DATE DUE

Microbial
Cell Surface
Hydrophobicity

Microbial Cell Surface Hydrophobicity

Editors

R. J. Doyle

Department of Microbiology,
Schools of Medicine and Dentistry,
University of Louisville,
Louisville, Kentucky

Mel Rosenberg

Maurice and Gabriela Goldschleger School of Dental
Medicine and Department of Human Microbiology,
Sackler School of Medicine, Tel Aviv University,
Ramat Aviv, Israel

American Society for Microbiology
Washington, D.C.

Copyright © 1990 American Society for Microbiology
1325 Massachusetts Avenue, N.W.
Washington, DC 20005

Library of Congress Cataloging-in-Publication Data

Microbial cell surface hydrophobicity / editors, R. J. Doyle, Mel Rosenberg.
　　p.　cm.
　　Includes index.
　　ISBN 1-55581-028-4
　　1. Bacteria—Adhesion.　2. Microorganisms—Adhesion.　3. Hydrophobic surfaces.
4. Bacterial cell walls. I. Doyle, R. J. II. Rosenberg, Mel.
　　[DNLM:　1. Bacterial Adhesion.　2. Cell Adhesion.　3. Microbiology.　4. Surface
Properties.　　QW 52 M6234]
QR96.8.M53　　1990
576'. 11—dc20.
DNLM/DLC
for Library of Congress　　　　　　　　　　　　　　　　　　　　　　　　90-1202
　　　　　　　　　　　　　　　　　　　　　　　　　　　　　　　　　　　　　CIP

Cover figure: Adhesion of *Bacteroides gingivalis* 2561 to hexadecane. Adhesion of *B. gingivalis* 2561 to hexadecane droplets was carried out by vigorously mixing washed cell suspensions with the liquid hydrocarbon in standard test tubes. After phase separation, samples of the upper layer were observed by light microscopy, using gentian violet to enhance visualization. Extrusion of hexadecane from areas on the droplets which are free of bound cells is evident. (From M. Rosenberg, I. A. Buivids, and R. P. Ellen, submitted for publication.)

CONTENTS

CONTRIBUTORS

Yeshaya Bar-Or • Division of Water Resources, Ministry of the Environment, P.O. Box 6234, 91061 Jerusalem, Israel

H. J. Busscher • Laboratory for Materia Technica, University of Groningen, Antonius Deusinglaan 1, 9713 AV Groningen, The Netherlands

Harry S. Courtney • Veterans Administration Medical Center and Department of Medicine, University of Tennessee, Memphis, TN 38163

R. J. Doyle • Academic Health Center, University of Louisville, Louisville, KY 40292

David Drake • College of Dentistry, University of Iowa, Iowa City, IA 52242

Wendy C. Duncan-Hewitt • Faculty of Pharmacy, University of Toronto, 19 Russell Street, Toronto, Ontario M5S 1A1, Canada

David L. Hasty • Veterans Administration Medical Center and Department of Anatomy, University of Tennessee, Memphis, TN 38163

Kevin C. Hazen • Department of Pathology, Box 214, University of Virginia Health Sciences Center, Charlottesville, VA 22908

Randall T. Irvin • Department of Medical Microbiology and Infectious Diseases, University of Alberta, Edmonton, Alberta T6G 2H7, Canada

Stephen A. Klotz • Departments of Medicine and Ophthalmology, Veterans Affairs and Louisiana State University Medical Centers, Shreveport, LA 71101-4295

R. Victor Lachica • U.S. Army Natick Research, Development and Engineering Center, Natick, MA 01760-5020

N. Mozes • Université Catholique de Louvain, Unité de Chimie des Interfaces, Place Croix du Sud 1, B-1348 Louvain-la-Neuve, Belgium

Itzhak Ofek • Department of Human Microbiology, Sackler School of Medicine, Tel-Aviv, Israel

Eugene Rosenberg • Department of Microbiology, Tel Aviv University, Ramat Aviv, Israel 69978

Mel Rosenberg • Maurice and Gabriela Goldschleger School of Dental Medicine and Department of Human Microbiology, Sackler School of Medicine, Tel Aviv University, Ramat Aviv, Israel 69978

P. G. Rouxhet • Université Catholique de Louvain, Unité de Chimie des Interfaces, Place Croix du Sud 1, B-1348 Louvain-la-Neuve, Belgium

Nechemia Sar • H. Steinitz Marine Biology Laboratory, P.O. Box 469, Eilat, Israel 88103

J. Sjollema • Laboratory for Materia Technica, University of Groningen, Antonius Deusinglaan 1, 9713 AV Groningen, The Netherlands

Gerrit Smit • Department of Plant Molecular Biology, Leiden University, Nonnensteeg 3, 2311 VJ Leiden, The Netherlands

Gary Stacey • Center for Legume Research, Department of Microbiology, and Graduate Program of Ecology, University of Tennessee, Knoxville, TN 37996

H. C. van der Mei • Laboratory for Materia Technica, University of Groningen, Antonius Deusinglaan 1, 9713 AV Groningen, The Netherlands

Torkel Wadström • Department of Medical Microbiology, University of Lund, Lund, Sweden

PREFACE

The past decade has witnessed the growth of a vast literature on microbial cell surface hydrophobicity. One reason for the emergence of this field is that new techniques have been developed to measure the association between microbes and nonpolar or hydrophobic surfaces. Furthermore, there is an increasing realization that hydrophobic interactions play a role in many, if not most, microbial adhesion phenomena. Microbial cell surface hydrophobicity is a factor in adhesion to soft host tissues, implants and prostheses, contact lenses, glass, oil, steel, teeth, submerged aquatic surfaces, plants, and fish. In many cases, the hydrophobic effect appears to stabilize so-called specific polar and stereochemical interactions. There is scarcely a microbial species of interest which has not been examined for hydrophobic surface properties.

Just as there is an increasing literature on the hydrophobic effect and how it influences microbial surface properties, there is also an awareness that the hydrophobic effect is difficult to define. One of the reasons may be our inability to fully understand the hydrophobic effect and how best to characterize it. Nevertheless, the techniques presently available have revealed several general principles and raised new questions for further research. Future techniques will expand the breadth and depth of this area of research. It is envisioned that the use of defined hydrophobic surface probes will increase in the coming decade. Novel, increasingly quantitative techniques and modifications will be introduced to detect the hydrophobic effect. Cooperative effects will be closely examined. The involvement of hydrophobic interactions alongside other interactions (electrostatic, stereochemical) will be clarified.

One of the lessons of elementary organic chemistry is that "like dissolves like." Similarly, water is attracted to water more than water is attracted to a hydrocarbon or nonpolar substance. Our definition of hydrophobicity, therefore, demands that hydrocarbons or other nonpolar compounds bind to each other in the presence or absence of water.

This is the first text dedicated solely to microbial cell surface hydrophobicity. The structure of this book reflects the experiences of its authors. We have selected the authors on the basis of their substantial contributions to microbial cell surface hydrophobicity. Chapter 2 was designed to provide a physical description of the hydrophobic effect. The understanding of this chapter is relevant to researchers in all areas of cell surface hydrophobicity.

The idea for this volume grew out of a joint grant from the United States-Israel Binational Science Foundation (68-00263), Jerusalem. M.R. thanks R. P. Ellen and the Faculty of Dentistry, University of Toronto, for hosting his sabbatical leave, during which portions of this book were edited.

R. J. Doyle
Mel Rosenberg

Chapter 1

Microbial Cell Surface Hydrophobicity: History, Measurement, and Significance

Mel Rosenberg and R. J. Doyle

Hydrophobic surface properties of microbial cells have been extensively investigated during the past decade. Whereas several years ago much of the pertinent research could be covered in review papers (88, 95, 151, 168; H. C. van der Mei, M. Rosenberg, and H. J. Busscher, *in* H. J. Busscher, N. Mozes, P. G. Rouxhet, and P. S. Handley, ed., *Analysis of Microbial Surfaces: Methodology and Applications*, in press), an entire book (such as this volume) is now required to review the vast literature that has accumulated on this subject. The purpose of this chapter is to present a critical overview, concentrating on the historical development of microbial cell surface hydrophobicity studies, the measurement techniques used, and the potential significance of the hydrophobic effect in understanding adhesion phenomena. In some instances, we have preferred to cite recent studies, earlier investigations having been covered in a 1986 review (151).

Over the past few years, investigators have implicated cell surface hydrophobicity in a wide variety of adhesion phenomena (some examples are listed in Table 1). Numerous tests have been proposed to measure microbial cell surface hydrophobicity. In many cases, cell surface components that promote or reduce hydrophobicity (termed, respectively, hydrophobins and hydrophilins [151]) have been identified (8, 9, 63, 86, 119, 126, 141, 147, 169). Yet despite the progress that has been made, the

Mel Rosenberg • Maurice and Gabriela Goldschleger School of Dental Medicine and Department of Human Microbiology, Sackler School of Medicine, Tel Aviv University, Ramat Aviv, Israel 69978.   **R. J. Doyle** • Academic Health Center, University of Louisville, Louisville, Kentucky 40292.

Table 1. Recent Studies Implicating Hydrophobic Interactions in
Microbial Cell Surface Phenomena

Phenomenon	Reference(s)
Adaptation to NaCl	60
Adhesion to fatty meat surfaces	29
Adhesion of fungal spores to host insect cuticle	13
Adhesion to biomaterials	76, 77, 127, 128, 136, 155
Adhesion to contact lenses	84, 105, 134
Adhesion to elemental sulfur	16
Adhesion to human foreskins	48
Adhesion to mineral particles	126, 172, 173
Adhesion to plant surfaces	28, 186
E. coli-mediated ulcerative colitis	18
Gliding motility	17, 191
Hemagglutination	50, 84, 163
Top vs bottom fermentation brewery strains	3
Opsonic requirements and phagocytosis	127, 170, 176, 185, 190
Plasmid-associated invasiveness of *Shigella flexneri*	125
Resistance of *Pseudomonas aeruginosa* to biocides	79
Virulence of coagulase-negative staphylococci	101
Virulence of fish pathogens	7, 15, 26

term "cell surface hydrophobicity" remains intuitively understood but poorly defined.

Much of the confusion lies with the term "hydrophobicity" itself. Physical chemists agree that hydrophobic interactions between surfaces depend in large part on the unique properties of water itself (see Duncan-Hewitt, this volume). Apolar moieties, immersed in an aqueous phase, are surrounded by structured layers of water (the degree of structure is inversely proportional to the distance from the apolar moiety). The water molecules in such shells, unable to freely undergo hydrogen bonding in all directions, are at a higher energy level than molecules in the bulk solution. If it is assumed that the apolar moiety is almost totally incapable of interacting with the water molecules (e.g., when the immersed moiety is a hydrocarbon molecule), then the energy involved in introducing this hydrophobic entity into the water phase is akin to the energy required to make a "cavity" in the water, the size of the immersed body (5, 12). The energy required to create this cavity is analogous to the energy required to increase the surface area of water. Thus, just as structured water molecules must assume constrained, energetically unfavorable arrays at the air surface (as indicated by the high energy required to increase the area of an air-water interface, i.e., surface tension), so must water molecules proximal to apolar moieties assume a configuration of higher energy. When two apolar moieties or surfaces approach one another, these constrained water molecules can be "squeezed out" or "freed" into the bulk aqueous phase, where their interactions with other water molecules are unencumbered. This energetically favorable process is associated with an increase in entropy concomitant with the decreased order of the system. Whereas attractive forces can draw apolar molecules toward one another, their relative contribution may well be a minor one (74, 129). Thus, one can look at adhesion mediated by hydrophobic interactions as primarily a process of exclusion from the water phase, a minimizing of energetically unfavorable interfaces, just as droplets of oil tend to coalesce in an aqueous environment. Such a viewpoint would predict that so-called hydrophobic microorganisms would not adhere to surfaces primarily by energetic bonds with the substratum but rather by exclusion from the bulk aqueous phase. Van Pelt et al. (184) describe lateral movement of bacteria "attached" to a polytetrafluoroethylene surface (see also Courtney et al., this volume). We have similarly observed that bacteria adhering to polystyrene are sometimes swept along the plastic surface by receding water droplets. This observation suggests that the attraction between the cells and the substratum is too weak to withstand the removal of the bulk water phase. Furthermore, cells with hydrophobic surface characteristics are often capable of adhering to wettable as well as nonwettable surfaces. Aqueous cell suspensions of *Serratia marcescens* RZ appear to stabilize both low-energy (e.g., hexadecane) and high-energy (mercury) liquid surfaces, whereas nonhydrophobic mutants do not exhibit either property (M. Barki, Y. Rishpon, and M. Rosenberg, unpublished data). Clearly, if water properties are the primary determinants in the hydrophobic interactions between microorganisms and surfaces, then the properties of the aqueous phase deserve a great deal of attention.

In the following chapter, Duncan-Hewitt provides a critical examination of the hydrophobic effect, concluding that it is mainly an "excess function" derived from anomalous properties of the solvent, water. As such, the term "hydrophobicity" lacks a definition on the mathematical level and is invoked only by descriptive analysis. Thus, in essence, how hydrophobicity is defined depends on how one goes about measuring it. The problem of defining hydrophobicity is not unique to the study of how it relates to microbial adhesion. Investigators have similar difficulties in assessing the contribution of the hydrophobic effect to protein structure and function. In one recent publication, 38 methods for measuring amino acid and resulting polypeptide hydrophobicity were compared (22).

One point that is sometimes overlooked when one discusses so-called hydrophobic microorganisms is the fact that most of them disperse quite readily in the bulk aqueous phase. Very few microbial strains studied to date are so hydrophobic that they can be wetted by oil but not by water and thus pass freely into the bulk oil phase (135). Most of the "hydrophobic" microorganisms described in this text are wetted by, dispersed, and suspended in water, suggesting that they should be called amphipathic rather than hydrophobic. It was previously suggested that the overall tendency of microorganisms to exhibit hydrophobic surface properties is determined by a complex interplay of polar and apolar outer surface components (152). In several microorganisms studied to date (e.g., *Acinetobacter calcoaceticus* and *S. marcescens*), surface moieties that promote hydrophobicity (hydrophobins) and reduce hydrophobicity (hydrophilins) appear to coexist on the cell surface (8, 9, 132, 141, 147). It is thus their relative concentration, distribution, configuration, and juxtaposition which determines the tendency of the cell to exhibit hydrophobic surface properties.

HISTORY

A chronology of some relevant hydrophobicity investigations and publications is summarized in Table 2. In 1924, the pioneering observations by Mudd and Mudd of bacterial partitioning at oil-water interfaces were published (110, 111). Seven years later, Reed and Rice (135) demonstrated quantitatively that nonwettable microorganisms (mycobacteria) are able to pass from the aqueous into the bulk oil phase. However, their attempts to quantify the observation by Mudd and Mudd of bacterial partitioning at the interface itself were not successful, presumably because of their different choice of strains (152). In the following decades, studies related to hydrophobicity were sporadic. Dyar (34), Dyar and Ordal (35), and Hill and co-workers (65) invoked nonpolar interactions in the binding of anionic surfactants to staphylococcal and streptococcal surfaces, proposing surface lipids as possible hydrophobins. In 1958, Boyles and Lincoln (14) initially described the interaction of endospores with the air-water interface. Subsequent studies have dealt with the partitioning of spores and other microbial cells at the air-water interface (25, 51, 63, 73) as well as partitioning of spores at liquid-liquid interfaces (4, 157). More recently, various investigators have addressed the relationship between hydrophobicity and spore morphogenesis (23, 31, 47, 87, 89). The involvement of the hydrophobic effect in aggregation of *Corynebacterium xeroxis* was considered by Stanley and Rose in 1967 (171).

In the early 1970s, considerable interest in microbial hydrophobicity arose from several directions. The studies by Marshall and co-workers on the polar adhesion of *Flexibacter* and *Hyphomicrobium* species at the

Table 2. Chronology of Selected Hydrophobicity Studies

Year	Studies	Reference(s)
1924	Mudd and Mudd study bacterial adhesion at the oil-water interface as a model for phagocytosis.	110, 111
1931	Reed and Rice measure partitioning of mycobacteria from aqueous phase into bulk oil phase.	135
1946–48	Dyar and Ordal and Dyar link surface hydrophobicity of *Staphylococcus aureus* with presence of surface lipids, as measured by binding of anionic surfactant.	34, 35
1958	Boyles and Lincoln report foam flotation of bacterial spores.	14
1958	Aqueous TPP described by Albertsson.	2
1959	Adhesion of spores to fluorocarbon reported.	4
1961	Partitioning of spores in aqueous two-phase system.	157
1963	Hill and co-workers study streptococcal hydrophobicity by binding of anionic probes.	65
1967	Hydrophobicity invoked in clumping of *Corynebacterium xeroxis*.	171
1971	Isolation of nonhydrophobic yeast mutant by foam flotation.	124
1972–78	Van Oss and co-workers revive the phagocytosis-hydrophobicity issue, using contact angles of bacterial films.	24, 182, 183
1972–81	Controversy over the possible role of adhesion in growth of bacteria and yeast cells on oil droplets.	80–82, 103, 106, 152
1973	Classic hydrophobicity text published by Tanford.	174
1973–75	Marshall and co-workers demonstrate oriented adsorption of microorganisms at the oil-water interface.	99–101
1974	Development of HIC by Hjertén and co-workers.	66
1976	Marshall publishes classic text on microbial adhesion, emphasizing possible role of hydrophobic interactions.	98
1976–	Fletcher and co-workers study microbial adhesion to polystyrene and other low-surface-free-energy substrata.	44–46, 133, 134
1977	Aqueous TPP method used to measure enterobacterial hydrophobicity.	96
1977	HIC first applied to the study of bacterial hydrophobicity.	169
1980–81	Introduction of several methods for measuring microbial hydrophobicity, including binding of dodecanoic acid, MATH, and SAT.	73, 83, 94, 96, 169
1981	High proportion of microorganisms on sea surface are hydrophobic.	25
1982–	Initial studies examining hydrophobic surface properties of oral microorganisms.	32, 53, 54, 116, 149, 150, 188, 189
1984	Hydrophobicity implicated in adhesion of benthic vs planktonic cyanobacteria.	39
1984–85	Studies on mechanisms of spore adhesion to hydrocarbon.	139
1985–	Increasing interest in fungal (particularly candidal) hydrophobicity.	62, 85, 105
1985–86	Improvements in MATH and SAT proposed.	92, 154, 165
1986–87	Additional tests proposed, including binding of nonionic surfactant and direction of spreading.	117, 158

oil-water interface were followed by a classic monograph on microbial adhesion (98) which emphasized the potential contribution of hydrophobic interactions. Concomitantly, the laboratory of C. J. van Oss undertook the reevaluation of the role of hydrophobicity in phagocytosis, being the first to use contact angle measurements (CAM) on bacterial layers (24, 182, 183). The role of hydrophobic interactions in bacterial adhesion to polystyrene and other nonwettable polymers has been examined since the 1970s by Fletcher and others (44–46, 133, 134). In the late 1970s and early 1980s, several hydrophobicity techniques were introduced by Scandinavian researchers, including application of P. A. Albertsson's aqueous two-phase partition (TPP) system (1, 96), hydrophobic interaction chromatography (HIC) (2, 169), binding of radiolabeled dodecanoic acid to bacterial cell surfaces (83), and the salt aggregation test (SAT) (94) (see below).

During the 1970s, much attention was directed to microbial oxidation of oil products, initially as a potential source of protein and subsequently as a possible environmental cleanup tool. Adhesion to hydrocarbon droplets was often observed during microbial growth on oils (103, 106) but was usually considered to result from stereospecific interactions characteristic of oil-degrading microorganisms (81, 82). In 1980, Iimura and co-workers implicated cell surface hydrophobicity as a pellicle formation factor in yeast cells, as determined by use of a two-phase hydrocarbon-water assay (74). In the same year, Rosenberg et al. (148) proposed bacterial adhesion to liquid hydrocarbons as a simple, general method for studying cell surface hydrophobicity.

During the 1980s, a wide variety of microbial adhesion and related phenomena have been explored with regard to the role of hydrophobic interactions. Some of these phenomena are listed in Table 3. Recent examples include adhesion to mineral surfaces (126, 172, 173), fish surfaces (7, 17, 26, 159), fatty meats (29), intrauterine contraceptive devices (76, 77, 136), bioprostheses (49), and host insect cuticle (14). Over the past decade, a great deal of attention has been directed at the potential role of hydrophobicity in mediating adhesion of oral microorganisms to teeth and tooth model surfaces (1, 32, 43, 53, 54, 67, 68, 78, 86, 107, 115, 116, 118, 121, 149, 178, 188, 189), buccal mucosa (140, 153), and dental restoration materials (85, 160). Although earlier papers raised the possibility that hydrophobic interactions might be of some importance in the oral cavity (91, 93), the first direct evidence that oral microorganisms possess hydrophobic outer surfaces was published in 1982 (115, 121, 188). Recent investigations into microbial adhesion to host tissues in general have often focused on the role of stereospecific, lectinlike interactions, frequently disregarding the potential contribution of hydrophobic effects.

Table 3. Agents That Inhibit Adhesion by Interfering with the Hydrophobic Effect[a]

Assay	Inhibitor(s)	Reference(s)
Adhesion to SHA	Thiocyanate anion, urea, tetramethyl urea, Li$^+$	115
	Sulfolane	Zhang et al., in press
Adhesion to hexadecane	Hydrophobic salivary proteins	6
Adhesion to polystyrene	Hydrophobic salivary components	37
Adhesion of *Bifidobacterium bifidum* to hydrocarbon	Albumin	122
Adhesion of *Clostridium difficile* to a human embryonic intestinal cell line	1% (but not 0.1%) Tween 80	192
Adsorption of *Actinomyces viscosus* to hydrophobic gels	Tween 80	21
Adhesion of *Candida* species to polystyrene	Tween 20	85
Hemagglutination by *Pseudomonas aeruginosa*	L-Tryptophan, L-leucine, 4-methylumbelliferone, *p*-nitrophenol	50
	Triton X-100, Nonidet P-40, Tween 20	84
Hemagglutination by *Gardnerella vaginalis*	Phosphatidylserine	164
MATH and adhesion to buccal epithelial cells	Emulsan	140
Adhesion of plaque microorganisms to buccal epithelial cells	Emulsan, Tween, SDS	153
HIC of *Bacillus subtilis* spores	SDS	31
Adhesion to sulfur granules	SDS	16
Coaggregation between *Actinomyces viscosus* and *Streptococcus sanguis*	SDS	102
Adhesion to SHA	Albumin	116, 196
Adhesion to polystyrene test tubes	Various graft copolymers	72
	Caseinoglycopeptides, BSA	114
Adhesion of *Escherichia coli* to mammalian cells	*p*-Nitrophenol, tyrosine, others	38
Adhesion of enteropathogenic *E. coli* to rabbit ileal microvillus membranes and to hydrophobic surfaces	Tetramethyl urea, propanol	33
Adhesion of *E. coli*, *Dictyostelium discoideum*, and erythrocytes to "hydrophobized" glass	Copolymer with hydrophobic and hydrophilic segments	123
Adhesion to latex, Teflon-coated, and vinyl catheters	Heparin and tridodecylmethylammonium chloride complex	155
Streptococcal adhesion to SHA	Fatty acids, normal human serum	93

(*Continued on next page*)

Table 3—*Continued*

Assay	Inhibitor(s)	Reference(s)
Adsorption of *E. coli* to HIC	Ethylene glycol	20
MATH of oral streptococci	Polyethylene glycol	67
Adhesion of group B streptococci to a polyethylene intrauterine device	Human plasma fibronectin, BSA	76
Adhesion of *Pseudomonas fluorescens* and *Acinetobacter* sp. to hydrogel and polystyrene surfaces	Lipopolysaccharide, BSA	85, 134

[a] Abbreviations: SHA, saliva-coated hydroxylapatite; SDS, sodium dodecyl sulfate; BSA, bovine serum albumin.

In those cases in which hydrophobicity has been considered, correlations have often (but not always) been observed (this point is discussed further toward the end of the chapter). The controversies over the extent of the hydrophobic effect in mediating adhesion within the oral cavity (19, 32), and to host surfaces in general (33, 59, 112, 113, 125, 156, 163, 164, 166; van der Mei et al., in press), will continue to fuel future investigations.

MEASUREMENT

Most microbial cell surface hydrophobicity assays can be divided into two broad categories. The first category measures the hydrophobic properties of the outer cell surface as a whole. Such assays include CAM, partitioning of cells into one or another liquid phase (TPP), and adsorption of individual hydrophobic molecular probes at the cell surface. The second category measures hydrophobicity in terms of adhesion. Such techniques include microbial adhesion to hydrocarbons (MATH), hydrophobic interaction chromatography (HIC), and adhesion to polystyrene and to other hydrophobic solid surfaces. In such instances, a hydrophobic "tip" may be sufficient to mediate adhesion, thus classifying a bacterium as hydrophobic (99, 100), although the remainder of the cell surface may be relatively hydrophilic. Inhibition of adhesion by agents that interfere with the hydrophobic effect (e.g., surfactants and chaotropes) is yet another adhesion-dependent approach. Techniques such as the SAT, in which a salting-out agent is used to induce aggregation of suspended cells, probably fall somewhere between these categories. Still other methods (e.g., direction of spreading for measuring colonial hydrophobicity) defy simplistic attempts at categorization.

Any measurement of microbial hydrophobicity has inherent technical difficulties. First, as mentioned above, hydrophobic interactions can often promote adhesion to a wide range of surfaces and interfaces. Thus, it is not uncommon to find high proportions of test microorganisms adhering to the walls of the vessel they happen to be in (or adsorbed to water films on surfaces such as glass). This problem emphasizes the need to use clean, standard vessels. Second, since most hydrophobicity assays use aqueous suspensions, attention must be paid to the properties of the aqueous phase. As already mentioned, the types and concentrations of salts and buffers and presence of amphipathic contaminants (e.g., cell surface components released by the cells themselves upon prolonged standing) can have a marked effect on results with a variety of hydrophobicity assays. Third, because microbial growth conditions can profoundly affect hydrophobic surface properties of a wide range of microorganisms, attention must be paid to differences in culturing conditions (8, 9, 34, 35, 39, 60, 119, 138, 142, 148, 193–195). In some cases, sequential subculturing of clinical isolates results in gradual disappearance of hydrophobic properties (7, 189), suggesting that fresh isolates (or even microorganisms obtained directly in situ [149, 151]) should be used whenever possible. Fourth, bacteria with hydrophobic surface properties sometimes have a tendency to adhere to one another, resulting in clumping (aggregation) (171). Fifth, adhesion of microorganisms to hydrophobic surfaces is, at least in some cases, cell density dependent (184; M. Barki, S. Goldberg, R. J. Doyle, and M. Rosenberg, unpublished data).

Finally, it is often difficult to implicate hydrophobic interactions in a given adhesion phenomenon. The hydrophobic effect cannot be inhibited by exquisitely stereospecific molecules; often the molecules that do inhibit hydrophobic interactions (e.g., surface-active agents) may also elicit structural changes in cell surface polymers (detergent molecules, for example). Conversely, it is just as difficult to rule out the contribution of hydrophobic interactions in adhesion phenomena. Researchers may erroneously interpret inhibition by cell surface components as evidence that these components take part in a given adhesion process. However, if the surface moieties are themselves amphipathic (e.g., lipopolysaccharide or lipoteichoic acid [LTA]), such components may be inhibiting adhesion by interfering with hydrophobic interactions (134). In a recent paper (20), a role was invoked for LTA in mediating adhesion of *Staphylococcus epidermidis* to fibrin clots on the basis of a fourfold inhibition by LTA. However, as the authors themselves reported, human serum albumin elicited similar inhibitory levels. Thus, LTA should not be implicated as an adhesion on the basis of its inhibition properties. Similarly, adhesion of *A. calcoaceticus* RAG-1 to hydrocarbons is inhibited by released exopolysaccharide (emulsan) (140). However, when emulsan is present on the cell surface, it does not mediate adhesion but rather prevents it (132, 141).

In the next section, we review various aspects of the methods commonly used for measuring microbial cell surface hydrophobicity.

HYDROPHOBICITY ASSAYS

MATH

In 1980, Rosenberg et al. showed that various bacterial strains thought to possess hydrophobic surface characteristics adhered to liquid hydrocarbons, whereas nonhydrophobic strains did not (148). This process could be inhibited and reversed by addition of surfactants. Adherent microorganisms could be observed adhering at the oil-water interface. Bacterial adhesion to hydrocarbons (BATH) was proposed as a simple, general technique for studying cell surface hydrophobicity. In recent years, the extensive use of this approach for the study of eucaryotic microorganisms (13, 85, 105, 108, 109, 195) suggests that MATH (microbial adhesion to hydrocarbons) is a more appropriate acronym.

The original experimental approach was based on mixing (vortexing) washed cell suspensions with test hydrocarbons (n-hexadecane, n-octane, and p-xylene) for a given time (2 min) and measuring adhesion simply as the decrease in turbidity in the aqueous phase after separation of the phases. Several aspects of the techniques, however, are frequently overlooked or misinterpreted. First, it is essential to observe the hydrocarbon droplets after phase separation, MATH being one of few assays in which the adhesion process itself can be readily observed in the microscope. The upper "cream," containing stable hydrocarbon droplets coated with adherent cells, can be transferred to a glass slide and readily observed in the phase microscope. Larger hydrocarbon droplets become pressed between the slide and cover slip, affording two planes of focus, at which the bacterial layer can be observed (Fig. 1). Adhesion to the droplets can also be observed in the light microscope after staining of the cells with a dye (e.g., gentian violet). Several investigators have erroneously reported that the cells partition "into" the hydrocarbon phase (1, 109, 130). This error could have been avoided by simple microscopic observation. Attention must also be paid to the possibility of clumping during the assay, which would result in a spurious reduction in optical density. Because the adherent cells rise after the mixing procedure, clumping can readily be observed in the aqueous phase.

In the original assay procedure, mixtures were vortexed for a given time in the presence of increasing hydrocarbon volumes, and adhesion

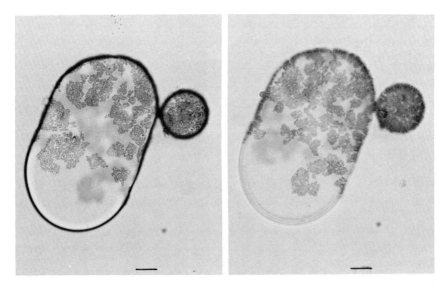

Figure 1. Microscopic observation of adhesion to hexadecane. When a large droplet of hydrocarbon, with adherent cells, is observed microscopically, the droplet is often pressed between the glass slide and the cover slip, affording two fields of focus. The figure shows the two fields of focus (left, upper field; right, lower field) for adhesion of *Bacteroides gingivalis* 2561, stained with gentian violet, to a large hexadecane droplet. Bars represent 10 μm. (Rosenberg, Buivids, and Ellen, unpublished data.)

was shown to increase with increasing hydrocarbon/volume ratio. However, steady-state adhesion levels were difficult to demonstrate. This problem prompted Lichtenberg et al. (92) to propose a kinetic approach, which provides quantitative data (removal coefficients) linking the rate of adhesion as a function of the hydrocarbon/water volume ratio. Sharon and co-workers (165) showed that kinetics of adhesion to hexadecane can be monitored directly in polystyrene cuvettes. (Busscher et al. [in press] have reported that hexadecane droplets can collect on the cuvette walls. In such cases, introduction of air bubbles with a Pasteur pipette can dislodge them.) Despite these advances in quantitation of MATH, care must be exercised, since we have recently found that changes in the initial cell density may affect the removal kinetics (M. Rosenberg, M. Barki, R. Bar-Ness, S. Goldberg, and R. J. Doyle, submitted for publication). Thus, in any comparison among closely related strains or treatments, the initial cell densities must be comparable to permit conclusions regarding relative hydrophobicities.

With certain strains, adhesion levels of over 99% to hexadecane can be observed. Under such conditions, nonhydrophobic mutants may be easily enriched from the lower aqueous phase. This process has been used to isolate hydrophobicity-deficient mutants of *A. calcoaceticus* RAG-1 (152), *S. marcescens* RZ (145), and *Streptococcus sanguis* strains (54, 78, 107).

MATH was originally developed to study the role of adhesion during growth of *A. calcoaceticus* RAG-1 on oil (152). The original MATH assay (148) used phosphate buffer, $MgSO_4$, and urea (PUM buffer), which enable growth of RAG-1 when hexadecane is present. This approach facilitated the measurement of adhesion under conditions approximating growth and the subsequent demonstration that adhesion fulfills an important role in growth of RAG-1 on hexadecane (147, 152). Generally, urea and magnesium sulfate appear to have no appreciable effect on MATH, and more recent studies have used the phosphate buffer alone (8, 9).

Many other additions to the aqueous phase do have a profound effect on MATH. Ammonium sulfate promotes adhesion of *Escherichia coli* to hydrocarbons and polystyrene, probably as a result of its salting-out properties (144, 146). Polyethylene glycol has been reported to decrease adhesion to hexadecane, allowing differences among adherent strains to be discerned (67, 68). It has recently been shown that certain amphipathic cations can "hydrophobize" a variety of microbial cells, thus mediating their partitioning at oil-water interfaces and promoting adhesion to polystyrene (57; S. Goldberg, R. J. Doyle, and M. Rosenberg, *J. Bacteriol*, in press).

The simple turbidimetric approach to MATH lends itself to the study of mixed populations. Thus, hydrophobic properties of bacteria derived directly from dental plaque (149) and human feces (151) were demonstrated. Separation of hydrophobic and nonhydrophobic cells (e.g., *Bacillus* spores versus vegetative cells) can be simply performed by partitioning at the oil-water (31, 143) interface.

One intriguing aspect of MATH is the effect of phase transition of the liquid hydrocarbon. When hexadecane droplets bearing adherent bacteria are cooled to solidification (below 16°C) and subsequently warmed to allow melting, the droplets coalesce and the bacteria are desorbed into the bulk aqueous phase (151). One possible explanation for this phenomenon is that bacteria partitioning at the interface extend somewhat into the oil phase (Mudd and Mudd [110, 111] first suggested that the bacteria deform the interface) and that the interface is "straightened out" after solidification, with concomitant desorption of cells. However, recent electron micrographs of cells of *Bacteroides gingivalis* adhering to hexadecane droplets (M. Rosenberg, I. Buivids, and R. P. Ellen, submitted for publication) show no indentation of the oil surface by the closely attached microbial cells. Clearly, if solidification of the hexadecane results in desorption of cells, any modification of the MATH assay that includes solidification of the hydrocarbon phase is to be avoided.

Recently, the possibility that MATH may involve cooperative interactions among the adhering cells themselves has been studied. In some strains (e.g., RAG-1), adhesion is high regardless of the cell density of the suspension (M. Barki, S. Goldberg, R. J. Doyle, and M. Rosenberg, unpublished data). However, clinical isolates of *S. marcescens* and *Candida albicans* exhibit apparent positive cooperativity: when the initial cell density is low, adhesion is barely observable; however, as the density of cells is increased, adhesion to hydrocarbons rises (despite the decreasing hexadecane/cell ratio) (Barki, Goldberg, Doyle, and Rosenberg, unpublished data). Interactions between cells on the oil surface can explain the "patchiness" sometimes observed in adhesion to oil droplets (Rosenberg, Buivids, and Ellen, submitted), as well as to other solid surfaces (184), and the observation that adherent cells may appear to act as a "girdle" for the droplet (see the cover of this volume); thus, areas free of adhering patches of cells tend to "bulge out" (Rosenberg, Buivids, and Ellen, submitted). It was previously proposed that MATH be performed with a wide range of initial cell densities so that overloading (i.e., too many cells mixed in the presence of too low a hexadecane volume) would not occur (143). Clearly, the opposite possibility (i.e., underloading) can also be the case if positive cooperativity is a factor.

Most easily performed techniques have potential limitations. One drawback of the MATH technique is the possibility that cells are damaged by vortexing in the presence of the liquid hydrocarbons. Vanhaeke and Pijck (179) recently showed that performing MATH with *n*-hexadecane has no effect on cell intactness and found a high correlation between optical density and ATP values. Xylene, however, caused cell lysis. These findings indicate that *n*-hexadecane should be preferred as the test hydrocarbon and that use of aromatic hydrocarbons should be avoided if possible. Similarly, the loss or extraction of cell surface components during the mixing procedure is probably rare and has not been shown to affect results. Nevertheless, any assay in which cells are harvested from liquid growth medium and are subjected to centrifugation, suspension, and mixing procedures may effect loss of loosely bound material, such as secreted polysaccharides, or even surface structures, such as fimbriae. Procedures such as that of Sar (158, 159), in which hydrophobicity of colonies, rather than washed cells, is measured, can circumvent this problem.

Other potential problems of MATH concern quantitation. The original assay was semiquantitative in nature. This limitation has been somewhat mitigated by the newer kinetic approaches (92, 165). Nevertheless, the surface area available for cell contact during the assay is extremely difficult to ascertain and varies greatly with the agitating instrument, vessel, and mixing intensity used. Moreover, in order for adhesion to be measured in MATH, the cell-coated droplets must remain stable over time. This stabilizing process may entail both cell-cell and cell-hydrocarbon interactions. We have on occasion observed momentary adhesion to hexadecane droplets, followed by droplet coalescence and concomitant release of cells. Such cells may be hydrophobic enough to initially adhere at the hydrocarbon-water interface but lack the cooperative cell-cell interactions necessary to form a stable cream of cell-covered droplets. This apparent disadvantage of MATH may ultimately prove to be of benefit, since cell-cell interactions may turn out to be a crucial factor in hydrophobicity-promoted adhesion events in vivo.

In the absence of added hydrocarbon, hydrophobic microorganisms tend to adhere at a water film that develops on the walls of the vessel during the vortexing procedure. Thus, controls consisting of cell suspensions vortexed in the absence of added hydrocarbons are not appropriate in the MATH assay. Results should always be compared against time zero measurement.

Modifications of MATH have potential applications. Examples include using oil droplets as a vehicle for desorbing oral microorganisms (150), flotation of larvicidal spores (146), cell immobilization on hydrocarbon droplets (57), and as a diagnostic tool. Martin and co-workers (101) recently reported that among various tests for pathogenicity of coagulase-negative staphylococci, MATH had the highest positive predictive value.

CAM

CAM constitutes a classic method for measurement of surface free energies (or degree of wettability by the droplet) of solid surfaces. The use of CAM to study hydrophobic surface properties of microorganisms was introduced in 1972 (183), and this technique has since been used in several laboratories (19, 24, 182; Busscher et al., in press). A recent review deals at length with the advantages and disadvantages of this procedure (Busscher et al., in press).

In general, CAM should be used for measurement of homogeneous, flat, smooth, dry surfaces. Obviously, these conditions cannot be met for microbial cells. Deviations from ideal conditions can lead to large variations in measurements as well as to differences between advancing and receding contact angles (hysteresis).

Several general aspects of the method include (i) obtaining a layer (as flat as possible) of microbial cells, either by filtration of washed cells or as lawns on agar; (ii) partial drying of the layer to remove "free" water, which could otherwise lead to deviations in measurement (A. W. Neumann, personal communication), but not desiccation of the cells themselves; (iii) meticulous application of a standard droplet of liquid (usually water or saline); and (iv) measurement of the contact angle directly or derivation of the contact angle via determination of the shape of the droplet.

Theoretically, CAM should give a definitive overall hydrophobicity value of the microbial cell surface. This has prompted investigators to calculate surface free energy values for various bacteria and to predict adhesion on the basis of relative surface free energies of the cells, the suspending liquid, and the substratum (19, 177, 178, 184). However, there are at least four possible ways to derive surface free energies from the contact angles themselves (Busscher et al., in press). Several additional potential problems may be mentioned. First, the measurement of contact angles on bacterial layers deviates greatly from the manner in which contact angles should theoretically be measured. Second, there is some disagreement on the technical aspects of measurement (e.g., how to deal with changes in contact angle immediately after application) as well as computational considerations (e.g., whether spreading pressures should be neglected). Third, when water is applied to the microbial layer, it tends to seep into the underlying layer. Moreover, measurements of water contact angles from only one profile may be misleading because they may not be axisymmetric (Neumann, personal communication). One alternative is to measure the contact angle by using nonaqueous solvents such as α-bromonaphthalene. However, these compounds often tend to disrupt the underlying cell layer. Finally, the fate of cell surface appendages in the drying-out process is unclear. Fimbriae which are thought to extend out from the cell surface into the bulk aqueous phase have often been proposed as mediators of bacterial hydrophobicity (43, 49, 75, 147). The necessity of evaporating "unbound water" may thus affect their normal configuration.

Despite the considerations mentioned above, CAM has proven to be an important technique for measurement of cell surface hydrophobicity and may indeed provide highly relevant data for its quantitation. Unfortunately, proper measurement of CAM requires special instrumentation and technical expertise that are beyond the capacities of most microbiology laboratories.

HIC

HIC measures microbial adsorption to octyl- or phenyl-Sepharose beads. This simple technique, originally developed for protein separation

(66), was first used by Smyth and co-workers (169) to test bacterial hydrophobicity. HIC can be easily performed by packing aqueous suspensions of Sepharose beads bearing covalently bound phenyl or octyl groups into small columns (e.g., in Pasteur pipettes). The percentage of bound cells can be determined by the loss in turbidity (or radioactivity) in the eluate as compared with the initial level. In some cases, salting-out agents (e.g., ammonium sulfate) have been added to promote adhesion to the gel. Desorption of bound cells can be demonstrated, for example, by elution in the presence of detergent. As a control, adsorption to untreated Sepharose beads (CL-4B) should also be determined; in some cases, a high percentage of cells adhere to the untreated beads (163). Although it is possible to calculate a hydrophobicity index on the basis of a comparison of adsorption to "hydrophobized" versus control beads (21), the meaning of this index is unclear, because hydrophobic interactions may play a role in adsorption to control beads as well (for example, Tween 80 inhibits binding of *Actinomyces viscosus* to phenyl-Sepharose as well as to untreated Sepharose beads [W. B. Clark, personal communication]). One potential problem of HIC is mechanical entrapment within the gel matrix or between the gel and glass support, which may occur with use of salting-out agents (which can cause aggregation) or in studies of chains or clumps of microorganisms or large cells. This problem may be overcome by using a batch procedure, i.e., by mixing the bacterial cells and beads freely in suspension (151).

As with other hydrophobicity assays, the nature of the suspending solution (e.g., pH, type, and concentration of salt) can have a large effect on the results (van der Mei et al., in press).

SAT

SAT is an extremely simple technique for studying the aggregative behavior of cells in increasing concentrations of salting-out agents (primarily ammonium sulfate). This technique, introduced by Lindahl et al. (94), is based on the premise that increasingly hydrophobic bacteria will aggregate at correspondingly lower salt concentrations. An improved test, in which stain is added to improve visualization of the aggregates, has been reported (154). As with MATH, cell density may be an important criterion (30).

The SAT technique has several limitations. First, many hydrophobic bacteria will clump in the absence of any added ammonium sulfate. Second, the method provides only a qualitative estimate of relative rank of hydrophobicity. Finally, charge-charge (electrostatic) interactions may play a greater role in influencing the results of SAT as compared with other hydrophobicity measurement techniques.

TPP

TPP is based on a separation technique originally proposed by Albertsson (2), measuring the distribution of cells and cell components between two aqueous phases (one containing dextran and the other containing polyethylene glycol) that are mutually immiscible. Hydrophobicity of the polyethylene glycol phase can be increased by covalent binding of fatty acid acyl groups (96). In this technique, a bacterial suspension is introduced to the two-phase system, and the phases are then mixed thoroughly and allowed to separate. Samples of each phase can then be taken for determining the distribution. The method is sensitive to changes in the concentration, batches, and molecular weight distribution of the two polymers; moreover, cells may partition in the interphase, complicating recovery and interpretation of results (151; van der Mei et al., in press). In principle, the technique appears to be an excellent one in that the cells are not subjected to harsh conditions and few artifacts are anticipated. Nevertheless, it has been used less frequently than the more common hydrophobicity tests (MATH, HIC, CAM, and SAT).

Binding of Molecular Probes

In principle, the degree of binding of hydrophobic molecular probes to the outermost cell surface appears to be a superior approach to the study of microbial hydrophobicity. Some examples are listed in Table 3. Earlier studies used small concentrations of ionic (usually anionic, to avoid electrostatic attraction) surfactants, and the level of adsorption to staphylococcal and streptococcal cells was ascertained by changes in electrophoretic mobilities (34, 35, 65). More recently, Kjelleberg et al. (83) measured binding of radiolabeled dodecanoic acid to bacterial cells and calculated the number and affinity of binding sites. A similar approach using radiolabeled palmitic acid was proposed (97). Nonionic surfactants have also been proposed as suitable probes for measuring bacterial hydrophobicity (117).

One potential problem associated with molecular probes is that a minimum degree of polarity is necessary to enable significant solvation and prevent their own aggregation. Thus, their hydrophobic properties must be compromised by introduction of polar or charged groups. In addition, such molecular probes may intercalate within, rather than on, the cell envelope and bind to sites that are not exposed at the outermost cell surface. Hydrophobic probes may also be internalized into the cells themselves via porins or as a result of outer surface damage. There do not seem to be any overall correlations between binding of hydrophobic dyes

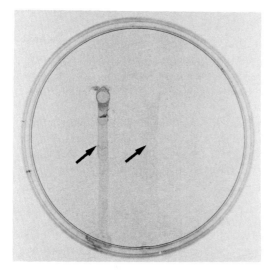

Figure 2. Instantaneous bacterial adhesion to polystyrene. Adhesion to polystyrene can be observed by allowing a droplet of cell suspension to run down a polystyrene petri dish. Samples (50 μl) of a dense (ca. 15 units of optical density at 400 nm), washed suspension of *Actinomyces viscosus* T14VJ1 in phosphate buffer, pH 7.1 (left), or containing 0.4 mg of bovine serum albumin per ml (right) were applied to the surface of an untreated polystyrene petri dish. The plate was tilted to allow the droplets to run down the plate. The petri dish was then washed with water and stained with gentian violet to visualize adhering microorganisms. A wake of bound cells is clearly observable (arrow, left) for the washed cell suspension but is almost completely absent in the presence of bovine serum albumin (arrow, right).

or susceptibility to antibiotics and cell surface hydrophobicity (27, 61, 190), although isolated examples have been reported (130).

Adhesion to Polystyrene and Other Hydrophobic Surfaces

Microbial adhesion to polystyrene represents a simple, versatile approach to studying hydrophobicity via adhesion to a rather uniform, nonwettable plastic surface. Commercial, non-tissue culture-treated polystyrene is readily obtained in a variety of forms, enabling a wide range of related assays. Adhering microorganisms may be counted microscopically, estimated by spectrophotometric measurements of stained cell layers, or estimated by outgrowth. The transparent nature of polystyrene is a great advantage in direct observation of the adhesion process. Adhesion of hydrophobic microorganisms to polystyrene takes place within a fraction of a second. This phenomenon can be simply demonstrated by allowing a droplet containing a turbid microbial suspension to run down a polystyrene dish and observing the adherent cell layer in the droplet's wake (Fig. 2). Polystyrene can also be used in a replica technique to identify colonies of adherent bacteria (142). Adhesion to polystyrene cuvettes can be conveniently used to study bacterial desorption (150).

Microspheres of polystyrene, usually carrying a net negative charge sufficient to enable their dispersion in aqueous suspension, have also been used to study hydrophobicity, e.g., by induced aggregation (90). The large size of yeast cells enables the number of bound polystyrene beads per cell to be enumerated (62), as described by Hazen (this volume).

One possible criticism of adhesion to polystyrene is that manufacturers include amphipathic plasticizing agents in production of polystyrene products. However, the hydrophobicity of flat, polystyrene surfaces can be monitored by measuring the water contact angles.

Several investigators have used solid surfaces of increasing surface energies to ascertain the role of hydrophobicity in mediating adhesion in various systems. Adhesion to hydrophobic, low-energy surfaces often prevails, especially when the strains used are hydrophobic (69, 136, 160, 161, 184). In any case, microorganisms that exhibit hydrophobic surface properties usually adhere better to most surfaces than do nonhydrophobic mutants or strains (108).

Additional Methods

Sar has reported a simple technique for ranking hydrophobicity of bacterial colonies and lawns that is based on direction of spreading of applied droplets (158). Indirect methods for studying hydrophobicity include use of agents or treatments that affect hydrophobicity. Some examples of agents that appear to block adhesion by inhibiting hydrophobic interactions are listed in Table 4. The main consideration in interpreting such effects is that the inhibitors are frequently protein-unfolding agents. For example, thiocyanate, Li^+, urea, and sodium dodecyl sulfate are well-known denaturants. Certain inhibitors, such as sulfolane, emulsan, stearic acid, and hydrophobic amino acids, may be superior to other surfactants and chaotropes as selective inhibitors of the hydrophobic effect (X. Zhang, M. Rosenberg, and R. J. Doyle, *FEMS Microbiol. Lett.*, in press). Other methods include growth in the presence of antibiotics (56, 128, 131, 162) and enzymatic treatments (118, 137, 176). Since hydrophobic interactions decrease at lowered temperatures, this may be an excellent way to ascertain the role of the hydrophobic effect in various adhesion assays (113). Finally, microbial strains that are closely related yet differ in hydrophobic surface properties have been compared for their ability to grow on hydrocarbons (133), adhere to saliva-coated hydroxylapatite (43, 54, 78, 87, 189), exhibit gliding motility (168), and foam (124).

Comparing Hydrophobicity Tests

A somewhat controversial issue is the degree of correlation between the various hydrophobicity techniques. Table 4 lists a variety of studies in which comparative data were presented. In general, most papers on

Table 4. Correlations among Various Hydrophobicity Tests

Strains used	Methods compared	Conclusions	Refer-ence(s)
Various	TPP vs CAM	Good, linear correlations.	52
Escherichia coli, *Serratia marcescens*	BATH vs adhesion to polystyrene microtiter plate wells	Good correlations.	144, 145
42 *Actinomyces* spp.	HIC vs SAT	Good correlations.	21
Renibacterium salmoninarum	BATH vs SAT	Some correlation evident.	7
16 different strains	CAM vs BATH	Good but nonlinear correlation.	180
	CAMS vs TPP	Good correlations for ca. half of the strains.	180
Streptococcus salivarius HB and mutants	CAM, HIC, BATH, adhesion to polymethylmethacrylate	Good overall correlations, with few exceptions.	178
Various streptococci	HIC, SAT, CAM, BATH	Poor to moderate correlations.	177
Various	Adhesion to untreated polystyrene, latex particle agglutination HIC, SAT	Good correlation between MATH and adhesion to untreated polystyrene; poor to moderate correlations among other tests.	30
E. coli strains	SAT, HIC	Good correlation.	166
Pasteurella multocida and *Actinobacillus lignieresii*	MATH, HIC	Good correlations for 15 *P. multocida* strains but poor correlations for 3 *A. lignieresii* strains.	27
Corynebacterium renale and *C. pilosum*	HIC, MATH, SAT	Strains adhered to hydrocarbon and to octyl-Sepharose but were not aggregated below 2 M ammonium sulfate in SAT.	75
Yersinia spp.	HIC, MATH, SAT	Some correlation between HIC and MATH but not with SAT.	163
Candida albicans, *Candida* spp.	MATH, CAM, adhesion to plastics of different surface free energies	Good correlations among all three methods except for *Candida krusei*.	85
Six *Candida* spp.	MATH, CAM	Good overall correlations.	105
Coagulase-negative staphylococci treated with various antibiotics	MATH, adhesion to tissue culture-treated polystyrene	Excellent ($r = 0.958$) correlations.	162

(*Continued on next page*)

Table 4—*Continued*

Strains used	Methods compared	Conclusions	Reference(s)
Eight species of marine bacteria	Adhesion to tissue culture-treated polystyrene vs untreated polystyrene	Good correlation.	133
14 mesophilic *Aeromonas* strains	Binding to nitrocellulose filters, SAT, HIC, uptake of gentian violet	Seven strains were more hydrophobic than counterparts in all tests.	130
Streptococcus sanguis wild type and mutants	MATH, HIC	Good correlation.	78
Five *Lactobacillus* strains	Modified CAM, adhesion to various plastics	High degree of correlation ($P = 0.994$).	136

hydrophobicity report good correlations among the methods tried, including those in which different classes of assays (e.g., MATH versus CAM) were compared. In some cases, researchers present the data obtained by two or more different tests but do not determine the degree of correlation. One such example is the study of Darnell et al. (27), who measured hydrophobicity of 15 strains of *Pasteurella multocida* by MATH and HIC. The high degree of correlation ($P = 0.91$) between the two sets of hydrophobicity data is shown in Fig. 3.

Interestingly, three independent comparative investigations were published in 1987 in a single volume of *Journal of Microbiological Methods*. All three comparisons reported a wide range of correlations between the various tests. Mozes and Rouxhet compared CAM, HIC, MATH, adhesion to polystyrene, and SAT (109). They found good agreement for the very hydrophobic strain (*Moniliella pollinis*) and the very hydrophilic strains (*Saccharomyces cerevisiae* and *Acetobacter*

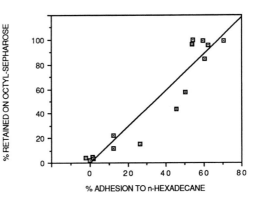

Figure 3. Comparison of two hydrophobicity methods. Cell surface hydrophobicity of 15 *Pasteurella multocida* strains was measured by percent adhesion to *n*-hexadecane and percent retained on an octyl-Sepharose column. A high degree of correlation ($r = 0.91$) between the two methods is evident. (Data taken from Darnell et al. [27].)

aceti). However, evaluation of the hydrophobicity of the others (*Entero-bacter aerogenes*, *Klebsiella oxytoca*, *Saccharomyces carlsbergensis*, and *Kluyveromyces fragilis*) varied widely according to the measurement technique. The authors recommended use of CAM in combination with HIC and adhesion to polystyrene and found MATH the "least adequate method." However, an examination of the experimental protocol shows that the authors did not adhere to the basic assay procedures. For example, toluene was used as the test hydrocarbon in the MATH test, and instead of small hydrocarbon/water ratios as suggested by Rosenberg et al. (148), equal volumes of hydrocarbon and water were used (73). HIC was performed without the untreated Sepharose CL-4B control. Further-more, three bacteria were compared alongside four fungal strains.

Dillon et al. (30) compared MATH, SAT, HIC, adhesion to polysty-rene, and polystyrene (latex) particle agglutination, using six bacterial strains. Bacteria were not grown in liquid culture but were scraped from agar plates. HIC was performed without the untreated Sepharose CL-4B control. Whereas both strains of *A. calcoaceticus* tested appeared hydro-phobic in all tests and *Streptococcus mitis* was relatively nonhydrophobic throughout, discrepancies were found for two *Staphylococcus aureus* strains and one *E. coli* strain.

Van der Mei and co-workers (177) compared a range of streptococcal isolates by using MATH, HIC, SAT, and CAM. Of the three comparison papers, this was the only one to analyze the data statistically. They found weak correlations among the various techniques, even when comparing tests from the same category (e.g., HIC versus MATH). This result may have been due to their use of the hydrophobicity index to determine HIC values. However, generally good correlations were found when related strains were studied.

It is not surprising that hydrophobicity tests sometimes fail to correlate. As previously emphasized (151), certain types of bacteria cannot be expected to give the same results in different tests. For example, a bacterium with a highly hydrophilic outer surface except for an adhesive hydrophobic tip (or a cell with a hydrophilic cell envelope surface but bearing fimbriae with hydrophobic binding properties) might be hydrophobic as measured by MATH, HIC, and adhesion to polysty-rene but nonhydrophobic as measured by CAM or TPP. Similarly, bacteria that tend to autoaggregate may give disproportionately high hydrophobicity values in SAT as compared with other techniques. Among bacterial species studied to date, perhaps *E. coli* is the most notorious in confounding hydrophobicity studies, since strains are often aggregated by low ammonium sulfate concentrations and sometimes bind to polystyrene but rarely adhere appreciably to hydrocarbons.

Electrostatic Interactions and the Hydrophobic Effect

Most microbial cells carry a net negative surface charge, the magnitude of which is affected by the type of strain, growth conditions, pH, and the presence and concentration of various inorganic ions (108). The role of electrostatic interactions in affecting microbial cell surface hydrophobicity has been addressed by many investigators during the past decade but remains somewhat unclear. Table 5 summarizes several recent studies that compared microbial charge and hydrophobicity. Intuitively, an inverse correlation should be expected between surface charge and hydrophobicity. Since surface charge increases the likelihood of polar interactions with proximal water molecules, the more charged surface groups, the lower the hydrophobicity. Indeed, treatments that tend to reduce the negative charge, e.g., covalent blocking (86; Goldberg et al., submitted) or addition of organic cations (57, 104; Goldberg et al., submitted), increase cell surface hydrophobicity. However, it is often found that the more electronegative the strain, the higher the observed hydrophobic properties (Table 5). This observation may depend on the techniques used. Weerkamp et al. compared MATH, CAM, and electrophoretic kinetics in six different oral species (187). Electrophoretic mobility (negative) correlated positively with water contact angles but negatively with MATH. This finding suggests either that MATH is relatively insensitive to charge effects or that water contact angle measurements on bacterial cell layers are more highly influenced by electrostatic interactions than was previously assumed.

Van Loosdrecht et al. (181) compared electrophoretic mobility and adhesion of various bacterial species to negatively charged polystyrene. Computer analyses of their data predicted that cells of high hydrophobicity and lowest electrophoretic mobility should be most adherent. However, relatively hydrophobic cells tended to also have high negative electrokinetic potentials.

Two explanations can be proffered for these apparent contradictions. First, the techniques used for measuring electrostatic interactions of microbial cells do not provide information regarding their distribution over the cell surface. Electrophoretic mobility measurements assess the overall net charge of the cell. Thus, a cell may have little or no net surface charge yet be covered by a large (but equivalent) number of negatively and positively charged moieties, all of which decrease the hydrophobic effect. Second, appendages such as fimbriae may simultaneously confer both hydrophobicity and electronegativity. Finally, it is conceivable that adhesion to positively charged gels involves a hydrophobic component and cannot be inferred as a direct indication of surface charge.

Table 5. Some Recent Studies Relating Hydrophobicity and Charge

Microorganisms studied	Hydrophobicity technique	Charge measurement	Result	Reference
Staphylococcus aureus	Binding of anionic surfactant		Cells "fattened" by growth in glycerol broth were more electronegative.	35
29 *Staphylococcus epidermidis* strains	CAM	Electrostatic interaction chromatography	No direct correlation.	40
Six *Lactobacillus* strains	Modified CAM	e.m.[2]	No direct correlation.	36
14 *Lactobacillus* strains	HIC	e.m.	Inverse correlation between hydrophobicity and negative surface charge.	58
Various streptococci	MATH, HIC, SAT, CAM	e.m.	Poor correlations with most hydrophobicity tests.	177
23 different strains	CAM	e.m.	Relatively hydrophobic cells tended to have higher negative electrokinetic potentials.	181
Five oral strains	CAM, MATH	e.m.	Direct correlation between CAM and e.m.; inverse correlation between MATH and e.m.	187
Rhizobium leguminosarum	CAM	e.m.	LPS mutants had both higher hydrophobicity and higher electrophoretic mobility.	28
Streptococcus sanguis and *S. mutans*	CAM	e.m.	Increasing contact angles correlating with decreasing zeta potentials.	160
Staphylococcus aureus	MATH	e.m.	Surface charge decreased with increasing hydrophobicity.	10

[a] e.m., Electrophoretic mobility.

BIOLOGICAL SIGNIFICANCE AND RELEVANCE

One general problem regarding microbial cell surface hydrophobicity may be stated as follows: Are hydrophobic surface properties actually important for microbial functions, or are they simply a measurement outcome reflecting the presence (or absence) of surface components? As an example, consider the function of bacterial exopolysaccharides and their effect on cell surface hydrophobicity. It has become increasingly clear that the presence of capsules in almost all instances (coagulase-negative staphylococci being one possible exception [69, 70]) results in decreased adhesion and hydrophobicity. The presence of capsule reduces adhesion to hydrocarbon of *P. multocida* (175), *Streptococcus pyogenes* (119), *A. calcoaceticus* (141), *Klebsiella pneumoniae* (11), and *Staphylococcus aureus* (137). Capsules similarly prevent adhesion of *Pseudomonas* strain S9 to siliconized glass (193, 194). Should the conclusion be drawn that capsules are elaborated as antiadhesive devices? This is a compelling argument (119), but exopolysaccharides likely serve many additional functions. Conversely, the presence of certain fimbriae has been shown to impart hydrophobic surface properties to various micro-organisms (49, 131, 147, 166, 173, 186). However, one cannot exclude the possibility that the fimbrial tips possess stereospecific adhesins and that whereas adhesion is correlated with hydrophobicity, the actual adhesion process occurs by a separate mechanism.

Role of Hydrophobic Interactions in So-Called Specific Adhesion

The term "specificity" in describing microbial adhesion is generally used to denote the stereospecific "fit" between cell surface molecules and stereospecific receptors on the surface of the substratum. This characteristic is usually inferred from inhibition data showing that only molecules of a certain stereochemical structure are able to inhibit the adhesion process. For example, adhesion of various enterobacterial strains (usually bearing type 1 fimbriae) to yeast cells and certain mammalian cells can be blocked by mannose and mannosides but not by other saccharides (41, 42). An additional example is the "lactose-specific" adhesion of oral microorganisms to one another (e.g., *Streptococcus sanguis* to *Actinomyces viscosus*) (102). On the other hand, adhesion mediated by hydrophobic interactions is often considered to be "nonspecific."

Certainly, this concept of specific versus nonspecific adhesion is misleading. First, cell surface hydrophobicity may be promoted by specific proteins with specific amino acid sequences. Similarly, adhesion may be specific to a single member of a given species, regardless of the mechanism involved. Finally and perhaps most importantly, we are far from understanding how specific the so-called specific adhesion mechanisms really are. In both of the examples cited above, (mannose- and lactose-inhibitable adhesion), hydrophobic molecules are more potent as inhibitors than the sugars or saccharides themselves. Firon and co-workers (41, 42) demonstrated that *p*-nitrophenyl-*o*-chlorophenylmanno-side was approximately 400 times more inhibitory than methylmannoside in blocking yeast aggregation by *E. coli* and in its adhesion to ileal epithelial cells. Ohman et al. (120) observed that mannose inhibited binding of *E. coli* to octyl-Sepharose beads only in those strains bearing mannose-sensitive adhesins. Harber and co-workers (59) reported inhibition of adhesion to polystyrene by mannose. Similarly, Falkowski and co-workers (38) have shown that small, hydrophobic molecules, structurally unrelated to mannose, are potent inhibitors of "mannose-specific" binding to epithelial cells. Similarly, surface-active molecules inhibit lactose-specific bacterial coadhesion at lower molar levels than does lactose itself (102). These investigations highlight the importance of hydrophobic interactions in promoting stereospecific interactions, but the precise mechanism is unclear. Does the hydrophobic effect enable an initial docking that facilitates subsequent, stereospecific linkages? Or do hydrophobic interactions proximal to the lectinlike binding sites increase the affinity of the receptor-ligand binding?

Conversely, how "nonspecific" is adhesion via hydrophobic interactions? The nonspecificity refers to a lack of understanding of the molecular texture at which the hydrophobic effect between microorganisms and interfaces takes place. For example, how does the surface array of microbial hydrophobins affect adhesion to mammalian cells? One might expect that for optimal adhesion to occur, there should be a topographical fit between interacting hydrophobins on both surfaces. This fit might involve a complementary, periodic surface distribution of hydrophobic and hydrophilic components (64).

SUMMARY

Certainly, the great body of literature summarized in this volume points toward a general role for the hydrophobic effect in mediating microbial adhesion and partitioning phenomena. Knowledge of hydrophobic surface properties and how they are mediated can facilitate manipulation of microorganisms, binding them to surfaces or removing them. Adhesion to relatively unexplored hydrophobic substrata, e.g., leaf surfaces (167) and gastric laminal mucosa (55), remains to be investigated. Microbial cell surface hydrophobicity is emerging as a diagnostic criterion (101). Future investigations will help to further our understanding of the measurement techniques now in use and how they are related, lead to development of novel assessment methods, aid in identification and probing of hydrophobins and hydrophilins and their surface topography, and help to further clarify the role of the hydrophobic effect in microbial ecology and pathogenesis.

Acknowledgments. Support from the United States-Israel Binational Science Foundation, Jerusalem, Israel (grant 86-00263), and Public Health Service grant DE07199 from the National Institutes of Health is gratefully acknowledged.

LITERATURE CITED

1. **Abbott, A., and M. L. Hayes.** 1984. The conditioning role of saliva in streptococcal attachment to hydroxyapatite surfaces. *J. Gen. Microbiol.* **130**:809–816.
2. **Albertsson, P. A.** 1958. Particle fractionation in liquid two-phase system. The composition of some phase systems and the behavior of some model particles in them. Application to the isolation of cell walls from microorganisms. *Biochim. Biophys. Acta* **27**:378–395.
3. **Amory, D. E., and P. G. Rouxhet.** 1988. Surface properties of *Saccharomyces cerevisiae* and *Saccharomyces carlsbergensis*: chemical composition, electrostatic charge and hydrophobicity. *Biochim. Biophys. Acta* **938**:61–70.
4. **Angus, T. A.** 1959. Separation of bacterial spores and parasporal bodies with a fluorocarbon. *J. Invertebr. Pathol.* **1**:97–98.
5. **Arakawa, K., K. Tokiwano, N. Ohtomo, and H. Kedaira.** 1979. A note on the nature of ionic hydrations and hydrophobic interactions in aqueous solutions. *Bull. Chem. Soc. Jpn.* **52**:2483–2488.
6. **Babu, J. P., E. H. Beachey, and W. A. Simpson.** 1986. Inhibition of the interaction of *Streptococcus sanguis* with hexadecane droplets by 55- and 66-kilodalton hydrophobic proteins of human saliva. *Infect. Immun.* **53**:278–284.
7. **Bandin, I., Y. Snatos, J. L. Barja, and A. E. Toranzo.** 1989. Influence of the growth conditions on the hydrophobicity of *Renibacterium salmoninarum* evaluated by different methods. *FEMS Microbiol. Lett.* **60**:71–78.
8. **Bar-Ness, R., N. Avrahamy, T. Matsuyama, and M. Rosenberg.** 1988. Increased cell surface hydrophobicity of a *Serratia marcescens* NS 38 mutant lacking wetting activity. *J. Bacteriol.* **170**:4361–4364.
9. **Bar-Ness, R., and M. Rosenberg.** 1989. Putative role of a 70 kd surface protein in mediating cell surface hydrophobicity of *Serratia marcescens*. *J. Gen. Microbiol.* **135**:2277–2281.
10. **Beck, G., E. Puchelle, C. Plotkowski, and R. Peslin.** 1988. Effect of growth on surface charge and hydrophobicity of *Staphylococcus aureus*. *Ann. Microbiol.* (Paris) **139**:655–664.
11. **Benedi, V. J., B. Ciurana, and J. M. Tomas.** 1989. Isolation and characterization of *Klebsiella pneumoniae* unencapsulated mutants. *J. Clin. Microbiol.* **27**:82–87.
12. **Bishop, W. H.** 1987. Molecular and macroscopic aspects of cavity formation in the hydrophobic effect. *Biophys. Chem.* **27**:197–210.
13. **Boucias, D. G., J. C. Pendland, and J. P. Latge.** 1988. Nonspecific factors involved in attachment of entomopathogenic deuteromycetes to host insect cuticle. *Appl. Environ. Microbiol.* **54**:1795–1805.
14. **Boyles, W. A., and R. E. Lincoln.** 1958. Separation and concentration of bacterial spores and vegetative cells by foam flotation. *Appl. Microbiol.* **6**:327–334.
15. **Bruno, D. W.** 1988. The relationship between auto-agglutination, cell surface hydrophobicity and virulence of the fish pathogen *Renibacterium salmoninarum*. *FEMS Microbiol. Lett.* **51**:135–140.

16. **Bryant, R. D., J. W. Costerton, and E. J. Laishley.** 1984. The role of *Thiobacillus albertis* glycocalyx in the adhesion of cells to elemental sulfur. *Can. J. Microbiol.* **30:**81–90.

17. **Burchard, R. P.** 1986. The effect of surfactants on the motility and adhesion of gliding bacteria. *Arch. Microbiol.* **146:**147–150.

18. **Burke, D. A., and A. T. Axon.** 1988. Hydrophobic adhesin of *E. coli* in ulcerative colitis. *Gut* **29:**41–43.

19. **Busscher, H. J., and A. H. Weerkamp.** 1987. Specific and nonspecific interactions in bacterial adhesion to solid substrata. *FEMS Microbiol. Rev.* **46:**165–173.

20. **Chugh, T. D., G. J. Burns, H. J. Shuhaiber, and G. M. Bahr.** 1990. Adherence of *Staphylococcus epidermidis* to fibrin-platelet clots in vitro mediated by lipoteichoic acid. *Infect. Immun.* **58:**315–319.

21. **Clark, W. B., M. D. Lane, J. E. Beem, S. L. Bragg, and T. T. Wheeler.** 1985. Relative hydrophobicities of *Actinomyces viscosus* and *Actinomyces naeslundii* strains and their adsorption to saliva-treated hydroxyapatite. *Infect. Immun.* **47:**730–736.

22. **Cornette, J. L., K. B. Cease, H. Margalit, J. L. Spouge, J. A. Berzofsky, and C. DeLisi.** 1987. Hydrophobicity scales and computational techniques for detecting amphipathic structures in proteins. *J. Mol. Biol.* **195:**659–685.

23. **Craven, S. E., and L. C. Blankenship.** 1987. Changes in the hydrophobic characteristics of *Clostridium perfringens* spores and spore coats by heat. *Can. J. Microbiol.* **33:**773–776.

24. **Cunningham, R. K., T. O. Soderstrom, C. F. Gillman, and C. J. van Oss.** 1975. Phagocytosis as a surface phenomenon. V. Contact angles and phagocytosis of rough and smooth strains of *Salmonella typhimurium*, and the influence of specific antiserum. *Immunol. Commun.* **4:**429–442.

25. **Dahlback, B., M. Hermansson, S. Kjelleberg, and B. Norkrans.** 1981. The hydrophobicity of bacteria—an important factor in their initial adhesion at the air-water interface. *Arch. Microbiol.* **128:**267–270.

26. **Daly, J. G., and R. M. W. Stevenson.** 1987. Hydrophobic and haemagglutinating properties of *Renibacterium salmoninarum*. *J. Gen. Microbiol.* **133:**3575–3580.

27. **Darnell, K. R., M. E. Hart, and F. R. Champlin.** 1987. Variability of cell surface hydrophobicity among *Pasteurella multocida* somatic serotype and *Actinobacillus lignieresii* strains. *J. Clin. Microbiol.* **25:**67–71.

28. **de Maagd, R. A., A. S. Rao, I. H. M. Mulders, L. Goosen-de Roo, M. C. M. van Loosdrecht, C. A. Wijffelman, and B. J. J. Lugtenberg.** 1989. Isolation and characterization of mutants of *Rhizobium leguminosarum* bv. *viciae* 248 with altered lipopolysaccharides: possible role of surface charge or hydrophobicity in bacterial release from infection thread. *J. Bacteriol.* **171:**1143–1150.

29. **Dickson, J. S., and M. Koohmaraie.** 1989. Cell surface charge characteristics and their relationship to bacterial attachment to meat surfaces. *Appl. Environ. Microbiol.* **55:**832–836.

30. **Dillon, J. K., J. A. Fuerst, A. C. Hayward, and G. H. G. Davis.** 1986. A comparison of five methods for assaying bacterial hydrophobicity. *J. Microbiol. Methods* **6:**13–19.

31. **Doyle, R. J., F. Nedjat-Halem, and J. S. Singh.** 1984. Hydrophobic characteristics of *Bacillus* spores. *Curr. Microbiol.* **10:**329–332.

32. **Doyle, R. J., W. E. Nesbitt, and K. G. Taylor.** 1982. On the mechanism of adherence of *Streptococcus sanguis* to hydroxylapatite. *FEMS Microbiol. Lett.* **15:**1–5.

33. **Drumm, B., A. W. Neumann, Z. Policova, and P. M. Sherman.** 1989. Bacterial cell surface hydrophobicity properties in the mediation of in vitro adhesion by the rabbit enteric pathogen *Escherichia coli* strain RDEC-1. *J. Clin. Invest.* **84:**1588–1594.

34. **Dyar, M. T.** 1948. Electrokinetical studies on bacterial surfaces. II. Studies on surface lipids, amphoteric material and some other surface properties. *J. Bacteriol.* **56**:821–834.

35. **Dyar, M. T., and E. J. Ordal.** 1946. Electrokinetic studies on bacterial surfaces. I. The effects of surface-active agents on the electrophoretic mobilities of bacteria. *J. Bacteriol.* **51**:149–167.

36. **Eisen, A., and G. Reid.** 1989. Effect of culture media on *Lactobacillus* hydrophobicity and electrophoretic mobility. *Microb. Ecol.* **17**:17–25.

37. **Eli, I., H. Judes, and M. Rosenberg.** 1989. Saliva-mediated inhibition on promotion of bacterial adhesion to polystyrene. *Biofouling* **1**:203–211.

38. **Falkowski, W., M. Edwards, and A. J. Schaeffer.** 1986. Inhibitory effect of substituted aromatic hydrocarbons on adherence of *Escherichia coli* to human epithelial cells. *Infect. Immun.* **52**:863–866.

39. **Fattom, A., and M. Shilo.** 1984. Hydrophobicity as an adhesion mechanism of benthic cyanobacteria. *Appl. Environ. Microbiol.* **47**:135–143.

40. **Ferreiros, C. M., J. Carballo, M. T. Criado, V. Sainz, and M. C. del Rio.** 1989. Surface free energy and interaction of *Staphylococcus epidermidis* with biomaterials. *FEMS Microbiol. Lett.* **60**:89–94.

41. **Firon, N., S. Ashkenazi, D. Mirelman, I. Ofek, and N. Sharon.** 1987. Aromatic alpha-glycosides of mannose are powerful inhibitors of the adherence of type 1 fimbriated *Escherichia coli* to yeast and intestinal epithelial cells. *Infect. Immun.* **55**:472–476.

42. **Firon, N., I. Ofek, and N. Sharon.** 1984. Carbohydrate-binding sites of the mannose-specific fimbrial lectins of enterobacteria. *Infect. Immun.* **43**:1088–1090.

43. **Fives-Taylor, P. M., and D. W. Thompson.** 1985. Surface properties of *Streptococcus sanguis* FW213 mutants non-adherent to saliva-coated hydroxyapatite. *Infect. Immun.* **47**:752–759.

44. **Fletcher, M.** 1976. The effects of proteins on bacterial attachment to polystyrene. *J. Gen. Microbiol.* **94**:400–404.

45. **Fletcher, M., and G. I. Loeb.** 1979. Influence of substratum characteristics on the attachment of a marine pseudomonad to solid surfaces. *Appl. Environ. Microbiol.* **44**:184–192.

46. **Fletcher, M., and J. H. Pringle.** 1983. The effect of surface free energy and medium surface tension on bacterial attachment to solid surfaces. *J. Colloid Interface Sci.* **104**:5–13.

47. **Foegeding, P. M., and M. L. Fulp.** 1988. Comparison of coats and surface-dependent properties of *Bacillus cereus* T prepared in two sporulation environments. *J. Appl. Bacteriol.* **65**:249–259.

48. **Fussell, E. N., M. B. Kaack, R. Cherry, and J. A. Roberts.** 1988. Adherence of bacteria to human foreskins. *J. Urol.* **140**:997–1001.

49. **Galdiero, F., M. Cotrufo, P. G. Catalanotti, T. S. De Luca, R. Ianniello, and E. Galdiero.** 1987. Adherence of bacteria to cardiac valve prostheses. *Eur. J. Epidemiol.* **3**:216–221.

50. **Garber, N., N. Sharon, D. Shohet, J. S. Lam, and R. J. Doyle.** 1985. Contribution of hydrophobicity to hemagglutination reactions of *Pseudomonas aeruginosa*. *Infect. Immun.* **50**:336–337.

51. **Gaudin, A. M., S. L. Mular, and R. F. O'Connor.** 1960. Separation of microorganisms by flotation. II. Flotation of spores of *Bacillus subtilis* var. *niger*. *Appl. Microbiol.* **8**:91–97.

52. **Gerson, D. F., and J. Akit.** 1980. Cell surface energy, contact angles and phase

partition. II. Bacterial cells in biphasic aqueous mixtures. *Biochim. Biophys. Acta* **602**:281–284.

53. **Gibbons, R. J., and I. Etherden.** 1983. Comparative hydrophobicities of oral bacteria and their adherence to salivary pellicles. *Infect. Immun.* **41**:1190–1196.

54. **Gibbons, R. J., I. Etherden, and Z. Skobe.** 1983. Association of fimbriae with the hydrophobicity of *Streptococcus sanguis* FC-1 and adherence to salivary pellicles. *Infect. Immun.* **41**:414–417.

55. **Goddard, P. J., Y. C. Kao, and L. M. Lichtenberger.** 1990. Luminal surface hydrophobicity of canine gastric mucosa is dependent on a surface mucous gel. *Gastroenterology* **98**:361–370.

56. **Godfrey, A. J., and L. E. Bryan.** 1989. Cell surface changes in *Pseudomonas aeruginosa* PAC4069 in response to treatment with 6-aminopenicillanic acid. *Antimicrob. Agents Chemother.* **33**:1435–1442.

57. **Goldberg, S., Y. Konis, and M. Rosenberg.** 1990. Effect of cetylpyridinium chloride on microbial adhesion to hexadecane and polystyrene. *Appl. Environ. Microbiol.* **56**:1678–1682.

58. **Gullmar, B., S. Hjertén, and T. Wadström.** 1988. Cell surface charge of lactobacilli and enterococci isolated from pig small intestine as studied by free zone electrophoresis: a methodological study. *Microbios* **55**:183–192.

59. **Harber, M. J., R. Mackenzie, and A. W. Asscher.** 1983. A rapid bioluminescence method for quantifying bacterial adhesion to polystyrene. *J. Gen. Microbiol.* **129**:621–632.

60. **Hart, D. J., and R. H. Vreeland.** 1988. Changes in the hydrophobic-hydrophilic cell surface character of *Halomonas elongata* in response to NaCl. *J. Bacteriol.* **170**:132–135.

61. **Hart, M. E., and F. R. Chaplin.** 1988. Susceptibility to hydrophobic molecules and phospholipid composition in *Pasteurella multocida* and *Actinobacillus lignieresii*. *Antimicrob. Agents Chemother.* **32**:1354–1359.

62. **Hazen, K. C., and B. W. Hazen.** 1987. A polystyrene microsphere assay for detecting surface hydrophobicity variations within *Candida albicans* populations. *J. Microbiol. Methods* **6**:289–299.

63. **Hermansson, M., S. Kjelleberg, T. K. Korhonen, and T.-A. Stenstrom.** 1982. Hydrophobic and electrostatic characterization of surface structures of bacteria and its relationship to adhesion at an air-water surface. *Arch. Microbiol.* **131**:308–312.

64. **Hewison, L. A., W. T. Coakley, and H. W. Meyer.** 1988. Spatially periodic discrete contact regions in polylysine-induced erythrocyte-yeast adhesion. *Cell Biophys.* **13**:151–157.

65. **Hill, M. J., A. M. James, and W. R. Maxted.** 1963. Some physical investigations of the behaviour of bacterial surfaces. X. The occurrence of lipids in the streptococcal cell wall. *Biochim. Biophys. Acta* **75**:414–424.

66. **Hjertén, S., J. Rosenbergen, and S. Pahlman.** 1974. Hydrophobic interaction chromatography. The synthesis and the use of some alkyl and aryl derivatives of agarose. *J. Chromatogr.* **101**:281–288.

67. **Hogg, S. D., and J. E. Manning.** 1987. The hydrophobicity of 'viridans' streptococci isolated from the human mouth. *J. Appl. Bacteriol.* **63**:311–318.

68. **Hogg, S. D., and L. A. Old.** 1987. The hydrophobicity of *Streptococcus salivarius* strain HB and mutants deficient in adhesion to saliva-coated hydroxyapatite. *Lett. Appl. Microbiol.* **4**:99–101.

69. **Hogt, A. H., J. Dankert, J. A. de Vries, and J. Feijen.** 1983. Adhesion of coagulase-negative staphylococci to biomaterials. *J. Gen. Microbiol.* **129**:2959–2968.

70. **Hogt, A. H., J. Dankert, C. E. Hulstaert, and J. Feijen.** 1986. Cell surface characteristics of coagulase-negative staphylococci and their adherence to fluorinated poly(ethylenepropylene). *Infect. Immun.* **51:**294–301.

71. **Humphries, M., J. F. Jaworzyn, and J. B. Cantwell.** 1986. The effect of a range of biological polymers and synthetic surfactants on the adhesion of a marine *Pseudomonas* sp. strain NCMB 2021 to hydrophilic and hydrophobic surfaces. *FEMS Microbiol. Ecol.* **38:**299–308.

72. **Humphries, M., J. F. Jaworzyn, J. B. Cantwell, and A. Eakin.** 1987. The use of non-ionic ethoxylated and propoxylated surfactants to prevent the adhesion of bacteria to solid surfaces. *FEMS Microbiol. Lett.* **42:**91–101.

73. **Iimura, Y., S. Hara, and K. Otsuka.** 1980. Cell surface hydrophobicity as a pellicle formation factor in film strain of *Saccharomyces*. *Agric. Biol. Chem.* **6:**1215–1222.

74. **Israelachivili, J. N., and P. M. McGuiggan.** 1988. Forces between surfaces in liquids. *Science* **241:**795–800.

75. **Ito, H., E. Ono, and R. Yanagawa.** 1987. Comparison of surface hydrophobicity of piliated and non-piliated clones of *Corynebacterium renale* and *Corynebacterium pilosum*. *Vet. Microbiol.* **14:**165–171.

76. **Jacques, M., and J. W. Costerton.** 1987. Adhesion of group B *Streptococcus* to a polyethylene intrauterine contraceptive device. *FEMS Microbiol. Lett.* **41:**23–28.

77. **Jacques, M., T. J. Marrie, and J. W. Costerton.** 1986. In vitro quantitative adherence of microorganisms to intrauterine contraceptive devices. *Curr. Microbiol.* **13:**133–137.

78. **Jenkinson, H. F., and D. A. Carter.** 1988. Cell surface mutants of *Streptococcus sanguis* with altered adherence properties. *Oral Microbiol. Immunol.* **3:**53–57.

79. **Jones, M. V., T. M. Herd, and H. J. Christie.** 1989. Resistance of *Pseudomonas aeruginosa* to amphoteric and quaternary ammonium biocides. *Microbios* **58:**49–61.

80. **Kappeli, O., and A. Fiechter.** 1976. The mode of interaction between the substrate and the cell surface of the hydrocarbon-utilizing yeast *Candida tropicalis*. *Biotechnol. Bioeng.* **18:**967–974.

81. **Kennedy, R. S., W. R. Finnerty, K. Sudarsanan, and R. A. Young.** 1975. Microbial assimilation of hydrocarbon. I. The fine structure of a hydrocarbon oxidizing *Acinetobacter* sp. *Arch. Microbiol.* **102:**75–83.

82. **Kirschner-Zilber, I., E. Rosenberg, and D. Gutnick.** 1980. Incorporation of ^{32}P and growth of pseudomonad UP-2 on *n*-tetracosane. *Appl. Environ. Microbiol.* **40:**1086–1093.

83. **Kjelleberg, S., C. Lagercrantz, and T. Larsson.** 1980. Quantitative analysis of bacterial hydrophobicity studied by the binding of dodecanoic acid. *FEMS Microbiol. Lett.* **7:**41–44.

84. **Klotz, S. A., S. I. Butrus, R. P. Misra, and M. S. Osato.** 1989. The contribution of bacterial surface hydrophobicity to the process of adherence of *Pseudomonas aeruginosa* to hydrophilic contact lenses. *Curr. Eye Res.* **8:**195–202.

85. **Klotz, S. A., D. J. Drutz, and J. E. Zajic.** 1985. Factors governing adherence of *Candida* species to plastic surfaces. *Infect. Immun.* **50:**97–101.

86. **Koga, T., N. Okahashi, I. Takahashi, T. Kanamoto, H. Asakawa, and M. Iwaki.** 1990. Surface hydrophobicity, adherence, and aggregation of cell surface protein antigen mutants of *Streptococcus mutans* serotype c. *Infect. Immun.* **58:**289–296.

87. **Koshikawa, T., M. Yamazaki, M. Yoshimi, S. Ogawa, A. Yamada, K. Watabe, and M. Torii.** 1989. Surface hydrophobicity of spores of *Bacillus* spp. *J. Gen. Microbiol.* **135:**2717–2722.

88. **Krekeler, C., H. Ziehr, and J. Klein.** 1989. Physical methods for characterization of microbial cell surfaces. *Experientia* **45:**1047–1055.

89. **Kutima, P. M., and P. M. Foegeding.** 1987. Involvement of the spore coat in germination of *Bacillus cereus* T spores. *Appl. Environ. Microbiol.* **53**:47–52.

90. **Lachica, R. V., and D. L. Zink.** 1984. Determination of plasmid-associated hydrophobicity of *Yersinia enterocolitica* by a latex particle agglutination test. *J. Clin. Microbiol.* **19**:660–663.

91. **Leach, S. A.** 1980. A biophysical approach to interactions associated with the formation of the matrix of dental plaque, p. 159–183. *In* S. A. Leach (ed.), *Dental Plaque and Surface Interactions in the Oral Cavity.* IRL Press, Oxford.

92. **Lichtenberg, D., M. Rosenberg, N. Sharfman, and I. Ofek.** 1985. A kinetic approach to bacterial adherence to hydrocarbon. *J. Microbiol. Methods* **4**:141–146.

93. **Liljemark, W. F., S. V. Schauer, and C. G. Bloomquist.** 1978. Compounds which affect the adherence of *Streptococcus sanguis* and *Streptococcus mutans* to hydroxyapatite. *J. Dent. Res.* **57**:373–379.

94. **Lindahl, M., A. Faris, T. Wadström, and S. Hjertén.** 1981. A new test based on 'salting out' to measure relative surface hydrophobicity of bacterial cells. *Biochim. Biophys. Acta* **677**:471–476.

95. **Magnusson, K.-E., C. Dahlgren, G. Maluszynska, E. Kihlstrom, T. Skogh, O. Stendahl, G. Soderlund, L. Ohman, and A. Walan.** 1985. Nonspecific and specific recognition mechanisms of bacterial and mammalian cell membranes. *J. Dispersion Sci. Technol.* **6**:69–89.

96. **Magnusson, K.-E., and G. Johansson.** 1977. Probing the surface of *Salmonella typhimurium* SR and R bacteria by aqueous biphasic partitioning in systems containing hydrophobic and charged polymers. *FEMS Microbiol. Lett.* **2**:225–228.

97. **Malmqvist, T.** 1983. Bacterial hydrophobicity measured as partition of palmitic acid between the two immiscible phases of cell surface and buffer. *Acta Pathol. Microbiol. Immunol. Scand. Sect. B* **91**:69–73.

98. **Marshall, K. C.** 1976. *Interfaces in Microbial Ecology.* Harvard University Press, Cambridge, Mass.

99. **Marshall, K. C., and R. H. Cruickshank.** 1973. Cell surface hydrophobicity and the orientation of certain bacteria at interfaces. *Arch. Mikrobiol.* **91**:29–40.

100. **Marshall, K. C., R. H. Cruickshank, and H. V. A. Bushby.** 1975. The orientation of certain root-nodule bacteria at interfaces, including root-hair surfaces. *J. Gen. Microbiol.* **91**:198–200.

101. **Martin, M. A., M. A. Pfaller, R. M. Massanari, and R. P. Wenzel.** 1989. Use of cellular hydrophobicity, slime production, and species identification markers for the clinical significance of coagulase-negative staphylococcal isolates. *Am. J. Infect. Control* **17**:130–135.

102. **McIntire, F. C., L. K. Crosby, and A. E. Vatter.** 1982. Inhibitors of coaggregation between *Actinomyces viscosus* T14V and *Streptococcus sanguis* 34: β-galactosides, related sugars, and anionic amphipathic compounds. *Infect. Immun.* **36**:371–378.

103. **McLee, A. G., and S. L. Davies.** 1972. Linear growth of a *Torulopsis* sp. on n-alkanes. *Can. J. Microbiol.* **18**:315–319.

104. **Miller, M. J., and D. G. Ahearn.** 1987. Adherence of *Pseudomonas aeruginosa* to hydrophilic contact lenses and other substrata. *J. Clin. Microbiol.* **25**:1392–1397.

105. **Minagi, S., Y. Miyake, Y. Fujioka, H. Tsuru, and H. Suginaka.** 1986. Cell-surface hydrophobicity of *Candida* species as determined by the contact-angle and hydrocarbon-adherence methods. *J. Gen. Microbiol.* **132**:1111–1115.

106. **Miura, Y., M. Okazaki, S.-I. Hamada, S.-I. Murakawa, and R. Yugen.** 1977. Assimilation of liquid hydrocarbon by microorganisms. I. Mechanism of hydrocarbon uptake. *Biotechnol. Bioeng.* **19**:701–714.

107. **Morris, E. J., N. Ganeshkumar, and B. C. McBride.** 1985. Cell surface components of *Streptococcus sanguis*: relationship to aggregation, adherence, and hydrophobicity. *J. Bacteriol.* **164:**255–262.

108. **Mozes, N., F. Marchal, M. P. Hermesse, J. L. van Haecht, L. Reuliaux, A. J. Leonard, and P. G. Rouxhet.** 1987. Immobilization of microorganisms by adhesion: interplay of electrostatic and nonelectrostatic interactions. *Biotechnol. Bioeng.* **30:**439–450.

109. **Mozes, N., and P. G. Rouxhet.** 1987. Methods for measuring hydrophobicity of microorganisms. *J. Microbiol. Methods* **6:**99–112.

110. **Mudd, S., and E. B. H. Mudd.** 1924. The penetration of bacteria through capillary spaces. IV. A kinetic mechanism in interfaces. *J. Exp. Med.* **40:**633–645.

111. **Mudd, S., and E. B. H. Mudd.** 1924. Certain interfacial tension relations and the behaviour of bacteria in films. *J. Exp. Med.* **40:**647–660.

112. **Nataro, J. P., I. C. Scaletsky, J. B. Kaper, M. M. Levine, and L. R. Trabulsi.** 1985. Plasmid-mediated factors conferring diffuse and localized adherence of enteropathogenic *Escherichia coli. Infect. Immun.* **48:**378–383.

113. **Nealon, T. J., and S. J. Mattingly.** 1984. Role of cellular lipoteichoic acids in mediating adherence of serotype III strains of group B streptococci to human embryonic, fetal, and adult epithelial cells. *Infect. Immun.* **43:**523–530.

114. **Neeser, J.-R., A. Chambaz, S. W. Vedovo, M.-J. Prigent, and B. Guggenheim.** 1988. Specific and nonspecific inhibition of adhesion of oral actinomyces and streptococci to erythrocytes and polystyrene by caseinoglycopeptide derivatives. *Immun. Infect.* **56:**3201–3208.

115. **Nesbitt, W. E., R. J. Doyle, and K. G. Taylor.** 1982. Hydrophobic interactions and the adherence of *Streptococcus sanguis* to hydroxylapatite. *Infect. Immun.* **38:**637–644.

116. **Nesbitt, W. E., R. J. Doyle, K. G. Taylor, R. H. Staat, and R. R. Arnold.** 1982. Positive cooperativity in the binding of *Streptococcus sanguis* to hydroxylapatite. *Infect. Immun.* **35:**157–165.

117. **Noda, Y., and Y. Kanemasa.** 1986. Determination of hydrophobicity on bacterial surfaces by nonionic surfactants. *J. Bacteriol.* **167:**1016–1019.

118. **Oakley, J. D., K. G. Taylor, and R. J. Doyle.** 1985. Trypsin-susceptible cell surface characteristics of *Streptococcus sanguis. Can. J. Microbiol.* **31:**1103–1107.

119. **Ofek, I., E. Whitnack, and E. H. Beachey.** 1983. Hydrophobic interactions of group A streptococci with hexadecane droplets. *J. Bacteriol.* **154:**139–145.

120. **Ohman, L., K.-E. Magnusson, and O. Stendahl.** 1985. Effect of monosaccharides and ethyleneglycol on the interaction between *Escherichia coli* bacteria and octyl-Sepharose. *Acta Pathol. Microbiol. Immunol. Scand. Sect. B* **93:**133–138.

121. **Olsson, J., and G. Westergren.** 1982. Hydrophobic surface properties of oral streptococci. *FEMS Microbiol. Lett.* **15:**319–323.

122. **Op den Camp, H. J. M., A. Oosterhof, and J. H. Veerkamp.** 1985. Cell surface hydrophobicity of *Bifidobacterium bifidum* subsp. *pennsylvanicum. Antonie van Leeuwenhoek J. Microbiol. Serol.* **51:**303–312.

123. **Owens, N. F., D. Gingell, and P. R. Rutter.** 1987. Inhibition of cell adhesion by a synthetic polymer adsorbed to glass shown under defined hydrodynamic stress. *J. Cell Sci.* **87:**667–675.

124. **Ouchi, K., and H. Akiyama.** 1971. Non-foaming mutants of Sake yeasts: selection by cell agglutination method and by froth flotation method. *Agric. Biol. Chem.* **7:**1024–1032.

125. **Pal, T., and T. L. Hale.** 1989. Plasmid-associated adherence of *Shigella flexneri* in a HeLa cell model. *Infect. Immun.* **57:**2580–2582.

126. **Parker, N. D., and C. B. Munn.** 1984. Increased cell surface hydrophobicity associated

with possession of an additional surface protein by *Aeromonas salmonicida*. *FEMS Microbiol. Lett.* **21**:233–237.

127. **Pascual, A., A. Fleer, N. A. Westerdaal, M. Berghuis, and J. Verhoef.** 1988. Surface hydrophobicity and opsonic requirements of coagulase-negative staphylococci in suspension and adhering to a polymer substratum. *Eur. J. Clin. Microbiol. Infect. Dis.* **7**:161–166.

128. **Pascual, A., A. Fleer, N. A. Westerdaal, and J. Verhoef.** 1986. Modulation of adherence of coagulase-negative staphylococci to Teflon catheters in vitro. *Eur. J. Clin. Microbiol.* **5**:518–522.

129. **Pashley, R. M., P. M. Mcguiggan, B. W. Ninham, and D. F. Evans.** 1985. Attractive forces between uncharged hydrophobic surfaces: direct measurements in aqueous solution. *Science* **229**:1088–1089.

130. **Paula, S. J., P. S. Duffey, S. L. Abbott, R. P. Kokka, L. S. Oshiro, J. M. Janda, T. Shimada, and R. Sakazaki.** 1988. Surface properties of autoagglutinating mesophilic aeromonads. *Infect. Immun.* **56**:2658–2665.

131. **Peros, W. J., I. Etherden, R. J. Gibbons, and Z. Skobe.** 1985. Alteration of fimbriation and cell hydrophobicity by sublethal concentrations of tetracycline. *J. Periodont. Res.* **20**:24–30.

132. **Pines, O., Y. Shoham, E. Rosenberg, and D. Gutnick.** 1988. Unmasking of surface components by removal of cell-associated emulsan from *Acinetobacter* sp. RAG-1. *Appl. Microbiol. Biotechnol.* **28**:93–99.

133. **Pringle, J. H., and M. Fletcher.** 1983. Influence of substratum wettability on attachment of freshwater bacteria to solid surfaces. *Appl. Environ. Microbiol.* **45**:811–817.

134. **Pringle, J. H., and M. Fletcher.** 1988. Influence of substratum hydration and adsorbed macromolecules on bacterial attachment to surfaces. *Appl. Environ. Microbiol.* **51**:1321–1325.

135. **Reed, G. B., and C. E. Rice.** 1931. The behaviour of acid-fast bacteria in oil and water systems. *J. Bacteriol.* **22**:239–247.

136. **Reid, G., L.-A. Hawthorn, R. Mandatori, R. L. Cook, and H. S. Beg.** 1988. Adhesion of lactobacilli to polymer surfaces in vivo and in vitro. *Microb. Ecol.* **16**:241–251.

137. **Reifsteck, F., S. Wee, and B. J. Wilkinson.** 1987. Hydrophobicity-hydrophilicity of staphylococci. *J. Med. Microbiol.* **24**:65–73.

138. **Robinson, A., A. R. Gorringe, L. I. Irons, and C. W. Keevil.** 1983. Antigenic modulation of *Bordetella pertussis* in continuous culture. *FEMS Microbiol. Lett.* **19**:105–109.

139. **Rosenberg, E., D. R. Brown, and A. L. Demain.** 1985. The influence of gramicidin S on hydrophobicity of germinating *Bacillus brevis* spores. *Arch. Microbiol.* **142**:51–54.

140. **Rosenberg, E., A. Gottlieb, and M. Rosenberg.** 1983. Inhibition of bacterial adherence to hydrocarbons and epithelial cells by emulsan. *Infect. Immun.* **39**:1024–1028.

141. **Rosenberg, E., N. Kaplan, O. Pines, M. Rosenberg, and D. Gutnick.** 1983. Capsular polysaccharides interfere with adherence of *Acinetobacter calcoaceticus* to hydrocarbons. *FEMS Microbiol. Lett.* **17**:157–160.

142. **Rosenberg, M.** 1981. Bacterial adherence to polystyrene: a replica method of screening for bacterial hydrophobicity. *Appl. Environ. Microbiol.* **42**:375–377.

143. **Rosenberg, M.** 1984. Bacterial adherence to hydrocarbons: a useful technique for studying cell surface hydrophobicity. *FEMS Microbiol. Lett.* **22**:289–295.

144. **Rosenberg, M.** 1984. Ammonium sulphate enhances adherence of *Escherichia coli* J-5 to hydrocarbon and polystyrene. *FEMS Microbiol. Lett.* **25**:41–45.

145. **Rosenberg, M.** 1984. Isolation of pigmented and nonpigmented mutants of *Serratia marcescens* with reduced cell surface hydrophobicity. *J. Bacteriol.* **160**:480–482.

146. **Rosenberg, M.** 1986. Concentration of larvicidal *Bacillus* spores at the water surface by adherence to oil droplets. *J. Microbiol. Methods* **5**:79–81.

147. **Rosenberg, M., E. A. Bayer, J. Delarea, and E. Rosenberg.** 1982. Role of thin fimbriae in adherence and growth of *Acinetobacter calcoaceticus* on hexadecane. *Appl. Environ. Microbiol.* **44**:929–937.

148. **Rosenberg, M., D. Gutnick, and E. Rosenberg.** 1980. Adherence of bacteria to hydrocarbons: a simple method for measuring cell-surface hydrophobicity. *FEMS Microbiol. Lett.* **9**:29–33.

149. **Rosenberg, M., H. Judes, and E. Weiss.** 1983. Cell surface hydrophobicity of dental plaque microorganisms in situ. *Infect. Immun.* **42**:831–834.

150. **Rosenberg, M., H. Judes, and E. Weiss.** 1983. Desorption of adherent bacteria from a solid hydrophobic surface by oil. *J. Microbiol. Methods* **1**:239–244.

151. **Rosenberg, M., and S. Kjelleberg.** 1986. Hydrophobic interactions: role in bacterial adhesion. *Adv. Microb. Ecol.* **9**:353–393.

152. **Rosenberg, M., and E. Rosenberg.** 1981. Role of adherence in growth of *Acinetobacter calcoaceticus* on hexadecane. *J. Bacteriol.* **148**:51–57.

153. **Rosenberg, M., M. Tal, E. Weiss, and S. Guendelman.** 1989. Adhesion of non-coccal dental plaque microorganisms to buccal epithelial cells: inhibition by saliva and amphipathic agents. *Microbial Ecol. Health Dis.* **2**:197–202.

154. **Rozgonyi, F., K. R. Szitha, A. Ljungh, S. B. Baloda, S. Hjertén, and T. Wadström.** 1985. Improvement of the salt aggregation test to study bacterial cell-surface hydrophobicity. *FEMS Microbiol. Lett.* **30**:131–138.

155. **Ruggieri, M. R., P. M. Hanno, and R. M. Levin.** 1987. Reduction of bacterial adherence to catheter surface with heparin. *J. Urol.* **138**:423–426.

156. **Runnels, P. L., and H. W. Moon.** 1984. Capsule reduces adherence of enterotoxigenic *Escherichia coli* to isolated intestinal epithelial cells of pigs. *Infect. Immun.* **45**:737–740.

157. **Sacks, L. E., and G. Alderton.** 1961. Behavior of bacterial spores in aqueous polymer two-phase systems. *J. Bacteriol.* **82**:331–340.

158. **Sar, N.** 1987. Direction of spreading (DOS): a simple method for measuring the hydrophobicity of bacterial lawns. *J. Microbiol. Methods* **6**:211–219.

159. **Sar, N., and E. Rosenberg.** 1987. Fish skin bacteria: colonial and cellular hydrophobicity. *Microb. Ecol.* **13**:193–202.

160. **Satou, J., A. Fukunaga, N. Satou, H. Shintani, and K. Okuda.** 1988. Streptococcal adherence on various restorative materials. *J. Dent. Res.* **67**:588–591.

161. **Satou, N., J. Satou, J. Shintani, and K. Okuda.** 1988. Adherence of streptococci to surface-modified glass. *J. Gen. Microbiol.* **134**:1299–1305.

162. **Schadow, K. H., W. A. Simpson, and G. D. Christensen.** 1988. Characteristics of adherence to plastic tissue culture plates of coagulase-negative staphylococci exposed to subinhibitory concentrations of antimicrobial agents. *J. Infect. Dis.* **157**:71–77.

163. **Schiemann, D. A., and P. J. Swanz.** 1985. Epithelial cell association and hydrophobicity of *Yersinia enterocolitica* and related species. *J. Med. Microbiol.* **19**:309–315.

164. **Scott, T. G., and C. J. Smyth.** 1987. Haemagglutination and tissue culture adhesion of *Gardnerella vaginalis*. *J. Gen. Microbiol.* **133**:1999–2005.

165. **Sharon, D., R. Bar-Ness, and M. Rosenberg.** 1986. Measurement of the kinetics of bacterial adherence to hexadecane in polystyrene cuvettes. *FEMS Microbiol. Lett.* **36**:115–118.

166. **Sherman, P. M., W. L. Houston, and E. C. Boedeker.** 1985. Functional heterogeneity of intestinal *Escherichia coli* strains expressing type 1 somatic pili (fimbriae): assess-

ment of bacterial adherence to intestinal membranes and surface hydrophobicity. *Infect. Immun.* **49**:797–804.

167. **Small, D. A., N. F. Moore, and P. F. Entwistle.** 1986. Hydrophobic interactions involved in attachment of a baculovirus to hydrophobic surfaces. *Appl. Environ. Microbiol.* **52**:220–223.

168. **Smyth, C. J.** 1988. Assays for fimbrial adhesins, p. 223–244. *In* P. Owen and T. J. Foster (ed.), *Immunochemical and Molecular Genetic Analysis of Bacterial Pathogens.* Elsevier Science Publishers, New York.

169. **Smyth, C. J., P. Jonsson, E. Olsson, O. Söderlind, J. Rosengren, S. Hjertén, and T. Wadström.** 1978. Differences in hydrophobic surface characteristics of porcine enteropathogenic *Escherichia coli* with or without K88 antigen as revealed by hydrophobic interaction chromatography. *Infect. Immun.* **22**:462–472.

170. **Speert, D. P., B. A. Loh, D. A. Cabral, and I. E. Salit.** 1986. Nonopsonic phagocytosis of nonmucoid *Pseudomonas aeruginosa* by human neutrophils and monocyte-derived macrophages is correlated with bacterial piliation and hydrophobicity. *Infect. Immun.* **53**:207–212.

171. **Stanley, S. O., and A. H. Rose.** 1967. On the clumping of *Corynebacterium xeroxis* as affected by temperature. *J. Gen. Microbiol.* **115**:509–512.

172. **Stenstrom, T.-A.** 1989. Bacterial hydrophobicity, an overall parameter for the measurement of adhesion potential to soil particles. *Appl. Environ. Microbiol.* **55**:142–147.

173. **Stenstrom, T.-A., and S. Kjelleberg.** 1985. Fimbriae mediated nonspecific adhesion of *Salmonella typhimurium* to mineral particles. *Arch. Microbiol.* **143**:6–10.

174. **Tanford, C.** 1973. The hydrophobic effect: formation of micelles and biological membranes. John Wiley & Sons, Inc., New York.

175. **Thies, K. L., and F. R. Champlin.** 1989. Compositional factors influencing cell surface hydrophobicity of *Pasteurella multocida* variants. *Curr. Microbiol.* **18**:385–390.

176. **van Bronswijk, H., H. A. Verbrugh, H. C. J. M. Heezius, N. H. M. Renders, A. Fleer, J. van der Meulen, P. L. Oe, and J. Verhoef.** 1989. Heterogeneity in opsonic requirements of *Staphylococcus epidermidis*: relative importance of surface hydrophobicity, capsules and slime. *Immunology* **67**:81–86.

177. **van der Mei, H. C., A. H. Weerkamp, and H. J. Busscher.** 1987. A comparison of various methods to determine hydrophobic properties of streptococcal cell surfaces. *J. Microbiol. Methods* **6**:277–287.

178. **van der Mei, H. C., A. H. Weerkamp, and H. J. Busscher.** 1987. Physico-chemical surface characteristics and adhesive properties of *Streptococcus salivarius* strains with defined cell surface structures. *FEMS Microbiol. Lett.* **40**:15–19.

179. **Vanhaecke, E., and J. Pijck.** 1988. Bioluminescence assay for measuring the number of bacteria adhering to the hydrocarbon phase in the BATH test. *Appl. Environ. Microbiol.* **54**:1436–1439.

180. **van Loosdrecht, M. C., J. Lyklema, W. Norde, G. Schraa, and A. J. Zehnder.** 1987. The role of bacterial cell wall hydrophobicity in adhesion. *Appl. Environ. Microbiol.* **53**:1893–1897.

181. **van Loosdrecht, M. C. M., J. Lyklema, W. Norde, G. Schraa, and A. J. B. Zehnder.** 1987. Electrophoretic mobility and hydrophobicity as a measure to predict the initial steps of bacterial adhesion. *Appl. Environ. Microbiol.* **53**:1898–1901.

182. **van Oss, C. J.** 1978. Phagocytosis as a surface phenomenon. *Annu. Rev. Microbiol.* **32**:19–39.

183. **van Oss, C. J., and C. F. Gillman.** 1972. Phagocytosis as a surface phenomenon. I. Contact angles and phagocytosis of non-opsonized bacteria. *RES J. Reticuloendothel. Soc.* **12**:283–292.

184. van Pelt, A. W. J., A. H. Weerkamp, M. H. W. J. C. Uyen, H. J. Busscher, H. P. deJong, and J. Arends. 1985. Adhesion of *Streptococcus sanguis* CH3 to polymers with different surface free energies. *Appl. Environ. Microbiol.* **49**:1270–1275.

185. van Steenbergen, T. J. M., F. Namavar, and J. de Graff. 1985. Chemiluminescence of human leukocytes by black-pigmented *Bacteroides* strains from dental plaque and other sites. *J. Periodontal Res.* **20**:58–71.

186. Vesper, S. J., N. S. A. Malik, and W. D. Bauer. 1987. Transposon mutants of *Bradyrhizobium japonicum* altered in attachment to host roots. *Appl. Environ. Microbiol.* **53**:1959–1961.

187. Weerkamp, A. H., H. M. Uyen, and H. J. Busscher. 1988. Effect of zeta potential and surface energy on bacterial adhesion to uncoated and saliva-coated human enamel and dentin. *J. Dent. Res.* **67**:1483–1487.

188. Weiss, E., M. Rosenberg, H. Judes, and E. Rosenberg. 1982. Cell-surface hydrophobicity of adherent oral bacteria. *Curr. Microbiol.* **7**:125–128.

189. Westergren, G., and J. Olsson. 1983. Hydrophobicity and adherence of oral streptococci after repeated subculture in vitro. *Infect. Immun.* **40**:432–435.

190. Williams, P., P. A. Lambert, C. G. Haigh, and M. R. W. Brown. 1986. The influence of the O and K antigens of *Klebsiella aerogenes* on surface hydrophobicity and susceptibility to phagocytosis and antimicrobial agents. *J. Med. Microbiol.* **21**:125–132.

191. Wolkin, R. H., and J. L. Pate. 1985. Selection for nonadherent or nonhydrophobic mutants co-selects for nonspreading mutants of *Cytophaga johnsonae* and other gliding bacteria. *J. Gen. Microbiol.* **131**:737–750.

192. Wood-Helie, S. J., H. P. Dalton, and S. Shadomy. 1986. Hydrophobic and adherence properties of *Clostridium difficile*. *Eur. J. Clin. Microbiol.* **5**:441–445.

193. Wrangstadh, M., P. L. Conway, and S. Kjelleberg. 1986. The production and release of an extracellular polysaccharide during starvation of a marine *Pseudomonas* sp. and the effect thereof on adhesion. *Arch. Microbiol.* **145**:220–227.

194. Wrangstadh, M., P. L. Conway, and S. Kjelleberg. 1988. The role of an extracellular polysaccharide produced by the marine *Pseudomonas* sp. S9 in cellular detachment during starvation. *Can. J. Microbiol.* **35**:309–312.

195. Wyroba, E., G. Bottiroli, and P. A. Giordano. 1987. Membrane region of increased hydrophobicity in dividing *Paramecium* cells revealed by cycloheptaamylose-dansyl chloride complex. *Eur. J. Cell Biol.* **44**:34–38.

196. Yen, S., and R. J. Gibbons. 1987. The influence of albumin on adsorption of bacteria on hydroxyapatite beads in vitro and human tooth surfaces in vivo. *Arch. Oral Biol.* **32**:531–533.

Microbial Cell Surface Hydrophobicity
Edited by R. J. Doyle and M. Rosenberg
© 1990 American Society for Microbiology, Washington, DC 20005

Chapter 2

Nature of the Hydrophobic Effect

Wendy C. Duncan-Hewitt

What's in a name?
W. Shakespeare, *Romeo and Juliet*, Act II, Scene II

Shuzan held out his short staff and said: "If you call this a short staff, you oppose its reality. If you do not call it a short staff, you ignore the fact. Now what do you wish to call this?"

Zen Koan

Even a cursory survey of the literature will reveal that hydrophobicity is considered to be an important determinant of interfacial processes in biology, including protein folding, antigenic site specification, assembly of phospholipid bilayers, micelle formation, phagocytosis, protein adsorption, and cell adhesion. However, the term "hydrophobicity," literally "water aversion," conveys only a vague impression of its fundamental meaning. It has been used to explain the relative insolubility of nonpolar substances in water, the propensity of nonpolar substances to aggregate in water, the inability of some substances to be wetted by water, and the tendency of some substances to partition unequally between water and another phase or at an interface. Dozens of "hydrophobicity tests" have been developed specifically to predict cell adhesion.

Hydrophobicity has been explained in a variety of ways: qualitatively, hydrophobic effects have been said to arise because "like prefers like" (49), suggesting that molecules will interact only if they possess similar structural and chemical characteristics, or to arise from a "solvent

Wendy C. Duncan-Hewitt • Faculty of Pharmacy, University of Toronto, 19 Russell Street, Toronto, Ontario M5S 1A1, Canada.

effect'' (23, 28, 39), which suggests that there is something special about the solvent, particularly water, which cannot be explained by the usual molecular interaction functions. Hydrophobicity has also been explained in terms of van der Waals interactions (discussed below), perhaps because at least one component of this group of forces acts between all molecules. Ben-Naim (5) considers hydrophobicity as a special case of ''solvophobicity.'' Water assumes no special status in this interpretation so that ''phobic'' phenomena should be predictable from a function of one or more molecular parameters that are common to all materials. Common experience appears to be at odds with this approach. In fact, the ubiquitous use of the term ''hydrophobicity'' implies that water possesses some molecular features that cause it to behave differently from all other materials.

The foregoing, essentially phenomenological definitions of hydrophobicity do not lend themselves to thermodynamic analyses. Ben-Naim (5) formally defines hydrophobicity in terms of the hydrophobic interaction process, in which two solute molecules are brought together within a solvent from infinite separation (Fig. 1). The free energy for this process is divided into two terms: one that quantitates the direct force between the two solute molecules and an indirect part that is mainly a function of the solvent, in our case water. It is this latter term that Ben-Naim equates with the hydrophobic interaction.

Hydrophobicity is an interfacial phenomenon. Interfaces are difficult to study both theoretically and experimentally because they involve so few molecules. If one considers any macroscopic system, such as a cell adhering to a polymer surface, only a minute fraction of the total volume of the system is taken up by the interface. Until very recently our measuring devices have been much too insensitive to detect the minute changes in a system that occur as a result of interfacial events. In fact, interfacial phenomena have been recognized historically as perturbations in bulk systems.

It is for this reason that interfacial thermodynamics was developed as an excess function. In this approach, the thermodynamic behaviors of contacting bulk phases are considered as though their properties are continuous and constant up to a mathematical ''dividing surface.'' Any differences between this abstraction and experimental measurements can conveniently be assigned to this surface. From thermodynamic and experimental standpoints, this device is extremely useful, since some phenomena can be quantified and predicted without considering the molecular events which give rise to them.

But great difficulties arise immediately if one wishes to interpret excess functions in terms of molecular interactions. The problem is that

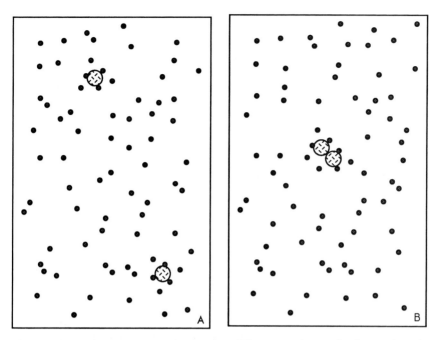

Figure 1. The hydrophobic interaction process. When two solute molecules are brought together from "infinite" separation, the associated free-energy change can be divided into two parts, one associated with the direct molecular interaction and an indirect part associated with changes in the solvent. It is the latter part that is linked with the hydrophobic effect.

the molecular world is so small that any observable phenomenon will be a composite result of many thousands of molecular interactions. To make things worse, every molecular phenomenon is statistical in nature. What we observe is the average of many distinctly different events. Our "molecular" explanations of bulk phenomena, our models of molecular interactions, are intimately linked to the statistical mean. If now an attempt is made to enlist our statistical molecular mechanisms to explain interfacial phenomena, which themselves are atypical (by the very fact that they occur in an interface where the properties of the system change drastically), excess functions will be required to account for the fact that we are no longer observing the bulk mean interactions. We require parameters such as hydrophobicity to account for the skew in the system.

The present understanding of hydrophobicity seems somewhat nebulous. To apprehend its role in biological adhesion and partitioning phenomena, the following questions must be asked:

(i) What role does the solvent (water) play in adhesion (adsorption) phenomena? What is the relative importance of this role? Is water in fact a unique solvent? How is it unique?

(ii) What fundamental information can be extracted from hydrophobicity assays? Do they really measure hydrophobicity? Can the results of the various tests be compared?

(iii) If the objective is to understand and predict adhesion, how far do the hydrophobicity tests take us?

The answer to these questions may be sought at three levels.

(i) Molecular approaches consider the combined effect of all possible molecular interactions. If we had the theoretical and analytical capability to relate the molecular interactions with macroscopic observations of wetting, adhesion, etc., the concept of hydrophobicity would be superfluous. At present, the total interaction potential for a large number of molecules is calculated as the sum of all the bimolecular interaction potentials in the region of interest. However, it is recognized that the total interaction potential for an assembly of molecules cannot be described in terms of the sum of its parts. For example, the attraction between two molecules separated by a few molecular diameters in a vacuum may be calculated in a straightforward manner. But as the molecules are brought closer together, some work must be expended in orienting the molecules to maximize their interaction. If a third molecule is added to the system, not only will its behavior be influenced by the two original molecules, but its presence will modify the interaction between the first two molecules. These problems make it impossible, at present, to adopt the molecular approach.

(ii) Modelistic approaches consider individual molecules to behave as microscopic versions of the bulk phase. In other words, they are assigned dielectric constants, densities, etc., that really have meaning only when a large number of molecules are considered. The relationships used in modelistic approaches can be conceptually misleading since they are described in terms of bimolecular interaction potentials in exactly the same way that the molecular approaches are. Although the interest may be principally in what happens to two molecules, it is the solution of the entire electrostatic field that will determine the bulk behavior of the system. Many body effects cause the force laws for larger entities (e.g., cells) to differ from those of their molecular constituents. Similarly, the total interaction potential for two molecules includes not only the solute-solute interactions but also any changes in the solute-solvent and solvent-solvent interaction energies that occur as the two molecules approach each other.

To circumvent this problem, the atomic structure is ignored completely in modelistic approaches: macroscopic bodies are viewed as continua, and the forces between these bodies are derived in terms of bulk properties (e.g., their dielectric constants and refractive indices).

The most familiar example of a modelistic approach is the manner in which charge interactions are assayed in colloid chemistry. The charge of a cell is measured by electrophoresis and similar methods which in reality measure only the net charge per unit volume or area. That a certain cell is found to possess a net negative charge does not imply that there are no positive charges on the surface. The approximation is often quite adequate: for example, the net charge determines the way in which the cell will respond to changes in the ionic composition of its environment. The approximation becomes inadequate at our level of interest; for example, a small positively charged region on a cell would make it possible for the cell to adhere to a negatively charged surface, an unexpected result if the net charge is used to predict adhesion.

(iii) Thermodynamic approaches rely on macroscopic measurements. These are slow on the atomic scale of time and coarse on the atomic scale of distance. Thermodynamics considers only a few parameters such as energy, entropy, volume, and number. This simplification is usually made at the expense of understanding, but the thermodynamic approach is predictive under suitable conditions. Usually, "suitable conditions" implies the attainment of equilibrium and knowledge of the relationship between the extensive variables in the fundamental equation (or, under more restricted conditions, the equation of state) of the system (13).

Most attempts to describe and explain adhesion are based on modelistic approaches which view a system not as a collection of discrete molecules but as a collection of featureless continua that contact each other in a variety of geometries. This is a helpful approximation because it avoids the complexity of the microscopic world. Macroscopic laws are thus relatively successful in many situations of practical importance. It can be shown, for example, that a continuum model describing the interaction between a monovalent ion and a dipolar solvent such as water provides an accurate description of the interaction behavior for any distance beyond the first shell of the surrounding solvent molecules. Unfortunately, the situations for which these laws are inadequate happen to be those of greatest interest: interfacial systems that contain biological cells in an aqueous environment.

To understand how interacting systems are affected by the aqueous environment, and in particular to determine the origins of the hydrophobic effect, the behavior of interfacial systems at all three levels of interpretation must be examined. In the process of comparing and contrasting the molecular, modelistic, and thermodynamic explanations of interfacial phenomena, a better understanding of hydrophobicity and the tests that purport to quantify it will hopefully emerge.

MOLECULAR ASPECTS

In order to know an object I must know not its external but all its internal qualities.

Ludwig Wittgenstein, *Tractatus Logico-Philosophicus*

The study of a physical system essentially consists in solving its time-independent Schrödinger equation.

A. Messiah (50)

A consistent theory of intermolecular forces can only be developed on the basis of quantum mechanical principles. Once the electron distributions are determined by solving the Schrödinger equation for the system, all intermolecular forces may be determined by classical electrostatics calculations. At present, however, the Schrödinger equation for systems as simple as a hydrogen molecule has still not been completely solved.

In the absence of such a unifying solution, it has been necessary to define a number of classes of intermolecular interactions. If two proximal particles in a vacuum are considered, each type of interaction can be characterized in terms of a simple potential. They can be distinguished by the manner in which the charge and geometry, both determined by quantum mechanical considerations, affect the distance dependence of the interaction energy (Table 1).

The electrostatic origins of dipole interactions are obvious, but the forces between nonpolar molecules seem, at first, to arise from nowhere. The van der Waals interactions, so frequently invoked in interfacial phenomena, actually include three distinct interactions, all with the same r^{-6} distance dependence (33). (i) The Keesom (orientation) component gives the angle-averaged interaction between a large number of dipoles whose kinetic energy is sufficient to allow them to rotate freely. (ii) The Debye (induction) component gives angle-averaged interaction between a nonpolar molecule and a dipole. The electrostatic field of the dipole causes the electrons surrounding the nonpolar molecule to become slightly displaced from their equilibrium positions around the nuclei. As a result, the centers of positive and negative charge within the molecule become slightly displaced with respect to each other, and a dipole moment results. The permanent and induced dipoles are now able to attract each other. (iii) The London (dispersion) component gives the

Table 1. Electrostatic Molecular Interactions

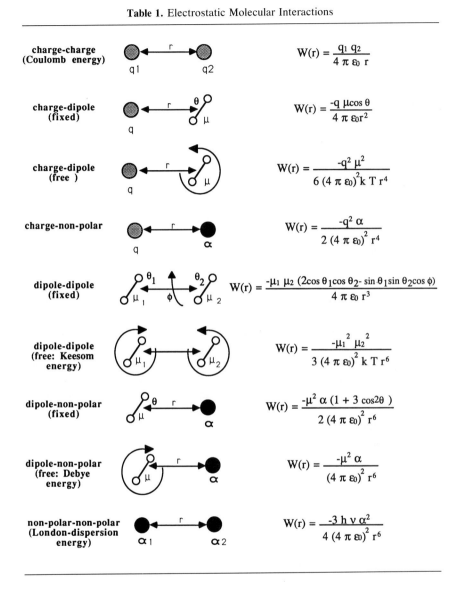

charge-charge **(Coulomb energy)**	$W(r) = \dfrac{q_1 q_2}{4 \pi \varepsilon_0 r}$
charge-dipole **(fixed)**	$W(r) = \dfrac{-q \mu \cos \theta}{4 \pi \varepsilon_0 r^2}$
charge-dipole **(free)**	$W(r) = \dfrac{-q^2 \mu^2}{6 (4 \pi \varepsilon_0)^2 k T r^4}$
charge-non-polar	$W(r) = \dfrac{-q^2 \alpha}{2 (4 \pi \varepsilon_0)^2 r^4}$
dipole-dipole **(fixed)**	$W(r) = \dfrac{-\mu_1 \mu_2 (2\cos \theta_1 \cos \theta_2 - \sin \theta_1 \sin \theta_2 \cos \phi)}{4 \pi \varepsilon_0 r^3}$
dipole-dipole **(free: Keesom** **energy)**	$W(r) = \dfrac{-\mu_1^2 \mu_2^2}{3 (4 \pi \varepsilon_0)^2 k T r^6}$
dipole-non-polar **(fixed)**	$W(r) = \dfrac{-\mu^2 \alpha (1 + 3 \cos 2\theta)}{2 (4 \pi \varepsilon_0)^2 r^6}$
dipole-non-polar **(free: Debye** **energy)**	$W(r) = \dfrac{-\mu^2 \alpha}{(4 \pi \varepsilon_0)^2 r^6}$
non-polar-non-polar **(London-dispersion** **energy)**	$W(r) = \dfrac{-3 h \nu \alpha^2}{4 (4 \pi \varepsilon_0)^2 r^6}$

interaction between two nonpolar molecules. The dispersion component is ubiquitous insofar as it acts between all molecules, and it often predominates. To understand dispersion interaction, one must consider the behavior of all molecules on a quantum mechanical level. Its origin may be grasped intuitively by considering the following. While the time-averaged dipole moment of a nonpolar molecule may be zero because the probability distribution for finding an electron anywhere around the molecule is symmetric, at any given moment a finite dipole moment exists because the electron can, in principle, be localized. The dipole moment is then given by the instantaneous positions of the electrons and protons. This dipole possesses an electric field which can polarize a nearby neutral molecule and induce a dipole moment in it. The two dipoles can then attract each other. The magnitude of the induced dipole depends on both the field strength and the polarizability of the molecule, which is often roughly proportional to its size. There is confusion regarding van der Waals forces, since this name is often given to the dispersion component alone.

In principle, one can calculate the equilibrium or steady-state configuration of a system if one knows the electron distribution functions for the molecules of interest. Monte Carlo simulations, made possible by modern computer technology, attempt to do just that (78). In this approach, the equilibrium thermodynamic functions and spatial correlations are evaluated directly from the assumed total intermolecular potential energy of n molecules, using a Monte Carlo sampling of points within the volume under consideration (in fact, the exact potential function is not known). Molecular dynamics calculations, an alternate simulational approach, also assume prior knowledge of the potential energy function for the molecules of interest (47). A system of several tens to a few thousand molecules is assembled in an arbitrary starting configuration and then allowed to evolve, using classical mechanics and electrostatics.

There are two drawbacks to these simulation approaches: (i) the exact potential functions for most molecules of interest are not known, and (ii) the computer time required is massive, so that it is practically impossible to study anything but the simplest systems. Therefore, for most interactions involving solvents, condensed phases, surfaces, and complex molecules, it has been necessary to simplify the calculations by using modelistic approximations.

Water Structure

The moon in the water;
Broken and broken again,
Still it is there.

Chosu

Before we discuss molecular and macroscopic body interactions in an aqueous environment and the origins of the hydrophobic effect, it will

Table 2. Comparison of the Physical Properties of Several Common Materials
with Those of Water[a]

Material	Mol wt (g)	Melting point (°C)	Boiling point (°C)	ΔH (vapor) (J/g)	Heat capacity (J/g · °C)	Surface tension (mJ/m², at 20°C)	Dielectric constant, at °C
Acetic acid	60	16.6	118.1	405	1.96, 0	27.8	16.9, 25
Acetone	58	−95.0	56.5	523	2.21, 20	23.7	6.2, 20
Ammonia	17	−77.7	−33.4	1,264	4.71, 20	18.1	1.9, 20
Benzene	78	5.5	80.1	394	1.70, 20	28.9	20.7, 25
Chloroform	120	−63.5	61.3	247	0.98, 20	27.1	2.3, 20
Diethyl ether	76	−116.0	34.6	373	2.29, 30	17.0	4.8, 20
Ethanol	46	−115.0	78.5	854	2.43, 20	22.8	24.3, 25
n-Octane	114	−56.5	125.8		2.42, 20	21.8	4.3, 20
Water	18	0.0	100.0	2,261	4.18, 20	72.5	78.0, 20

[a] Data taken from reference 29a.

be useful to survey the properties of this ubiquitous and remarkable substance, the presence of which is essential for life. Indeed, water is no ordinary liquid: given that its intermolecular bonds are not metallic or ionic, it has an unexpectedly high melting point, boiling point, latent heat of vaporization, dielectric constant, and heat capacity (Table 2) and a low isothermal compressibility (36, 70). It has a leveling effect: electrostatic interactions are greatly reduced in it, and it acts as a buffer against temperature changes.

Many of the special effects associated with water can be explained, in part, by bulk properties such as the dielectric constant. These are classified in continuum theories as solvent effects. But the continuum models that adequately account for the behavior of nonpolar solutes or surfaces in relatively nonpolar media (e.g., hexane) continually break down for aqueous media. Simulation approaches indicate that these failures can be attributed to changes in the structure of water.

Despite its importance, investigations into the structure of water have begun only recently, and most of the pertinent investigations have taken place since the 1950s. The water molecule can be considered to have a tetrahedral structure made up of its two hydrogen atoms and two lobes associated with its nonbonding electron pairs (Fig. 2). It is this arrangement which gives rise to the tetrahedral hydrogen-bonded structure that is observed both in the solid and in the liquid state. As a result, although spherically symmetrical molecules pack with a coordination number of 12, water has only four nearest neighbors in the solid state and only about five in the liquid state.

Water can act as both a hydrogen donor and a hydrogen acceptor;

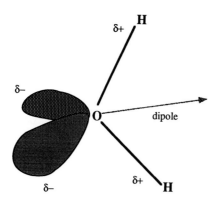

Figure 2. The ST2 model of water. The water molecule appears tetrahedral, with hydrogen atoms forming two of the lobes and the nonbonding electrons accounting for the other two lobes. Partial charges of 0.24e (e is the proton charge) and −0.24e, respectively.

many other polar molecules are capable of forming only one type of hydrogen bond and are unable to form hydrogen-bonded structures with themselves. Liquid water is structured by the hydrogen bonds, but the structure is very labile. The typical lifetime of a hydrogen bond in liquid water is 10^{-11} s. This property, in turn, helps to explain the relatively low viscosity of water.

Hydrogen-bonded systems are particularly complex (relative to simpler electrostatic systems) because hydrogen bonding interactions cannot be assumed to be additive, even as a first approximation. For example, acetone dissolves readily in water because it can hydrogen bond strongly with the solvent, yet pure acetone contains only weak hydrogen bonds (between the ketone and C—H groups). It is a very strong hydrogen bond donor and a very weak hydrogen bond acceptor. Although hydrogen bonds can be considered in many instances to be essentially electrostatic in nature, some phenomena cannot be explained by such a simple model (61). For one thing, the strength of hydrogen bonds is extremely orientation dependent. It is this feature which gives rise to many of the effects attributed to water structure.

The structure of water has been investigated extensively by using computer simulation approaches (14, 16–18, 29, 46, 74, 75). Some theoretical results have been supported at least qualitatively by X-ray and neutron diffraction studies (78). These studies indicate that when water contacts certain surfaces and molecules, its structure is altered. In particular, the propensity of water molecules to form hydrogen bonds with each other strongly influences their behavior when they approach a molecule that is incapable of forming these bonds. To minimize the number of "dangling" hydrogen bonds (a situation that would occur if the nonpolar molecule is situated in a potential hydrogen bonding direction), the water molecules tend to reorient themselves. If the nonpolar solute is small enough, it may be possible to do this without the sacrifice of any hydrogen bonds. But the choice of one or more configurations that maximize the number of hydrogen bonds also decreases the total number of configurations available to a given water molecule: the water becomes more "ordered." The decrease in entropy associated with this ordering is thermodynamically unfavorable. This can be seen immediately from the equation for Gibbs free energy: since $\Delta G = \Delta H - T\Delta S$, any decrease in ΔS would cause ΔG to increase.

Molecular dynamic and Monte Carlo simulations have shown that the sizes and shapes of the nonpolar molecules are very important in determining the structure that the water adopts. The formation of clathrates of water (cagelike structures) around nonpolar molecules has been simulated extensively (9), as has the water structure against a nonpolar wall (45). The results of these studies showed that the orientational structure adopted next to small solute molecules was distinctly different from that adopted next to extended surfaces. The thermodynamic consequences of this behavior will be discussed in detail below.

MODELISTIC APPROACHES

The fact that accurate interaction potential functions are not known for most molecules and the sheer magnitude of the computer time that would be required for even simple many-body calculations make it practical and indeed necessary to model the behavior of molecules in terms of the average observed for a large number of molecules. By critically examining some of the models and the consequences of their assumptions, it will become apparent how hydrophobicity and other solvation effects arise naturally as excess functions.

Charge-Charge Interactions

Most ionic interactions take place within a medium, the net effect of which is to screen the electrostatic field associated with the ions so that their net interaction energy is reduced. To avoid the complexity of calculating the effect of each solvent molecule, the system is modeled as the interaction of point charges in a homogeneous solvent continuum (33). The extent of the screening effect is given by the dielectric constant of the solvent.

It can be shown that this approximation is adequate unless one is considering interactions within the first few solvent layers, i.e., when the ions approach each other to within a few angstroms. Within the primary hydration shell, the water molecules are much more restricted in their motion, which can substantially change the local dielectric behavior. The concept of a dielectric constant is truly macroscopic, and it is probably inappropriate even to consider such a parameter on the molecular scale. Nevertheless, for the sake of argument, the usual net effect of the local ordering of water around an ion is an apparent decrease in the dielectric constant. Furthermore, when the solvation zones of two molecules overlap, a short-range force arises that cannot be treated by continuum theories.

What is the net effect of assuming that water behaves as a continuum? Consider the relative magnitude of the dissociation constant for an ionic solid in a vacuum and within an aqueous medium (33). While the continuum approach correctly predicts that dissociation occurs much more readily in water because the dielectric constant of the solvent is large, the predicted solubility is much greater than that measured experimentally. In practice, it has been found necessary to assume that the molecular radii of the ions are 6 to 200% larger to account for the hydration effect (33). Conceptually, the effect may be seen to arise because as the ions are hydrated, they restrict the motion of the water molecules in the first few layers, decreasing their entropy. The free-energy decrease is smaller as a result.

Charge-Dipole and Other Orientation-Dependent Interactions

It is important to note that ion-dipole, dipole-dipole, and dipole-nonpolar interaction energies are orientation dependent (Table 1). At room temperature, many molecules in the gaseous and liquid states possess sufficient kinetic energy to permit them to rotate more or less freely. If no particular configurations were favored, the total interaction would average to zero, but in fact certain orientations are preferred because they minimize the total free energy of the system and so are favored statistically. The equations in Table 2 are satisfactory as long as the net interaction is determined only by these statistical thermodynamic considerations, as is the case for interactions that occur beyond the first few hydration shells. Within this region, e.g., in interfaces or at the surfaces of protein molecules, orientational effects become important and the error of predictions based on macroscopic models can become large.

For example, Israelachvili (33) has shown that the surface tension of nonpolar and weakly polar liquids can be predicted with surprising accuracy on the basis of continuum theories of van der Waals interactions. However, if the surface tension of water is calculated by this method, the value (18 mJ m^{-2}) is only 25% of the experimental value (72.5

mJ m^{-2}). The difference can be shown to arise almost completely from a constrained orientational effect extrinsic to van der Waals considerations.

Thus, the apparent hydrophobicity of nonpolar molecules in an aqueous environment is to a large degree an orientational effect. To maximize the hydrogen bonding around a nonpolar molecule (and so decrease the enthalpy), the water molecules sacrifice some of their rotational degrees of freedom (thus decreasing the entropy). The hydrophobic effect, while perhaps largely determined by the entropic term, is in reality a function of the sum of these positive and negative contributions to the free energy of the system.

Van der Waals (Polarization) Interactions

Polarization interactions, in which an initially nonpolar molecule is polarized by an electric field (the centers of positive and negative charge are displaced relative to each other), become extremely difficult to predict when many-body interactions are considered or a third medium such as a solvent is involved. Beyond the immediate difficulties of assuming that the medium is homogeneous with a constant dielectric constant (a problem which is common to all of the modelistic approaches), the polarizability of a molecule is not independent of the medium for two reasons: (i) the moment of a permanent dipole may be changed in a complex manner that depends on the local interactions, so that total polarizability can be determined only by experimentation, and (ii) a dissolved molecule can move (i.e., be polarized) only if it can displace the solvent. If, for example, the field affects each molecule within a medium equally, no net polarization can occur and the molecules appear "invisible" to the field. Thus, it is not the polarizability but the excess polarizability of the molecule which must be quantified.

The manner in which the excess polarizability is calculated depends very much on the structure of the system being considered. For example, the excess polarizability of a molecule or small particle is calculated in the following way. The particle is modeled as a sphere of volume v_i and dielectric constant ε_i. In a medium which possesses a dielectric constant of ε, this sphere can be shown to possess the following effective excess polarizability (α_i) of (41): $\alpha_i = 3\varepsilon_0\varepsilon \, (\varepsilon_i - \varepsilon/\varepsilon_i + 2\varepsilon)v_i$. If, on the other hand, the interactions of interest are no longer molecular, then it becomes more meaningful to consider the interactions across a larger interface. The continuum approach ignores the atomic structure altogether and evaluates the excess polarizability in terms of bulk properties such as the dielectric constant and the refractive index. The apparent excess polarizability is decreased because molecules deep within the bulk of each phase have zero excess polarizability with respect to each other (41): $\alpha_i = 2\varepsilon_0\varepsilon\ (\varepsilon_i - \varepsilon/\varepsilon_i + \varepsilon)v_i$.

This last concept is extremely useful theoretically because it shows that dispersion forces can be positive, negative, or zero, depending on the relative magnitudes of the dielectric constants of the interacting particles and the medium. Note that the two expressions differ.

The predictions of the interactions of molecules and particles based on excess polarizabilities are quantitatively accurate for interactions between surfaces at large distances and for nonpolar molecules at small separations. It is not certain whether the discrepancy between theory and experiment for polar molecules is due to the inadequacy of the simple model (especially hydrogen-bonding molecules, because the forces are complex and directional) or to a change in the dipole moment of molecules in the condensed state. Hydrophobic (and hydrophilic) effects could be considered to be the excess functions which arise from the application of continuum theories.

What are the practical consequences of the approximations associated with continuum theories? It is well known that in water, two identical nonpolar molecules will attract each other much more strongly than in the vapor phase. For example, the van der Waals interaction energy between two contacting methane molecules in vacuum is $-2.5e^{-21}$ J, whereas the interaction energy in water for these two molecules is actually $-14e^{-21}$ J. Continuum theories, which only consider the fact that the dielectric medium will act to shield the molecules from each other, predict that this interaction should be decreased!

THERMODYNAMICS OF THE HYDROPHOBIC EFFECT

It is the insensitivity to specific structural or mechanical details that underlies the universality and simplicity of thermodynamics.

Herbert Callen

Let us reiterate Ben-Naim's definition of the hydrophobic interaction (5), which arises when two solute molecules are brought together within water from infinite separation: hydrophobicity is associated with the part of the free energy for this process that is determined by the solvent (water). Is it possible to study this hydrophobic interaction in isolation? Experimentally, two fundamentally different processes are considered: (i) the transfer of a solute from water to a nonaqueous solvent (e.g., octanol), represented by the standard free energy of transfer, and (ii) the tendency of two molecules to aggregate in aqueous solutions, for example the noncovalent dimerization of carboxylic acids in solution. The aggregation behavior is not additive: as the aggregates get bigger, it becomes more and more likely that two or more of the immediate neighbors of a given solute molecule will also be solute molecules. Quantitatively speaking, some solute molecules will be transferred to an interface (resembling adsorption), but many more will be transferred from the solvent to a pure, bulk solute phase. In the limit, the formation of a large aggregate can be seen to be approximated by the transfer process, because the surface effect will become negligible with respect to that of the bulk. The two processes are quite different, and in general one cannot infer anything about one from the other. Adsorption processes involve the formation of more than one solute-solute contact. At low coverage adsorption resembles a pairwise interaction, whereas at high coverage it resembles a transfer process more closely.

Pairwise Hydrophobic Interaction

The pairwise hydrophobic interaction can be used to model adsorption at low surface coverage. The free-energy change associated with the hydrophobic interaction between two molecules in water is considered by Ben-Naim (5) to be equal to the total free-energy change for the system minus the interaction potential between the molecules (which is assumed to be well known). It can be shown that this value is equal to the difference between the standard free energy of solution of the noncovalent dimer and twice that of the monomer (5).

Unfortunately, this quantity is not experimentally accessible, so Ben-Naim assumed that the free-energy change for the hydrophobic interaction could be approximated by the difference between the free energy of solution of a covalent molecule that resembles the noncovalent dimer and twice the free energy of solution of the monomer (Fig. 3). For example, the free-energy change for the hydrophobic interaction between two methane molecules in water is approximated by the difference between the free energy of solution for ethane and twice the free energy of solution for methane. This type of analysis can be extended to more complex molecules in order to estimate their hydrophobic behavior.

It may be shocking to think that two contacting methane molecules would be approximated in this way by a single ethane molecule: obviously, the intermolecular separation represented by this approach is unrealistic, but there are as yet no good ways to estimate the magnitude of the hydrophobic interaction for the correct separation (i.e., molecular contact). But the approximate method is perfectly acceptable from a thermodynamic standpoint as long as it is realized that some inaccuracies

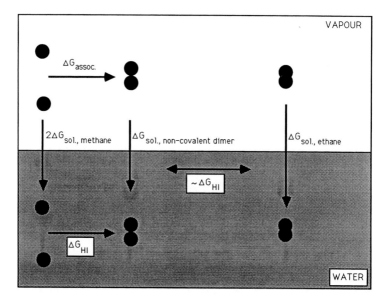

Figure 3. Calculation of free-energy change. The free-energy change of the hydrophobic interaction process can be calculated from the difference in the solubilities of the single molecules and the dimer formed in the gaseous state, shown in the left portion of the diagram. Since in reality the concentration of the gaseous dimer is extremely small, this process is experimentally inaccessible. Therefore, the solubility of the dimer is approximated by the process shown on the right: a covalent molecule that resembles the dimer is substituted (e.g., ethane replaces the methane dimer).

may result because such close contact could never be achieved experimentally. Furthermore, it is a convenient way of characterizing the hydrophobic effect by changing various thermodynamic parameters such as the temperature, pressure, and solvent composition.

The pairwise hydrophobic interaction has several important features.

(i) The reaction is often endothermic. Contrary to the popular attitude that hydrophobicity implies stronger solute-solute rather than solute-solvent van der Waals interactions, this observation implies that the opposite is the case. The solute-solvent interaction energy is often greater.

(ii) The hydrophobic interaction is highly temperature dependent, suggesting that entropic factors play a major role in the process. The entropic effect may, in fact, be anticipated from energetic considerations: since the reaction is spontaneous (negative ΔG) and endothermic (positive ΔH), it must be strongly favorable entropically (positive $T\Delta S$). The fact that the entropy of the system increases as a result of the hydrophobic

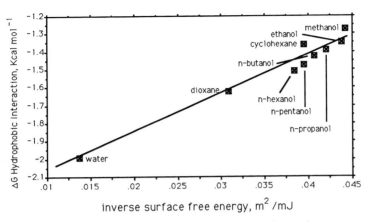

Figure 4. Free-energy change for the approximate hydrophobic interaction process versus the inverse surface tension of the solvent. Although it appears that a fairly simple relationship exists between these two parameters, it cannot replace molecular understanding. The exact form of the relationship changes with temperature, pressure, and chemical composition. (Data from reference 5.)

interaction (where the entropy of the solute molecules certainly decreases) suggests that the solvent becomes more disordered. This hypothesis has been supported by recent molecular dynamic studies showing that water becomes more structured at a hydrophobic interface. The hydrophobic interaction decreases the total surface area accessible to water and thus frees some molecules.

It appears at first that a simple relationship exists between this hydrophobic interaction energy and the surface tension of the solvent (Fig. 4). However, the explicit form of the relationship changes with temperature. Specifically, the value for cyclohexane increases with temperature, whereas that for water decreases drastically over the same temperature range. The temperature does not seem to affect the behavior of the other liquids significantly.

It is interesting to note that the free-energy change for the hydrophobic interaction process defined in this way for hydrocarbon chains of increasing molecular weight increases linearly with the number of carbon residues at 25°C (Fig. 5). This series has been used to model the aggregation effect, but it must be noted that the pairwise free-energy change for the dimerization of two methane molecules (-2.16 kcal mol^{-1}) is substantially different than the change in free energy per residue (-1.88 kcal [1 kcal $= 4.184$ kJ] mol^{-1}). Furthermore, it is probable that only nearest-neighbor interactions contribute to this behavior. The free energy associated with the hydrophobic interaction for cyclopropane, where

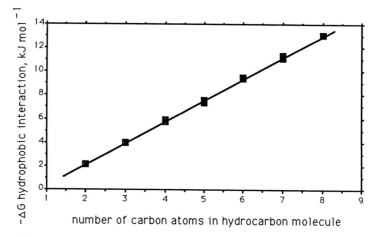

Figure 5. Free-energy change for the approximate hydrophobic interaction process versus the chain length of the aliphatic hydrocarbon solute. The linear relationship provides evidence supporting the hypothesis that formation of the solute-solvent interface is the dominant factor that determines the energetics of the hydrophobic interaction. (Data from reference 5.)

three-body interactions occur, is -3.49 kcal mol^{-1}, which seems to indicate that the nonadditivity is appreciable.

In many cases the hydrophobic interaction increases when electrolytes are added to the solution (6). This salting-out effect often increases almost linearly with the electrolyte concentration. However, caution must be exercised when one is attempting to generalize; for example, some tetraalkylammonium salts have a salting-in effect on the solubility of methane and ethane.

The effect of nonelectrolytes is more complicated. The addition of a very small amount of ethanol (mole fraction less than 0.03) to water decreases the hydrophobic interaction. From a concentration ranging from 0.03 to 0.2 mole fraction, the hydrophobic interaction increases. After this, the interaction decreases monotonically up to 100% ethanol, where the hydrophobic interaction is relatively weak.

A salt aggregation test is used frequently to characterize the hydrophobicity of bacteria (68). When one considers the interactions between two essentially nonpolar solute molecules the effect is already complex, since the hydrophobic effect does not always increase monotonically with the salt concentration. With bacteria, the system is much more complicated, since their surfaces are charged. Increasing the salt concentration decreases electrostatic repulsion by compressing the double layer, thus increasing the relative importance of the van der Waals forces in addition to changing the hydrophobic interaction.

Hydrophobicity, Solubility, and Partitioning

Solubility, partitioning, and other transfer phenomena resemble adhesion processes at high (especially multilayer) surface coverage because in all of these processes most of the solute-water interactions are exchanged for solute-nonaqueous solvent interactions (a neighboring solute molecule can be equated to a nonaqueous solvent). Partition coefficients and relative solubilities quantitate the relative phobia of the solute for two solvents. It has been shown that these quantities are highly temperature dependent, indicating a large entropic component in agreement with the results of the hydrophobic interaction method.

The solubility or partition approach is often used in an attempt to predict the hydrophobicity of a protein from the sum of the hydrophobicities of its constituent amino acids, and a large number of hydrophobicity scales are described in the literature (15). However, the assumption that there is a correspondence between them is erroneous: the effect of the close proximity of the amino acid groups and the tertiary structure of the protein are not usually considered.

Nevertheless, the hydrophobicity profile of a protein has been used with variable success to predict: (i) peptide chain turns (65), (ii) interior and exterior regions (66, 67), (iii) antigenic sites (30), (iv) membrane-spanning segments (40), and (v) "sticky" segments of molecules. The partitioning of amino acids between ethanol and water has been used extensively in models that attempt to predict protein folding (58). However, if a different scale is used, the predictions can be strikingly different (44). Lawson and co-workers suggest that the system N-cyclohexyl-2-pyrrolidone is a better model of the protein interior since it contains a substantial apolar region as well as a peptide bond-like moiety. The hydrophobic contribution is much smaller for this system, but its predictive capability appears greater.

Once again, it is safest to assume that transfer processes such as dissolution and partitioning have little in common with the hydrophobic interaction process. For example, if one transfers methane from water to a mixture of ethanol and water, the free-energy change is -0.17 kcal mol^{-1}; the methane appears to prefer the mixture (it is more hydro-*phobic*). On the other hand, the hydrophobic interaction (approximated by the difference in the solubilities of methane and ethane described above) appears to be less favorable in pure water (-1.99 kcal mol^{-1}) than in a mixture of water and ethanol (-2.11 kcal mol^{-1}); methane appears to be more "phobic" of the mixture than of water! One is tempted to conclude that these results conflict. But we must remember that these tests are showing essentially two different things: in the first, the entire surroundings of the molecule are being changed (i.e., from water to methanol and water), whereas in the latter, only part of the environment is changed (from solvent to solvent and solute).

The energetics of the transfer process cannot be inferred from the energetics of the pairwise hydrophobic interaction because the molecular events which give rise to hydrophobicity are not additive.

Relationship of Hydrophobicity to Surface Free Energy

The free energy of transfer of a solute molecule (e.g., methane) into a solvent (e.g., water) can be divided into two parts: one that represents an attractive term and another that represents a repulsive term. If the repulsive potential is steep, this is akin to the work required to introduce a hard-sphere (rigid) solute into a liquid in a fixed place. Let us consider a hypothetical microorganism that does not interact with water. In such an instance, we measure the work required to create a complementary cavity that has the same size, shape, and surface area as the microbe (7). Since the surface tension or free energy gives the work required to expand the liquid surface by unit area, this derivation immediately suggests that there could be a relationship between the free energy of transfer and the surface tension (surface free energy) of the liquid, and indeed, this has been shown to be true in some cases. For example, the standard free energy of solution of argon in solvents as diverse in structure and behavior as simple hydrocarbons, aromatics, and water changes almost linearly with the surface tension. Similarly, the difference in the solubility of hydrocarbons in water and in an arbitrary nonpolar solvent is seen to vary in a strictly linear fashion with chain length in a homologous series of substances (48).

It must be stressed, however, that finding a relationship between the two parameters is not equivalent to understanding the hydrophobic effect. Both are derived quantities, and a molecular theory should be able to predict both. Furthermore, there are conceptual problems associated with defining an explicit relationship between the two: How large should the cavity be? This will probably be a complex function of the local intermolecular interactions.

What will happen when both hydrophobic and hydrophilic groups are present? Some experimental evidence seems to suggest that the hydrophobic effect is neutralized if hydrophilic groups are nearby (59). If this is the case, then the surface free energy-hydrophobicity relationship will no longer apply. This reaffirms the assertion that it is dangerous to assume that a simple relationship between parameters such as solubility and surface free energy can be found, since the effect of adding a hydrophilic group is far from additive.

Temperature Dependence of the Hydrophobic Interaction

The partial derivative of the free-energy change with temperature for the hydrophobic interaction gives the entropy change for the process (5), whereas the change of the enthalpy with temperature gives the heat capacity change. Because the dispersion interaction between nonpolar solutes is essentially independent of temperature, the entropy change calculated from the free energy change for the interaction process must be that associated with the hydrophobic interaction. The steep decrease of the hydrophobic interaction free energy in water indicates that the entropy change is significantly more negative for this solvent than for the others. This observation implies that water is unique. Conversely, the nearly linear relationship between the surface free energy and the free energy of the pairwise hydrophobic interaction indicates that water is special only insofar as it possesses a surface tension which is greater than that of most other solvents. Since most solvents with high surface tensions also form strong hydrogen bonds, these observations may indicate simply that hydrophobicity is really a function of hydrogen bonding.

The heat capacity change for the hydrophobic interaction process is negative. This fact implies that although the hydrophobic interaction increases with temperature, its rate of increase decreases with temperature. Regardless of the type of model or process that is used to evaluate interaction processes such as aggregation and adhesion, it is obvious that the sign and magnitude of these two parameters are very important indicators of whether hydrophobicity is an important factor to consider.

Most of the nonpolar groups in water-soluble proteins are not exposed to the solvent (35, 66), and thus it is generally assumed that hydrophobic interactions play a major role in the formation of the compact structure of proteins. However, if the results of protein denaturation and hydrocarbon dissolution in water are compared, there are a number of surprising similarities and differences among the processes (62).

(i) When nonpolar molecules are dissolved in water, volume and isothermal compressibility decrease because the solute molecules can "fit" into the water lattice, so that the water molecules have less freedom to move in response to compressive stress. The opposite behavior is observed during the hydration of nonpolar chains in a protein. This behavior can be explained if the interior of a protein is not "oily", as is normally assumed. In fact, the interior is more crystal-like, with a packing density of about 0.75 for most proteins. In contrast, organic liquids have packing densities which do not exceed 0.44.

(ii) Whereas the entropy change for the so-called hydrophobic interaction of hydrocarbons is zero in the temperature range of 130 to 160°C (because the rate of change of entropy decreases with increasing temperature in classical hydrophobic phenomena), the entropy change for protein folding is about 17.6 J K^{-1} mol^{-1} (this is close to the value expected for a helix-coil transition for a protein) in this temperature range.

(iii) The heat capacity difference between native and unfolded states of a protein was shown to be directly related to the number of possible hydrophobic contacts in the protein interior, which correlates with the microscopic stability of a protein, i.e., its "breathing." This measure of the mobility of the protein structure is almost linearly inversely proportional to the heat capacity change.

Privalov and Gill (62) concluded that the roles of water, entropy, hydrogen bonds, and van der Waals forces in protein stability cannot be separated. Proteins are only marginally stable, and only a small change in any of these variables could lead to the denaturation of the entire structure. The conclusions for proteins have important implications in many adhesion phenomena, since adhesion is often mediated by proteins on the cell surface. We cannot allow ourselves the luxury of assuming that adhesion to surfaces in the absence of electrostatic effects is solely a function of hydrophobicity without probing more deeply into the behavior by quantifying the effect of temperature changes. For example, it may be informative to investigate the adhesion of bacteria to n-octane droplets between 0 and 50°C, having previously characterized the changes in densities and surface tensions of all phases and the water–n-octane interfacial tension within this range.

Thermodynamic Interpretation of Adhesion Tests

The preceding discussions should convey the impression that hydrophobicity, as an independent concept, arises only because we prefer to study the interactions between two particles as though they were essentially isolated from their environment, apart from a shielding factor such as a dielectric constant. The only instance in which the assumption of isolation is warranted on a molecular scale, however, is when interactions occur in the gaseous state; in the liquid state, the energetics of a given process will be determined by all molecular interactions. Thus, hydrophobicity arises naturally as an excess function, just as the surface tension is an excess function that arises because it is mathematically convenient to assume that the characteristics of a given phase are invariant right up its interface with another phase. Thermodynamics is well suited to the task of quantifying excess functions.

Many of the tests that are assumed to evaluate the hydrophobicity of particles, surfaces, proteins (51, 56), and biological cells (69) are actually adhesion tests. For example, hydrophobic interaction chromatography is an adhesion test that considers kinetic phenomena (see, for example, reference 63) complicated by the effects of mechanical trapping.

At the outset, the reader must be warned that it is very difficult to evaluate the results of most adhesion tests solely on the basis of hydrophobicity (as it is formally defined) because so many parameters are involved in most interfacial systems of interest. The factors that have been shown to play a significant role include pH, ionic strength, type of nonpolar solvent, the presence of added hydrophobic compounds (21), chemical constitution of the surfaces (52), distribution and orientation of surface groups (71, 73), and contamination or the presence of monolayers on the surfaces (42).

To interpret an adhesion test, one must scrutinize its inherent assumptions in order to define the thermodynamic variables that will determine the equilibrium configuration. Consider the following system. A suitable substrate (an immiscible liquid of known chemical composition or a solid of known chemical composition and wetting behavior) is prepared and placed within a fluid of known composition. If the fluid is a liquid, its surface tension is usually known as well. A number of model adherents (cells, particles, etc.) are then allowed to adhere in some specified way to the substrate. If the results are to be analyzed thermodynamically, the system must be allowed to reach equilibrium. This restriction is problematic enough, since adhesion is often mediated by protein adsorption, which may require hours to reach steady state.

Second, in the attempt to study the role of hydrophobicity alone, it is often assumed that holding the pH, ionic strength, or other ionic variables constant is sufficient to permit ionic interactions to be ignored. But hydrophobicity per se is rarely if ever quantified in adhesion tests.

The following experimentally accessible parameters provide information that ultimately may be useful in understanding the thermodynamic aspects of adhesion. Certain properties of materials that quantitate the macroscopic manifestations of the short-range attractive intermolecular forces may be evaluated by measuring melting points, boiling points, enthalpies of vaporization and fusion, adsorption isotherms, surface potentials, solubility, partitioning, surface tensions, or osmotic pressure. Viscosity and compressibility measurements are typical of the types of experiments that quantitate the short-range repulsive forces.

Of these parameters, the surface tensions are, in principle, the most useful because the reversible work of adhesion is given by (19) $W_{ad} = \gamma_{sl} + \gamma_{pl} - \gamma_{sp}$, where γ_{sl} is the interfacial free energy between the solid surface and the liquid, γ_{pl} is the interfacial free energy between the particle and the liquid, and γ_{sp} is the interfacial free energy between the particle and the solid. Therefore, if the appropriate free energies are known, then the free-energy balance determines both whether the reaction can occur and, theoretically, the equilibrium constant of the reaction. The test does not provide information about the mechanisms which give rise to the behavior.

The interfacial free energy is defined as the reversible work required to form a unit area of new interface at constant temperature, volume, and mass of the system, and in a typical biological system it can be shown to be a function of the surface entropy, internal energy, and number densities of the interfacial components (10). Other independent variables may be added as required in order to fully represent the system. For example, if the role of charge is not considered explicitly in the electrochemical potential, then in systems in which ions are present, the charge may be considered as an independent variable.

The fundamental problem of obtaining quantitative information about an adhesive system is therefore reduced to one of determining the interfacial free energies. In practice, information about the interfacial free energies is derived from tests of wettability (e.g., contact angles). The fundamental problem of obtaining predictive information requires a generalized scheme that relates wettability and interfacial free energy.

These statements apply generally. Thus, even though a test may be based on adhesion to one surface in a pure aqueous environment, the test is still based on the assumption that particular surface is typical of some characteristic. The characteristic that is usually assumed to be possessed by the surface is hydrophobicity.

This hydrophobicity is not equivalent to that described by the formal definition. Hydrophobicity, for the purposes of the adhesion test, is defined in terms of wettability. This fact does not pose a problem to the investigator who is attempting to derive quantitative or predictive information from the test. It does pose a problem, however, if the results of the test are to be interpreted mechanistically.

The solution of adhesion problems is not straightforward because few methods exist which evaluate interfacial free energies. Although several techniques are available which evaluate the surface free energy directly (fracture mechanics [43], zero creep rate method [76], and crystal nucleation theories [8, 79]), these are not useful in the present situation.

Contact Angles

Because of the often insurmountable difficulties associated with measuring solid surface and interfacial tensions directly, considerable attention has been paid in recent years to alternate methods that quantitate wettability on the basis of measurement of contact angles (2).

Contact angles can be measured readily on most solids and are often used simply as empirical parameters to quantify wettability (64). More than 180 years ago, Thomas Young proposed that the contact angle between a liquid and a solid arises from the requirement that the interfacial tensions acting at the three-phase line be balanced at equilibrium (80): $\gamma_{sv} - \gamma_{sl} = \gamma_{lv}\cos(\theta)$.

More recently, general thermodynamic derivations of the Young equation have been published (37, 53). From the point of view of hydrostatics, Buff (11) showed that Young's equation provides the condition for hydrostatic equilibrium locally at the three-phase line. The underlying assumptions of all these derivations include (i) that the surface is flat, smooth, and rigid (72) and (ii) that the system is at equilibrium. Under normal conditions, the effects of surface deformation will be small.

The thermodynamic status of contact angles has been a subject of considerable debate (1, 27, 53, 81). Underlying this debate is the fact that it is difficult to establish whether the system is in equilibrium, a problem which arises because solids are able to sustain shear stresses.

The interpretation of contact angles is further complicated by contact angle hysteresis, a phenomenon in which the contact angle formed by a liquid advancing across an unwetted surface is generally larger than the contact angle of the same liquid as it recedes across the previously wetted surface. Several investigators have advanced models which predict the existence of metastable states and contact angle hysteresis arising from two factors: (i) surface heterogeneity and (ii) roughness (38, 54). Whereas the presence of heterogeneity does not preclude the application of the Young equation, the presence of roughness does, and it is often difficult to determine which case prevails.

In 1937, Bangham and Razouk (3, 4) called attention to the fact that the solid-vapor interfacial tension, i.e., the quantity measured in most experiments, is smaller than the solid surface tension due to adsorption effects. While it is generally agreed that this factor must be considered, there is disagreement about the magnitude of the spreading pressure (the difference between the solid surface free energy in a vacuum and the value measured experimentally in air or other gaseous environments). For example, in the work of Good (25) and Fowkes (22), the spreading pressure is assumed to be negligible. On the other hand, many recent European studies claim that the spreading pressure is considerable (12).

Given that a surface meets the requirements listed above, many investigators have become convinced that contact angles are thermodynamically significant quantities. Moreover, they propose that a fundamental relationship exists which relates the liquid surface tension with the solid-liquid interfacial tension and the solid-vapor interfacial tension, the difference of which is calculated by using Young's equation (24, 26, 55). At present, however, it is safest to assume that contact angles give only relative measures (77).

For contact angle information to give more that just a qualitative indication of the expected adhesion behavior, one requires another explicit relationship between the interfacial free energies. That such a relationship exists has been demonstrated in two ways by J. K. Spelt (Ph.D. thesis, University of Toronto, Toronto, Canada, 1983). This relationship has been termed an equation of state and has been applied to the evaluation of many different adhesion systems.

However, a problem of generality may arise in any such an approach. Consider the familiar example of the equation of state for ideal gases. It does not even apply generally; in fact, it does not adequately predict the behavior of most real gases. This outcome can be readily understood if the microscopic behavior of real molecules is considered: whereas ideal gases do not interact except by collision, real molecules attract each other by van der Waals forces. As the molecules in the gas become less and less ideal, the more the equation of state for gases needs to be modified to be predictive. It is often necessary to incorporate "material constants."

It would not be surprising if a given equation of state for contact angles was also a function of specific molecular parameters (such as hydrogen bonding propensity!). In other words, a set of equations of state, the precise forms of which will be dependent on the chemical composition and other microscopic variables, is probably needed. In other words, useful though the thermodynamic approach may be for quantitation, it cannot be used independently of more fundamental studies that consider the intermolecular forces explicitly.

A number of investigators attempt to allow for the fact that different materials possess different molecular behaviors, by dividing the interfacial tensions into components that are functions of a number of molecular interaction mechanisms (e.g., dispersion, polar, hydrogen bonding, etc.). These approaches are valid to a point. Consider the force relationships shown in Table 1. Most of them are functions of the products of the relevant material variables such as the dipole moment and polarizability. The combining laws commonly used in the surface tension expressions take this behavior into account. The surface free energies are divided into components that are the square roots of the products of dispersion, polar, and other forces (the induction forces that arise from interaction between dipolar and nonpolar molecules are usually ignored under the assumption that they play a relatively minor role in most molecular interactions). The system breaks down only when additivity cannot be assumed or if important parameters are ignored.

Unfortunately, the systems of greatest interest are aqueous. Furthermore, cell surfaces are very complex. The free-energy changes associated with hydrogen bonds are not additive (recall the case of acetone and water). Also, hydrophilic and hydrophobic alterations in the structure of water induced by the proximity of both the cell surface and the surface to which it adheres may play a significant role in an aqueous environment; these too are not additive. Until these problems are addressed explicitly, each experimental situation must be considered carefully so that, on the basis of knowledge of the molecular characteristics of the system, it will be possible to decide to what extent predictions may be in error, given that a satisfactory understanding of the interfacial free energies does not exist.

It may be tempting to study materials that interact solely by dispersion forces and then attempt to extrapolate from these observations. Consider, for example, the contact angle of water on fluoropolymers. The latter are considered to be purely dispersive: they are assumed to interact with other materials only through dispersion forces. But this assumption is incorrect if the wetting fluid is water. The polar and induction contribution to the total van der Waals interaction may be considerable for this material. But this is obviously not the end of the story, because contact with the surface will alter the structure of water in the vicinity of the interface, and this hydrophobic effect cannot be ignored. Finally, the fluoro groups have the capability of forming weak hydrogen bonds in the presence of a material that is a strong hydrogen bonder. Water is such a material (water may tend to hydrogen bond with itself because the strongest bond will be favored energetically; nevertheless, statistical thermodynamics ensures that the equilibrium will be shifted).

DLVO THEORY AND ADHESION

Although this chapter focuses on hydrophobicity, it is useful to comment on the DLVO theory, which is used so frequently in explanations of adhesion phenomena (57). According to this classical approach, the interaction free energy comprises two additive terms, one quantifying the contribution from the van der Waals interactions and the other resulting from the overlap of electrical double layers that build up around a charged species (e.g., a cell) in an electrolyte solution. In evaluating the utility of the DLVO approach, it is helpful to return to the commonly held hypothesis that the van der Waals forces determine wetting and adhesion in the absence of electrostatic effects. As was discussed earlier, theories of van de Waals forces consider only the angle-averaged attractions, but only the dispersion component would be expected to be relatively insensitive to the presence of an interface. The orientation and induction components will be modified greatly by the proximity of the surface, and it is the deviations from the expected van der Waals interactions, not the interactions themselves, that give rise to the hydrophobic effect. Normally it is hydrophobicity, not the van der Waals interaction between the particles per se, that controls the observed behavior.

It has been argued that Hamaker constants (ostensibly the parameter that gives the intensity of the van der Waals interaction) calculated from contact angles (i.e., wettability measurements) are preferable to those which quantify van der Waals forces alone (S. N. Omenyi, Ph.D. thesis, University of Toronto, Toronto, Canada, 1978). From a predictive standpoint alone this may indeed be preferable, since wettability is a function of the interfacial free energies and incorporates the hydrophobic effect. From a theoretical and interpretive standpoint, though, it is dangerous. The inaccurate nomenclature implies that the force laws for the hydrophobic interaction and van der Waals interaction are of the same form, and this has not been established.

ADHESION FORCE MEASUREMENT

Experimental methods such as steady-state adhesion measurements and contact angle determinations provide thermodynamic information but do not permit the intermolecular potential functions to be quantified. This information is accessible only through direct force measurements. Although several different procedures have been described, some are more amenable to theoretical evaluation.

A spinning disk configuration has been used by several groups. The centrifugal force imposed by the motion of the disk effectively shears the weakest bonds attaching microorganisms to the substrate (31). Unfortunately, this method suffers several drawbacks: (i) the shear force varies across the radius of the disk; (ii) microscopy shows that often debris is left behind, indicating the cells are torn rather than detached; and (iii) the stresses acting on a given particle are not well defined because the adherent cells cause the surface to be rough and induce turbulent flow.

Flow cells pose the same difficulties. However, the fact that attachment under flow conditions can be observed is a distinct practical advantage, since most adhesion phenomena do not take place under static conditions.

The most sophisticated approach measures the force between two surfaces as they are brought together to distances within 1 nm. Pashley and co-workers (32, 60) measured the force between two mica surfaces covered with surfactant monolayers in water and various electrolyte solutions in the range of 0 to 10 nm in a direct study of the hydrophobic effect. Below 6 nm, the measured attraction was observed to be 10- to 100-fold stronger than that predicted by van der Waals theory alone. They attributed this observation to the hydrophobic effect. They were able to conclude that the hydrophobic interaction acted over long distances on the atomic scale (characteristic decay length of 6 to 15 Å [0.6 to 1.5 nm]). Conversely, surfaces with groups that are water soluble and therefore hydrophilic were found to repel each other (34). Both of these observations confirm the predictions of molecular theories of water structure.

HYDROPHILICITY

Hydrophilicity is sometimes used as an antonym for hydrophobicity. There is no phenomenon actually known as the hydrophilic interaction, but it is used to describe substances that either are wetted by water or are soluble in it. Hydrophilic materials tend to repel each other in an aqueous environment (while hydrophobic materials aggregate).

The structures of aqueous solutions of electrolytes (generally considered to be hydrophilic) have been studied by simulational approaches, some aspects of which have been verified by X-ray and neutron diffraction studies (20). These studies show that ions are strongly hydrated. Hydrophilic substances such as ions are believed to increase the disorder in liquid water. This view at first seems to disagree with experimental fact, since the mobility of water in the primary hydration shell around an ion is restricted. To account for hydrophilic behavior, it is believed that a second layer of water that is extremely disordered exists around the primary shell. The disorder in this layer is assumed to arise from a competition between the structures of water in the bulk and in the hydration layer for the intervening molecules. The molecular dynamics simulations might provide evidence that supports this hypothesis. Unfortunately, only the ion-ion and ion-water correlation functions have been reported. These data yield no information about water structuring.

SUMMARY

The antipathy of the paraffin-chain for water is, however, frequently misunderstood. There is no question of actual repulsion between individual water molecules and paraffin chains, nor is there any very strong attraction of paraffin chains for one another. There is, however, a very strong attraction of water molecules for one another in comparison with which the paraffin-paraffin or paraffin-water interactions are very slight.

G. S. Hartley (28)

A working definition for hydrophobicity is elusive. It is formally defined in terms of the hydrophobic interaction process, in which two solute molecules are brought together within a solvent from infinite separation. The free energy for this process is divided into two terms, one that quantitates the direct force between the two solute molecules and an indirect part that is mainly a function of the solvent (the hydrophobic interaction).

The fact that hydrophobic interactions are often endothermic indicates that the solute-solvent interaction energy is often greater than the solute-solute van der Waals interaction, which is contrary to the popular conception that hydrophobicity is a manifestation of van der Waals forces. Furthermore, the hydrophobic interaction is highly temperature dependent, suggesting that entropic factors play a major role in the process. Indeed, the hypothesis that it is largely the structure of the solvent that determines the hydrophobic interaction has been supported by recent molecular dynamic studies showing that water becomes more ordered at a hydrophobic interface.

Two fundamentally different processes that are used commonly to characterize hydrophobicity were considered: (i) the transfer of a solute from water to a nonaqueous solvent, (represented by the standard free energy of transfer), and (ii) the tendency of two molecules to aggregate in aqueous solutions (pairwise interaction). Adsorption processes involve the formation of more than one solute-solute contact. At low coverage, adsorption resembles a pairwise interaction; at high coverage, it resembles a transfer process more closely. Experimental evidence suggests that there could be a relationship between the free energy of transfer and the surface tension (surface free energy) of the liquid (water). Therefore, it should be possible to correlate surface tension with the relative phobia of a solute for two different solvents and adsorption phenomena at high surface coverage.

Although these correlations are encouraging vis à vis predicting adhesion from surface tension data, it is very difficult to evaluate the results of most adhesion tests solely on the basis of hydrophobicity (as it is formally defined), since so many parameters are involved in most interfacial systems of interest. The fundamental problem of obtaining quantitative information about an adhesive system is determining the interfacial free energies. In practice, information about the interfacial free energies is derived from tests of wettability. The fundamental problem of obtaining predictive information requires a generalized scheme that relates wettability and interfacial free energy. This fact does not pose a problem to the investigator who is attempting to derive quantitative or qualitatively predictive information from an adhesion test. It does pose a problem, however, if the results of the test are to be interpreted mechanistically.

LITERATURE CITED

1. **Adamson, A. W., and I. Ling.** 1964. The status of contact angle as a thermodynamic property. *Adv. Chem. Ser.* **43**:57–73.
2. **American Chemical Society.** 1964. *Contact Angles, Wettability and Adhesion.* Advances in Chemistry Series, vol. 43. American Chemical Society, Washington D.C.
3. **Bangham, D. H.** 1937. The Gibbs adsorption equation and adsorption on solids. *Trans. Faraday Soc.* **33**:805–811.
4. **Bangham, D. H., and R. I. Razouk.** 1937. Adsorption and the wettability of solid surfaces. *Trans. Faraday Soc.* **33**:1459–1463.
5. **Ben-Naim, A.** 1980. *Hydrophobic Interactions.* Plenum Publishing Corp., New York.
6. **Ben-Naim, A., and M. Yaacobi.** 1974. Effects of solutes on the strength of hydrophobic interaction and its temperature dependence. *J. Phys. Chem.* **78**:170–175.
7. **Bishop, W. H.** 1987. Molecular and macroscopic aspects of cavity formation in the hydrophobic effect. *Biophys. Chem.* **27**:197–210.
8. **Blakely, J. M.** 1973. *Introduction to the Properties of Crystal Surfaces*, p. 7. Pergamon Press, Oxford.
9. **Bolis, E., and E. Clementi.** 1981. Methane in aqueous solution at 300 K. *Chem. Phys. Lett.* **82**:147–152.
10. **Boruvka, L., and A. W. Neumann.** 1977. Generalization of the classical theory of capillarity. *J. Chem. Phys.* **66**:5464–5469.
11. **Buff, F. P.** 1960. *Handbook der Physik*, vol. 10. Springer-Verlag, Berlin.
12. **Busscher, H. J., A. W. J. van Pelt, H. P. de Jong, and J. Arends.** 1983. Effects of spreading pressure on surface free energy determinations by means of contact angle measurements. *J. Colloid Interface Sci.* **95**:23–27.
13. **Callan, H. B.** 1985. *Thermodynamics and an Introduction to Thermostatistics*, 2nd ed., p. 5. John Wiley & Sons, Inc., New York.

14. **Christou, N. I., J. S. Whitehouse, D. Nicholson, and N. G. Parsonage.** 1985. Studies of high density water films by computer simulation. *Mol. Phys.* **55:**397–410.

15. **Cornette, J. L., K. B. Cease, H. Margalit, J. L. Spouge, J. A. Berzofsky, and C. DeLisi.** 1987. Hydrophobicity scales and computational techniques for detecting amphipathic structures in proteins. *J. Mol. Biol.* **195:**659–685.

16. **Croxton, C. A.** 1979. Molecular orientation and surface potential at the liquid water interface. *Phys. Lett.* **74A:**325–328.

17. **Croxton, C. A.** 1980. *Statistical Mechanics of Liquid Surfaces*, p. 198–218. John Wiley & Sons, Inc., New York.

18. **Croxton, C. A.** 1981. Molecular orientation and interfacial properties of liquid water. *Physica* **106A:**239–259.

19. **Dupré, A.** 1869. *Théorie Mechanique de la Chaleur*, p. 369. Gauthier-Villars, Paris.

20. **Enderby, J. E., and C. W. Nielson.** 1979. X-ray and neutron scattering by aqueous solutions of electrolytes, p. 1–46. *In* F. Franks (ed.), *Water, a Comprehensive Treatise*, vol. 6. *Recent Advances.* Plenum Publishing Corp., New York.

21. **Falkowski, W., M. Edwards, and A. J. Schaeffer.** 1986. Inhibitory effect of substituted aromatic hydrocarbons on adherence of *Escherichia coli* to human epithelial cells. *Infect. Immun.* **52:**863–866.

22. **Fowkes, F. M.** 1963. Additivity of intermolecular forces at interfaces. I. Determination of the contribution to surface and interfacial tensions of dispersion forces in various liquids. *J. Phys. Chem.* **67:**2538–2541.

23. **Frank, H. S., and M. W. Evans.** 1945. Free volume and entropy in condensed systems. III. Entropy in binary liquid mixtures; partial molal entropy in dilute solutions; structure and thermodynamics in aqueous electrolytes. *J. Chem. Phys.* **13:**507–532.

24. **Girifalco, L. A., and R. J. Good.** 1957. A theory for the estimation of surface and interfacial energies. I. Derivation and application to interfacial tension. *J. Phys. Chem.* **61:**904–909.

25. **Good, R. J.** 1964. Theory for estimation of surface and interfacial energies. VI. Surface energies of some fluorocarbon surfaces from contact angle measurements. *Adv. Chem. Ser.* **43:**74–87.

26. **Good, R. J., L. A. Girifalco, and G. Kraus.** 1958. A theory for the estimation of surface and interfacial energies. II. Application to surface thermodynamics of Teflon and graphite. *J. Phys. Chem.* **62:**1418–1421.

27. **Gray, V. R.** 1967. *Wetting*, p. 99. Monograph 25. Society of Chemical Industries, London.

28. **Hartley, G. S.** 1936. *Aqueous Solutions of Paraffin-Chain Salts.* Hermann & Cie., Paris.

29. **Henderson, D., L. Blum, and M. Lozada-Cassou.** 1983. The statistical mechanics of the electrical double layer. *J. Electroanal. Chem.* **150:**291–303.

29a.**Hodgman, C. D., R. C. Weast, R. S. Shankland, and S. M. Selby (ed.).** 1963. *The Handbook of Chemistry and Physics*, 44th ed. The Chemical Rubber Publishing Co., Cleveland.

30. **Hopp, T. P., and K. R. Woods.** 1981. Prediction of protein antigenic determinants from amino acid sequences. *Proc. Natl. Acad. Sci. USA* **78:**3824–3828.

31. **Horbett, T. A., J. J. Waldburger, B. D. Ratner, and A. S. Hoffman.** 1988. Cell adhesion to a series of hydrophilic-hydrophobic copolymers studied with a spinning disc apparatus. *J. Biomed. Mater. Res.* **22:**383–404.

32. **Israelachvili, J., and R. Pashley.** 1982. The hydrophobic interaction is long range, decaying exponentially with distance. *Nature* (London) **300:**341–342.

33. **Israelachvili, J. N.** 1989. *Intermolecular and Surface Forces with Applications to Colloidal and Biological Systems*, p. 28–156. Academic Press, Inc., New York.

34. **Israelachvili, J. N., and P. M. McGuiggan.** 1988. Forces between surfaces in liquids. *Science* **241**:795–800.

35. **Janin, J.** 1979. Surface and inside volumes in globular proteins. *Nature* (London) **277**:491–492.

36. **Joesten, M. D., and L. J. Schaad.** 1974. *Hydrogen Bonding.* Marcel Dekker, Inc., New York.

37. **Johnson, R. E.** 1959. Conflicts between Gibbsian thermodynamics and recent treatments of interfacial energies in solid-liquid-vapor systems. *J. Phys. Chem.* **63**:1655–1658.

38. **Johnson, R. E., and R. J. Dettre.** 1964. Contact angle hysteresis. 1. Study of an idealized rough surface. *Adv. Chem. Ser.* **43**:112–135.

39. **Kauzmann, W.** 1959. Some factors in the interpretation of protein denaturation. *Adv. Protein Chem.* **14**:1–63.

40. **Kyte, J., and R. F. Doolittle.** 1982. A simple method for displaying the hydropathic character of a protein. *J. Mol. Biol.* **157**:105–132.

41. **Landau, L. D., and E. M. Lifshitz (ed.).** 1963. *Electrodynamics of Continuous Media,* vol. 8, 2nd ed. Pergamon Press, Oxford.

42. **Langmuir, I.** 1916. The constitution and fundamental properties of solids and liquids. *J. Am. Chem. Soc.* **38**:2221–2295.

43. **Lawn, B. R., and T. R. Wilshaw.** 1975. *The Fracture of Brittle Solids.* Cambridge University Press, London.

44. **Lawson, E. Q., A. J. Sadler, D. Harmatz, D. T. Brandau, R. Micanovic, R. D. MacElroy, and C. R. Middaugh.** 1984. A simple experimental model for hydrophobic interactions in proteins. *J. Biol. Chem.* **259**:2910–2912.

45. **Lee, C. Y., J. A. McCammon, and P. J. Rossky.** 1984. The structure of liquid water at an extended hydrophobic surface. *J. Chem. Phys.* **80**:4448–4455.

46. **Lee, C. Y., and H. L. Scott.** 1980. The surface tension of water: a Monte Carlo calculation using an umbrella sampling algorithm. *J. Chem. Phys.* **73**:4591–4596.

47. **Marchesi, M.** 1983. Molecular dynamics simulation of liquid water between two walls. *Chem. Phys. Lett.* **97**:224–230.

48. **McAuliffe, C.** 1966. Solubility in water of paraffin, cycloparaffin, olefin, acetylene, cycloolefin, and aromatic hydrocarbons. *J. Phys. Chem.* **70**:1267–1275.

49. **McBain, J. W.** 1950. *Colloid Science.* D. C. Heath and Co., Boston.

50. **Messiah, A.** 1976. *Quantum Mechanics,* vol. 1, p. 343. John Wiley & Sons, Inc., New York.

51. **Morrissey, B. W., and C. C. Han.** 1978. The conformation of γ-globulin adsorbed on polystyrene lattices determined by quasielastic light scattering. *J. Colloid Interface Sci.* **65**:423–431.

52. **Mozes, N., F. Marchal, M. P. Hermesse, J. L. Van Haecht, L. Reuliaux, A. J. Leonard, and P. G. Rouxhet.** 1987. Immobilization of microorganisms by adhesion; interplay of electrostatic and nonelectrostatic interactions. *Biotechnol. Bioeng.* **30**:439–450.

53. **Neumann, A. W.** 1974. Contact angles and their temperature dependence: thermodynamic status, measurement, interpretation, and application. *Adv. Colloid Interface Sci.* **4**:105–127.

54. **Neumann, A. W., and R. J. Good.** 1972. Thermodynamics of contact angles. I. Heterogeneous solid surfaces. *J. Colloid Interface Sci.* **38**:341–358.

55. **Neumann, A. W., R. J. Good, C. J. Hope, and M. Sejpal.** 1974. An equation of state approach to determine surface tensions of low-energy solids from contact angles. *J. Colloid Interface Sci.* **49**:291–304.

56. **Norde, W., and J. Lyklema.** 1978. The adsorption of human plasma albumin and bovine pancreas ribonuclease at negatively charged polystyrene surfaces. I. Adsorption iso-

therms, effects of charge, ionic strength, and temperature. *J. Colloid Interface Sci.* **66**:257–302.

57. **Norde, W., and J. Lyklema.** 1989. Protein adsorption and bacterial adhesion to solid surfaces: a colloid-chemical approach. *Colloids Surf.* **38**:1–13.

58. **Nozaki, Y., and C. Tanford.** 1971. The solubility of amino acids and two glycine peptides in aqueous ethanol and dioxane solutions. Establishment of a hydrophobicity scale. *J. Biol. Chem.* **246**:2211–2217.

59. **Ohshima, H., Y. Inoko, and T. Mitsui.** 1982. Hamaker constant and binding constants of Ca^{2+} and Mg^{2+} in dipalmitoyl phosphatidyl choline/water system. *J. Colloid Interface Sci.* **86**:57–72.

60. **Pashley, R. M., P. M. McGuiggan, B. W. Ninham, and D. F. Evans.** 1985. Attractive forces between uncharged hydrophobic surfaces: direct measurements in aqueous solution. *Science* **229**:1088–1089.

61. **Pimentel, G. C., and A. L. McClellan.** 1960. *The Hydrogen Bond*, p. 229–241. W. H. Freeman and Co., San Francisco.

62. **Privalov, P. L., and S. J. Gill.** 1988. Stability of protein structure and hydrophobic interaction. *Adv. Protein Chem.* **39**:191–234.

63. **Raynes, J. G., and K. P. W. J. McAdam.** 1988. Purification of serum amyloid A and other high density apolipoproteins by hydrophobic interaction chromatography. *Anal. Biochem.* **173**:116–124.

64. **Rivollet, I., D. Chatain, and M. Eustathopoulos.** 1987. Mouillabilité de l'alumine monocrystalline par l'or et l'etain entre leur point de fusion et 1673K. *Acta Metall.* **35**:835–844.

65. **Rose, G. D.** 1978. Prediction of chain turns in globular proteins on a hydrophobic basis. *Nature* (London) **272**:586–590.

66. **Rose, G. D., A. R. Geselowitz, S. J. Gleen, G. J. Lesser, R. H. Lee, and M. H. Zehfus.** 1985. Hydrophobicity of amino acid residues in globular proteins. *Science* **229**:834–838.

67. **Rose, G. D., and S. Roy.** 1980. Hydrophobic basis of packing in globular proteins. *Proc. Natl. Acad. Sci. USA* **77**:4643–4647.

68. **Rozgonyi, F., K. R. Szitha, A. Ljungh, S. B. Baloda, S. Hjertén, and T. Wadström.** 1985. Improvement of the salt aggregation test to study bacterial cell-surface hydrophobicity. *FEMS Microbiol. Lett.* **30**:131–138.

69. **Sar, N., and E. Rosenberg.** 1987. Fish skin bacteria: colonial and cellular hydrophobicity. *Microb. Ecol.* **13**:193–202.

70. **Schuster, P., G. Zundel, and C. Sandorfy.** 1976. *The Hydrogen Bond*, vol. I, II, and III. North Holland Publishing Co., Amsterdam.

71. **Shafrin, E. G., and W. A. Zisman.** 1957. The adsorption on platinum and wettability of monolayers of terminally fluorinated octadecyl derivatives. *J. Phys. Chem.* **61**:1046–1053.

72. **Shanahan, M. E. R.** 1985. Contact angle equilibrium on thin elastic solids. *J. Adhesion* **18**:247–267.

73. **Shapira, R.** 1959. Adsorption of ribonuclease on glass. *Biochem. Biophys. Res. Commun.* **1**:236–237.

74. **Stillinger, F. H., and A. Ben Naim.** 1967. Liquid-vapour interface potential for water. *J. Chem. Phys.* **47**:4431–4437.

75. **Stillinger, F. H., and A. Rahman.** 1974. The molecular dynamic simulation of liquid water. *J. Chem. Phys.* **60**:1545–1555.

76. **Udin, H.** 1952. Measurement of solid:gas and solid:liquid interfacial energies, p. 114–133. *In Metal Interfaces.* American Society for Metals, Ohio.

77. **Whitesides, G. M., and P. E. Laibinis.** 1990. Wet chemical approaches to the charac-

terization of organic surfaces: self-assembled monolayers, wetting, and the physical-organic chemistry of the solid-liquid interface. *Langmuir* **6**:87–96.

78. **Wood, D. W.** 1979. Computer simulation of water and aqueous solutions, p. 279–436. *In* F. Franks (ed.), *Water, a Comprehensive Treatise*, vol. 6. *Recent Advances*. Plenum Publishing Corp., New York.

79. **Woodruff, D. P.** 1973. *The Solid-Liquid Interface*, p. 21. Cambridge University Press, London.

80. **Young, T.** 1855. *Miscellaneous works*, vol. 1. G. Peacock (ed.). J. Murray, London.

81. **Zisman, W. A.** 1964. Relation of the equilibrium contact angle to liquid and solid constitution. *Adv. Chem. Ser.* **43**:1–51.

Microbial Cell Surface Hydrophobicity
Edited by R. J. Doyle and M. Rosenberg
© 1990 American Society for Microbiology, Washington, DC 20005

Chapter 3

Microbial Hydrophobicity and Fermentation Technology

N. Mozes and P. G. Rouxhet

INTRODUCTION

General Scope

Surface hydrophobicity is involved in interfacial interactions of microbial cells with other microbial cells (flocculation), with liquids and solids (adhesion), or with air (flotation) (69).

The role of hydrophobicity in adhesion of bacteria in natural systems (inanimate, such as aquatic surfaces, rocks, and soil particles, or animate, such as plant roots, animal tissues, and organs) is described in various chapters of this volume. The adverse effects of hydrophobicity in various technological fields (fouling of marine submerged surfaces, fermentation equipments, industrial membranes, and implants) are well documented and addressed elsewhere in this volume. This chapter describes beneficial aspects of microbial hydrophobicity in relation to fermentation technology, namely, microbial cell immobilization and separation techniques.

Our concern in this context is the term "hydrophobicity" in its broadest sense: "the tendency to avoid water" (51). This is a rather ambiguous definition, as are the various methods used to measure hydrophobicity and the parameters used to quantify it (69). Hydrophobicity is conferred to a cell by combination of certain hydrophobic structures or molecules present on the cell surface. Rosenberg and Kjelleberg (69) proposed that these molecules or structures be called

---

N. Mozes and P. G. Rouxhet • Université Catholique de Louvain, Unité de Chimie des Interfaces, Place Croix du Sud 1, B-1348, Louvain-la-Neuve, Belgium.

hydrophobins. In this chapter, we shall refer to hydrophobicity as an overall property of a population of whole cells.

Chemical Aspects of Cell Hydrophobicity

Chemical analysis of surfaces

X-ray photoelectron spectroscopy (XPS) is a powerful tool in investigations of surfaces. It involves irradiation of a sample by an X-ray beam, which induces the ejection of photoelectrons. The kinetic energy of the photoelectrons is analyzed and permits determination of their binding energy. Each peak is characteristic of a given energy level of a given element. Because of inelastic scattering of the electrons in a solid, the information obtained relates to the outermost molecular layers and thus provides an elemental analysis of the surface over a thickness of 2 to 5 nm. The precise binding energy of the photoelectrons depends on the oxidation state of the analyzed element and on the electronegativity of bound atoms. Therefore, the peak position, or the decomposition of a complex peak into various components, may provide information on the chemical functions present at the surface layer.

Application of the XPS technique to microorganisms is delicate. Since the analysis is performed under high vacuum, the sample must be dehydrated, and this fact raises questions concerning the representativity of the surface analyzed. The procedure developed by Amory et al. (1) allows one to gather information about the surfaces of yeasts and bacteria, ensuring at least to some extent the relevancy of the XPS data to the hydrated state of the cells.

Relationship between surface chemical composition and surface hydrophobicity of microorganisms

The presence of polar molecular groups on the surface of a microbial cell favors interactions with water molecules via H bonds and therefore is expected to reduce the overall hydrophobicity of that surface. Indeed, analyses of various bacteria showed that their hydrophobicity (determined by hydrophobic interaction chromatography [HIC]) was inversely related to the oxygen concentration on the cell surface (determined by XPS). Hydrophobicity was also directly correlated with the fraction of surface carbon that was attributed to hydrocarbon moieties of proteins, lipoproteins, phospholipids, or lipopolysaccharides (55, 56). An interesting difference between yeasts and bacteria was noted when their surface N/P atomic concentration ratios were correlated with hydrophobicity parameters (contact angle of water and retention by HIC): bacteria were more hydrophilic as their N/P ratio increased, whereas yeasts were more hydrophobic (2, 55). This finding was hypothetically explained by the fact that bacterial nitrogen is found in various types of surface molecules, including numerous polar ones, whereas yeast nitrogen is found mainly in proteins.

The surface energy of streptococci (78) and staphylococci (77) was found to increase with the surface concentration ratio O/C and to decrease with N/C.

Hydrophobicity And Surface Charge Properties

Surface charge properties of microbial cells

The surfaces of living cells, including microorganisms, carry a net negative charge at physiological pH. The cells acquire that charge by ionization of surface groups (amino, carboxylate, and phosphate), which is a function of pH. The surface charge neutralization is governed by competition between electrical and chemical forces tending to bind counterions to the charged groups on one hand and molecular motion on the other. An electrical double layer is thus established at the interface. According to the classical model usually used to describe the surface, the part of the double layer extending into the aqueous phase is separated in two regions. In the first region, called the Stern layer, ions of sign opposite that of the surface are retained in close contact with the latter, and the potential falls linearly across this layer. The second region extends much farther into the liquid phase and is characterized by an exponential fall in potential. Anions and cations have different distributions through this region, hence the term "diffuse double layer." The thickness of the diffuse layer depends on the ionic strength of the electrolyte solution. Long-distance electrostatic interactions between a cell and any other solid (another particle or a macroscopic body) are due to overlap of the diffuse layers associated with the two surfaces.

The surface charge properties of microbial cells are usually characterized by the zeta potential, which is close to the potential at the beginning of the diffuse layer. The zeta potential is deduced from electrophoretic mobility measurements.

Influence of electrostatic interactions on hydrophobicity measurements

Measurements of hydrophobicity of microbial cells in suspension may be influenced by electrostatic interactions. These interactions, being determined by the potential and the thickness of the diffuse layer, can be modulated by variations of pH and of the ionic strength of the suspension. At pH values close to the isoelectric point of the cells, the zeta potential is minimal. At another pH, if the electrolyte concentration increases, the surface charge is neutralized at shorter distance because of increased retention of counterions in the Stern layer and compaction of the diffuse layer. In both situations, electrostatic interactions are reduced and hydrophobic interactions are revealed best.

HIC is usually carried out at neutral pH (6.8) and at very high ionic strength [4 M NaCl or 1 M $(NH_4)_2SO_4$], as originally proposed by Smyth et al. (74). It has been demonstrated (58) that performing the test at different ionic strengths and pH values provides more information: subtle differences between hydrophobicity of various strains are revealed, and distinction between cells that otherwise behave similarly becomes possible.

Measurement of adhesion to polystyrene, proposed for testing hydrophobicity of colonies (66) and individual cells (58), is also subject to various degrees of interference of electrostatic interactions, as shown by the dependence of the results on pH and ionic strength (58).

The sensitivity of microbial adhesion to hydrocarbons (MATH) test (68) was greatly increased by addition of ammonium sulfate (67). The same effect could probably be obtained by lowering the pH to a value close to the isoelectric point of the cells. Actually, it was reported that a shift of the pH of the test from 4.2 to 5.5 led to a decrease in the measured hydrophobicity of a bottom fermentation yeast (37) and that raising the pH of the test in the range of 3 to 6.5 caused a regular reduction in apparent hydrophobicity of a floating yeast (34).

Finally, the salt aggregation test (48) measures the minimal concentration of ammonium sulfate that allows sufficient screening of the surface charge to provoke cell association, besides modifying water structure to favor the hydrophobic effect.

In contrast to the aforementioned methods, evaluation of hydrophobicity from measurements of contact angles on dried cellular films is not influenced by interactions between electrically charged surfaces.

Microbial Interfacial Interactions in Fermentation Processes

The association of a microbial cell with another phase is the basis of a few processes that are of major interest in fermentation technology.

Biofilm formation consists of an initial step in which a primary monolayer of cells adheres to a solid support and a subsequent step in which many other layers accumulate on top of each other. The first step involves cell-support interactions. The second step could result from cell-cell interactions that are also dependent on surface properties, from embedment of the cells in a polymeric matrix, or from failure of cell progeny to separate from each other. The two last mechanisms are not related to the physicochemical properties of the cells and will not be discussed here.

When formation of cell aggregates consists of association of previously isolated cells, it is called flocculation and depends on the surface properties of the cells. Other possible mechanisms for aggregate formation are embedments in a polymer matrix and lack of separation after cellular multiplication, as described above for biofilm formation.

Flotation of cells on top of a liquid phase may involve isolated cells or aggregates. In both cases, interfacial interactions are determinants in the process.

Acquiring information on the zeta potential and on surface energy is essential for understanding the phenomenon of accumulation of microbial cells at interfaces and eventually interfering with it. The aim of this chapter is to present information, ideas, and examples related to the effects of microbial hydrophobicity in fermentation technology, specifically with respect to microbial adhesion and accumulation at solid, gas, or liquid interfaces.

IMMOBILIZATION OF MICROBIAL CELLS BY ADHESION TO A SOLID SUPPORT

Immobilization of Biocatalysts

The use of biocatalysts (enzymes, bacteria, and fungi) is a common practice in various industrial processes. The performance of biocatalyst-based systems can be improved by immobilization (31). The term is defined as spatial confinement, or control of localization, of the biocatalyst in order to use it in a continuous process or to recover it easily for repeated utilization. Immobilization thus allows one to achieve a high catalyst concentration and consequently increased productivity (amount produced per time unit and reactor volume). In certain cases, immobilization also improves biocatalyst stability. Additional advantages are simplification of operation, possibility for automation, and ease of product separation.

Immobilization is achieved by numerous techniques. These can be classified into three categories: (i) covalent coupling, (ii) interfacial interaction, and (iii) physical retention.

Covalent coupling can be involved in cross-linking or in attachment to a carrier. In the process of cross-linking, the individual biocatalyst units are joined to one another with the help of bi- or multifunctional reagents. The most commonly used bifunctional reagent is glutaric dialdehyde; others are hexamethylene and toluene diisocyanates. Otherwise, the catalyst may be bound to a carrier directly or via a spacer arm, the carrier being inorganic material or natural or synthetic polymer. The coupling agents are glutaric dialdehyde, cyanogen bromide, thionyl chloride, and others. Covalent binding to a carrier is used only for enzymes and is not suitable for whole-cell immobilization.

Immobilization via interfacial interactions implies association of the catalyst with a support surface as a result of favorable energetic balance. The forces involved are physicochemical (electrostatic and van der Waals forces and hydrogen bonds). The term "adsorption" is used for enzymes; "adhesion" refers to whole-cell attachment to a support surface. The supports used are a wide variety of inorganic and organic materials as well as synthetic polymers. Formation of cell aggregates with no fixed support is another way of immobilizing microbial cells.

Physical retention of biocatalysts may involve their embedment in polymer of a gel-like structure or confinement within a semipermeable membrane. Entrapment in a porous polymeric matrix is the immobilization technique most widely used for whole cells. Natural and synthetic polymers can be used. Membrane immobilization is achieved by the liposome technique, by microencapsulation, or by use of a membrane reactor with hollow fibers, ultrafiltration, or dialysis membranes.

Microbial Immobilization by Retention on a Support

Adhesion is the oldest method for immobilizing microbial cells; the production of vinegar by acetic bacteria adhering to wood shavings (the trickling process) has been practiced since 1815 (31). The technique is inexpensive and easy to carry out, and it does not rely on nonphysiological pretreatment conditions as do some of the other methods. Adhesion provides a more efficient system than entrapment in a matrix, since it ensures easy contact and free mass transfer (nutrients and products) between the immobilized cells and the circulating liquid medium. Practically speaking, the aim is to create thick microbial films adhering on a fixed solid support. Biofilm formation allows accumulation of large amounts of cells that can be regenerated by cellular growth. Its drawback is the risk of diffusional limitations. The carrier may be a porous matrix filling the entire reactor volume (fixed bed); it may also be particles packed or maintained in suspension by gas or liquid flow (fluidized bed). Cell aggregates used in fluidized bed reactors are also considered in the latter category; they will be treated in a separate section (see below, Flocculation as an Immobilization Technique).

For each specific production system, adhesion must be examined in terms of amount of immobilized cells, initial activity, storage stability, and operational stability.

The association of a bacterium (considered as a solid particle suspended in a polar medium) with another solid phase (in contact with the same polar medium) can be described as a multistep process. It is convenient to distinguish a generic step that depends on overall physicochemical properties (electrical charge and hydrophobicity) and may result in reversible or irreversible adhesion. A possible subsequent step, more specific, depends on molecular complementarity and stereochemistry. It involves only interactions acting at short distances and leads to irreversible adhesion. Colonization and biofilm formation may occur at later stages.

Busscher and Weerkamp (11) proposed a model in which at separation distances greater than 50 nm, only van der Waals forces are operative. At separation distances between 10 and 20 nm, both van der Waals and electrical double-layer forces occur; their sum can result in a secondary minimum at which reversible adhesion takes place. Certain surface appendages could overcome the repulsive electrostatic forces and bridge the entire cell with the support. Hydrophobic regions (molecules, structures, and appendages) of the surface may play a major role at this point in allowing the removal of the water film from between the interacting surfaces and enabling short-range interactions to occur. At very short separation distances, when any potential barrier has been overcome, a variety of specific and nonspecific interactions operate, leading to irreversible adhesion.

Immobilization of Hydrophobic Microorganisms

In this section, several examples will be described to illustrate the role of cell surface hydrophobicity in the first step of biofilm formation, namely, adhesion to a solid support. All of the information presented concerns laboratory-scale experiments.

Moniliella pollinis

The fungus *Moniliella pollinis* is used for the production of erythritol and other polyols or of a viscous polysaccharide of the glucan type. In its yeastlike form, this fungus is an extremely hydrophobic particle, scoring the highest values by any method used for evaluating hydrophobicity (contact angle of water, >90°; 100% retention by HIC; 0.05 M in the salt aggregation test; 99% exclusion from water phase in MATH) (58). Such hydrophobic cells are expected to adhere to any solid that is less polar than the aqueous medium. *M. pollinis* could indeed be immobilized irreversibly on a variety of supports ranging from glass to polyvinyl chloride, with preference toward low-energy materials (57).

Table 1. Surface Properties of Phosphate-Saturated and -Depleted Cells
of *C. glutamicum*[a]

	Parameter							
Cells	Surface concn ratio			Electrophoretic mobility at pH 6 $(10^{-8}$ $m^2 s^{-1}$ $V^{-1})$	Water contact angle (°)	Retention by HIC (%)[b]	Concn of DEAE-dextran for provoking maximum flocculation (g/liter)	Amt of cells adsorbed (mg/g of carrier)
	O/C	N/C	P/C					
P saturated	0.2100	0.0400	0.0047	3.1	49	91	0.3	10
P depleted	0.2300	0.4400	0.0011	2.7	28	26	0.8	4

[a] Adapted from Büchs et al. (10).
[b] Hydrophobic interaction chromatography, performed at pH 7 with 4 M NaCl.

Corynebacterium glutamicum

Corynebacterium glutamicum was studied by Büchs et al. (10) for transformation of α-ketoisocaproic acid into L-leucine. Phosphate supply to growing cells was found to have an influence on cell wall properties, resulting in variation of adhesion behavior. The main results are summarized in Table 1. Cells saturated with phosphate were hydrophobic and showed a high tendency to adsorb on treated porous glass surfaces. The glass treatment consisted of coating with DEAE-dextran or with aminosilane; both coatings caused substantial decrease of the surface energy of the glass and made its electrokinetic potential less negative. Phosphate depletion during culture led to a decrease in surface hydrophobicity and consequently to a smaller tendency to associate with the support. Since phosphate depletion (down to a phosphorus content of about 20% of the saturation value) had no influence on growth rate and on specific L-leucine production, a strategy for obtaining optimal functioning of a continuous amino acid fermentation was proposed. The strategy is based on the inducible variation of the surface properties of *C. glutamicum*: by adequate phosphate feeding, one can obtain initially hydrophobic cells that will tend to adhere to the support and associate with each other and then hydrophilic cells that will prevent plugging of the carrier by excessive accumulation.

Hyphomicrobium sp. and Xanthobacter autotrophicus

Hyphomicrobium sp. and *Xanthobacter autotrophicus* are bacteria capable of growing on dichloromethane and 1,2-dichloromethane, respectively, as sole carbon sources. They are exploited for degradation of the environment-polluting halogenated hydrocarbons. M. P. Hermesse (per-

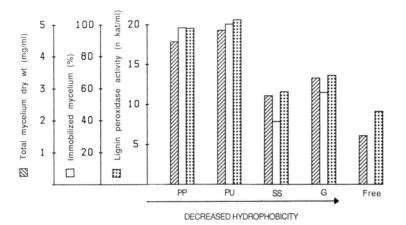

Figure 1. Total mycelium dry weight, percentage of immobilized mycelium, and lignin peroxidase activity for *P. chrysosporium* INA-12 immobilized on polypropylene (PP), polyurethane (PU), stainless steel (SS), and grey (G) or as free pellets after 5 days of total incubation. (Adapted from Asther et al. [4].)

sonal communication) investigated their immobilization by adhesion to active carbon during growth (in the culture medium and under continuous agitation) or by sedimentation of harvested bacteria suspended in water. *X. autotrophicus* was found to be more hydrophobic than the *Hyphomicrobium* sp. (contact angles of water were 55° and 30°, respectively). The former adhered only to a carbon carrier with an apolar surface, whereas the hydrophilic *Hyphomicrobium* sp. adhered to both apolar and polar carbon (the latter obtained by air oxidation), with somewhat better results on the polar carbon.

Phanerochaete chrysosporium

Asther et al. (4) studied immobilization of *Phanerochaete chrysosporium*, a lignin peroxidase-producing fungus, to carriers of different hydrophobicities. They determined the surface free energy of the cells and the carriers, using contact angles of mercury and α-bromonaphthalene, and showed that the fungus is hydrophobic, the mycelium being less hydrophobic than the conidiospores. Experimental results agreed with predictions based on a thermodynamic model: immobilization was higher for hydrophobic than for hydrophilic carriers (Fig. 1). More efficient retention by the carrier led to an increase of biomass, presumably as a result of protection of the cells from agitation, and therefore to an enhanced production of lignin peroxidase. It was concluded that hydrophobic interactions, in addition to the rugosity, porosity, and surface area of the carrier, are crucial in the immobilization process.

Methanogenic bacteria

In the continuous process of anaerobic digestion and methane production by mixed microbial populations, retention of certain species by the carrier is followed by an increase in the number of species and in the total biomass held on the support surface. The installation of the first bacterial layer is thus of major importance for the formation of biofilms in anaerobic digesters. Verrier et al. (79, 80) studied the short-term adhesion of four methanogenic bacteria on polymer surfaces with different hydrophobicities. *Methanobrevibacter arboriphilicus* (66% remaining in the aqueous phase in the MATH test) adhered to all of the supports, with highest density on the intermediately hydrophobic ones. *Methanothrix soehngenii* (79% in the MATH test) adhered preferentially to hydrophobic polymers. *Methanospirillum hungatei* (>92% in the MATH test) preferred hydrophilic surfaces. *Methanosarcina mazei* did not adhere to any support, a finding attributed to excretion of extracellular polymer of high hydrophilic nature. The data were used to define criteria for selection of support materials.

Interplay of Hydrophobic and Electrostatic Interactions

Although hydrophobic interactions are predominant in adhesion of hydrophobic cells, electrostatic interactions are often present and influence the adhesion. In the case of *M. pollinis* mentioned above, adhesion was influenced by the pH (Fig. 2). Coverage of any support was most dense around pH 4, slightly above the isoelectric point of the cells. At higher pH values, as the electrokinetic potential increased, electrostatic cell-cell repulsion prevented dense coverage. At pH 3.5, the electrokinetic potential was zero, and the cells flocculated as a result of lack of cell-cell repulsion and the extreme surface hydrophobicity. Below this pH, the cells remained flocculated. The flocs adhered to the support but did not resist the shear force of washing and detached before numeration (57).

The adhesion of *C. glutamicum* from culture without phosphate limitation (see above) was better from suspension in NaCl solution than from suspension in pure water. The phosphate-depleted bacteria (the more hydrophilic ones) adhered only from the electrolyte solution (10).

Examination of adhesion of a collection of microorganisms to polystyrene at pH 3 to 7 and at low (10^{-3} M) and high (10^{-1} M) ionic strengths (58) revealed that for cells of moderate hydrophobicity, adhesion oc-

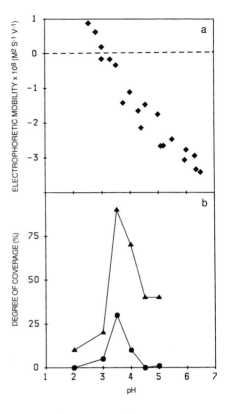

Figure 2. Variation of the surface properties and adhesion of *M. pollinis* as a function of pH. (a) Electrophoretic mobility; (b) coverage of solid supports in adhesion tests. Symbols: ●, glass; ▲, polystyrene. (Adapted from Mozes et al. [57].)

curred at low ionic strength only at pH 3, i.e., close to the isoelectric points. Under conditions of high ionic strength, adhesion took place over all pH values tested. In both cases, adhesion was weak. For very hydrophobic cells, irreversible adhesion occurred in the whole pH range and at both high and low ionic strengths; the density of the adhering cells diminished only at the low ionic strength and pH 6 to 7. Hydrophilic cells did not adhere at all on the native support. Such cells could adhere only if the electrostatic repulsions were decreased by modifying the surface of the support (13, 57).

FLOCCULATION

The Phenomenon of Flocculation

Flocculation is a process of gathering particles to form an aggregate that is a transportable entity, with no need for substratum to attach to. The aggregate is called floc, flock, clump, cluster, or pellet. The process is active, involving the coming together of originally dispersed cells, in contrast to the passive association that results from failure of cellular progeny to separate from each other or from embedment of cells in a matrix of excreted polymer. It could be stochastic (as by fluid motion and random collisions) or directed (as by chemotactic signals). The genesis of flocs in certain cases is part of the natural history of the organism and depends on its genetic makeup and stage in the life cycle. In other cases, formation of flocs occurs under certain nonphysiological conditions, leading to the expression of competence to flocculate; thus, flocculation can be provoked by man-made intervention (12).

The mechanisms by which dispersed cells are repelled and the forces that cause them to aggregate depend on the molecular architecture of the cell surface (65). Flocculation can be considered in the same terms as adhesion. Interfacial interactions are the primary driving force. They are determined by the overall physicochemical properties of the cell surface and can be modulated by the presence of polymers and cell appendages (76). Specific interactions play a significant role at short separation distances.

Microbial flocculation and aggregate formation are of particular importance in many industrial processes, such as brewing, biomass production, waste water treatment, and bioconversions. The implication of flocculation for biotechnology is discussed by Atkinson and Daoud (5), Esser and Kues (21), and Unz (76). For fermentors containing organisms in suspension, the association of single cells into flocs is a key factor in the overall process. The performance characteristics and productivity of the fermentor may be affected by the presence of cell aggregates due to increased microbial hold-up, ease of separation of biomass from the fermented broth, and possibilities for continuous operations beyond normal washout flow rate. Against these advantages stands the fact that overall rates of growth and metabolism are likely to be reduced as a result of diffusional limitations on the transfer of substrate, product, oxygen, and protons through the floc.

Flocculation of Brewer's Yeasts

Importance and proposed mechanisms

The ability of selected yeast strains to flocculate has traditionally been exploited in the manufacture of beer. The yeast cells remain dispersed during the fermentation of the wort; as a result, the conversion of sugar to ethanol and CO_2 is efficient and free of mass transfer limitations. Once the sugar is consumed, flocculation of the cells (followed by sedimentation or flotation) helps in the separation of the yeast cells from the fermented beer. According to the site of cell accumulation at the end of fermentation, brewing yeasts are distinguished as bottom (sedimenting) or top (floating) strains.

Kamada and Murata (37) explained that flocculation of brewer's yeast occurs easily and reversibly, depending on the environmental conditions. According to these authors, there is no evidence that the yeast excretes a certain substance from the cell when it flocculates (contrary to the situations reported for some other microorganisms). It therefore seems reasonable to describe the mechanism of the process, at least initially, in terms of the general theory of colloid stability (the DLVO theory). Yeasts remain dispersed as a result of repulsion of the negative cell surfaces (65); they associate when attractive forces overcome that repulsion.

Investigations of flocculation concern two aspects: (i) the inherent tendency of the cells to flocculate (flocculence) and (ii) the actual interactions between potentially flocculent cells to form flocs (54). The ability of yeasts to flocculate is genetically and physiologically determined (18, 19, 50). Expression of this ability depends on environmental factors such as the presence of proper inducers and rate of cellular collisions.

The nature of the yeast surface molecular components responsible for flocculation has been widely investigated. A few authors showed that it is mannoprotein (37, 54, 60), of which the native stereospecific structure is indispensable for the reversibility of the dispersion-flocculation process. The role of phosphate, in the phosphomannan-protein complex, was often evoked, but the reports are controversial. Whereas one study (30) could not demonstrate any significant difference in phosphate content between flocculent and powdery strains, other authors (50) claimed that flocculent yeast walls have more exposed phosphate groups than do nonflocculent yeast walls, and still others (36) showed that removal of phosphate by HF treatment enhanced flocculation. Such disagreement could result from differences in defining the yeast surface and in methodologies used for measuring surface phosphate groups. Extraction of surface lipids resulted in decreased flocculation of flocculent strains (36). Enhanced flocculation of top yeasts (washed cells) was observed upon addition of organic solvents (54). This was attributed to increased strength of hydrogen bonds as a result of decreased dielectric constant of the liquid medium, for cell flocculation occurred at about the same dielectric constant with different solvents tested. Participation of cell wall mannan in interactions leading to flocculation was revealed by selective staining and electron microscopy (52, 53) and by the inhibitory effects of mannose derivatives (18, 39, 52) and of mannose-specific lectins (53).

The evident specific role of Ca^{2+} ions in promoting flocculation has led to the suggestion that anionic groups of cell wall components of adjacent cells are bridged by the cation. There is no unanimity as to what type of anions would be involved, carboxyl groups of cell wall proteins (36) or phosphate groups of the phosphomannan skeleton of the cell wall (50). Miki et al. (53) proposed a more elaborate model that accounts for the role of Ca^{2+} ions, mannan, and protein in the flocculation mechanism: "a specific lectin-like component of the cell wall recognizes and adheres to alpha-mannan carbohydrate on an adjoining cell, Ca^{2+} ions acting as cofactors activating the binding capacity." Kihn et al. (39) have shown that calcium specificity is linked to the ionic radius of the element.

Flocculence of top and bottom yeast strains

Saccharomyces carlsbergensis is a bottom strain, the cells sedimenting at the end of fermentation. *Saccharomyces cerevisiae* is a top strain; the yeast cells associate with CO_2 bubbles and gather at the surface of the broth.

Amory et al. (3) tested a bottom and a top fermentation yeast strain cultivated on malt extract or on yeast extract. The test consisted of measuring the clearing rate of suspension of washed cells after provoking flocculation by addition of calcium and ethanol. The top fermentation strain (*S. cerevisiae* MUCL28733) did not flocculate in the presence of calcium alone, even at high concentrations. It did, however, flocculate in the presence of ethanol, alone or together with calcium. The bottom fermentation strain (*S. carlsbergensis* MUCL28285) flocculated only under the synergistic effect of ethanol and calcium or under very high calcium concentrations (and this only for cells coming from yeast extract cultures). In both strains, cells grown on yeast extract were more flocculent than those derived from malt extract cultures, as shown by the threshold concentrations of ethanol or calcium inducing flocculation (Fig. 3). In this way, it was demonstrated that the *S. carlsbergensis* strain tested was less flocculent than the *S. cerevisiae* strain and that cultures in malt extract produced less flocculent cells of both strains.

Relationship between surface properties and flocculence

The variation between the sensitivities of *S. carlsbergensis* and *S. cerevisiae* to calcium and ethanol indicates that their surface properties are different. Indeed, when the surface chemical composition, the zeta potential, and the surface hydrophobicity of a series of bottom and top fermentation strains were investigated, clear distinctions were observed (2). The values of the P/C atomic concentration ratios of *S. carlsbergensis*

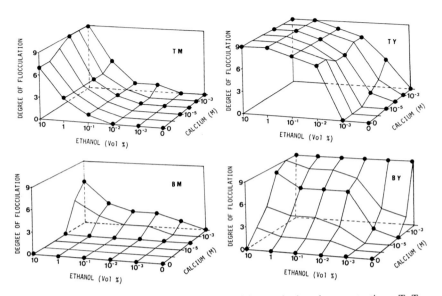

Figure 3. Comparison of flocculation at various calcium and ethanol concentrations. T, Top fermentation *S. cerevisiae*; B, bottom fermentation *S. carlesbergensis*; Y, culture on yeast extract; M, culture on malt extract. (Mozes et al. [57]. Reproduced with permission.)

strains were between 0.0027 and 0.0034, the values of the zeta potential at pH 4 were in the range of -29 to -40 mV, and contact angles of water measured on lawns of freezed-dried pressed cells were between 52° and 65°. For *S. cerevisiae* strains, the values were as follows: P/C, 0.0011 to 0.0018; zeta potential, -6 to -29 mV; and water contact angle, 72° to 90°. Higher surface hydrophobicity and lower surface charge were thus shown to be related to higher flocculence of the latter strains. Moreover, their high hydrophobicity accounts for their association with CO_2 bubbles and their floating on the top of the broth at the end of fermentation. These observations agree with the ideas discussed by Jayatissa and Rose (36) that the negative charge of the phosphate surface groups creates a repulsive force that helps to keep the cells dispersed and that hydrophobicity influences the process.

Changes in flocculation as a culture develops with time occur both in wort and in simple culture media. These changes have been attributed to changes in the composition of the cell wall (19). Amory et al. (3) showed that cell age affects the surface properties of bottom fermentation yeast cells. The most important change was observed at the level of the P/C atomic concentration ratio; it decreased gradually from a value of 0.0042 at day 1 to 0.0018 after 10 days of fermentation. Hydrophobicity (estimated by testing adhesion to polystyrene) increased during the first 4 days and remained high thereafter. Flocculence (detected by the test with calcium and ethanol as described above) was high from day 4, but flocculation in the fermentation medium was observed only after 7 days.

The influence of nutritional source on surface properties of the fermentation yeasts was also investigated (3). Here the correlations were not always straightforward. Difference in flocculence of the bottom strain cultivated on yeast or on malt extract could be predicted better from the P/C atomic ratio than from the zeta potential, the latter being practically identical for the cells from the two media. Hydrophobicity (estimated by testing adhesion to polystyrene) was also the same for cells derived from both types of media. As for the top strain, whereas P/C and zeta potential seemed consistent with the measured flocculence (increased electrostatic repulsion is related to reduced flocculence), the hydrophobicity of the top fermentation yeast cells grown on malt extract was higher than that of cells grown on yeast extract, although the former were less flocculent.

Kamada and Murata (37) showed that the hydrophobicity of the bottom yeast cell surface increased under conditions inducing flocculation; no such change was observed with top strains. In the froth flotation test used for evaluating hydrophobicity (affinity to bubbles), top and bottom strains behaved similarly in the absence of sugar. When sugar was present, the hydrophobicity of the bottom strains decreased considerably, whereas that of the top strains did not vary much. Surprisingly, high electrolyte concentration (1 M) caused flocculated yeast cells to redisperse. Treatment of flocculent yeast cells with proteolytic enzymes led to simultaneous loss of their ability to flocculate and reduction of hydrophobicity (affinity to benzene compared with water). The effect of denaturing cell surface proteins (by urea, guanidine hydrochloride, detergents, or heat) was complex. Figure 4 illustrates the influence of ionic detergent concentration on flocculation. The suggested explanation was that because of partial unfolding of proteins, interior hydrophobic groups are increasingly exposed to other yeast cells as the detergent concentration increases slightly, and cell-cell hydrophobic interactions are established. At excessive concentrations, the detergents not only provoke changes of surface proteins conformation but adsorb completely on the cell surface, causing a dispersion and dissociation of the aggregates. The ensemble of observations led the authors to conclude that reversible dispersion-flocculation changes are based on variation of hydrophobic and ionic interactions between cell surfaces.

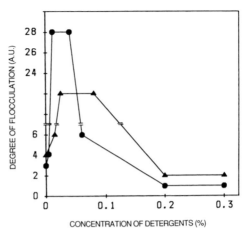

Figure 4. Degree of yeast flocculation, determined from sedimentation height, as a function of detergent concentration. Symbols: ●, sodium dodecyl sulfate; ▲, cetylethyl ammonium bromide. (Adapted from Kamada and Murata [37].)

Cell Separation by Flocculation

Separation of microorganisms by flocculation is described in numerous reports. Bautista et al. (7) described a process of cell removal based on flocculation and sedimentation. A flocculating agent (polyelectrolyte) is added to the broth after fermentation (production of hyaluronate lyase by *Streptococcus equisimilis*). Within 2 h, a compact cell sediment is obtained, leaving a clear supernatant that contains the produced enzyme, maintaining 100% of the enzyme activity. The authors calculated a power saving of 80% in comparison with a nonsedimentation process based on continuous centrifugation.

Eriksson and Hardin (20) studied flocculation of *Escherichia coli* with the cationic polyelectrolyte chitosan. They showed that flocculation is related to the amount of extracellular polymers released by the cells into the solution and that the hydrophobicity of the cells (estimated by partition in a two-phase system) depends on the bound extracellular polymers on the cell surface.

The same approach to facilitate separation was used by Kim et al. (40). They used positively charged submicron polymeric particles to interact with cells and to form flocs. The sedimentation velocities of flocculated *E. coli* and *S. cerevisiae* increased 4 and 3 orders of magnitude, respectively. Sedimentation of *E. coli* cells and cell debris was completed within 10 min under gravity, which represents an appreciable energy saving. Bioproducts (hexokinase from *S. cerevisiae*, creatininase from *Flavobacterium filamentosum*, and glycerophosphate oxidase from *Aerococcus viridans*) remained in the supernatant without adsorption or inactivation.

These three examples show that exploiting cell surface charge properties to provoke flocculation is easier and more common than interfering with hydrophobicity.

In various processes of biomass production, flocculation or aggregate formation followed by sedimentation of the flocs or the aggregates in a settler allows the produced cells to be harvested easily. Barreto et al. (6) reported production of a starter culture in a fluidized bed reactor with a flocculent strain of *Lactobacillus plantarum*. Greenshields and Smith (27) described the production of biomass by flocculent strains of brewer's yeast (*S. cerevisiae*) by using a tower fermentor.

Flocculation as an Immobilization Technique

It must be pointed out that there is a certain ambiguity in the definition of the term "flocculation." Some authors use the term to refer to any aggregate formation. Probably certain examples cited below do not involve flocculation in the sense defined earlier (see above, The Phenomenon of Flocculation).

Flocculation of industrially important microorganisms is frequently used as a means for their retention in bioreactors (see above, Microbial Immobilization by Retention on a Support). Natural or artificially provoked flocculation is used to allow high cell concentrations in continuous fermentors or to facilitate cell separation in batch fermentors. A common installation for continuous fermentation is an upflow tower, the operation of which relies either on the ability of the flocculated cells to settle against the fluid upflow or on collecting the flocculated cells by sedimentation in a settler at the outlet of the tower and recycling them back to the bottom of the reactor.

Flocculation is the only immobilization method that is well adapted to production processes requiring a high rate of cellular multiplication (70). A limitation of this method is that cells in flocs may respire, metabolize, or grow at lower rates as a result of internal gradients within the floc, whereas dispersed cells have greater access to substrate and less exposure to their own products.

Typical examples are use of *Arthrobacter* cells flocculated by polyelectrolyte for production of fructose-enriched syrup (17), anaerobic process of ethanol production by flocs of bacteria (*Zymomonas mobilis* [44, 45, 63]) or yeasts (*S. cerevisiae* [33, 38, 42, 59, 64]), aerobic fermentation by flocs of *Bacillus* sp. (43), and lactic acid production by flocculating mutants of *Lactobacillus bulgaricus* (32).

Information on the relevance of surface properties of microbial cells to their immobilization by flocculation is sparse. Büchs et al. (10) showed that *C. glutamicum* cells from culture with no phosphate limitation, which were more hydrophobic (by HIC and contact angle measurements), were also more flocculent than cells from phosphate- or sulfur-limited cultures (Table 1). The flocculation test in that case was based on addition of polycation and measurement of the relative volume of the sedimented flocs. A low additive concentration required to achieve flocculation was ascribed to high flocculence of the bacteria.

Development of industrially feasible methods for manipulating non-flocculent suspensions to produce stable and manageable aggregates requires further research in the field of microbial flocculation (12).

Role of Flocculation in Activated Sludge Processes

Biological processes are used extensively in the treatment of domestic and industrial wastewater. The quality of the effluent water depends on effective removal of the pollutants by metabolic activity of the microorganisms and effective removal of the latter after they have fulfilled their metabolic role. The bacteria and the majority of the protozoa, of which the sludge is composed, usually exist in aggregated masses (22). According to Tenney and Verhoff (75), the process of aggregate formation consists of attachment and subsequent bridging between cell surfaces by polymeric species to form a large lattice. However, the aggregates may contain filamentous cells that constitute a more intricate matrix.

The surface structures and properties of the sludge organisms determine whether they flocculate. However, relevant information is rare. Sludge organisms growing on nitrogen- and phosphorus-restricted media were reported to have exceptionally large capsules and a high surface charge (82). Singh and Vincent (73) described the properties of a clump-forming bacterial strain (designated KEWA-1) isolated from sludge of a wastewater treatment plant. In a diluted medium, the cells were hydrophobic (tested by MATH [68]). They grew slowly and produced extensive capsules that fused with other cells to form clumps; this process is not considered true flocculation (see above, The Phenomenon of Flocculation). In a rich medium, growth was faster, and bacteria swam freely, did not demonstrate hydrophobicity, did not have a capsule, and did not clump.

FLOTATION

General Aspects

The interaction between bacteria and the air-water interface is a well-documented phenomenon. Adsubble (adsorptive bubble separation) processes are defined as phenomena and techniques in which dissolved or suspended material is segregated within, or removed from, a liquid by adsorption or attachment at the surface of rising bubbles (47).

The natural occurrence of high numbers of microorganisms, predominantly bacteria, at the air-water interfaces of aquatic systems and its environmental implications were described and discussed by Kjelleberg (41). The bacteria populating the sea surface microlayer consist of a higher percentage of hydrophobic cells than can be found in subsurface water (14, 49). The spontaneous tendency of metabolizing cells to separate from the medium by accumulating at the liquid surface is encountered in certain industrial systems: the wine industry, breweries (see above, Flocculence of top and bottom yeast strains), water treatment plants, and biomass production facilities. For certain applications, flotation is provoked by addition of a surface-active substance.

According to froth flotation theory (46), both frother surfactant and hydrophobic particles are required for generation of stable froths. Studies conducted on the surface properties of floating microorganisms show generally that the cells involved in flotation are highly hydrophobic. The surfactant, whether natural, produced by the cells themselves (biosurfactants [25]), or added, is an essential factor for achieving flotation. Its main role is to reduce the surface tension of the aqueous medium. Moreover, adsorption of the surfactant and its orientation at the cell surface, with the hydrocarbon portion exposed to the solution phase, may make cells hydrophobic.

Cell Concentration and Separation by Flotation

Typical examples

Dognon and Dumontet (16) were the first to observe the enrichment of microorganisms in foam and proposed the use of flotation process for separation and concentration of cells. The separation could be improved by use of quaternary ammonium ions (28, 29) or flocculation additives (72). Koch bacilli (strain BCG) could be distinguished from all other species under study (yeasts and bacteria) as being easily concentrated at the froth in the form of voluminous packs (16). The other species could be separated from the suspension and concentrated in the foam only by addition of surfactants or electrolytes (NaCl, $CaCl_2$, or Na_2SO_4). The authors suggested application in the field of medical bacteriological analysis.

Boyles and Lincoln (9) described a selective method for removing or concentrating bacterial spores and vegetative cells from the culture growth medium by foam flotation. They efficiently collected spores of *Bacillus anthracis* and *Bacillus subtilis* as well as cells of *Serratia marcescens*, less successfully collected various strains of *Brucella suis*, and failed to collect cells of *Pasteurella tularensis*. The authors concluded that whether a cell is carried by foam or not depends on the nature of the cell surface, mainly its hydrophobicity.

Grieves and Wang (28, 29) suggested that the most effective foam separation agent should be a cationic surfactant. They investigated foam separation of six bacterial strains. The efficiency of separation depended on the strain. This finding was attributed to differences in cell size and in the nature of the cell surfaces, presumably with respect to hydration and charge magnitude. For usual lyophobic colloidal particles, surfactant ions of charges opposite that of the particle are adsorbed with the head group attached to the particle surface and with the hydrocarbon chain lying perpendicularly to the surface. For bacteria, the situation is more complex since the constituting material is hydrated. The surfactant ions may not only be retained by oppositely charged surface groups but may also react chemically with surface molecules; they may even be retained within the membrane or be absorbed into the cytoplasm.

Rubin et al. (71, 72) introduced the technique of microflotation, whereby very low gas flow rates and insoluble collector surfactants are used to produce a surface phase. Adsorption and orientation of the collector surfactant (lauric acid or laurylamine) at the cell surface led the bacteria to float and accumulate at the air-liquid interface. The method was applied to harvesting bacteria (*E. coli* and *Aerobacter aerogenes*) and algae (*Chlamydomonas reinhardtii* and *Chlorella ellipsoidea*).

Industrial applications

In industrial processes in which the goal is production of biomass, flotation offers a means of collecting the cells. When the microorganism has only a catalytic role, flotation can help in recovering the active biocatalyst at the end of the process.

A process of cell recovery by continuous foam flotation was developed and described by Viehweg and Schügerl (81) and Gehle et al. (24). Separation of *Hansenula polymorpha* cells from the growth medium was achieved by permitting the broth to flow down a column, at the bottom of which nitrogen bubbles were produced. The separator is illustrated in Fig. 5. Cell concentration in the foam (C_f) was very high (65 g/liter) compared with the concentration in the effluent at the bottom outlet of the column (0.08 g/liter) or the concentration of the feed broth (4.8 g/liter), thus achieving an efficient process (enrichment factor of 812). The authors did not discuss the surface properties of the cells. They investigated the influence of culture conditions and found that the enrichment factor was

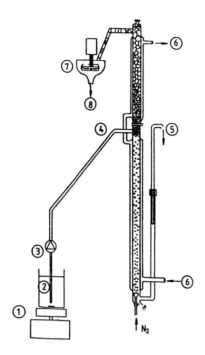

Figure 5. Cell flotation column. 1, Magnetic stirrer; 2, cell suspension; 3, pump; 4, feed; 5, liquid outlet with cell residue; 6, thermostat; 7, foam breaker; 8, foam liquid. (Gehle et al. [24]. Reproduced with permission.)

reduced considerably if cell growth occurred under nitrogen limitation. In their effort to optimize separation, they tested the effects of added electrolytes. When "structure maker salts" (such as Na_2SO_4) were added to the medium, C_f was reduced. With use of chaotropic salts (such as $NaClO_4$), C_f was increased. These results point to the role of hydrophobic interactions in the process of foam flotation, since hydrophobicity is manifested best in nonstructured water.

Recently, R. Tybussek (personal communication) used a similar approach for collecting baker's yeast (*S. cerevisiae*). A distinct difference between the flotation behavior of two strains was noticed. Analysis of the surface properties of these strains revealed that the one with the higher flotation index was more hydrophobic (water contact angle of 69°) and had a lower zeta potential (electrophoretic mobility of $-1.35 \times 10^{-8}\,m^2\,s^{-1}\,V^{-1}$) than did the strain with the lower flotation index (27° and $-1.85 \times 10^{-8}\,m^2\,s^{-1}\,V^{-1}$). Carbon or phosphorus limitation during growth caused alteration of flotation concomitantly with modification in hydrophobicity (Fig. 6) and surface proportion of apolar/polar moieties (determined by XPS analysis).

In many industrial fermentation processes, gas bubbles are either formed by intensive agitation (air) or produced as a result of microbial

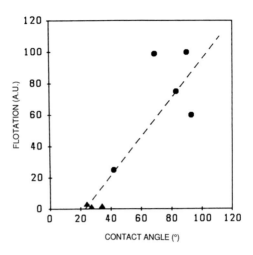

Figure 6. Flotation of *S. cerevisiae* cells as a function of hydrophobicity (water contact angle). The two symbols refer to different strains. Each datum point represents yeast cells grown under different culture conditions. (R. Tybussek and N. Mozes, unpublished results.)

metabolism (CO_2, CO, CH_4, and H_2). If the bacteria have a hydrophobic surface, they will attach to the bubbles. An adequate reactor design (generally a tower with a foam breaker) allows one to separate the biocatalyst at the top of the fermented broth.

Tool in biological research

The differences between flotation of vegetative bacterial cells, germinating spores, and dormant spores suggest an easy method for their separation. Dobias and Vinter (15) reviewed differences in the surfaces of vegetative cells and spores; tests of separation in an aqueous polymer two-phase system indicated that the spores are more hydrophobic.

Gaudin et al. (23) studied flotation of *B. subtilis* var. *niger* to obtain information on the surface properties of the bacteria and to develop a method for separating and concentrating bacterial cells according to their physiological states. Flotation of spores or vegetative cells could be targeted by adequate selection of surfactant. Thus, the addition of secondary amines resulted in a preferential flotation of spores as compared with the rest of the cells. On the other hand, the addition of carboxylic acid increased flotation of vegetative cells as compared with the spores. This difference was attributed to the presence of different polar groups on the surface of the spores and the vegetative cells. Both cationic and anionic sites are available at the surface of both forms of cells, but the relative abundance of the sites is different. The selective flotation could be due to the fact that the spores have an excess of carboxylate groups and the vegetative cells have an excess of amino groups. Factors other than the physiological state of the cells that were found to influence flotation behavior were the electrolyte concentration and the pH.

Grieves and Wang (28, 29) proposed that cells be concentrated from diluted suspensions by flotation in order to obtain more accurate counts.

Implications of Flotation in the Wine Industry

Film-forming yeasts

Propagation of certain yeasts, called film yeasts, on the surface of wine causes formation of a pellicle. These yeasts often continue to live in the wine during storage, resulting in an inferior product. Iimura et al. (34), in their studies on film yeasts of wine, showed that the factor which keeps yeast cells floating on the liquid surface to form a pellicle is cell surface hydrophobicity. The method for quantitative estimation of hydrophobicity was the distribution ratio of cells between aqueous and organic solvent phases. The nonfilm strain (*S. cerevisiae*) showed low hydrophobicity. The film strain (*Saccharomyces bayanus*) did not form a pellicle at the early period of fermentation but did so upon respiration at the postfermentation period of wine making. The change from nonfilm to film stage was shown to be due to a change of the cell surface from hydrophilic to hydrophobic.

The influence of growth conditions on pellicle formation and hydrophobicity was further examined with various strains and genera. The conditions tested were provision of ethanol as sole carbon source, inhibition of ethanol assimilation, repression of respiratory activity, use of a respiratory-deficient mutant, biotin limitation, and increase of pH. It was shown that the tendency to form a pellicle was positively related to cell surface hydrophobicity and that both were promoted by oxidative growth.

The following mechanism of pellicle formation was proposed. The film yeasts that survive at the post-fermentation period will regenerate, with ethanol assimilation near the medium-air interface. The surface of the resultant cells possesses high hydrophobicity in the medium with low pH. Consequently, these cells will be supported strongly on the medium surface and exposed to aerobic conditions. It is likely that this behavior will ease oxygen uptake by the yeast cells and result in an even more hydrophobic surface.

The same authors (35) also investigated the substance responsible for the hydrophobicity of the cell surface of film yeasts. They concluded that hydrophobicity depends on fatty acids rather than on proteins.

Sake yeast

Most sake mash develops a high froth at an early stage of fermentation. The froth head, being about 1.5 times the volume of the mash itself, markedly reduces the available capacity of the fermentation tank (62). The froth head of sake mash is produced abundantly without aeration or agitation. It is maintained only in the earlier stage of fermentation (a week or so) and then breaks down and disappears. Nevertheless, from a practical point of view, a frothless mutant is quite desirable because approximately one-third of the capacity of a fermentation tank can be saved; if the parent (frothing) yeast is used, the same space must be reserved for froth formation.

Nunokawa et al. (61) isolated some frothless mutants from Kyokai, a strain of *S. cerevisiae* used for sake fermentation. The usefulness of those mutants was demonstrated by fermentation tests performed on industrial scale; some of them were put into practical use (62).

Flotability (degree of affinity to the gas-liquid interface) of the parental and mutant cells was measured parallel to analysis of their physicochemical properties (61, 62). Two main differences between the parent and mutants could be discerned: hydrophobicity and electrokinetic potential of the cell surface. Test of distribution of the cells in a two-phase system showed that the parental yeast cells were more hydrophobic than the mutants. Electrophoresis showed a higher isoelectric point for the parental cells. Examination of the changes of the cell flotability after enzymatic and chemical treatments suggested that the hydrophobic substance on the cell surface is proteinaceous. The presence of surfactants enhanced flotability of both parental and mutant cells, the effect being stronger for the former. Though the mechanism of froth formation in yeast cells is not yet clearly understood, it seems that both hydrophobic and electrostatic interactions are involved.

Flotation in Activated Sludge

The formation of a stable foam on the surface of aeration tanks in activated sludge plants is a source of continuous problems (8, 26). Actinomycetes are key organisms in the production of stable foam in the activated sludge process.

Goddard and Forster (26) studied the role of *Nocardia amarae* and *Microthrix paravicella*. They showed that substances produced by *M. paravicella* are responsible for formation and stabilization of the foam. Examination of two oxidation ditches in a sewage plant, one of which had a foam problem, revealed that (i) when foam was present, the dominant species in both the foam itself and the mixed liquors was *N. amarae*, whereas the nonfoaming samples were dominated by other filamentous species, and (ii) the surface tension of the mixed liquor from the ditch having a foam problem was distinctly lower than that of the liquor from the other ditch. This finding was attributed to specific surface-active compounds produced by *N. amarae*. It was suggested that a polar foam stabilizer is produced in addition to the biosurfactants.

Blackall and Marshall (8) showed that the formation of foam in cultures of *N. amarae* at laboratory scales conforms to the adsubble process theory: the *N. amarae* cells are hydrophobic (hydrocarbon affinity test), and they produce a surface-active substance (reduced surface tension). Appearance of foam is dependent on simultaneous presence of the hydrophobic cells and the surfactant. Stable foam could not be formed in culture filtrate (absence of cells) or by washed cells (absence of surfactant). Prevention of stable foam formation during culture was achieved by addition of montmorillonite, a hydrophilic clay. The authors postulate that a salt-dependent, reversible bacterium-montmorillonite complex which confers hydrophilicity to the otherwise hydrophobic actinomycetes is readily formed. This property prevents cells from entering and stabilizing the foam phase in an appropriately aerated pure culture.

CONCLUDING REMARKS

This chapter points to the fact that hydrophobic interactions, along with electrostatic interactions, play a determinant role in all phenomena of cellular accumulation at interfaces. In industrial operations, acting through the charge properties is a more common practice than interfering with hydrophobic interactions. However, the increasing body of data that support a role for hydrophobic interactions strongly suggests that these interactions should receive more attention. Indeed, hydrophobic interactions can be altered through genetics and physiology as well as by addition of suitable compounds, thus providing several avenues for modifying their effects. Therefore, additional studies are needed to refine methods for evaluating cell surface hydrophobicity and to relate this property to culture conditions, to other surface properties, and to overall cell behavior.

Acknowledgments. This laboratory is affiliated with the Faculty of Agriculture and Research Center for Advanced Materials. We thank the Commission of the European Communities (Biotechnology Action Programme) and Service de Progammation de la Politique Scientifique (Concerted Action Physical Chemistry of Interfaces and Biotechnology) for financial support.

LITERATURE CITED

1. **Amory, D. E., M. J. Genet, and P. G. Rouxhet.** 1988. Application of XPS to the surface analysis of yeast cells. *Surface Interface Anal.* **11**:478–486.

2. **Amory, D. E., and P. G. Rouxhet.** 1988. Surface properties of *Saccharomyces cerevisiae* and *Saccharomyces carlsbergensis*: chemical composition, electrostatic charge and hydrophobicity. *Biochim. Biophys. Acta.* **938**:61–70.

3. **Amory, D. E., P. G. Rouxhet, and J. P. Dufour.** 1988. Flocculence of brewery yeasts and their surface properties: chemical composition, electrostatic charge and hydrophobicity. *J. Inst. Brew.* **94**:79–84.

4. **Asther, M., M. N. Bellon-Fontaine, C. Capdevila, and G. Corrieu.** 1990. A thermodynamic model to predict *Phanerochaete chrysosporium* INA-2 adhesion to various solid carriers in relation to lignin peroxidase production. *Biotechnol. Bioeng.* **35**:477–482.

5. **Atkinson, B., and I. S. Daoud.** 1976. Microbial flocs and flocculation in fermentation process engineering. *Adv. Biochem. Eng.* **4**:41–124.

6. **Barreto, M. T. O., E. P. Melo, and M. J. T. Carrondo.** 1989. Starter culture production in fluidized bed reactor with a flocculent strain of *Lactobacillus plantarum*. *Biotechnol. Lett.* **11**:337–342.

7. **Bautista, J., E. Chico, and A. Machado.** 1987. Cell removal from fermentation broth by flocculation and sedimentation. *Biotechnol. Lett.* **8**:315–318.

8. **Blackall, L. L., and K. C. Marshall.** 1989. The mechanism of stabilization of actinomycete foams and the prevention of foaming under laboratory conditions. *J. Ind. Microbiol.* **4**:181–188.

9. **Boyles, W. A., and R. E. Lincoln.** 1958. Separation and concentration of bacterial spores and vegetative cells by foam flotation. *Appl. Microbiol.* **6**:327–334.

10. **Büchs, J., N. Mozes, C. Wandrey, and P. G. Rouxhet.** 1988. Cell adsorption control by culture conditions: influence of phosphate on surface properties, flocculation and adsorption behaviour of *Corynebacterium glutamicum*. *Appl. Microbiol. Biotechnol.* **29**:119–128.

11. **Busscher, H. J., and A. H. Weerkamp.** 1987. Specific and non-specific interactions in bacterial adhesion to solid substrata. *FEMS Microbiol. Rev.* **46**:165–173.

12. **Callega, G. B.** 1984. Aggregation. Report of the aggregation group in the Dahlem workshop, p. 303–321. *In* K. C. Marshall (ed.), *Microbial Adhesion and Aggregation—1984*. Springer-Verlag KG, Berlin.

13. **Changui, C., A. Doren, W. E. E. Stone, N. Mozes, and P. G. Rouxhet.** 1987. Surface properties of polycarbonate and promotion of yeast cells adhesion. *J. Chem. Phys.* **84**:275–281.

14. **Dahlback, B., M. Hermansson, S. Kjelleberg, and B. Norkrans.** 1981. The hydrophobicity of bacteria—an important factor in their initial adhesion at the air-water interface. *Arch. Microbiol.* **128**:267–270.

15. **Dobias, B., and V. Vinter.** 1966. Flotation of microorganisms. *Folia Microbiol.* **11**:314–322.

16. **Dognon, A., and A. Dumontet.** 1941. Concentration et séparation des microorganismes par moussage. *C.R. Soc. Biol.* **135**:884–887.

17. **Dunnill, P.** 1980. Immobilized cell and enzyme technology, p. 131–142. *In* S. Brenner, B. S. Hartley, and P. J. Rodgers (ed.), *New Horizons in Industrial Microbiology*. The Royal Society, London.

18. **Eddy, A. A., and D. Phil.** 1955. Flocculation characteristics of yeasts. II. Sugars as dispersing agents. *J. Inst. Brew.* **61**:313–317.

19. **Eddy, A. A., and D. Phil.** 1958. Composite nature of the flocculation process of top and bottom strains of *Saccharomyces*. *J. Inst. Brew.* **64:**143–151.
20. **Eriksson, L. B., and A. M. Hardin.** 1987. Flocculation of *E. coli* bacteria with cationic polyelectrolytes, p. 441–455. *In* Y. A. Attia (ed.), *Flocculation in Biotechnology and Separation Systems*. Elsevier, Amsterdam.
21. **Esser, K., and U. Kues.** 1983. Flocculation and its implication for biotechnology. *Process Biochem.* **18:**21–23.
22. **Forster, C. F., J. B. Knight, and A. D. J. Wase.** 1985. Flocculating agents of microbial origin. *Adv. Biotechnol. Processes* **4:**211–240.
23. **Gaudin, A. M., A. L. Mular, and R. F. O'Connor.** 1960. Separation of microorganisms by flotation. *Appl. Microbiol.* **8:**84–97.
24. **Gehle, R. D., T. L. Sie, H. Viehweg, T. Kramer, and K. Schügerl.** 1984. Cell recovery by continuous foam flotation in bench and pilot plant scale equipment, p. 545–548. *In Proceedings of the Third European Congress on Biotechnology*, vol. III. Verlag Chemie, Weinheim.
25. **Gerson, D. F., and J. E. Zajic.** 1977. Microbial surfactants. *Process Biochem.* **1979:**20–29.
26. **Goddard, A. J., and C. F. Forster.** 1986. Surface tension of activated sludges in relation to the formation of stable foams. *Microbios* **46:**29–43.
27. **Greenshields, R. N., and E. L. Smith.** 1971. Tower fermentation systems and their applications. *Chem. Eng.* **249:**182–190.
28. **Grieves, R. B., and S. L. Wang.** 1966. Foam separation of *Escherichia coli* with a cationic surfactant. *Biotechnol. Bioeng.* **8:**323–336.
29. **Grieves, R. B., and S. L. Wang.** 1967. Foam separation of bacteria with a cationic surfactant. *Biotechnol. Bioeng.* **9:**187–194.
30. **Griffin, S. R., and I. C. MacWilliam.** 1969. Variation of cell wall content in flocculent and non-flocculent yeast strains. *J. Inst. Brew.* **75:**355–358.
31. **Hartmeier, W.** 1988. *Immobilized Biocatalysts*. Springer-Verlag KG, Berlin.
32. **Hartmeier, W., G. Rüsch, C. Sieger, and C. Bücker.** 1987. Continuous production of lactic acid using carrier-fixed and flocculating bacteria, p. THP 57. *In* O. M. Neijssel, R. R. van der Meer, and K. A. M. Luyben (ed.), *Proceedings of the Fourth European Congress of Biotechnology*, vol. II. Elsevier, Amsterdam.
33. **Hoshino, K., M. Taniguchi, H. Marumoto, and M. Fuju.** 1989. Repeated batch conversion of raw starch to ethanol using amylase immobilized on a reversible soluble-autoprecipitating carrier and flocculating yeast cells. *Agric. Biol. Chem.* **53:**1961–1969.
34. **Iimura, Y., S. Hara, and K. Otsuka.** 1980. Cell surface hydrophobicity as a pellicle formation factor in film strain of *Saccharomyces*. *Agric. Biol. Chem.* **44:**1215–1222.
35. **Iimura, Y., S. Hara, and K. Otsuka.** 1980. Fatty acids as hydrophobic substance on cell surface of film strain of *Saccharomyces*. *Agric. Biol. Chem.* **44:**1223–1229.
36. **Jayatissa, P. M., and A. H. Rose.** 1976. Role of wall phosphomannan in flocculation of *Saccharomyces cerevisiae*. *J. Gen. Microbiol.* **96:**165–174.
37. **Kamada, K., and M. Murata.** 1984. On the mechanism of brewer's yeast flocculation. *Agric. Biol. Chem.* **48:**2423–2433.
38. **Kida, K., H. Kamayashi, S. Asano, T. Nakata, and Y. Sonoda.** 1989. The effect of aeration on stability of continuous ethanol fermentation by a flocculating yeast. *J. Ferment. Bioeng.* **68:**107–111.
39. **Kihn, J. C., C. L. Masy, and M. M. Mestdagh.** 1988. Yeast flocculation: competition between non-specific repulsion and specific bonding in cell adhesion. *Can. J. Microbiol.* **34:**773–778.

40. **Kim, C. W., S. K. Kim, R. Chokyum, and E. Robinson.** 1987. Removal of cell and cell debris by electrostatic adsorption of positively charged polymeric particles, p. 429–439. *In* Y. A. Attia (ed.), *Flocculation in Biotechnology and Separation Systems.* Elsevier, Amsterdam.
41. **Kjelleberg, S.** 1984. Mechanisms of bacterial adhesion at gas-liquid interfaces, p. 163–194. *In* D. C. Savage and M. Fletcher (ed.), *Bacterial Adhesion and Physiological Significance.* Plenum Publishing Corp., New York.
42. **Kuriyama, H., Y. Seiko, T. Murakami, H. Kobayashi, and Y. Sonoda.** 1985. Continuous ethanol fermentation with cell recycling using flocculating yeast. *J. Ferment. Technol.* **63:**159–165.
43. **Kwok, K. H., and I. G. Prince.** 1989. Flocculation of *Bacillus* species for use in high-productivity fermentation. *Enzyme Microb. Technol.* **11:**597–603.
44. **Lee, J. H., R. J. Pagan, and P. L. Rogers.** 1983. Continuous simultaneous saccharification and fermentation of starch using *Zymomonas mobilis. Biotechnol. Bioeng.* **25:**659–669.
45. **Lee, J., L. Skotnicki, and P. L. Rogers.** 1982. Kinetic studies on a flocculent strain of *Zymomonas mobilis. Biotechnol. Lett.* **4:**615–620.
46. **Leja, A.** 1982. *Surface Chemistry of Froth Flotation.* Plenum Publishing Corp., New York.
47. **Lemlich, R.** 1972. Adsubble processes: foam fractionation and bubble fractionation. *J. Geophys. Res.* **77:**5204–5210.
48. **Lindahl, M., A. Faris, T. Wadström, and S. Hjertén.** 1981. A new test based on "salting out" to measure relative hydrophobicity of bacterial cells. *Biochim. Biophys. Acta* **677:**461–476.
49. **Lion, L. W., and J. O. Leckie.** 1981. The biogeochemistry of the air-sea interface. *Annu. Rev. Earth Planet Sci.* **9:**449–486.
50. **Lyons, T. P., and J. S. Hough.** 1970. Flocculation of brewer's yeast. *J. Inst. Brew.* **76:**564–571.
51. **Magnusson, K. E.** 1980. The hydrophobic effect and how can it be measured with relevance for cell-cell interactions. *Scand. J. Infect. Dis. Suppl.* **24:**131–134.
52. **Miki, B. L. A., N. H. Poon, A. P. James, and V. L. Seligy.** 1980. Flocculation in *Saccharomyces cerevisiae*: mechanism of cell-cell interactions, p. 165–170. *In* G. Stewart and I. Russel (ed.), *Current Developments in Yeast Research.* Pergamon Press, Toronto.
53. **Miki, B. L. A., N. H. Poon, A. P. James, and V. L. Seligy.** 1982. Possible mechanism for flocculation interactions governed by gene *FLO1* in *Saccharomyces cerevisiae. J. Bacteriol.* **150:**878–889.
54. **Mill, P. J.** 1964. The nature of the interactions between flocculent cells in the flocculation of *Saccharomyces cerevisiae. J. Gen. Microbiol.* **35:**61–68.
55. **Mozes, N., D. E. Amory, A. J. Léonard, and P. G. Rouxhet.** 1989. Surface properties of microbial cells and their role in adhesion and flocculation. *Colloids Surfaces* **42:**313–329.
56. **Mozes, N., A. J. Léonard, and P. G. Rouxhet.** 1988. On the relations between the elemental surface compositions of yeasts and bacteria and their charge and hydrophobicity. *Biochim. Biophys. Acta* **945:**324–334.
57. **Mozes, N., F. Marchal, M. P. Hermesse, J. L. Van Haecht, L. Reuliaux, A. J. Léonard, and P. G. Rouxhet.** 1987. Immobilization of microorganisms by adhesion: interplay of electrostatic and non-electrostatic interactions. *Biotechnol. Bioeng.* **30:**439–450.
58. **Mozes, N., and P. G. Rouxhet.** 1987. Methods for measuring hydrophobicity of microorganisms. *J. Microbiol. Methods* **6:**99–112.
59. **Netto, C. B., A. Destruhaut, and G. Goma.** 1985. Ethanol fermentation by flocculating

yeast: performance and stability dependence on a critical fermentation rate. *Biotechnol. Lett.* 7:355–360.

60. **Nishihara, H., T. Toraya, and S. Fukui.** 1977. Effect of chemical modification of cell surface components of a brewer's yeast on the floc forming ability. *Arch. Microbiol.* 115:19–23.

61. **Nunokawa, Y., H. Toba, and K. Ouchi.** 1971. Froth flotation of yeast cells. *J. Ferment. Technol.* 49:959–967.

62. **Ouchi, K., and Y. Nunokawa.** 1973. Non-foaming mutants of sake-yeast: their physicochemical characteristics. *J. Ferment. Technol.* 51:85–95.

63. **Prince, J. G., and J. P. Barford.** 1982. Tower fermentation using *Zymomonas mobilis* for ethanol production. *Biotechnol. Lett.* 4:525–530.

64. **Prince, J. G., and J. P. Barford.** 1982. Induced flocculation of yeasts for use in the tower fermenter. *Biotechnol. Lett.* 4:621–626.

65. **Rose, A. H.** 1984. Physiology of cell aggregation: flocculation by *Saccharomyces cerevisiae* as a model system, p. 323–335. *In* K. C. Marshall (ed.), *Microbial Adhesion and Aggregation—1984*. Springer-Verlag KG, Berlin.

66. **Rosenberg, M.** 1981. Bacterial adhesion to polystyrene: a replica method of screening for bacterial hydrophobicity. *Appl. Environ. Microbiol.* 42:375–377.

67. **Rosenberg, M.** 1984. Ammonium sulfate enhances adherence of *Escherichia coli* J-5 to hydrocarbon and polystyrene. *FEMS Microbiol. Lett.* 25:41–45.

68. **Rosenberg, M., D. Gutnick, and E. Rosenberg.** 1980. Adherence of bacteria to hydrocarbons: a simple method for measuring cell-surface hydrophobicity. *FEMS Microbiol. Lett.* 9:29–33.

69. **Rosenberg, M., and S. Kjelleberg.** 1986. Hydrophobic interactions: role in bacterial adhesion. *Adv. Microb. Ecol.* 9:353–393.

70. **Rouxhet, P. G.** 1985. New ways of enzyme utilization—immobilization of catalysts. *Cerevisia* 4:167–184.

71. **Rubin, A. J.** 1968. Microflotation: coagulation and foam separation of *Aerobacter aerogenes*. *Biotechnol. Bioeng.* 10:89–98.

72. **Rubin, A. J., E. A. Cassel, D. Henderson, J. D. Jonhson, and J. C. Lamb III.** 1966. Microflotation: new low gas-flow rate foam separation technique for bacteria and algae. *Biotechnol. Bioeng.* 8:135–151.

73. **Singh, K. K., and W. S. Vincent.** 1987. Clumping characteristics and hydrophobic behaviour of an isolated bacterial strain from sewage sludge. *Appl. Microbiol. Biotechnol.* 25:396–398.

74. **Smyth, C. J., P. Jonsson, E. Olsson, O. Soderlind, J. Rosengren, S. Hjertén, and T. Wadström.** 1978. Differences in hydrophobic surface characteristics of porcine enteropathogenic *Escherichia coli* with or without K88 antigen as revealed by hydrophobic interaction chromatography. *Infect. Immun.* 22:462–472.

75. **Tenney, M. W., and F. H. Verhoff.** 1973. Chemical and autoflocculation of microorganisms in biological wastewater treatment. *Biotechnol. Bioeng.* 15:1045–1073.

76. **Unz, R. F.** 1987. Aspects of bioflocculation: an overview, p. 351–358. *In* Y. A. Attia (ed.), *Flocculation in Biotechnology and Separation Systems*. Elsevier, Amsterdam.

77. **van der Mei, H. J., P. Brokke, J. Dankert, J. Feijen, P. G. Rouxhet, and H. J. Busscher.** 1989. Physicochemical surface properties of nonencapsulated and encapsulated coagulase-negative staphylococci. *Appl. Environ. Microbiol.* 55:2806–2814.

78. **van der Mei, H. J., M. J. Genet, A. H. Weerkamp, P. G. Rouxhet, and H. J. Busscher.** 1989. A comparison between the elemental surface compositions and electrokinetic properties of oral streptococci with and without adsorbed salivary constituents. *Arch. Oral Biol.* 11:889–894.

79. **Verrier, D., B. Mortier, and G. Albagnac.** 1987. Initial adhesion of methanogenic bacteria to polymers. *Biotechnol. Lett.* **9:**735–740.

80. **Verrier, D., B. Mortier, H. C. Dubourguies, and G. Albagnac.** 1988. Adhesion of anaerobic bacteria to inert supports and development of methanogenic biofilms, p. 61–69. *In* E. R. Hall and P. N. Hobson (ed.), *Anaerobic Digestion: Proceedings of the International Symposium.* Pergamon Press, Oxford.

81. **Viehweg, H., and K. Schügerl.** 1983. Cell recovery by continuous flotation. *Eur. J. Appl. Microbiol. Biotechnol.* **17:**96–102.

82. **Wu, Y. C.** 1978. Chemical flocculability of sludge organisms in response to growth conditions. *Biotechnol. Bioeng.* **20:**677–696.

Microbial Cell Surface Hydrophobicity
Edited by R. J. Doyle and M. Rosenberg
© 1990 American Society for Microbiology, Washington, DC 20005

Chapter 4

Role of Hydrophobic Interactions in Microbial Adhesion to Plastics Used in Medical Devices

Stephen A. Klotz

Microbial adhesion to plastics is a major health and economic concern. The microfouling of plastics that follows upon microbial adhesion to these surfaces results in inefficiency and destruction of the underlying materials (66). Paradoxically, adhesion to plastics may be a desirable and necessary step in the process of microbial biodegradation of polluting polymers. Many authorities believe that hydrophobic interactions between microorganisms and plastic devices are important factors in the early steps of microbial adhesion to these devices. This chapter addresses microbial adhesion to plastics found in medical devices and the role of hydrophobic interactions in this process.

The potential for infections arising from microbial adhesion to plastics is enormous. For example, more than 4 million Americans wear plastic extended-wear contact lenses, often for weeks at a time. It is estimated that 1 in 500 wearers will contract infectious ulcerative keratitis each year as a consequence of wearing these lenses (88). Foley catheters are inserted into the bladders of more than 5 million patients annually in the United States, and approximately 1 million of these patients will contract a urinary tract infection secondary to the catheter insertion (37). More than 20 million patients receive fluid therapy through plastic intravenous catheters yearly in the United States. About 0.2% of these patients will develop purulent thrombophlebitis as a complication of this

Stephen A. Klotz • Departments of Medicine and Ophthalmology, Veterans Affairs and Louisiana State University Medical Centers, Shreveport, Louisiana 71101-4295.

percutaneous cannulation (106). Thus, the potential for device-related infections in humans is calculated in the millions.

The microbial adhesion to plastics, which is considered an early and necessary event that leads to colonization or infection of indwelling plastic devices, is a contributing factor to many, if not most, nosocomial infections. Therefore, determining factors that are important in the process of microbial adhesion to these hydrophobic surfaces is important, since a better understanding of the microbe-plastic interaction may lead to strategies to interrupt colonization.

Although hydrophobicity is likely to play a substantial role in the infection of plastic devices, it is unlikely that every infection associated with an indwelling plastic device occurs from direct interaction of a microbe with the plastic surface of the device. This is because each indwelling plastic device elicits a host response that alters the normal anatomy and histology, forming a *locus minoris resistentiae* centered around the plastic device. The abnormal tissue that is formed in response to the foreign object either walls off or integrates with the plastic surface and is always subject to infection from almost any transient or sustained bacteremia.

GENERAL CONSIDERATIONS OF PLASTIC AS A SUBSTRATUM FOR MICROBIAL ADHESION

The target for microbial adhesion to be considered in this chapter is the plastic surface. The term "plastic" will refer to extruded or molded organic polymers, which are typically hydrophobic and thus have low surface free energy or surface tension. The terms "hydrophobic," "low surface free energy," and "low surface tension" will be used interchangeably (see chapter by Rosenberg and Doyle and chapter by Duncan-Hewitt, this volume). It must be borne in mind that the various proprietary polymers vary greatly in surface features because of additives such as plasticizers and because of proprietary manipulations of surface charge. It should also be appreciated that local topography of a plastic surface may be important in determining microbial adhesion. Although these products appear smooth surfaced macroscopically, at the ultrastructural level there are numerous irregularities that may serve as a nidus for bacterial adhesion (25, 56, 87). Other poorly defined properties of these polymers may greatly affect microbial adhesion. For example, some of the products are highly porous, polymeric knits used for such procedures as vascular grafting, and the degree of porosity of these products may bear a direct relationship to the number of microbes that adhere to them (22, 41).

If adhesion of a microorganism to a plastic medical device is the necessary step in establishment of colonization or infection of such a device, as it is believed to be for infectious diseases in general (4), then characterizing the interaction between the microorganism and the device is important. One difficulty in characterizing this interaction is determining the relevant plastic surface for experimental purposes; i.e., is the adhesion target the native plastic surface as it is machined and then purchased, or is it a plastic surface covered with adsorbed host-derived proteins, cells and tissue debris, or a combination of both? The difficulty in determining the relevant target is illustrated by the following example. Early prosthetic valve endocarditis occurs within 2 months after surgery and is believed to occur as a result of contamination at the time of surgery. Investigators have shown that air in the operative field, the plastic perfusion lines, the exposed tissues, the prosthetic valves themselves, and plastic peripheral intravenous lines are all contaminated with microorganisms during the operative procedure (42). However, it is not known which contaminants adhering to one of the several plastic surfaces are responsible for the ensuing infection, whether adhesion occurred to coated or to machined plastic, or whether adhesion occurred before or after insertion.

This uncertainty concerning the appropriate target surface is reflected in experimental models of microbial adhesion to plastic devices. For example, investigators will often demonstrate microbial adhesion to a plastic surface that has been treated with serum or plasma on the presumption that the substances coating the plastic form the relevant target. However, it seems to this author that the compositions of the coating substances on plastics when exposed in vivo are poorly characterized or unknown in most circumstances. For example, the insertion of an intravenous catheter through the skin or a Foley catheter into the bladder may provide ample opportunities for microorganisms to adhere to "uncoated plastic" as well as to "coated plastic." However, the types of host coatings on plastics may be due to the composition of the polymer, and thus the composition of the plastic device may affect both the host tissue acceptance of the foreign material and the rate of infection.

What constitutes the relevant target surface for microbial adhesion may be difficult to characterize; therefore, conclusions regarding the wisdom of using one plastic or another may have to be drawn entirely from data on use of various plastics in patients. An example of this approach is the work of Sheth et al. (99), who demonstrated a highly significant increase in bacteremias related to the use of polyvinyl chloride central venous catheters, much in excess of the number of bacteremias associated with the use of polytetrafluorethylene central venous catheters. Another is the study of Karlan et al. (56), who found that silicone myringotomy tubes were associated with more postinsertion ear infections than were polytetrafluorethylene myringotomy tubes. Until reliable predicative models of microbe-medical device interactions are developed, clinical studies such as these are needed to judge whether one plastic or another should be used. It is also evident that until models that mimic the in vivo state are available, no firm conclusions about the role of hydrophobicity or any other virulence factor can be made; only correlations will be possible.

MICROORGANISMS INVOLVED IN INFECTIONS OF PLASTIC MEDICAL DEVICES

Colonization and infection of plastic medical devices involve a unique interaction of a microbe with a nonbiologic surface. That this is a special interaction is supported by the facts that (i) the overwhelming number of microorganisms implicated in these infections are part of the normal cutaneous, mucosal, or gastrointestinal microbiota and are not traditional pathogens and (ii) each device is subject to infection with microorganisms usually resident at the site of insertion of the device. Some plastic devices and their commonly associated infecting microorganisms are listed in Table 1. These microorganisms are neither virulent nor pathogenic to normal hosts. They are implicated in infections in compromised hosts and in normal hosts when host barriers have been breached by the insertion of catheters or other medical devices.

Investigators have demonstrated the hydrophobic nature of the cell surface of most of these microorganisms. Some of the microorganisms, such as the coagulase-negative staphylococci and *Candida* spp., have been subject to extensive investigations of their capability to adhere to plastics. The coagulase-negative staphylococci are the leading microbes associated with infections of plastic medical devices at almost any site in the body or with any therapeutic modality. Hydrophobic interactions may explain the unique hydrophobic niche and medical device-related infections of two of these microorganisms. *Propionibacterium acnes*, for instance, normally resides in the highly hydrophobic pilosebaceous unit. This microorganism occasionally infects patients with indwelling plastic devices that remain exposed to the environment and the skin, such as intravenous catheters and cerebrospinal fluid reservoirs (53). The affinity of this bacterium for hydrophobic surfaces such as catheters is demonstrated in vitro by the fact that the microorganism adheres best to low-surface-energy (i.e., hydrophobic) substrates (acrylic > Plexiglas

Table 1. Some Plastic Medical Devices and the Microorganisms
That Commonly Infect Them

Medical device	Pathogen	Reference(s)
Central nervous system reservoirs	Coagulase-negative staphylococci, *Propionibacterium acnes*, *Corynebacterium* spp.	13 53
Cerebrospinal fluid shunts	Coagulase-negative staphylococci	35, 80
Intraocular lenses	Coagulase-negative staphylococci, *P. acnes*, *Corynebacterium* spp.	113
Extended-wear contact lenses	*Pseudomonas aeruginosa*, *Acanthamoeba* spp.	88
Dentures	*Candida albicans*	10
Total artificial heart	Polymicrobial infections with coagulase-negative staphylococci, *P. aeruginosa*, *Candida albicans*	39 64
Arteriovenous shunt	Coagulase-negative staphylococci	13, 55
Central vein catheters	Coagulase-negative staphylococci, *Candida* spp., *Malassezia furfur*, *Corynebacterium* spp.	98 57
Peripheral catheters	Coagulase-negative staphylococci, *Staphylococcus aureus*	13
Bladder catheters	Coagulase-negative staphylococci, *Candida* spp., urea-splitting *Enterobacteriaceae*	13
Biliary stents	*Escherichia coli*, *Citrobacter freundii*, coagulase-negative staphylococci	104
Tracheal tubes	Polymicrobial, coagulase-negative staphylococci	49 21
Prosthetic joints	Coagulase-negative staphylococci	50
Intrauterine contraceptive devices	Group B streptococci, coagulase-negative staphylococci, *Corynebacterium* spp., *Actinomyces israelii*	51

> > glass) (J. E. Zajic and S. Bhaduri, unpublished data). Furthermore, when the whole bacterial cell is suspended in saline, it acts as a surface-active agent and reduces the surface tension of the suspending liquid (Zajic and Bhaduri, unpublished data). This surfactant property is due to the cell surface hydrophobicity of the bacterium.

Malassezia furfur is another hydrophobic cutaneous saprophyte with a close association with the pilosebaceous unit. *M. furfur* is a highly hydrophobic yeast that requires the presence of exogenous lipids of the C_{12} to C_{24} series for in vitro growth. It causes a trivial scaling dermatosis and pustular folliculitis (60) but is associated with a distinctive sepsis syndrome in patients receiving intravenous lipid therapy (57). In this setting, the organism is known to adhere to plastic infusion catheters, giving rise to fungemia. Systemic infections with *P. acnes* or *M. furfur* are documented infrequently but were not known to occur prior to the use of indwelling plastic devices.

The microorganisms used to study adhesion to plastic surfaces can be categorized into three groups: (i) marine organisms that may contribute to biofouling of solid objects submerged in ocean waters, such as marine pseudomonads (31) and *Vibrio proteolytica* (85); (ii) microorganisms implicated in plastic device-related infections, such as the coagulase-negative staphylococci (15), *Staphylococcus aureus* (111), *Candida* spp. (61, 76), and *Pseudomonas aeruginosa* (12); and (iii) human pathogens such as mycoplasmas (9), *Clostridium difficile* (114), *Legionella pneumophila* (115), and *Bordetella pertussis* (27), which do not cause plastic device-related infections per se. However, measurement of their adhesion to a plastic surface provides an estimate of their cell surface hydrophobicity, and this characteristic may relate to their pathogenicity. I will include in this survey the adhesion of marine microorganisms, since only limited studies have been performed with the pathogens. Much of the theory and empiric data regarding microbial adhesion to plastic have been gained from studies of marine bacteria, and results of these studies have been applied to the subject of microorganism adhesion to plastic medical devices as well.

PHYSICOCHEMICAL CONCEPTS TO EXPLAIN MICROBIAL ADHESION TO PLASTICS

From a purely physicochemical standpoint, microbial adhesion to plastic medical devices is likely to depend on several factors. First, the microorganism must physically approach and make contact with the surface. If a relatively hydrophobic microorganism is suspended in an aqueous environment (for example, at mucosal surfaces), nonpolar, hydrophobic forces may enhance the concentration of a relatively hydrophobic microorganism at liquid-air and liquid-solid interfaces (70). This is true for any relatively hydrophobic microorganism in aqueous suspension because a hydrophobic particle would tend to be excluded from the bulk liquid by the great attraction of water for itself rather than for nonpolar suspended particles (107). It is also important that once a microbe is adjacent to a plastic surface, it remain in close proximity and not be dislodged. In this situation, hydrophobic forces and electrostatic forces would be important. At this stage, the possession of surface projections, perhaps hydrophobic proteins capable of penetrating through repulsive electrostatic forces, would enhance the chances of microbial adhesion. Bacterial pili or fibrillae may subserve this purpose. The last step, which is the irreversible adhesion of the microorganism to a substrate, may be accomplished by the elaboration of a water-soluble polysaccharide glycocalyx (40).

Although adhesion of a microorganism to a solid polymer is commonly used to quantitate cell surface hydrophobicity (92), other methods are more easily performed and as a rule correlate well with adhesion to a plastic surface. Examples of such methods are microbial adhesion to hydrocarbons (MATH) (93) or biphasic separation methods using polyethylene glycol and dextran (36). These methods may be even better estimates of the contribution of hydrophobicity to the adhesion of microorganisms to a plastic surface, since they measure the exclusion of the microorganisms from the bulk liquid as well as immobilization at a hydrophobic interface.

Because plastics are a nonbiological surface, the adhesion of microorganisms to them is often treated in physicochemical terms. This approach is conceptually appealing since the plastic surface is essentially inert. The physicochemical treatment, however, requires investigators to treat the microorganisms as colloidal particles. This may be a valid assumption for many microorganisms, since viability is often not required for bacterial adhesion to occur to plastic. For example, Fletcher found that formaldehyde-treated marine pseudomonads adhered to plastic as well as did viable bacteria (29). Others have found that *Escherichia coli*, *Aeromonas liquefaciens*, and *Pseudomonas fluorescens* killed by UV irradiation adhered as well as viable cells to a solid surface (73). However, not all microorganisms adhere well when killed. Nonviable *P. aeruginosa*, for instance, does not adhere well to polystyrene (75). The adhesion of some pathogenic microorganisms to plastic is an active metabolic process. For example, physiologically inactivated *V. proteolytica* does not adhere to polystyrene (85). However, some surfaces allow no adhesion to occur. For example, urinary catheters impregnated with silver, which functions as an antimicrobial agent, will not allow the adhesion of viable or nonviable *P. aeruginosa* to the catheter surface (67).

The characteristic feature of plastics is their nonwettable, hydrophobic surfaces. This nonwettable characteristic is a prominent part of all cell adhesion models. The most encompassing theory to explain the adhesion of microorganisms to solid surfaces is the concept of interfacial free energy as the controlling force in cell adhesion (2). According to this theory, cell adhesion is favored when interfacial tension between the solid, the particle, and the suspending liquid is reduced. For most microorganisms, this translates into greater microbial adhesion to solid surfaces in direct proportion to the hydrophobicity of the surface, provided that the suspending medium is water or a simple buffer.

However, even with these reservations, the interfacial free energy formula cannot explain the adhesion phenomenon entirely. For instance, Uyen et al. (108) compared the adhesion of polystyrene particles to a low-surface-energy substrate with that of *Streptococcus mitis* of approximately equal size and surface free energy. It was found that in suspensions of low ionic strength, *S. mitis* adhered 30 times greater than did the polystyrene particles, whereas at higher ionic strengths the adhesion ratios were nearly equal. Thus, it appears that in solutions of low ionic strengths, factors in addition to surface free energy of the microbe and the target are responsible for adhesion. The authors speculate that surface projections of the bacteria probably account for this difference and that these projections may collapse or fold around the bacterial surface in high-ionic-strength solutions (108).

When aqueous suspensions are more complex, such as in experiments measuring adhesion of microorganisms to submerged plastic planchettes in seawater, the preferential colonization of the substrate will depend on the specific microorganisms present at the time and on their own unique mode of attachment (68). Furthermore, the presence of proteins in solution will modify microbial adhesion in ways peculiar to each microorganism, sometimes increasing adhesion and at other times decreasing adhesion.

Plastics exert an important influence on the nature of protein layers that adsorb to them. Baier (5) has demonstrated the existence of a thromboresistant and adhesion-resistant range of critical surface tension for plastics used in medical devices. This adhesion-resistant range falls within 20 to 30 dynes/cm. This phenomenon has been observed to occur experimentally. For example, plastic planchettes submerged in ocean water became covered by microorganisms in numbers that correlated with the critical surface tension for wetting. The fewest number of organisms adhered to substrates with surface energy in the range of 20 to 30 dynes/cm (20) (Fig. 1). This same relationship of the critical surface tension for wetting and adhesion of marine microorganisms was confirmed in different coastal waters (19). An earlier study of submerged metals, wood, glass, and polymethylmethacrylate (PMMA) demonstrated little differences among these surfaces and the number of adhering microbes (97). In contrast, in a study of adhesion of marine microorganisms to glass slides or PMMA, the plastic always had more adherent microorganisms than did glass at all times of submersion studied (81). However, the relationship of critical surface tension of the substrate to adhesion has been shown to explain the adhesion and retention of oral microorganisms (16). In these experiments, the adhesion minimum occurred on plastics possessing a critical surface tension of 20 to 30

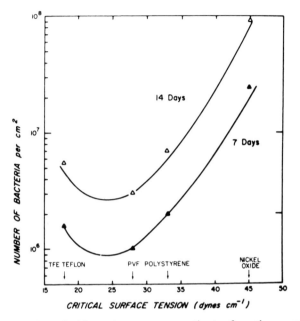

Figure 1. Number of attached bacteria per square centimeter from the scanning electron microscopy data as a function of the critical surface tension for wetting. (From Dexter et al. [20] with permission.)

dynes/cm (16), as had been observed earlier (20). These results, which are so divergent from those of experiments performed with microorganisms suspended in buffer alone, are best explained by protein coating of the plastic substrates. This may be a general phenomenon of microbial adhesion in the presence of some proteins but probably does not explain results for situations in which specific interactions between the microorganisms and the protein may occur, such as is the case for *S. aureus* with fibronectin.

All theories, however, must take into account what Fletcher and Loeb (31) and others (86) have shown, i.e., that a given microorganism adheres differently to a hydrophilic substrate than to a hydrophobic substrate. For example, Triton X-100 inhibited adhesion of *V. proteolytica* to hydrophobic polystyrene but not to tissue culture dishes with a hydrophilic surface (Fig. 2). Similarly, addition of propanol to the suspending medium reduced *E. coli* adhesion to polystyrene but not to a hydrophilic surface (23).

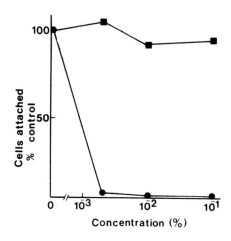

Figure 2. Effect of Triton X-100 on the attachment of *V. proteolytica* to polystyrene (●) or tissue culture dishes (■). (From Paul and Jeffrey [86] with permission.)

GENERAL TREATMENT OF MICROBIAL ADHESION TO PLASTICS

In practice, investigations into the contribution of hydrophobic interactions to the adhesion of microorganisms to plastic surfaces have been performed with the microorganisms suspended in physiologic buffers. In aqueous solutions, electrostatic interactions clearly play an important role in the process of microbial adhesion to plastics. Mozes et al. (79) have demonstrated that electrostatic interactions may be the predominant physicochemical force influencing the adhesion of "relatively hydrophilic" microorganisms to high-surface-energy surfaces such as glass. They also showed that electrostatic interactions contribute to the adhesion of "relatively hydrophobic" microorganisms to low-surface-energy surfaces. Van Loosdrecht et al. (109) even observed that with a decrease in hydrophobic interactions between microbes and a target substrate, there was a corresponding increase in the amount of electrostatic interactions. The reverse of this phenomenon was found with *Neisseria meningitidis*: a reduction of negative surface charge of the bacteria or neutralization of both positive and negative surface charges resulted in a corresponding increase in cell surface hydrophobicity and an increase in adhesion to buccal cells (17). Therefore, electrostatic interactions not only play a direct role in the process of microbial adhesion to plastic but also may affect the adhesion process by influencing the magnitude of hydrophobic interactions.

Cell surface hydrophobicity is highly dependent on the growth medium used. For example, McEldowney and Fletcher (72) demonstrated that the cell surface hydrophobicity of *P. fluorescens, Enterobacter*

Figure 3. End result of microbial adhesion on a plastic medical device. Shown are coagulase-negative staphylococci on the lumen of an endotracheal tube removed from a patient. (Scanning electron photomicrograph courtesy of S. Manocha and R. C. Clawson.)

cloacae, *Chromobacter* sp., and *Flexibacter* sp., as determined by the contact angle method, varied according to whether the organisms were in exponential or stationary phase growth and according to the growth medium used. Therefore, the ability of bacteria to adhere to polystyrene was affected (72). The temperature at which growth occurs also affects cell surface hydrophobicity of microorganisms. When *Candida albicans* is grown at room temperature, cell surface hydrophobicity is greater than when it is grown at 37°C (43). Unrecognized characteristics such as these may account for the fact that one investigator has reported *C. albicans* to be hydrophobic in Sabouraud dextrose broth (61), whereas another reported *C. albicans* to be hydrophilic when grown in tryptic soy broth (105).

Finally, the period of incubation of microbe with plastic undoubtedly affects adhesion; the longer the incubation, the greater the opportunity for reversibly bound microorganisms to elaborate extracellular polymers capable of acting as holdfasts (Fig. 3).

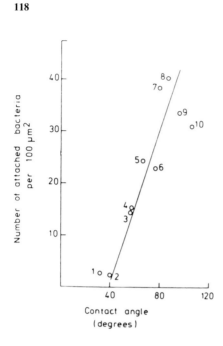

Figure 4. Relationship between numbers of attached bacteria and advancing contact angles of water on different surfaces. Substrata: 1, radiofrequency plasma-cleaned polystyrene; 2, radiofrequency plasma-cleaned polyethylene terephthalate; 3, gold-treated nylon; 4, epoxy; 5, nylon 6.6; 6, gold-treated polyethylene; 7, polyethylene terephthalate); 8, polystyrene; 9, polyethylene; 10, polytetrafluorethylene. (From Fletcher and Loeb [31] with permission.)

ADHESION OF MICROORGANISMS TO PLASTICS

The most powerful demonstrations of the influence of hydrophobic interactions on microbial adhesion to plastics are experiments demonstrating the direct correlation of microbial adhesion with increasing substrate hydrophobicity. Fletcher and Loeb (31) demonstrated the adhesion of a marine pseudomonad, suspended in artificial seawater, to surfaces of increasing hydrophobicity. These surfaces ranged from a hydrophilic polystyrene (contact angle of water = 32°) to the very hydrophobic polytetrafluorethylene (contact angle of water = 105°). The results (Fig. 4) demonstrate the direct correlation between the hydrophobicity of the substrate and the number of microorganisms adhering to the substrate. This direct correlation between greater microbial adhesion and increasing substrate hydrophobicity has also been observed for *Candida* spp. (61), *Streptococcus sanguis* (110), and coagulase-negative staphylococci (69). This phenomenon even describes the adhesion of plant cells of *Catharanthus roseus* to polymeric surfaces (26). Further investigations of microbial adhesion to hydrophobic surfaces performed in the absence of proteins are summarized in Table 2.

When placed in a solution containing proteins, plastics are rapidly coated by the proteins (8). The effect on microbial adhesion that a protein monolayer will have is partly determined by the microorganism in

Table 2. Some Reports Emphasizing the Predominant Role of Hydrophobicity in Adhesion of Microorganisms to Hydrophobic Substrates

Microorganism	Substrate	Blocking	Comments	Reference
Coagulase-negative staphylococci	Fluorinated polyethylenepropylene	Partial blocking with lipoteichoic acid.	Bacteria aggregated on substrate; adhesion was decreased by phenol treatment of bacteria, thus decreasing lipoteichoic acid.	46
Flexibacter sp.	Air-water interface	Blocked with 0.05% Tween 80.	Orientation of bacteria at interface with hydrophobic portion into interface.	71
Coagulase-negative staphylococci	Fluorinated polyethylenepropylene	Exogenous slime blocked adhesion.	Enzyme-treated bacteria did not adhere to fluorinated polyethylpropylene or xylene.	47
Candida spp.	Acrylic	Saliva coating of acrylic decreased adhesion.	Adhesion to acrylic correlated with that to hexadecane, $r = 0.849$.	77
Coagulase-negative staphylococci	Polyethylene	Adhesion decreased with pretreatment with lyzozyme or slime.	Adhesion to low-surface-energy substrates increased.	69
16 different bacterial species	Sulfated polystyrene		Bacterial cell surface hydrophobicity measured with drops of water and α-bromonaphthalene correlated directly with adhesion, $r = 0.8$.	109

question, specifically whether it possesses receptors for the immobilized protein. The outcome is also determined by the degree of hydrophilicity that the immobilized protein confers upon the substrate surface.

For example, Fletcher (28) demonstrated that bovine serum albumin, gelatin, fibrinogen, and pepsin inhibited the ability of a marine pseudomonad to adhere to polystyrene. Inhibition occurred whether the proteins were added simultaneously with the bacteria or by precoating the polystyrene with the proteins. Even bacteria preincubated with albumin and washed thoroughly could not adhere to polystyrene. It was thought that the principal action of the proteins may have been to decrease the surface tension of the aqueous suspension, thus reducing bacterial adhesion. PMMA precoated with serum and saliva (33) decreased the subsequent adhesion of *Streptococcus mutans*, and serum also inhibited the adhesion of coagulase-negative staphylococci and *P. aeruginosa* to polyvinyl chloride and polyurethane (78).

Coating PMMA with fibronectin, however, enhances the adhesion of *Staphylococcus aureus* (112). Fibronectin immobilized on the surface of plastics has also been reported to enhance the adhesion of coagulase-negative staphylococci and *P. aeruginosa* (78). The differences between Fletcher's work with proteins (28) and investigations with *S. aureus* are probably explained by the presence of a well-documented receptor on *S. aureus* for fibronectin (65). This specific interaction appears to overpower the less specific hydrophobic interactions that are responsible for the adhesion of *S. aureus* to PMMA. Presumably, in the case of the marine pseudomonad studied by Fletcher, no such specific interaction between microorganism and immobilized proteins occurred (28).

DISRUPTION OF HYDROPHOBIC INTERACTIONS AND ADHESION

Surface-active agents reduce hydrophobic interactions and by doing so reduce microbial adhesion to plastics. The decrease in adhesion is explained by the reduction in the surface tension of the suspending medium caused by the surfactant, creating a situation in which interfacial tensions (surface free energies) of the microbe, substrate, and suspending medium are such that microbial adhesion cannot occur (1).

An interesting application of nonionic surfactants was found by Humphries et al. (48), who coated polystyrene with alcohol propoxylates, surfactants whose hydrophobin is cetyl alcohol and whose hydrophilin is propoxylate. Polystyrene precoated with these alcohol propoxylates inhibited adhesion of *Pseudomonas* sp. and *Serratia marcescens* by 99%. This inhibition of adhesion remained intact for days. Precoating with other nonionic surfactants was not as effective, but other surfactants did possess considerable inhibitory power when added simultaneously with the bacteria at 10 parts per million. In this situation, the familiar NP and Tween series of surfactants were effective inhibitors of microbial adhesion (48).

Antibiotic treatment of bacteria will in many instances reduce microbial adhesion to plastic surfaces. This reduction in adhesion after exposure to antibiotics has been linked to a corresponding decrease in cell surface hydrophobicity for some organisms. For example, treatment of *Bifidobacterium bifidum* with greater than minimum inhibitory concentrations of penicillin G resulted in a release of lipoteichoic acid into the growth medium and a concomitant loss of cell surface hydrophobicity (82). Schadow et al. (96) demonstrated that even subinhibitory concentrations of some antibiotics resulted in diminished adhesion of coagulase-negative staphylococci to polystyrene. This diminished adhesion correlated with a corresponding decrease in cell surface hydrophobicity as determined by MATH (see Rosenberg and Doyle, this volume). Pretreatment of silicone surfaces with bacitracin A has been reported to inhibit the adhesion of coagulase-negative staphylococci to the same degree as does pretreatment of the same surfaces with Triton X-100. This effect was believed to be due to the surface-active properties of the antibiotic rather than to its antimicrobial properties (38).

Alcohols such as methanol, ethanol, propanol, and butanol will inhibit *Pseudomonas* adhesion to polystyrene, probably as a result of a change in surface tension of the suspending medium (30). These results are similar to those obtained when dimethyl sulfoxide was added to the suspending medium (1). As mentioned above (see Physicochemical Concepts To Explain Microbial Adhesion to Plastics), the addition of propanol to the suspending medium reduced *E. coli* adhesion to polystyrene but did not reduce adhesion to a hydrophilic surface (Fig. 5). This phenomenon was attributed to the change in surface tension of the suspending medium (23).

The clinical utility of hydrophobic interaction inhibitors such as the nonionic surfactants or alcohols has yet to be demonstrated. However, it is possible that use of such agents may prove advantageous in attempts to reduce adhesion and thus infection of plastic devices.

ADHESION OF MICROORGANISMS TO MEDICAL DEVICES

Adhesion of microorganisms to medical devices is often a prelude to infection of the device, ultimately giving rise to distant sites of infection.

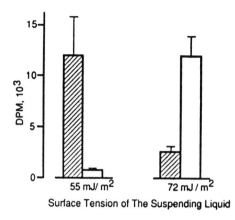

Figure 5. Adhesion of piliated RDEC-1 when organisms were suspended in either phosphate-buffered saline with 3.5% propanol (surface tension, 55 mJ/m^2) or phosphate-buffered saline alone (surface tension, 72 mJ/m^2). Symbols: □, bacterial binding to polystyrene; ▨, adhesion of bacteria to the relatively more hydrophilic sulfonated polystyrene. Bars indicate standard errors. The increase in adhesion to sulfonated polystyrene ($P <$ 0.01) and decrease in adhesion to polystyrene ($P <$ 0.001) after suspension of bacteria in a 3.5% propanol solution are both significant. (From Drumm et al. [23] with permission.)

Once infection of an indwelling plastic device occurs, it is generally believed that the only effective treatment is removal of the device (116). If the device remains in situ, it is virtually impossible to sterilize it, perhaps because of the elaboration of an antiphagocytic biofilm (40) that impedes penetration of antibiotics (100).

Little attention has been devoted to specifically investigating the relationship of microorganism adhesion to plastic medical devices and the role of hydrophobicity in this process. This section reviews the work with intravenous catheters, contact lenses, and intrauterine contraceptive devices (IUCDs).

Adhesion to Intravenous Catheters

Because adhesion of microorganisms to intravenous catheters may result in life-threatening bloodstream infections, there has been a great deal of clinical attention to this problem. In a series of publications, Sheth and co-workers have systematically investigated microbial adhesion to intravenous catheters. Their first observation was that polyvinyl chloride catheters were more often colonized with microorganisms than were Teflon catheters. For example, of 687 Teflon catheters inserted into patients, 6.9% were positive by culture, compared with 24.6% of 77 polyvinyl chloride catheters (99). The vast majority of these colonizing microorganisms were coagulase-negative staphylococci. The preferential adhesion of coagulase-negative staphylococci to polyvinyl chloride catheters has been confirmed by a separate group studying another hospital outbreak of coagulase-negative staphylococcal bacteremia (14). In the study by Sheth et al. (99), gram-negative bacteria, yeasts, and *S. aureus* appeared to adhere to Teflon catheters more than to polyvinyl chloride catheters, suggesting that different adhesion mechanisms were operative with these microorganisms. However, a preference of *Candida* spp. to adhere to polyvinyl chloride catheters rather than to Teflon catheters has been reported for an in vitro model (94). In a study of the adhesion of coagulase-negative staphylococci to Teflon catheters, it was found that in vitro, hydrophobic interactions explained adhesion; the more hydrophobic strains adhered in greater numbers than the less hydrophobic strains (84). The authors also found that pepsin treatment of the bacteria decreased adhesion and that adding serum or albumin to the catheter (or in suspension) decreased adhesion (84). Franson et al. (32) were able to inhibit by one-half the adhesion of coagulase-negative staphylococci to unused polyvinyl chloride catheters with a 10-mg/ml solution of α-D-methylmannoside and almost complete inhibition occurred with mannosamine, suggesting that the bacterial cell wall constituents were critical to adhesion of the bacteria to the catheter surface. The mechanisms of bacterial adhesion to catheters removed from patients and adhesion to unused polyvinyl chloride catheters were identical, both involving the production of extracellular material (32). Both slime-producing and non-slime-producing isolates are capable of survival on the catheter surface for extended periods of time in the absence of nutrients (32), perhaps by degrading the catheter surface itself (87).

Ashkenazi et al. (3) noted that adhesion of *Staphylococcus aureus*, *Serratia marcescens*, and *E. coli* to intravenous catheters paralleled the degree of bacterial cell surface hydrophobicity and that the adhesion could be inhibited with detergents. A particularly interesting experiment with *S. marcescens* demonstrated the role of hydrophobicity in the adhesion process. Hydrophilic and partially hydrophobic mutant strains were derived from a hydrophobic isolate of *S. marcescens*, and their adhesion to Teflon catheters occurred in proportion to the cell surface hydrophobicity of each strain (3).

A specific type of adhesion mechanism, however, has been suggested for the adhesion of *Staphylococcus aureus* to fibronectin-coated strips of PMMA (112) and to intravenous catheters in vitro (95). Intravenous catheters removed from patients were demonstrated to be coated with fibrin and fibronectin, and coated catheters demonstrated greater staphylococcal adhesion than did unused catheters (111). The potential role of hydrophobic interactions in the latter model is obscured by the methodology, however, since the bacteria were suspended in a buffer containing albumin. Albumin and serum are known to inhibit adhesion of coagulase-negative staphylococci to plastic (45). The inhibition of adhesion by albumin may be due to the ability of albumin to increase the hydrophilicity of the plastic surface.

Barrett (7) compared the adhesion of coagulase-negative staphylococci, *S. aureus*, *E. coli*, *Klebsiella aerogenes*, and *P. aeruginosa* to
intravenous catheters that had been previously implanted subcutaneously
in rabbits and to unused catheters. It was found that in unused catheters,
bacteria adhered greatest to low-surface-energy substrates. However, the
adhesion of the microorganisms to catheters removed from the subcuta-
neous implantation sites demonstrated different results. Both gram-
positive cocci and gram-negative bacilli adhered in much greater numbers
to the catheters that had been implanted in the animal. This result was
irrespective of the Gram reaction (7). This observation would argue
against the specificity of the adhesion as suggested by the work of
Vaudaux et al. (111), who have implicated fibronectin as the target protein
for *S. aureus* adhesion.

In a recent investigation, the role of hydrophobic interactions in the
adhesion of coagulase-negative staphylococci to three types of intravenous catheters, polyurethane, silicone, and hydrogel coated, was studied
(63). The polyurethane and silicone catheters are hydrophobic, whereas
the hydrogel-coated catheters absorb water from the environment, resulting in a hydrophilic surface. It was found that bacterial adhesion to
unused and serum-coated catheters was not influenced by bacterial cell
hydrophobicity. However, adhesion to hydrogel-coated catheters was
less than adhesion to polyurethane or silicone catheters, the implication
being that hydrogel-coated catheters possessed a higher surface free
energy, i.e., were less hydrophobic, and therefore had a marked reduction in microbial adhesion (63).

The pathogenesis of catheter-associated bacteremias may not be the
same in every circumstance. For example, it is generally believed that
infecting bacteria track down the extraluminal surface of the catheter to
engage the fibrin sleeve that forms about the cannula as it protrudes into
the lumen of the vein. Sitges-Serra et al. have challenged this tenet and
claim that bacteria introduced through the hub of the catheter are
responsible for the bacteremias (101, 102). If this theory is correct, the
likelihood of a microbe interacting with an uncoated plastic surface is
greater. In this context, it is interesting to note that the inner surfaces of
intravenous catheters are apparently more adherent for coagulase-negative staphylococci than are the outer surfaces (25).

Adhesion to Contact Lenses

The role that hydrophobicity plays in the adhesion of *P. aeruginosa*
to contact lenses can be stated with more precision. The process of
adhesion of this organism to the lens surface is an active one, although

nonviable organisms, killed by heat or hydrogen peroxide, can adhere to a hydrogel lens surface at about 10% the number of normal viable bacteria (54). Significant adhesion occurs within 5 min of incubation of lenses with bacteria, since about one-third of maximum adhesion will occur within that period of time (54). Dart and Badenoch, in an interesting investigation of bacterial adhesion to hard PMMA and soft hydrogel lenses, found that *S. aureus* and *P. aeruginosa* adhered in greater numbers to the more hydrophobic PMMA lenses than to the soft hydrogel lenses (18). These authors demonstrated that lenses subjected to routine measures such as cleaning and rinsing in tap water, insertion into the eye, or being worn for 3 h showed an astounding number of adherent bacteria. Up to 130 CFU of gram-positive cocci per lens were found after PMMA lenses were rinsed under the tap. About 400 CFU of coagulase-negative staphylococci and viridans streptococci were adherent after the PMMA lenses had been placed in the mouth for wetting, a practice that should be discouraged but is often performed in order to wet the lens before insertion in the eye (18). However, *P. aeruginosa* was not found in any of the cultures despite the fact that this microorganism is the leading cause of ulcerative keratitis in soft contact lens wearers (88).

The adhesion of microorganisms to hard lenses such as those made of PMMA is very likely controlled to a major extent by hydrophobic interactions. For example, *C. albicans* adheres to hard lenses in direct proportion to the water contact angle of the lens (11), and adhesion is greatly enhanced if tear components such as albumin, lactoferrin, lysozyme, and fibronectin are placed in the bacterial suspension. The mechanism of enhancement presumably occurs through the adsorption of the proteins onto the lens surfaces. This marked enhancement could be partially accounted for by the fact that *C. albicans* possesses receptors for albumin (83) and fibronectin (103). This in vitro coating of lenses which results in enhanced microbial adhesion is consistent with the finding that both *C. albicans* (11) and *P. aeruginosa* (12) adhere in greater numbers to used than to unused soft contact lenses. In the case of *P. aeruginosa*, the explanation for increased adhesion to used lenses may be the specific interaction of the bacteria with *N*-acetylneuraminic acid or sialic acid (12). Sialic acid has been reported to be a receptor for *P. aeruginosa* adhesion to the injured murine tracheobronchial tree (90) and can inhibit *P. aeruginosa* adhesion to the highly hydrophobic murine cornea (44). Sialic acid is found in abundance on worn contact lenses (62) and clearly inhibits *P. aeruginosa* adhesion to unworn and worn soft contact lenses (12). If sialic acid plays a unique role in the adhesion of *P. aeruginosa* to contact lenses, however, it is likely to be a complex one. For example, mucin, a glycoprotein in the tear film responsible for wetting the hydro-

Table 3. Effect of Various Additives on the Adhesion of *P. aeruginosa* to
Unworn Extended-Wear Soft Contact Lenses[a]

Additive	n	Mean no. of bacteria/lens (10^8)	Adherence inhibition (%)	P
EBSS,[b] pH 7.0	4	2.1 ± 0.2		
EBSS, pH 3.0	4	1.2 ± 0.2	43	0.01
EBSS, pH 7.0	6	4.5 ± 1.3		
Mucin, 0.1%	6	1.3 ± 0.3	72	0.001
EBSS, pH 7.0	4	2.2 ± 0.6		
Sodium hyaluronate, 1%	4	1.9 ± 0.8	14	NS[c]
Methylcellulose, 1%	4	2.1 ± 0.2	5	NS
Gum guar, 0.1%	4	2.8 ± 0.6		NS

[a] From Butrus et al. (12) with permission.
[b] EBSS, Earle balanced salt solution.
[c] NS, Not significant.

phobic cornea, contains abundant sialic acid. Mucin, in suspension with
P. aeruginosa, caused a pronounced inhibition of *Pseudomonas* adhesion
to soft contact lenses (Table 3). The specificity of this interaction was
suggested by the fact that other extensively cross-linked substances such
as gum guar and sodium hyaluronate had no inhibitory activity against
Pseudomonas adhesion. This interaction may be related to the surface-
active powers of mucin (74), which reduces the surface tension of the
suspension and thus the wetting angle of the bulk liquid on the contact
lens surface. This effect may result in diminished adhesion of bacteria to
the lens surface.

Miller and Ahearn investigated the adhesion of 20 clinical isolates of
P. aeruginosa to 12 different hydrogel contact lenses (75). They demonstrated that viability was critical for adhesion to these products, since
heat-killed microorganisms adhered at 10% the number of viable microorganisms to the soft contact lens surface. In their model, maximum
adhesion occurred at 3 to 4 h, suggesting that the formation of an
extracellular polymer adhesive was important. In a previous study of
Pseudomonas adhesion to soft contact lenses, it was shown that with
prolonged incubation, a bacterial glycocalyx began to be formed (24). In
general, the findings of Miller and Ahearn agreed with previously reported
results of bacteria adhering to uncoated plastics; i.e., most isolates
adhered to more hydrophobic surfaces than to relatively hydrophilic
surfaces (75). However, some strains adhered to hydrogels as well as they
did to polystyrene. They further observed that nonionic polymers supported greater *Pseudomonas* adhesion than did ionic polymers, indicating

Table 4. Attachment of *P. fluorescens* and *Acinetobacter* sp. to Substrata[a]

Surface[b]	Water content (%, wt/wt)[c]	Contact angle (°)	10^6 bacteria/cm^2 (SD)	
			P. fluorescens	*Acinetobacter* sp.
SP 85	79	11	6.2 (1.5)	0.54 (0.22)
SP 77	77	21	9.0 (1.0)	0.69 (0.30)
SP 70	70	27	9.3 (1.1)	1.9 (0.4)
SP 55	55	34	12.0 (1.0)	2.3 (0.42)
SP	1	0	21.0 (1.3)	17.0 (2.1)
TC-PS	1	66	29.0 (1.2)	63.0 (3.1)
PS	1	90	43.0 (4.0)	66.0 (3.0)

[a] From Pringle and Fletcher (89) with permission.
[b] SP, Trade name of hydrogel lenses; TC-PS, tissue culture polystyrene; PS, polystyrene.
[c] Values published by Special Polymers, London.

the importance of electrostatic interactions in the adhesion of *P. aeruginosa* to hydrogel lenses. Their results are more or less in agreement with those of Pringle and Fletcher (89), who demonstrated that *P. fluorescens* and an *Acinetobacter* sp. showed a decrease in adhesion of bacteria as the water content of the hydrogel soft contact lenses was increased (Table 4).

If we accept that the clean unused surface of a soft contact lens may be a target for *P. aeruginosa* adhesion, then hydrophobic interactions may clearly be important in this process. For example, *P. aeruginosa* isolates obtained from patients with ulcerative keratitis were grown at 37°C statically and at 26°C with shaking, resulting in organisms that were hydrophobic and hydrophilic, respectively, as shown by MATH and polyethylene glycol-dextran biphasic separation (59). The *P. aeruginosa* isolates at 37°C grew as a pellicle on the surface of the broth and were piliated. Ten isolates grown in this fashion were tested for their ability to adhere to unused soft contact lenses, and in each instance the more hydrophobic bacterium adhered more to the lens surface than did the hydrophilic bacterium (Table 5).

Erythrocyte agglutination by *P. aeruginosa* has been attributed to hydrophobic interactions (34) and occurred to a greater extent in *Pseudomonas* isolates grown at 37°C than it did in those grown at 26°C (59). Hemagglutination did not occur when nonionic surfactants such as Triton X-100, Nonidet P-40, and Tween 20 were added at concentrations as low as 0.01% (vol/vol). The inhibition of hemagglutination by surfactants correlated with bacterial adhesion to contact lenses as well. For example, these same surfactants were capable of reducing *Pseudomonas* adhesion to the lens surface as well; $3.5 \times 10^7 \pm 1.6 \times 10^7$ bacteria adhered to each lens without surfactant, whereas with a 0.1% solution of Triton X-100, adhesion was reduced to $1.7 \times 10^7 \pm 0.3 \times 10^7$ bacteria per lens, a 52% reduction (59).

Table 5. Adhesion of *P. aeruginosa* to Soft Contact Lenses when Grown at 37°C Statically or at 26°C with Shaking[a]

Isolate	No. of adherent bacteria/lens ± SD[b]	
	37°C	26°C
1	$4.4 \times 10^8 \pm 0.1 \times 10^8$	$8.0 \times 10^7 \pm 1.4 \times 10^7$
2	$2.4 \times 10^7 \pm 0.8 \times 10^7$	$8.4 \times 10^6 \pm 2.4 \times 10^6$
3	$1.5 \times 10^8 \pm 0.5 \times 10^8$	$5.0 \times 10^7 \pm 0.1 \times 10^7$

[a] From Klotz et al. (59) with permission.
[b] Adhesion of organisms grown at 37°C compared with 26°C was significantly different for each isolate ($P < 0.05$ by two-tailed Student t test). For each isolate, three lenses were tested.

In subsequent studies, it was shown that piliated *Pseudomonas* strains grown at 37°C statically adhered to previously worn soft contact lenses more than did isolates grown at 26°C. This adhesion appeared to occur in direct proportion to the amount of lens deposits (S. I. Butrus and S. A. Klotz, presented, at the 15th Cornea Research Conference, Boston, Mass., 1987). Therefore, it appears that hydrophobic interactions are important in the process of adhesion of *P. aeruginosa* to unworn and worn soft contact lenses alike, a process that may lead to vision-threatening infectious keratitis.

Adhesion to IUCDs

The role of hydrophobic interactions in the adhesion of microorganisms to IUCDs has received some attention. Lactobacilli were isolated from IUCDs removed from patients, and their ability to adhere to hydrophobic surfaces was investigated. The lactobacilli adhered in direct proportion to the surface hydrophobicity of the substrate (91). In another study, seven different microbial species were tested for their ability to adhere to three different plastic IUCDs (52). Group B streptococci were found to be the most adherent bacterial species, and the group B cocci were also the most hydrophobic microorganisms as determined by MATH. Scanning electron microscopy demonstrated that adhesion of group B streptococci to IUCDs was greatest on surface irregularities of the plastic devices.

In a follow-up report, it was shown that the adhesion of group B streptococci to polyethylene IUCDs may involve a bacterial cell surface protein-lipid complex consisting of nonfibrillar surface proteins and cellular lipoteichoic acids (51). Treatments of the bacterial cells that changed or diminished the amount of surface protein led to decreased cell surface hydrophobicity as determined by MATH, and this reduction

Table 6. Effects of Different Pretreatments on Viability, Affinity toward Xylene, and Adhesion to Polyethylene of a Group B Streptococcus Type II Isolate[a]

Pretreatment	Viability (% of control)	Affinity toward xylene (% of cells bound to xylene)[b]	Adhesion to polyethylene[c]
None	100	93.9 ± 5.2	432 ± 21
Heat (60°C, 1 h)	0	94.8 ± 6.4	416 ± 26
Formaldehyde (2%)	0	1.3 ± 3.1	127 ± 14[d]
Hot aqueous phenol (65°C, 30 min)	0	7.5 ± 2.8	17 ± 4[d]
Pepsin (1 mg/ml, 37°C, 1 h)	100	38.4 ± 4.3	28 ± 9[d]
Trypsin (1 mg/ml, 37°C, 1 h)	100	31.7 ± 3.9	56 ± 8[d]

[a] From Jacques and Costerson (51) with permission.
[b] Twenty minutes after 0.5 ml of xylene was added and mixed with the bacterial cell suspension.
[c] Expressed as the mean number of bacterial cells per 0.01 mm^2.
[d] Significant difference ($P < 0.001$) between control and test values.

correlated with decreased adhesion of the bacteria to polyethylene. Decreased adhesion of group B streptococci also occurred after treatment with formaldehyde, which changed protein configuration, hot phenol, which reduced lipoteichoic acid, and pepsin and trypsin, which reduced surface proteins (Table 6). These various treatments reduced both cell hydrophobicity and adhesion.

CONCLUSION

All plastic medical devices are subject to infection after implantation in or application on the human body. Infections are usually due to resident normal microbiota. These devices (some of which are listed in Table 1) are hydrophobic and thus nonwettable, and as a general rule they favor the adhesion of microorganisms to their surfaces. This phenomenon has been convincingly demonstrated for diverse microorganisms such as bacteria, mycoplasmas, fungi, and plant cells. The cell constituents responsible for these hydrophobic interactions have not been characterized, but recent studies with coagulase-negative staphylococci (13) and group B streptococci (51) appear to implicate lipoteichoic acid in the hydrophobic interactions that occur.

The actual adhesion target surface for these medical devices, i.e., plastic or host-coated plastic, has never been adequately determined, nor has the time of adhesion of microorganisms to medical devices been determined with any accuracy. It has been recently shown that vitronectin, more so than fibronectin, adsorbs on plastics, and this protein may deserve study from the standpoint of microbial adhesion to plastic surfaces (6). Furthermore, depending on the microorganism studied, the addition of host proteins that coat the plastic surface may enhance or decrease microbial adhesion. If microorganisms possess receptors for the protein in question, adhesion may increase. Conversely, protein adsorbed to the plastic surface which makes the surface more wettable or hydrophilic will decrease microbial adhesion. In the latter case, the critical surface tension model, which describes the presence of a nadir of microbial adhesion to low-surface-energy plastics, may be an accurate predictor of microbial adhesion in the presence of protein molecules.

Our knowledge of the extent of hydrophobic interactions involved in the plastic device-microorganism interaction will be improved as investigators characterize the cell surface moieties responsible for cell surface hydrophobicity, determine the surface characteristics of inserted devices, and determine what, if any, biologically compatible substance inhibits or abolishes adhesion of microorganisms to these plastic medical devices. Future advancements in medical care and technology to a large extent will be measured by our ability to prevent microbial infections of these devices.

Acknowledgment. I thank Ann Shows for her help in preparing the manuscript.

LITERATURE CITED

1. **Absolom, D. R., F. V. Lamberti, Z. Policova, W. Zingg, C. J. van Oss, and A. W. Neumann.** 1983. Surface thermodynamics of bacterial adhesion. *Appl. Environ. Microbiol.* **46:**90–97.

2. **Andrade, J. D.** 1973. Interfacial phenomena and biomaterials. *Med. Instrum.* **7:**110–120.

3. **Ashkenazi, S., E. Weiss, and M. M. Drucker.** 1986. Bacterial adherence to intravenous catheters and needles and its influence by cannula type and bacterial surface hydrophobicity. *J. Lab. Clin. Med.* **107:**136–140.

4. **Baddour, L. M., G. D. Christensen, W. A. Simpson, and E. H. Beachey.** 1990. Microbial adherence, p. 9–25. *In* G. L. Mandell, R. G. Douglas, Jr., and J. E. Bennett (ed.), *Principles and Practice of Infectious Diseases.* Churchill Livingstone, New York.

5. **Baier, R. E.** 1972. Influence of the initial surface condition of materials on bioadhesion, p. 633–639. *In* R. F. Acker, B. F. Brown, J. R. De Palma, and W. P. Iverson (ed.), *Proceedings of the Third International Congress on Marine Corrosion and Fouling.* National Bureau of Standards, Gaithersburg, Md.

6. **Bale, M. D., L. A. Wohlfahrt, D. F. Mosher, B. Tomasini, and R. C. Sutton.** 1989. Identification of vitronectin as a major plasma protein adsorbed on polymer surfaces of different copolymer composition. *Blood* **74:**2698–2706.

7. **Barrett, S. P.** 1988. Bacterial adhesion to intravenous cannulae: influence of implantation in the rabbit and of enzyme treatments. *Epidemiol. Infect.* **100:**91–100.

8. **Brash, J. L., and D. J. Lyman.** 1969. Adsorption of plasma proteins in solution to uncharged, hydrophobic polymer surfaces. *J. Biomed. Mater. Res.* **3:**175–189.

9. **Bredt, W., J. Feldner, and I. Kahane.** 1981. Adherence of mycoplasmas to cells and

inert surfaces: phenomena, experimental models and possible mechanisms. *Isr. J. Med. Sci.* **17**:586–588.

10. **Budts-Jorgensen, E.** 1974. The significance of *Candida albicans* in denture stomatitis. *Scand. J. Dent. Res.* **82**:151–190.

11. **Butrus, S. I., and S. A. Klotz.** 1986. Blocking *Candida* adherence to contact lenses. *Curr. Eye Res.* **5**:745–750.

12. **Butrus, S. I., S. A. Klotz, and R. P. Misra.** 1987. The adherence of *Pseudomonas aeruginosa* to soft contact lenses. *Ophthalmology* **94**:1310–1314.

13. **Christensen, G. D., L. M. Baddour, D. L. Hasty, J. H. Lowrance, and W. A. Simpson.** 1989. Microbial and foreign body factors in the pathogenesis of medical device infections, p. 27–59. *In* A. L. Bisno and F. A. Waldvogel (ed.), *Infections Associated with Indwelling Medical Devices.* American Society for Microbiology, Washington, D.C.

14. **Christensen, G. D., A. L. Bisno, J. T. Parisi, R. N. McLaughlin, M. G. Hester, and R. W. Luther.** 1982. Nosocomial septicemia due to multiple antibiotic resistant *Staphylococcus epidermidis. Ann. Intern. Med.* **96**:1–10.

15. **Christensen, G. D., W. A. Simpson, J. J. Younger, L. M. Baddour, F. F. Barrett, D. M. Melton, and E. H. Beachey.** 1985. Adherence of coagulase-negative staphylococci to plastic tissue culture plates: a quantitative model for the adherence of staphylococci to medical devices. *J. Clin. Microbiol.* **22**:996–1006.

16. **Christersson, C. E., R. G. Dunford, P.-O. J. Glantz, and R. E. Baier.** 1989. Effect of critical surface tension on retention of oral microorganisms. *Scand. J. Dent. Res.* **97**:247–256.

17. **Criado, M. T., C. M. Ferreiros, and V. Sainz.** 1985. Adherence and hydrophobicity in *Neisseria meningitidis* and their relationship with surface charge. *Med. Microbiol. Immunol.* **174**:151–156.

18. **Dart, J. K., and P. R. Badenoch.** 1986. Bacterial adherence to contact lenses. *CLAO J.* **12**:220–224.

19. **Dexter, S. C.** 1979. Influence of substratum critical surface tension on bacterial adhesion—*in situ* studies. *J. Colloid Interface Sci.* **70**:346–354.

20. **Dexter, S. C., J. D. Sullivan, Jr., J. Williams, and S. W. Watson.** 1975. Influence of substrate wettability on the attachment of marine bacteria to various surfaces. *Appl. Microbiol.* **30**:298–308.

21. **Diaz-Blanco, J., R. C. Clawson, S. M. Roberson, C. B. Sanders, A. K. Pramanik, and J. J. Herbst.** 1989. Electron microscopic evaluation of bacterial adherence to polyvinyl chloride endotracheal tubes used in neonates. *Crit. Care Med.* **17**:1335–1340.

22. **Dougherty, S. H.** 1986. Implant infections, p. 276–289. *In* A. F. von Recum (ed.), *Handbook of Biomaterials Evaluation. Scientific, Technical and Clinical Testing of Implant Materials.* Macmillan Publishing Co., New York.

23. **Drumm, B., A. W. Neumann, Z. Policova, and P. M. Sherman.** 1989. Bacterial cell surface hydrophobicity properties in the mediation of in vitro adhesion by the rabbit enteric pathogen *Escherichia coli* strain RDEC-1. *J. Clin. Invest.* **84**:1588–1594.

24. **Duran, J. A., M. F. Refojo, I. K. Gipson, and K. R. Kenyon.** 1987. *Pseudomonas* attachment to new hydrogel contact lenses. *Arch. Ophthalmol.* **106**:106–109.

25. **Elliott, T. S. J., V. C. D'Abrera, and S. Dutton.** 1988. The effect of antibiotics on bacterial colonization of vascular cannulae in a novel in-vitro model. *J. Med. Microbiol.* **26**:229–235.

26. **Facchini, P. J., F. DiCosmo, L. G. Radvanyi, and Y. Giguere.** 1988. Adhesion of *Catharanthus roseus* cells to surfaces: effect of substrate hydrophobicity. *Biotechnol. Bioeng.* **32**:935–938.

27. **Fish, F., Y. Navon, and S. Goldman.** 1987. Hydrophobic adherence and phase variation in *Bordetella pertussis. Med. Microbiol. Immunol.* **176:**37–46.

28. **Fletcher, M.** 1976. The effects of proteins on bacterial attachment to polystyrene. *J. Gen. Microbiol.* **94:**400–404.

29. **Fletcher, M.** 1980. The question of passive versus active attachment mechanisms in non-specific bacterial adhesion, p. 197–210. *In* R. C. W. Berkeley, J. M. Lynch, J. Melling, P. R. Rutter, and B. Vincent (ed.), *Microbial Adhesion to Surfaces.* Ellis Horwood, Chichester, England.

30. **Fletcher, M.** 1983. The effects of methanol, ethanol, propanol, and butanol on bacterial attachment to surfaces. *J. Gen. Microbiol.* **129:**633–641.

31. **Fletcher, M., and G. I. Loeb.** 1979. Influence of substratum characteristics on the attachment of a marine pseudomonad to solid surfaces. *Appl. Environ. Microbiol.* **37:**67–72.

32. **Franson, T. R., N. K. Sheth, H. D. Rose, and P. G. Sohnle.** 1984. Quantitative adherence *in vitro* of coagulase-negative staphylococci to intravascular cathers: inhibition with D-mannosamine. *J. Infect. Dis.* **149:**116.

33. **Fujioka-Hirai, Y., Y. Akagawa, S. Minagi, H. Tsuru, Y. Miyake, and H. Suginaka.** 1987. Adherence of *Streptococcus mutans* to implant materials. *J. Biomed. Mater. Res.* **21:**913–920.

34. **Garber, N., N. Sharon, D. Shohet, J. S. Lam, and R. J. Doyle.** 1985. Contribution of hydrophobicity to hemagglutination reactions of *Pseudomonas aeruginosa. Infect. Immun.* **50:**336–357.

35. **George, R., L. Leibrock, and M. Epstein.** 1979. Long-term analysis of cerebrospinal fluid shunt infections. *J. Neurosurg.* **51:**804–811.

36. **Gerson, D. F., and J. Akit.** 1980. Cell surface energy, contact angles and phase partition. II. Bacterial cells in biphasic aqueous mixtures. *Biochim. Biophys. Acta.* **602:**281–284.

37. **Givens, C. D., and R. P. Wenzel.** 1980. Catheter-associated urinary tract infections in surgical patients: a controlled study on the excess-morbidity and cost. *J. Urol.* **124:**646–648.

38. **Gower, D. J., V. C. Gower, S. H. Richardson, and D. L. Kelly, Jr.** 1986. Reduced bacterial adherence to silicone plastic neurosurgical prosthesis. *Pediatr. Neurosci.* **12:**127–133.

39. **Gristina, A. G., J. J. Dobbins, B. Giammara, J. C. Lewis, and W. C. DeVries.** 1988. Biomaterial-centered sepsis and the total artificial heart. Microbial adhesion vs tissue integration. *J. Am. Med. Assoc.* **259:**870–874.

40. **Gristina, A. G., and P. T. Naylor.** 1988. The race for the surface: microbes, tissue cells, and biomaterials, p. 177–211. *In* L. Switalski, M. Hook, and E. Beachey (ed.), *Molecular Mechanisms of Microbial Adhesion.* Springer-Verlag, New York.

41. **Harris, J. M., and L. F. Martin.** 1987. An *in vitro* study of the properties influencing *Staphylococcus epidermidis* adhesion to prosthetic vascular graft materials. *Ann. Surg.* **206:**612–620.

42. **Harris, R. L., W. R. Wilson, and T. W. Williams, Jr.** 1984. Infections associated with prosthetic heart valves, p. 90–112. *In* B. Sugarman and E. J. Young (ed.), *Infections Associated with Prosthetic Devices.* CRC Press, Inc., Boca Raton, Fla.

43. **Hazen, K. C., and B. W. Hazen.** 1987. Temperature-modulated physiological characteristics of *Candida albicans. Microbiol. Immunol.* **31:**497–508.

44. **Hazlett, L. D., M. Moon, and R. S. Berk.** 1986. In vivo identification of sialic acid as the ocular receptor for *Pseudomonas aeruginosa. Infect. Immun.* **51:**687–689.

45. Herrmann, M., P. Vaudaux, D. Pittet, R. Auckenthaler, R. D. Lew, F. Schumacher-Perdreau, G. Peters, and F. A. Waldvogel. 1988. Fibronectin, fibrinogen, and laminin act as mediators of adherence of clinical staphylococcal isolates to foreign material. *J. Infect. Dis.* **158**:693–701.

46. Hogt, A. H., J. Dankert, J. A. de Vries, and J. Feijen. 1983. Adhesion of coagulase-negative staphylococci to biomaterials. *J. Gen. Microbiol.* **129**:2959–2968.

47. Hogt, A. H., J. Dankert, C. E. Hulstaert, and J. Feijen. 1986. Cell surface characteristics of coagulase-negative staphylococci and their adherence to fluorinated poly(ethylenepropylene). *Infect. Immun.* **51**:294–301.

48. Humphries, M., J. F. Jaworzyn, J. B. Cantwell, and A. Eakin. 1987. The use of non-ionic ethoxylated and propoxylated surfactants to prevent the adhesion of bacteria to solid surfaces. *FEMS Microbiol. Lett.* **42**:91–101.

49. Inglis, T. J. J., M. R. Miller, J. G. Jones, and D. A. Robinson. 1989. Tracheal tube biofilm as a source of bacterial colonization of the lung. *J. Clin. Microbiol.* **27**:2014–2018.

50. Inman, R. D., K. V. Gallegos, B. D. Branse, P. B. Redecha, and C. L. Christian. 1984. Clinical and microbial features of prosthetic joint infection. *Am. J. Med.* **77**:47–53.

51. Jacques, M., and J. W. Costerton. 1987. Adhesion of group B *Streptococcus* to a polyethylene intrauterine contraceptive device. *FEMS Microbiol. Lett.* **41**:23–28.

52. Jacques, M., T. J. Marrie, and J. W. Costerton. 1986. In vitro quantitative adherence of microorganisms to intrauterine contraceptive devices. *Curr. Microbiol.* **13**:133–137.

53. James, H. 1984. Infections associated with cerebrospinal fluid prosthetic devices, p. 24–41. *In* B. Sugarman and E. J. Young (ed.), *Infections Associated with Prosthetic Devices*. CRC Press, Inc., Boca Raton, Fla.

54. John, T., M. F. Refojo, L. Hanninen, F. L. Leong, A. Medina, and K. R. Kenyon. 1989. Adherence of viable and non-viable bacteria to soft contact lenses. *Cornea* **8**:21–23.

55. Jones, P. G., R. L. Hopfer, L. Elting, J. A. Jackson, V. Fainstein, and G. P. Bodey. 1986. Semiquantitative cultures of intravascular catheters from cancer patients. *Diagn. Microbiol. Infect. Dis.* **4**:299–306.

56. Karlan, M. S., B. Skobel, M. Grizzard, N. J. Cassisi, G. T. Singleton, P. Buscemi, and E. P. Goldberg. 1980. Myringotomy tube materials: bacterial adhesion and infection. *Otolaryngol. Head Neck Surg.* **88**:783–795.

57. Klotz, S. A. 1989. *Malassezia furfur. Infect. Dis. Clin. North Am.* **3**:53–64.

58. Klotz, S. A. 1989. Surface-active properties of *Candida albicans. Appl. Environ. Microbiol.* **55**:2119–2122.

59. Klotz, S. A., S. I. Butrus, R. P. Misra, and M. S. Osato. 1989. The contribution of bacterial surface hydrophobicity to the process of adherence of *Pseudomonas aeruginosa* to hydrophilic contact lenses. *Curr. Eye Res.* **8**:195–202.

60. Klotz, S. A., D. J. Drutz, M. Huppert, and J. E. Johnson. 1982. *Pityrosporum* folliculitis: its potential for confusion with skin lesions of systemic candidiasis. *Arch. Intern. Med.* **142**:2126–2129.

61. Klotz, S. A., D. J. Drutz, and J. E. Zajic. 1985. Factors governing the adherence of *Candida* species to plastic surfaces. *Infect. Immun.* **50**:97–101.

62. Klotz, S. A., R. P. Misra, and S. I. Butrus. 1987. Carbohydrate deposits on the surfaces of worn extended-wear soft contact lenses. *Arch. Ophthalmol.* **105**:974–977.

63. Kristinsson, K. G. 1989. Adherence of staphylococci to intravascular catheters. *J. Med. Microbiol.* **28**:249–257.

64. Kunin, C. M., J. J. Dobbins, J. C. Melo, M. M. Levinson, K. Love, L. D. Joyce, and W. C. DeVries. 1988. Infectious complications in four long-term recipients of the Jarvik-7 artificial heart. *J. Am. Med. Assoc.* **259**:860–864.

65. **Kuusela, P.** 1978. Fibronectin binds to *Staphylococcus aureus*. *Nature* (London) **276**:718–720.

66. **Lewin, R.** 1984. Microbial adhesion is a sticky problem. *Science* **224**:375–377.

67. **Liedberg, H., and T. Lundeberg.** 1989. Silver coating of urinary catheters prevents adherence and growth of *Pseudomonas aeruginosa*. Urol. Res. **17**:357–358.

68. **Loeb, G., C. Bailey, and S. Wajagrass.** 1983. Variation in marine microbial attachment preference. *Int. Biodeterior. Bull.* **19**:79–82.

69. **Ludwicka, A., B. Jansen, T. Wadström, and G. Pulverer.** 1984. Attachment of staphylococci to various synthetic polymers. *Zentralbl. Bakteriol. Parasitenkd. Infektionskr. Hyg. Abt. 1 Reihe A* **256**:479–489.

70. **Marshall, K. C.** 1972. Mechanism of adhesion of marine bacteria to surfaces, p. 625–632. In R. F. Acker, B. F. Brown, J. R. De Palma, and W. P. Iverson (ed.), *Proceedings of the Third International Congress on Marine Corrosion and Fouling*. National Bureau of Standards, Gaithersburg, Md.

71. **Marshall, K. C., and R. H. Cruickshank.** 1973. Cell surface hydrophobicity and the orientation of certain bacteria at interfaces. *Arch. Mikrobiol.* **91**:29–40.

72. **McEldowney, S., and M. Fletcher.** 1986. Effect of growth conditions and surface characteristics of aquatic bacteria on their attachment to solid surfaces. *J. Gen. Microbiol.* **132**:513–523.

73. **Meadows, P. S.** 1971. The attachment of bacteria to solid surfaces. *Arch. Mikrobiol.* **75**:374–381.

74. **Meyer, F. A.** 1976. Mucus structure: relation to biological transport function. *Biorheology* **13**:49–58.

75. **Miller, M. J., and D. G. Ahearn.** 1987. Adherence of *Pseudomonas aeruginosa* to hydrophilic contact lenses and other substrata. *J. Clin. Microbiol.* **25**:1392–1397.

76. **Minagi, S., Y. Miyake, K. Inagaki, H. Tsuru, and H. Suginaka.** 1985. Hydrophobic interaction in *Candida albicans* and *Candida tropicalis* adherence to various denture base resin materials. *Infect. Immun.* **47**:11–14.

77. **Miyake, Y., Y. Fujita, S. Minagi, and H. Suginaka.** 1986. Surface hydrophobicity and adherence of *Candida* to acrylic surfaces. *Microbios* **46**:7–14.

78. **Mohammed, S. F., N. S. Topham, G. L. Burns, and D. B. Olsen.** 1988. Enhanced bacterial adhesion on surfaces pretreated with fibrinogen and fibronectin. *Trans. Am. Soc. Artif. Organs* **34**:573–577.

79. **Mozes, N., F. Marchal, M. P. Hermesse, J. L. Van Haecht, L. Reuliaux, A. J. Leonard, and P. G. Rouxhet.** 1987. Immobilization of microorganisms by adhesion: interplay of electrostatic and nonelectrostatic interactions. *Biotechnol. Bioeng.* **30**:439–450.

80. **Odio, C., G. H. McCraken, and J. D. Nelson.** 1984. CSF shunt infections in pediatrics. *Am. J. Dis. Child.* **38**:1103–1108.

81. **O'Neill, T. B., and G. L. Wilcox.** 1971. The formation of a "primary film" on materials submerged in the sea at Port Hueneme, California. *Pacific Sci.* **25**:1–12.

82. **Op den Camp, H. J. M., A. Ossterhof, and J. H. Veercamp.** 1985. Cell surface hydrophobicity of *Bifidobacterium bifidum* subsp. *pennsylvanicum*. *Antonie van Leeuwenhoek J. Microbiol. Serol.* **51**:303–312.

83. **Page, S., and F. C. Odds.** 1988. Binding of plasma proteins to *Candida* species *in vitro*. *J. Gen. Microbiol.* **134**:2693–2702.

84. **Pascual, A., A. Fleer, N. A. C. Westerdaal, and J. Verhoef.** 1986. Modulation of adherence of coagulase-negative staphylococci to Teflon catheters *in vitro*. *Eur. J. Clin. Microbiol.* **5**:518–522.

85. **Paul, J. H.** 1984. Effects of antimetabolites on the adhesion of an estuarine *Vibrio* sp. to polystyrene. *Appl. Environ. Microbiol.* **48**:924–929.

86. **Paul, J. H., and W. H. Jeffrey.** 1985. Evidence for separate adhesion mechanisms for hydrophilic and hydrophobic surfaces in *Vibrio proteolytica*. *Appl. Environ. Microbiol.* **50:**431–437.

87. **Peters, G., R. Locci, and G. Pulverer.** 1982. Adherence and growth of coagulase-negative staphylococci on surfaces of intravenous catheters. *J. Infect. Dis.* **146:**479–482.

88. **Poggio, E. C., R. J. Glynn, O. D. Schein, J. M. Seddon, M. J. Shannon, V. A. Seardino, and K. R. Kenyon.** 1989. The incidence of ulcerative keratitis among users of daily-wear and extended-wear soft contact lenses. *N. Engl. J. Med.* **321:**779–783.

89. **Pringle, J. H., and M. Fletcher.** 1986. Influence of substratum hydration and adsorbed macromolecules on bacterial attachment to surfaces. *Appl. Environ. Microbiol.* **51:**1321–1325.

90. **Ramphal, R., and M. Pyle.** 1983. Evidence for mucins and sialic acid as receptors for *Pseudomonas aeruginosa* in the lower respiratory tract. *Infect. Immun.* **41:**339–344.

91. **Reid, G., L.-A. Hawthorn, R. Mandatori, R. L. Cook, and H. S. Beg.** 1988. Adhesion of lactobacilli to polymer surfaces in vivo and in vitro. *Microb. Ecol.* **16:**241–251.

92. **Rosenberg, M.** 1981. Bacterial adherence to polystyrene: a replica method of screening for bacterial hydrophobicity. *Appl. Environ. Microbiol.* **42:**375–377.

93. **Rosenberg, M., D. Gutnick, and E. Rosenberg.** 1980. Adherence of bacteria to hydrocarbons: a simple method for measuring cell-surface hydrophobicity. *FEMS Microbiol. Lett.* **9:**29–33.

94. **Rotrosen, D., T. R. Gibson, and J. E. Edwards.** 1983. Adherence of *Candida* species to intravenous catheters. *J. Infect. Dis.* **147:**599.

95. **Russell, P. B., J. Kline, M. C. Yoder, and R. A. Polin.** 1987. Staphylococcal adherence to polyvinyl chloride and heparin-bonded polyurethane catheters is species dependent and enhanced by fibronectin. *J. Clin. Microbiol.* **25:**1083–1087.

96. **Schadow, K. H., W. A. Simpson, and G. D. Christensen.** 1988. Characteristics of adherence to plastic tissue culture plates of coagulase-negative staphylococci exposed to subinhibitory concentrations of antimicrobial agents. *J. Infect. Dis.* **157:**71–77.

97. **Sechler, G. E., and K. Gunderson.** 1972. Role of surface chemical composition of the microbial contribution to primary films, p. 610–616. *In* R. F. Acker, B. F. Brown, J. R. De Palma, and W. P. Iverson (ed.), *Proceedings of the Third International Congress on Marine Corrosion and Fouling.* National Bureau of Standards, Gaithersburg, Md.

98. **Sherertz, R. J., R. J. Folk, K. A. Huffman, C. A. Thomann, and W. D. Mattern.** 1983. Infections associated with subclavian Uldall catheters. *Arch. Intern. Med.* **143:**52–56.

99. **Sheth, N. K., T. R. Franson, H. D. Rose, F. L. A. Buckmire, J. A. Cooper, and P. G. Sohnle.** 1983. Colonization of bacteria on polyvinyl chloride and Teflon intravascular catheters in hospitalized patients. *J. Clin. Microbiol.* **18:**1061–1063.

100. **Sheth, N. K., T. R. Franson, and P. G. Sohnle.** 1985. Influence of bacterial adherence to intravascular catheters on *in vitro* antibiotic susceptibility. *Lancet* **ii:**1266–1268.

101. **Sitges-Serra, A., E. Juarrieta, J. Linares, J. L. Perez, and J. Garau.** 1983. Bacteria in total parenteral nutrition catheters: where do they come from? *Lancet* **i:**531.

102. **Sitges-Serra, A., J. Linares, and J. Garau.** 1985. Catheter sepsis: the clue is the hub. *Surgery* **97:**355–357.

103. **Skerl, K. G., R. A. Calderone, E. Segal, T. Sreevalson, and W. M. Scheld.** 1984. *In vitro* binding of *Candida albicans* yeast cells to human fibronectin. *Can. J. Microbiol.* **30:**221–227.

104. **Speer, A. G., P. B. Cotton, J. Rode, A. M. Seddon, C. R. Neal, J. Holton, and J. W. Costerton.** 1988. Biliary stint blockage with bacterial biofilm. A light and electron microscopy study. *Ann. Intern. Med.* **108:**546–553.

105. **Sugarman, B.** 1982. *In vitro* adherence of bacteria to prosthetic vascular grafts. *Infection* **10**:9–12.
106. **Sugarman, B., and E. J. Young.** 1989. Infections associated with prosthetic devices: magnitude of the problem. *Infect. Dis. Clin. North Am.* **3**:187–198.
107. **Tanford, C.** 1980. *The Hydrophobic Effect: Formation of Micelles and Biological Membranes*, p. 1–4. John Wiley & Sons, Inc., New York.
108. **Uyen, H. M., H. C. van der Mei, A. H. Weerkamp, and H. J. Busscher.** 1988. Comparison between the adhesion to solid substrata of *Streptococcus mitis* and that of polystyrene particles. *Appl. Environ. Microbiol.* **54**:837–838.
109. **van Loosdrecht, M. C. M., J. Lyklema, W. Norde, G. Schraa, and A. J. B. Zehnder.** 1987. The role of bacterial cell wall hydrophobicity in adhesion. *Appl. Environ. Microbiol.* **53**:1893–1897.
110. **van Pelt, A. W. J., A. H. Weerkamp, M. H. W. J. C. Uyen, H. J. Busscher, H. P. deJong, and J. Arends.** 1985. Adhesion of *Streptococcus sanguis* CH3 to polymers with different surface free energies. *Appl. Environ. Microbiol.* **49**:1270–1275.
111. **Vaudaux, P., D. Pittet, A. Haeberli, E. Huggler, U. E. Nydegger, D. P. Lew, and F. A. Waldvogel.** 1989. Host factors selectively increase staphylococcal adherence on inserted catheters: a role for fibronectin and fibrinogen or fibrin. *J. Infect. Dis.* **160**:865–875.
112. **Vaudaux, P. E., F. A. Waldvogel, J. J. Morgenthaler, and U. E. Nydegger.** 1984. Adsorption of fibronectin onto polymethylmethacrylate and promotion of *Staphylococcus aureus* adherence. *Infect. Immun.* **45**:768–774.
113. **Weber, D. J., K. L. Hoffman, R. A. Thoft, and A. S. Baker.** 1986. Endophthalmitis following intraocular lens implantation: report of 30 cases and review of the literature. *Rev. Infect. Dis.* **8**:12–20.
114. **Wood-Helie, S. J., H. P. Dalton, and S. Shadomy.** 1986. Hydrophobic and adherence properties of *Clostridium difficile*. *Eur. J. Clin. Microbiol.* **5**:441–445.
115. **Wright, J. B., I. Ruseska, M. A. Athar, S. Corbett, and J. W. Costerton.** 1989. *Legionella pneumophila* grows adherent to surfaces *in vitro* and *in situ*. *Infect. Control Hosp. Epidemiol.* **10**:408–415.
116. **Young, E. J., and B. Sugarman.** 1984. Introduction to prosthetic devices and their regulation in the United States, p. 2–10. *In* B. Sugarman and E. J. Young (ed.), *Infections Associated with Prosthetic Devices*. CRC Press, Inc., Boca Raton, Fla.

Microbial Cell Surface Hydrophobicity
Edited by R. J. Doyle and M. Rosenberg
© 1990 American Society for Microbiology, Washington, DC 20005

Chapter 5

Hydrophobicity of Proteins and Bacterial Fimbriae

Randall T. Irvin

INTRODUCTION

The structure and function of the bacterial surface appendages termed pili or fimbriae have been extensively reviewed (93, 96, 104, 105, 107, 129), but neither the hydrophobic nature of these structures nor the contribution of hydrophobicity to their function has been previously examined in depth. This chapter will examine the role of the hydrophobic effect in both fimbrial structure and function, particularly in relation to bacterial adhesion.

Fimbriae or pili are bacterial cell surface appendages composed primarily of a self-assembling protein subunit that is generally called pilin or, more awkwardly, a fimbrial structural subunit. Fimbriae are helical assemblages of pilin which form a linear unbranching rod-shaped structure, 2 to 12 nm in diameter and of variable length (generally from 0.2 to 10 μm), that extends outwards from the cell surface. The degree of fimbriation of a fimbriated bacterial cell varies from one or two to more than several hundred fimbriae per cell, with the number and length of the fimbriae being regulated by various mechanisms. Bacterial fimbriae have generally been associated with conjugative or adhesive functions. The significance of the adhesive function in initiating an infection is now widely recognized (5), and antifimbrial vaccines have proven to be efficacious in preventing homologous infections (15, 37, 93, 113, 115, 128, 135).

Randall T. Irvin • Department of Medical Microbiology and Infectious Diseases, University of Alberta, Edmonton, Alberta T6G 2H7, Canada.

The terminology used to describe bacterial fimbriae or pili varies substantially, but these structures are now generally referred to as fimbriae (a Latin term meaning fibers or threads that was introduced by Duguid et al. [33]) or pili (a term introduced by Brinton [13] that is Latin for hairlike structure). A number of nomenclature systems have been proposed (see reference 107 for a discussion of these systems and their limitations), but unfortunately there has been no generally accepted terminology. For consistency in this book, I will use the term "fimbriae" to describe proteinaceous nonprostheceal, nonflagellar structures and the term "pilin" to describe the protein that constitutes the dominant structural subunit of the fimbriae, although I personally feel that the two terms can and perhaps should be used interchangeably.

Fimbrial Types

Bacterial fimbriae can be roughly split into two groups, conjugative fimbriae and nonconjugative or "adhesive" fimbriae, on the basis of aspects of their functions (107). Whereas fimbriae are widely distributed in both gram-negative and gram-positive species, conjugative fimbriae have to date been described only for gram-negative species. Conjugative plasmids have an extensive transfer operon that codes for conjugative fimbriae and a genetic transfer system, whereas nonconjugative or adhesive fimbriae are encoded either chromosomally or on a plasmid and are associated with a more limited operon.

Conjugative fimbriae

Conjugative fimbriae mediate the initial interaction between donor and recipient cells. The conjugative fimbriae are then retracted or disassembled, with the pilin subunits being stored in the outer membrane to bring the recipient cell into close contact with the donor cell, where subsequent fusion of the cell envelopes occurs to result in the formation of a stable mating pair (107). The extensive work on the conjugative plasmids has resulted in numerous studies on the nature of the fimbria-recipient cell interaction. In the best-studied system, the F conjugative fimbriae of *Escherichia coli*, the pilin structural subunits at the tip of the fimbriae mediate the interaction with the recipient cell without the assistance of a tip- or fimbria-associated adhesin pilin (107). There is, however, evidence that the pilin subunits at the fimbrial tip constitute a unique tip-associated antigen that presumably is associated with the adhesive function of these pilin subunits, which is normally occluded in the fimbrial shaft as a result of the interaction of the pilins (107). The recipient cell surface receptor for the F pilus is the major outer membrane protein OmpA (82, 94). Whereas the F conjugative fimbriae have received considerable study, there have been few studies on the other conjugative fimbriae (see reference 107 for a comprehensive review on conjugative fimbriae). This chapter will focus on the nonconjugative or adhesive fimbriae of bacteria.

Adhesive fimbriae

As the term "adhesive fimbriae" implies, these structures have been ascribed a central role in the adhesion of bacterial cells to various biological surfaces, including erythrocytes (102), mammalian epithelial cells, fungi, and plant roots (32, 33). Indeed, the initial morphological descriptions of bacterial fimbriae suggested a potential role in bacterial adhesion for these very interesting structures (58). The literature concerning the role of fimbriae in mediating bacterial adhesion to various cells has obviously confirmed these initial observations and established the significance of these structures. Bacterial fimbriae are now clearly recognized as significant virulence factors for a wide variety of pathogens (5).

Distribution and examples of fimbriae in procaryotes

Fimbriae have been described for a wide-ranging group of gram-negative species and for an increasing number of gram-positive organisms (see reference 92 for a substantial list of fimbriated organisms; even this list is incomplete, and the number of fimbriated organisms that have been identified is still increasing). It is very likely that the number of fimbriated organisms and the variety of fimbriae that they express will expand considerably as a wider variety of organisms are examined more closely by both morphological and genetic means. It is technically difficult to detect the presence of small numbers of fimbriae on the surfaces of bacteria, particularly if synthesis of the fimbriae is tightly regulated and the growth conditions are such that fimbrial synthesis is repressed. The available evidence indicates that the synthesis of fimbriae is tightly regulated and responsive to specific environmental signals (19, 31, 107, 140).

E. coli fimbriae represent perhaps the best-defined systems with respect to genetics. The advanced genetic systems available in *E. coli* have enabled the identification of very low copy number fimbria-associated proteins, which are tip localized for the *pap* fimbriae (78, 80, 81). These proteins have been demonstrated to function as the fimbria-associated adhesins rather than the pilin structural protein (80, 81). A number of the *E. coli* fimbrial systems have been found to utilize a similar system.

Evidence is now mounting that some *E. coli* fimbriae do not have a separate adhesin protein and that the pilin structural subunit functions as the adhesin (66, 100). There is also evidence that for fimbriae possessing a separate fimbria-associated adhesin, there is substantial homology of the pilin structural subunit with the fimbria-associated adhesin and with afimbrial adhesins (1, 26). The methylphenylalanine (MePhe) or type 4 fimbriae produced by *Neisseria gonorrhoeae*, *Neisseria meningitidis*, *Pseudomonas aeruginosa*, *Bacteroides nodosus*, and *Moraxella bovis* are characterized by the presence of an unusual N-terminal amino acid consisting of *N*-methylphenylalanine and a region of extensive homology at the N-terminal region of the pilin (107). The MePhe fimbriae are perhaps the next-best-defined fimbrial system. The MePhe fimbriae to date have not been demonstrated to have a distinct fimbria-associated adhesin. For *P. aeruginosa* fimbriae, the pilin structural subunit contains an epithelial cell-binding domain that binds to the same receptor moieties and sites on human epithelial cells as do intact pili (28, 59). *N. gonorrhoeae* fimbriae recently were shown to have low copy numbers of fimbria-associated proteins (109), but an epithelial cell-binding domain has been identified in the pilin structural subunit (122, 129). Although negative experimental evidence does not prove a point, it appears that there are two general types of fimbrial systems, one in which the pilin structural subunit contains a binding domain and functions as the adhesin and one in which the adhesin is a discrete, low-copy-number, fimbria-associated protein.

Description of Fimbriae

Bacterial fimbriae are a helical assembly of a single protein termed pilin that forms a fiberlike structure. Pili vary considerably in diameter (2 nm for CS3 fimbriae [76] to over 12 nm [107]); some fimbriae have an axial hole within the structure (47), whereas others do not (132). The presence of an axial hole appears to correlate with the diameter of the fimbriae and the packing of pilin of monomers within the fimbrial filament. Morphologically, fimbriae vary considerably in that some are fairly rigid (the fimbriae are visualized by negative staining as straight unbending linear structures, as for *E. coli* type 1 fimbriae [14]), whereas others are much more flexible (the fimbriae, while generally straight in micrographs, are capable of bending or distorting without breaking, as is observed in *P. aeruginosa* [Fig. 1], and still others appear to be flexible and somewhat helical or corkscrewlike in structure, as for *Bacteroides gingivalis* 381 [149]). Fimbriae are remarkably stable structures that are resistant to both proteolysis and environmental insult, being stable to a wide range of pHs

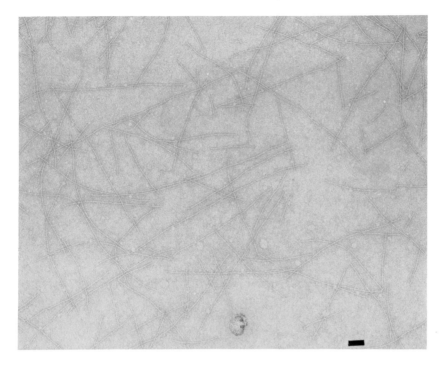

Figure 1. Negative stain of dispersed fimbriae of *P. aeruginosa* K. Bar = 100 nm.

and ionic strengths. Fine-structure analysis of bacterial fimbriae has been limited to a few fimbrial types, and as yet no high-resolution structure has been determined for any pilin or fimbrial structure.

Structure of fimbriae

Fine-structure analysis of bacterial fimbriae has been limited to a very small number of fimbrial types: *Bordetella pertussis* fimbriae (132), *E. coli* type 1 fimbriae (14), *E. coli* F pilus (fimbriae) (85), and *P. aeruginosa* fimbriae (47) (see reference 107 for a detailed discussion of fimbrial structure). These studies have indicated that the fimbriae are helical assemblies of pilin that spontaneously self-assemble into the defined morphological structure. The diameter, pitch of the helix, and whether there is an axial hole in the filament appear to depend on the particular pilin protein and its packing characteristics. The *B. pertussis* fimbriae have a diameter of ~5.5 nm, with a packing arrangement of pilin subunits that results in a pitch of 6.5 nm consisting of 2.5 repeating units per helical turn (132). This fimbrial structure varies somewhat from what appears to be a more typical packing arrangement of five repeating subunits per helical turn found in the F pilus (46, 85), the *E. coli* type 1 fimbriae (14), and the *P. aeruginosa* fimbriae (47, 107). The fivefold rotational symmetry found in these fimbriae is a common feature of a number of filamentous structures (85) and may prove to be a fairly constant structural feature of bacterial fimbriae.

Synthesis and assembly of fimbriae

Bacterial fimbriae are assembled from a prepilin that is normally processed during protein export to produce the final pilin (for the details of *P. aeruginosa* pilin synthesis, see references 40, 64, 65, 101, 110, 111, 126, 133, and 134). The synthesized pilin appears to be stored in the outer membrane of the cell before assembly into an intact fimbria (146). Growth or assembly of fimbriae occurs from the base of the fimbriae (79), consistent with the presence of pilin monomer pools in the outer membrane (146). It appears that for *P. aeruginosa* and *N. gonorrhoeae*, the fimbrial filament is actually assembled from dimers of pilin (109, 143, 145).

The assembled pilin proteins constituting the fimbriae are capable of presenting unique antigenic epitopes associated with the shaft and tip regions of the fimbriae as a result of packing constraints of the assembled pilin (148). Similarly, separate bacteriophage receptor sites reside on pilin molecules but are found to be surface accessible only in specific areas of the fimbriae (9–12). The length of bacterial fimbriae is variable and in the case of *E. coli* type 1 fimbriae is genetically regulated (67). For the MePhe structural class of fimbriae (the classification is based on the novel N-terminal amino acid residue that is found on these pilins, *N*-methylphenylalanine [49]), there is reasonable evidence that fimbrial length can be varied substantially by retraction or extension of the fimbriae (9–12), similar to the retraction ability of conjugative fimbriae (see reference 107). Fimbrial retraction has been associated with twitching motility and the erosion of agar surfaces by colonies of *Acinetobacter*, *Alteromonas*, *Bacteroides*, *Eikenella*, *Moraxella*, *Neisseria*, and *Pseudomonas* bacteria that possess this class of fimbriae (56, 105). Fimbrial retraction has thus far been established only for the MePhe class of fimbriae and conjugative fimbriae (107).

Structural variability in pilins

Bacterial cells frequently express multiple types of fimbriae simultaneously, and there is mounting evidence that several pathogens utilize phase variation or equivalent genetic manipulations to alter the fimbriae that are surface expressed at any one time on the bacterial cell surface

(19, 31, 50, 140). The variation of fimbrial structure in pathogenic bacteria is presumably due to substantial selective pressure from the host immune systems against a significant and critical surface antigen of the pathogen. It is generally possible to identify a number of immunologically related pilins for a given bacterial species. For several species, pilin serotyping schemes have been developed and used in epidemiological studies. A variety of strategies have been evolved by organisms to elude the selective pressure and yet maintain the functionality of the fimbriae.

N. gonorrhoeae demonstrates tremendous immunological variability in its fimbriae (55, 129, 138). N. gonorrhoeae cells have multiple genomic copies or versions of pilin that can be expressed (114). Whereas the immunodominant region of the N. gonorrhoeae pilin is varied considerably, portions of the pilin protein are immunologically conserved, presumably as a result of constraints for either adhesive function or assembly of the pilin subunits into intact fimbriae (122).

The strategy of varying the immunodominant region of the pilin protein while retaining conserved but nondominant immunologically conserved regions is also utilized by M. bovis and P. aeruginosa. The immunologically dominant region of P. aeruginosa pilin is found in the central region of the protein (127), whereas the N- and C-terminal portions of the protein are reasonably conserved in the different P. aeruginosa pilins (24, 107). This appears to be an effective strategy, since the N-terminal region is highly conserved as a result of processing and assembly constraints (101, 111). The C-terminal region is semiconserved (24, 107). Only antibodies directed against C-terminal antigenic epitopes either near or within the disulfide loop of P. aeruginosa pilin inhibit fimbria-mediated adhesion to human epithelial cells (28, 73). The C-terminal region of pilin has an epithelial cell-binding domain that mediates attachment to human buccal and ciliated tracheal epithelial cells, since synthetic peptides of the same sequence as the disulfide loop region of pilin bind to the same sites and receptors as do intact fimbriae (27, 59). The large majority of antibodies directed against the P. aeruginosa fimbriae do not interfere with the fimbrial adhesin.

M. bovis utilizes an interesting genetic inversion mechanism to vary the central immunodominant region of its pilins (50). Presumably, the M. bovis pilin is similar to the P. aeruginosa pilin, and these significant immunological changes (which would significantly alter the immune response of a host) have limited effects on the assembly or function of the intact fimbriae. The inversion mechanism described by Fulks et al. (50) thus offers an elegant mechanism for avoidance of a host response.

E. coli appears to use a significantly different strategy in dealing with host immune selective pressures. The extensive variety of fimbriae that E.

coli can utilize or express has presumably led to the development of phase variation mechanisms for high-frequency switching between the fimbrial type(s) expressed at a single time (31). The synthesis and production of fimbriae in *E. coli* also appear to be dependent on factors such as pH, nutrient status of the environment, and growth rate of the cells (141).

Fimbrial morphogenesis

A single bacterium can simultaneously possess more than one type of fimbriae; however, a single fimbria is (with very few exceptions) composed of a single type of pilin. The specificity of synthesis, processing, and assembly into intact fimbriae is thus extraordinarily specific. The association of the fimbria-associated adhesins with their appropriate fimbriae thus presumably requires a high degree of sequence or structural similarity so that the adhesin can become a functional part of the fimbriae. Substantial homology between fimbria-associated adhesins and the pilin structural proteins has been noted for both the tip-associated *papG* adhesin (26, 78, 80, 81) and the type 1 adhesin that is laterally associated with the filament of the type 1 fimbriae (1, 54, 70, 91). The degree of homology or structural similarity required for a protein to be processed and assembled into a fimbria is uncertain but is presumably considerable.

There is a high degree of conservation in both amino acid and DNA sequences among the various MePhe fimbriae (24, 107). The conservation in the MePhe fimbriae allows for the synthesis and assembly of intact fimbriae from MePhe pilins of various species in *P. aeruginosa* (6, 34, 35, 86). MePhe pilins have been cloned and expressed in *E. coli*, but to date pilin assembly into functional fimbriae has not been found, likely because of a basic difference in the processing and assembly of fimbriae in *E. coli* (40, 110). A number of other fimbrial processing and assembly systems will undoubtedly be recognized.

The MePhe fimbriae have provided considerable information concerning the nature of the specificity of the assembly of pilin monomers into intact fimbriae. *P. aeruginosa* and *N. gonorrhoeae* pilin monomers appear to initially form dimers, which are then polymerized into the final filamentous structure (109, 143, 145). The equilibrium constant for the association of pilin monomers appears to strongly favor the formation of dimers even in the presence of detergents. Crystallization studies with *N. gonorrhoeae* pilin indicate that the unit cell is composed of a dimer and that at a pH of >9.5, dimer formation and polymerization are strongly inhibited (109). The purification of *N. gonorrhoeae* pilin takes advantage of a depolymerization of the fimbriae at pH 9.5, which suggests that one or more lysine or tyrosine (or potentially histidine) residues play significant roles in the pilin-pilin assembly process (109).

Figure 2. Crystals of *P. aeruginosa* PAK pilin crystallized from a detergent solution of pilin dimers. (Courtesy of R. Read, University of Alberta.)

P. aeruginosa pilin also forms very stable dimers (even in the presence of detergents it is very difficult to obtain any significant concentration of dissociated monomers), and a role for tyrosine residues in pilin-pilin assembly has been suggested (143, 145). Interestingly, *P. aeruginosa* PAK pilin readily crystallizes from a solution of dimers (Fig. 2). There is considerable sequence homology in both the N- and C-terminal regions of MePhe pilins (24, 107), which raises the possibility that both regions are conserved because they are involved in subunit-subunit interactions (24, 107). An interesting feature of the MePhe pilins is that the C-terminal portion of the protein is semiconserved and contains an intrachain disulfide bridge that forms a loop structure. Reduction of the disulfide bridge results in at least partial disassembly of fimbriae into dimers and monomers of pilin (unpublished observations from this laboratory) and reduced affinity of a synthetic peptide containing the *P. aeruginosa* PAK fimbrial epithelial cell-binding domain (60). This finding suggests that both the C- and N-terminal regions of pilin are involved in subunit-subunit interactions.

There is now evidence to support a role for the C-terminal region of pilin in the pilin assembly process in addition to the role in adhesion. A

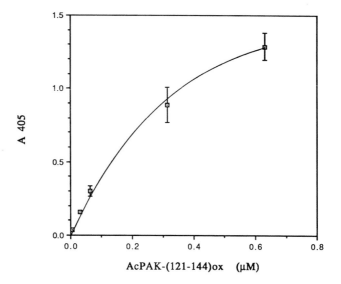

Figure 3. Enzyme-linked assay of the binding of the synthetic peptide AcPAK-(121-144)$_{ox}$ (which constitutes the 24 C-terminal amino acid residues of the PAK pilin protein, with the intrachain disulfide bridge being in the oxidized state [i.e., the bridge is formed]) to immobilized PAK fimbriae. The K_m of binding of AcPAK-(121-144)$_{ox}$ to PAK fimbriae is 0.25 μM.

synthetic peptide consisting of the 24 C-terminal amino acid residues of PAK pilin binds with high affinity (K_m = 0.25 μM) to intact PAK fimbriae (Fig. 3). There is clearly a domain in this synthetic peptide that likely constitutes part of the subunit interfacial region. This domain appears to be primarily on the N-terminal side of the intrachain disulfide bridge because a synthetic peptide consisting of the terminal 17 amino acid residues (the intact disulfide loop region) does bind to pili, but substantially less effectively (the maximal amount of binding to PAK fimbriae is 10-fold lower than that of the larger peptide).

The assembly of pilin monomers into an intact fimbria is a very interesting process, since it produces an oriented fiber that does not have equivalent ends or tips. Dispersed PAK fimbriae do not agglutinate epithelial cells, which indicates that the adhesin function of the fimbriae is localized to only one end of these structures. Immunochemical localization studies with monoclonal antibodies PK3B and PK99H have established that both antibodies recognize tip-localized epitopes (Fig. 4). Neither antibody binds to the sides of the fimbrial filament, but both are highly specific for *P. aeruginosa* PAK pilin (28). These results indicate that during the assembly process, the C-terminal region is exposed at only one tip of the fimbriae and that normally this region of PAK pilin is not accessible to antibodies as a result of its involvement in subunit assembly. It is tempting to speculate that the base of the fimbriae may express a unique antigenic epitope recognized by monoclonal antibody PK3B. However, it remains to be determined whether monoclonal antibodies PK3B and PK99H bind to the same or different ends of the fimbriae.

Specificity of the fimbrial assembly process

Cloning of the *B. nodosus* pilin structural gene into *P. aeruginosa* results in the expression and assembly of functional fimbriae that are immunologically indistinguishable from those synthesized by *B. nodosus* (34, 37). An immunocytochemical study of *P. aeruginosa* cells that were capable of synthesizing both the *P. aeruginosa* pilin and a *B. nodosus* pilin indicated that generally the *P. aeruginosa* fimbriae are inhibited but that with partial induction of the *B. nodosus* pilin, cells with both types of fimbriae could be observed (35). Elleman and Peterson (35) observed hybrid fimbriae (~0.1% of the fimbrial population) in which the *Pseudomonas* and *Bacteroides* pilin subunits were observed in homogeneous but discrete segments of a single fimbria. Elleman and Peterson (35) suggested that only one pilin is synthesized at any time and is thus incorporated into functional fimbriae as a homogeneous population.

Recently, Beard et al. (6) examined the expression of *M. bovis* fimbriae in *P. aeruginosa* cells and observed in rare fimbriae that heterologous subunits were mixed and not necessarily found in homogeneous discrete segments of the fimbriae. Beard et al. (6) have suggested that the pilin-pilin assembly process is very specific, that the affinity constant for homologous pilin assembly is greater than that of heterologous pilin assembly, and that heterologous pilin-pilin assembly occurs as a result of fimbrial extension using heterologous pilin pools formed from previous retraction of segments of both pilin types.

HYDROPHOBICITY OF BACTERIAL FIMBRIAE

Bacterial pilins in general are fairly hydrophobic, with hydrophobic amino acids constituting ~30% or more of the pilin protein. Dalrymple and Mattick (24) have suggested that a considerable portion of the hydrophobic amino acid residues constitute a basic secondary-structure framework for the organization of the pilin protein similar to the framework-region structures of immunoglobulins (25). Secondary-structure predictions based on deduced amino acid sequences for the pilins of *Actinomyces viscosus* T14V and *Streptococcus sanguis* FW213 suggest

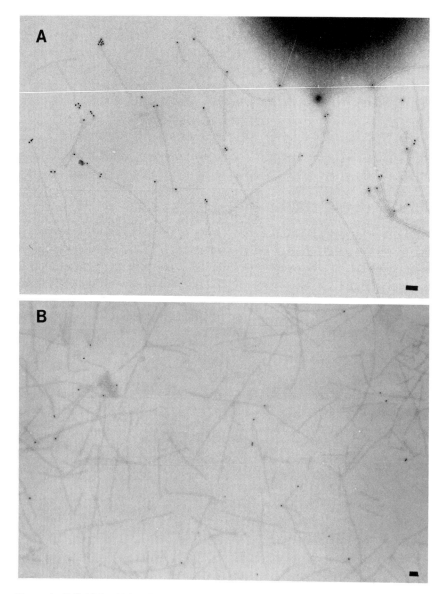

Figure 4. Colloidal gold localization of the PAK fimbrial antigenic epitopes recognized by the murine monoclonal antibodies PK3B (A) and PK99H (B) and a control (C). PK99H and PK3B both recognize a tip-associated fimbrial antigenic epitope found only at one end of the fimbriae. Both monoclonal antibodies are specific for PAK fimbriae (28), but only PK99H inhibits *P. aeruginosa* whole-cell binding or PAK fimbrial binding to human epithelial cells (28). PK99H specifically binds to an antigenic epitope located in the C-terminal region of PAK pilin (28). Bars = 100 nm.

Figure 4—*Continued*

that a common framework structure for pilins of specific families may be a general phenomenon (39). The hydrophobic amino acid residues of pilin, however, are not completely buried; in fact, bacterial fimbriae are reasonably hydrophobic structures. The surface hydrophobicity of bacteria is increased when fimbriae are expressed (17, 57, 87, 142). The increased bacterial cell surface hydrophobicity is particularly associated with the *E. coli* type 1 fimbriae (14, 63, 102).

The dramatic increase in cell surface hydrophobicity associated with the type 1 fimbriae is presumably associated with pellicle formation that was originally noted as an aspect of fimbriation (102). Aside from the tendency toward pellicle formation (likely due to the interaction of the hydrophobic fimbriae with the air-liquid interface and with cell-cell association [52]), increased cell surface hydrophobicity also results in increased binding to polystyrene (121), enhanced interaction with hydrophobic matrices (36, 131), and cell agglutination or salting out in high-salt solutions (77). Elevated surface hydrophobicity has also been associated with a number of cell surface interactions, including increased binding to mammalian epithelial cells (17, 63, 142) and yeast cells (63, 68, 69).

Aggregation of a number of bacterial fimbriae in paracrystalline bundles occurs in relatively high salt solutions that are of appropriate pH (14, 109, 132). Aggregation of bacterial fimbriae occurs to a certain extent for most fimbriae (Fig. 5). There appear to be a specific pH optimum and optimal salt concentration for each fimbria in order for fimbrial self-association to occur. Fimbrial aggregation appears to require an appropriate fimbrial surface charge and to be mediated by a hydrophobic effect. Fimbrial aggregation may also explain in part why purified fimbriae that have a tip-associated fimbrial antigen (presumably localized solely on one tip of a fimbria) readily agglutinate erythrocytes possessing the appropriate cell surface receptor structures. Aggregated fimbriae could then interact with more than one erythrocyte.

NATURE OF THE HYDROPHOBIC EFFECT

It is relevant to discuss the nature of the hydrophobic effect and its implications for the structure and function of bacterial fimbriae. Generally, the hydrophobic effect is considered to be a nonspecific interaction that is driven by solvent or hydration energetics (see Duncan-Hewitt, this volume). Although this is certainly true, a number of other factors can be and frequently are involved. It is also important to realize that a hydrophobic environment does not preclude interactions that are normally associated with a hydrophilic environment (e.g., H-bond formation or dipole-dipole interactions). A hydrophobic environment also need not be extensive in size or well isolated from polar environments and may in fact be found as regions or domains on the surfaces of single proteins. Similarly, hydrophobic domains may be grouped or ordered such that there is a discrete three-dimensional pattern that is hydrophobic, and thus specific "lock and key" interactions may be generated by a "nonspecific" mechanism.

Energetics

Solvent structure and energetics

The energy required to remove one molecule of H_2O from pure liquid at 25°C is 10.5 kcal (1 kcal = 4.184 kJ)/mol, which is a reflection of both the number of H bonds that one H_2O molecule is involved in and the dipole-dipole interactions between neighboring H_2O molecules (72). Water is a small molecule (radius of <1.5 Å [1 Å = 0.1 nm]) that has a rotational relaxation time of 2 to 3 ps (72), which is ample time for the formation and breakage of H bonds, which requires ~1 ps (120). Thus, physically water is an H-bonded and electrostatically interacting transient network of molecules whose individual components rapidly shift positions. Indeed, H_2O molecules further apart than ~8 to 10 Å do not

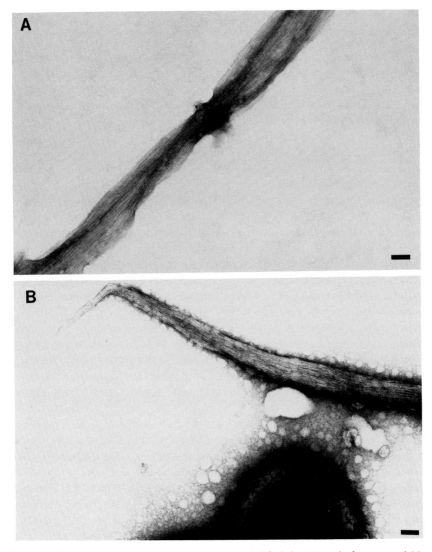

Figure 5. Negative stain of *P. aeruginosa* K aggregated fimbriae (A) and of aggregated *M. catarrahlis* fimbriae (B). Note that the fimbriae align themselves longitudinally and that the interaction between fimbriae occurs continually along the length of the filament. The extent of fimbrial aggregation can be extensive in high salt concentrations. Bars = 100 nm.

statistically interact but rather behave as though they were independent molecules (72, 120).

Hydration energetics

When a protein is dissolved in an aqueous environment, it is to a certain extent hydrated and thus interacts with a number of H_2O molecules. Adsorption isotherms clearly indicate that proteins bind H_2O molecules to form a hydration shell of water. The H_2O that is bound to proteins is almost exclusively surface associated, since X-ray diffraction studies have indicated that the interiors of proteins are generally devoid of water (71, 72). The amount of H_2O bound on the surface of a protein is not uniform; ionic amino acid residues may bind up to about six H_2O molecules per residue, compared with about one molecule per polar amino acid residue, whereas hydrophobic amino acid residues do not bind any H_2O molecules (72). A thermodynamic analysis of the hydration of a number of proteins indicates that both enthalpy and entropy components are involved in determining hydration, with the enthalpy component giving rise to the most significant change in the free energy (~ -11 to -15 kcal/mol of H_2O bound). A rigorous thermodynamic treatment of the hydration of proteins examines both the enthalpy and entropy contributions of all of the solvent-accessible amino acid residues and deals with the specific amino acid residues individually (120). Intriguingly, a rigorous thermodynamic treatment of hydration also includes an additional term for a surface tension factor (72, 120), a reflection of the physicochemical properties of water. The degree of hydrophobicity of a protein will thus affect the solubility of the protein and its degree of hydration in an aqueous solution.

The surfaces of most proteins should be viewed as mosaics consisting of (i) surface regions (i.e., solvent-accessible areas of the protein or regions where the folding of the protein does not exclude the accessibility of the solvent) in which interactions with H_2O occur (i.e., polar regions or hydrophilic domains) and (ii) surface regions in which H_2O-protein interactions are minimal (i.e., nonpolar regions or hydrophobic domains) and solvent-solvent interactions are maximized. The degree of hydration of a protein is very susceptible to the environment, since a change in surface charge of a protein or the binding of charged species to the surface of a protein will change the amount of water bound to the protein surface. Increasing the salt concentration of a buffer generally results in increased salt binding to the protein, which in turn results in significantly less water being bound to the protein (71), causing an apparent increase in the hydrophobicity of the protein. High salt concentrations thus maximize hydrophobic interactions. However, an increased salt concentration also modifies the bulk water structure significantly by altering the lattice of H bonding found in water, thus altering the surface tension of the solution. Molecular interactions that are ionic in nature are strongly dehydrating in nature and result in significantly less water in the hydration shell of the protein (71). Similarly, H bonding between solutes reduces the amount of water in the hydration shell whereas hydrophobic interactions between solutes have limited effects, since those regions of the protein would not have been interacting with water.

Hydrophobic effect

The hydrophobic effect or hydrophobic "bond" arises from the energetics of interaction of two hydrophobic solutes in an aqueous environment such that interactions between the hydrophobic portions of the solutes and the water are minimized (see Duncan-Hewitt, this volume). Hydrophobically mediated interactions would be expected to occur initially as random collisions whose frequency is dictated by the concentrations of the two solutes, their diffusion rates, and the temperature of the system. The fate of such random collisions would be determined by the hydrophobic effect; i.e., the enthalpy contribution from the avoidance of interaction of the hydrophobic domains with water would counter the entropy component of the interaction.

For Tanford's early description of the hydrophobic effect (136, 137), a thermodynamic approach was used to analyze this very interesting type of interaction; his analysis clearly demonstrated that the free-energy change associated with the interaction is proportional to the surface area of the hydrophobic regions. The free-energy change associated with this type of interaction arises largely from occluding water from these regions, thus minimizing the interaction between water and the hydrophobic portions of the solutes. This type of interaction is thus viewed as a nonspecific type of interaction (since there need be no obligate interaction of the two solutes through the classical bonding). Similarly, a hydrophobic effect is viewed as a flexible bond that allows for ready deformation and yet will maintain the integrity of the interaction, a bond that is very useful in generating flexible and dynamic biological structures (137). This approach maximizes the solvent contribution and neglects the interaction of the solutes, which can be quite significant.

In a hydrophobic environment, one may still have the full range of interactions between solutes. There has been a heavy emphasis on the van der Waals interactions in the hydrophobic effect, since (i) there are no requirements for specific structural features in order for the components to interact and (ii) the energetic contribution is proportional to the surface area of the components that interact. Van der Waals interactions or the London dispersion forces arise from the transitory induction of dipoles in atoms as a result of the movement of electrons within their orbits. These forces are significant only over very limited distances, and the strength of the interaction rapidly decreases in intensity as a function of $1/d^6$ (where d is the distance between the two atoms) and are functionally negligible beyond ~5 Å (16).

There is a tendency to overlook the other interactions that occur in a hydrophobic environment and which may greatly enhance the interaction in a highly specific manner. A number of interactions, such as the H bonding between the hydroxyl groups of carbohydrates, exclude water from the interaction site, and the appropriate amino acid residues in the binding domains of the appropriate lectins (both as donors and as acceptors for H-bond formation) effectively exclude water from the binding domain to generate a hydrophobic domain or interaction site (116). The exclusion of water from the interaction site yields an energy contribution to the interaction as a result of a hydrophobic effect. Water need not be totally excluded from an interaction mediated by a hydrophobic effect; indeed, one may have a bound water molecule in the interaction site and still have a hydrophobic effect (4). Similarly, ionic interactions between positively and negatively charged groups on the surfaces of two proteins result in a substantial decrease in the amount of water bound to the proteins, increasing their net hydrophobicity and thus providing a further energy contribution to the interaction through the hydrophobic effect (4). Both H bonding and ionic interactions occur readily in a hydrophobic environment and thus may further enhance the favorable energetics of what was initially a hydrophobic interaction. Similarly, dipoles of polar groups will interact in a hydrophobic milieu and thus can readily potentiate an initial hydrophobic effect. Clearly, electrostatic interactions or H bonding can potentiate hydrophobic effects and vice versa, resulting in significantly more favorable or specific molecular interactions (i.e., increasing the apparent association constant of the interaction for components that are complementary in nature). It is also prudent to recall that because of molecular motion any structure is dynamic and that molecular interactions are not necessarily static.

Factors affecting a hydrophobic effect

The magnitude of a hydrophobic effect that mediates an interaction involving proteins depends not only on the surface area of the hydrophobic domain (4, 103) but also on the particular amino acid residues that constitute that domain (120) and the salt concentration of the environment. The surface area of the solvent-accessible hydrophobic domain will largely define the magnitude of the solvent contribution to any interaction mediated by a hydrophobic effect for obvious reasons. The surface area of the hydrophobic domain of a protein can be modified by conformational changes in proteins, resulting in significant changes in the potential for a hydrophobic effect-mediated interaction (150). The amino acid residues that constitute the hydrophobic domain influence the strength of the hydrophobic effect because the polarities of the various amino acid side groups in their particular microenvironments vary significantly (4). The variable polarity of the side chains of the amino acid residues that constitute the hydrophobic domain will modify the energy contribution of the hydrophobic effect significantly as a result of the variability in chemical potential of hydration (120). Increasing the salt concentration generally dehydrates a protein (71), resulting in an increased surface area for hydrophobic domains. Thus, higher salt concentrations should increase the magnitude of hydrophobic interactions. Changes in the surface tension of the environment will also affect the hydrophobic effect as a result of changes in the bulk structure of the water (either by disrupting the H-bonding network of the bulk water or by minimizing the dipole-dipole interactions of the water). Thus, decreasing the surface tension of the aqueous solution will decrease the magnitude of the hydrophobic effect.

Specificity and the Hydrophobic Effect: Lessons from Protein Structure

Protein chemistry and our understanding of hydrophobic effects have been considerably enhanced by high-resolution structural determination of a wide variety of proteins as a result of the ability to determine which molecular interactions are occurring in the protein crystal. X-ray diffraction studies have revealed that the interiors of proteins are generally devoid of water; in fact, the peptide chains are packed so closely that water cannot penetrate past a definable surface region of the protein (62). Thus, the interior regions of proteins are hydrophobic in nature and are enriched in hydrophobic amino acid residues relative to the surface regions or solvent-accessible regions of the proteins (62). There is actually a very strong correlation between the mass of a protein and its solvent-accessible surface area ($A_s = 6.3M_r^{0.73}$ for monomeric proteins and $A_s = 5.3M_r^{0.76}$ for oligomeric proteins [A_s is the solvent-accessible surface area in square angstroms], the difference reflecting the subunit interfacial domains), which indicates that the hydrophobic core of the protein plays a major role in protein function (62). It is becoming apparent that a hydrophobic environment is required to induce protein secondary structure, particularly for helix formation (to eliminate solvent-protein side group H bonding and allow intrachain H bonding to generate a helix) (150). In fact, a hydrophobic environment has been observed to induce a conformational change in a peptide, resulting in the formation of an amphipathic helix that can interact with the hydrophobic component of a reverse-phase column (150). Thus, it is likely that the protein hydrophobic core allows for the correct folding of the peptide chain with the appropriate secondary structure to generate the final protein conformation. As the size of the hydrophobic core of a protein increases, one would expect that because of the hydrophobic effect, the protein may in fact be more stable. Hodges et al. (R. S. Hodges, N. E. Zhou, C. M. Kay, and P. D. Semchuk, submitted for publication) have provided experimental evidence that the greater the hydrophobicity of a protein, the greater its stability. Janin et al. (62) observed that cysteine residues (and thus high numbers of disulfide bridges) are frequently found in the interiors of small proteins, whereas larger proteins generally do not have buried disulfide bridges. This observation is consistent with the greater hydrophobic interior of the protein greatly stabilizing the structure; in fact, there would be a hydrophobic contribution of ~25 cal/mol per Å^2 of accessible surface that is buried (there would be ~32 Å^2 of surface area buried per amino acid residue in the interior of the protein) (62). Thus, the hydrophobic contribution to the energetics of folding and stability of a protein conformation can be extensive.

Whereas an interaction mediated by a hydrophobic effect is considered to be nonspecific, there are numerous examples indicating that hydrophobic effect-mediated interactions can be extremely specific high-affinity interactions. Oligomeric proteins provide some of the best examples of the potential specificity of hydrophobic interactions. Mismatched oligomeric proteins are extremely rare in cells, and unassembled monomers are rarely detected in cellular extracts. Analyses of numerous X-ray diffraction studies indicate that the subunit-subunit interactions are primarily due to hydrophobic interactions in well-defined interfacial regions or domains and that the interfacial area involved is >700 Å^2 of previously solvent-accessible surface area, which is large enough to stabilize the interaction and drive the assembly of subunits (62). Janin et al. (62) also observed that H bonding (generally involving charged species), ionic salt bridges, and other polar interactions contribute significantly to the stabilization of subunit-subunit interactions. In general, there is a linear relationship between the size of the interfacial domain and the number of H bonds that are formed (62). The interfacial domains differ significantly in amino acid composition from both the interior and the surface of the proteins and are found to be enriched for the amino acid residues leucine, arginine, isoleucine, valine, cysteine, and methionine, with the arginine residues being involved in ~42% of the polar interactions (62). Thus, the high specificity and affinity of the subunit interactions found in oligomeric proteins are due primarily to a hydrophobic effect, with significant contributions from H bonding, salt bridges, and other polar interactions. Clearly, a complementary structure consisting of hydrophobic domains with charged species, a structure reminiscent of a "lock and key," is necessary to mediate the specific and high-affinity assembly of oligomeric proteins.

MOLECULAR BASIS OF PROTEIN-MEDIATED INTERACTIONS

Somewhat surprisingly, relatively little is known about the molecular basis for the interactions of proteins or, more specifically, of fimbriae (or fimbria-associated adhesins) with their specific receptors or ligands, which explains in part why rational drug design has not yet borne substantial results. In general, structural studies on lectins cocrystallized with their appropriate sugars, immunoglobulins cocrystallized with their appropriate antigens, and proteases cocrystallized with various inhibitors suggest several interesting hypotheses concerning molecular interactions to be synthesized and tested in more complex systems.

Lectins

Lectinlike saccharide-binding specificities have been associated with bacterial fimbriae in that their specific receptors appear to have very specific saccharide components, although other receptors appear to be primarily proteinaceous in nature. The best details concerning the molecular basis for these protein-mediated interactions have been derived from X-ray crystallographic studies of various lectins. It is likely that at the very least some of the features of the binding domains of already characterized lectins will be found as features of the fimbrial or fimbria-associated adhesins.

Quiocho (116) observed that arginine residues are frequently found in the carbohydrate-binding domains of lectins in addition to a reasonably high frequency of charged and aromatic amino acid residues. It is interesting to note that the carbohydrate-binding domains are normally clefts with a few hydrophobic amino acid residues that are generally not directly interacting with bound carbohydrate. There are normally a number of van der Waals interactions and H bonds between aromatic amino acid residues and bound carbohydrate (116). The binding of the carbohydrate to the cleft in the lectin then facilitates the formation of a significant hydrophobic domain that is stabilized not only by van der Waals interactions but also by H bonding and dipole-dipole interactions.

Refined structural studies of carbohydrate-lectin interactions have revealed that there is an extensive network of H bonds, both as acceptors and as donors, between the bound carbohydrate and the numerous amino acid side chains that constitute the carbohydrate-binding domain of the lectin (116). It is particularly intriguing that water is effectively excluded from the bulk of the binding domain once the carbohydrate is bound to the lectin, with carbohydrate being made largely solvent inaccessible. This provides a significant thermodynamic contribution to the interaction through a hydrophobic effect (116). The specificity of a sugar binding to its appropriate lectin appears to be due to the interaction of complementary structures which are capable of forming numerous H bonds (with the carbohydrate acting as both donor and acceptor) such that water is largely excluded from the binding domain, since all of the water hydration shell has been exchanged in order to form H bonds and van der Waals-mediated interactions.

H bonding (both as acceptors and as donors) with both charged and uncharged groups in lectin-carbohydrate interactions has been confirmed by binding studies using derivatized carbohydrates (i.e., fluoro, deoxy, and thio sugars) (7). Thus one would expect to find very extensive H bonding between the fimbrial adhesins and their appropriate receptors in a binding domain that would be reasonably hydrophobic. One would also anticipate that factors influencing hydrophobic interactions would influence the lectinlike binding activity of bacterial fimbriae, and this is in fact generally the case.

Immunoglobulins

Immunoglobulins are one of the more interesting classes of proteins, since their primary function is to specifically interact with unique three-dimensional structures. It is now reasonable to assume that virtually any surface-accessible region of a protein can serve as an antigenic epitope (25), yet the specificity of a particular antibody for its respective antigenic structure is impressive. High-resolution structures have been determined for several Fab fragments of monoclonal antibodies cocrystallized with their appropriate antigens. The amino acid residues of immunoglobulins that are found in the complementary determining region of antibodies are quite distinct from those found in other regions of the antibody and are enriched for aromatic amino acid residues, histidine, asparagine, and arginine (106). Padlan (106) presents considerable evidence (based on the refined structures of a number of antibody-antigen complexes) for the hypothesis that these amino acid residues may be found in higher frequencies as a result of their ability to interact with antigens through multiple mechanisms (i.e., H bonding, ionic pairing, dipole-dipole interactions, and van der Waals interactions). Padlan (106) has also suggested that because of the relatively large size of the aromatic side chains, interactions with an antigen or hapten can occur through a wider range of angles and slightly differing positions of antigen binding to the antibody. Antigen binding to antibodies is obviously facilitated by a hydrophobic effect contribution (both from a solvent and from van der Waals interactions). The hydrophobicity of most immunoglobulins leads to a general "stickiness" of antibodies (the nonspecific antibody interactions due to hydrophobic effects are generally eliminated in most assays through the use of high concentrations of bovine serum albumin or by the use of detergents), and thus the specific interactions must involve a complementary three-dimensional structure, ionic interactions, and H-bond formation between the antigen and antibody.

Protease Inhibitors

Analysis of the interaction of proteases and their polypeptide inhibitors provides a useful insight into the molecular basis of protein-protein interactions. Conceptually, the function of several proteinase inhibitors is to inhibit as many proteases as possible in order to protect sensitive components from degradation or microbial attack. Obviously, the various proteases vary considerably in their three-dimensional structures, particularly in the active sites, yet their active sites have many structural features in common (119). The protease inhibitors thus utilize these similar features to bind to and inhibit the proteases. I would suggest that the interaction of fimbrial or fimbria-associated adhesins with their appropriate receptors is an analogous system in that there are a limited set of potential cell surface receptors present on the surface of epithelial cells with which fimbriae may interact. I would also suggest that individual variation of these cell surface receptors requires that the fimbrial adhesins be somewhat versatile with respect to the nature of the receptors to which they can bind, extending the analogy to the proteases and their inhibitors.

The molecular details of protease-protease inhibitor interactions are largely consistent with the observations found for antibody-antigen interactions and for lectin-carbohydrate interactions. The molecular interaction of a protease with an appropriate inhibitor is driven to a certain extent by a hydrophobic effect (with energy contributions from both solvent effects and van der Waals interactions) but also by ionic interactions, H bonding, and dipole-dipole interactions (119). In particular, aromatic amino acid residues (tyrosine and phenylanine in particular) appear to play a significant role in mediating contact (through both H bonding and van der Waals interactions) (119). In addition, the aromatic amino acid residues do not pack as well and thus restrict the mobility of the protease inhibitors (which leads to better inhibitor activity) while still allowing for a reasonable amount of contact surface (119). Padlan (106) has suggested that the H-bonding potential and the increased surface area of a number of the aromatic amino acid residues allow for a greater number of potential interactions yet can facilitate specificity because of their poor packing characteristics. It is clear that the interaction of a protease inhibitor with a protease is mediated by both a hydrophobic effect and polar interactions.

ROLE OF FIMBRIAE IN BACTERIAL ADHESION

Bacteria adhere to an extraordinary wide range of tissues and surfaces during their life cycles in their natural environments (22, 23). It is now clear that fimbriae are centrally involved with the adhesion mechanisms used by most bacteria to attach to viable cell surfaces. Fimbriae are now believed to play a limited role in mediating attachment to inert surfaces (61); capsular components have been ascribed the central role in mediating attachment to inert surfaces (43–45, 83, 84, 89, 90) and a minor role in mediating attachment to epithelial surfaces (29, 117). Our understanding of fimbrial involvement in adhesion to inert surfaces is limited and needs to be reevaluated. Fimbriae are reasonably hydrophobic structures, and the significance of hydrophobic interactions in mediating bacterial attachment to surfaces is well established (2, 3, 18, 95, 121).

ROLE OF FIMBRIAE IN ADHESION TO SOLID SURFACES

The initial colonization of a solid surface is thought to be mediated by polymer cross-bridging (the capsular components) of the bacterial cell surface with the inert surface (43–45, 84, 89, 91). The adhesion of bacteria to surfaces has been modeled on the basis of either surface tension effects (2, 3, 18) or the DLVO model, which is based on early studies in colloidal chemistry (18, 123). The DLVO model, however, has several significant difficulties, the primary one being that van der Waals forces are ascribed a long range (> >10 Å in distance; see reference 123) whereas they are significant only for very short distances (<6 Å) (16). There is a significant

Table 1. Affinities of *P. aeruginosa* PAK Fimbriae
for Various Surfaces[a]

Surface	K_m (μg/ml)
Stainless steel	0.33
Polystyrene	21
Bovine serum albumin	25.6

[a] PAK fimbriae were suspended in phosphate-buffered saline, pH 7.2. The two antibodies tested (PK99H and PK34C) had no effect on fimbrial binding. These antibodies specifically inhibit the PAK fimbrial epithelial cell-binding domain (28, 59).

role for polymer cross-bridging in both models, and a role of fimbriae in this process has not been closely examined.

The role of fimbriae in mediating attachment to solid surfaces should be anticipated because (i) they have a widespread distribution, (ii) they are exposed at the surface of the cell and extend beyond the capsule, and (iii) they are reasonably hydrophobic. My laboratory has recently examined the role of fimbriae in mediating attachment of *P. aeruginosa* to stainless-steel and polystyrene surfaces. *P. aeruginosa* fimbriae bind to both stainless steel and polystyrene in a time-dependent, saturable manner with defined but differing affinities (Table 1). We have also established, on the basis of monoclonal antibody inhibition of fimbrial binding, that the fimbriae mediate attachment to stainless steel and polystyrene through a domain or region of the pilin structural protein that differs from the epithelial cell-binding domain used to mediate attachment to human respiratory epithelial cells.

The mechanism of *P. aeruginosa* fimbrial binding to stainless steel is now under investigation, but our initial studies suggest a role for both an electrostatic effect and a hydrophobic effect. Fimbrial binding is significantly enhanced at high salt concentrations (as a result of increasing affinity for the stainless steel) and relatively insensitive to the presence of methanol (which significantly alters the surface tension of the environment), which indicates a significant hydrophobic effect (Table 2). Fimbrial binding to stainless steel is relatively insensitive to pH. However, above pH 9.5 binding to the stainless steel is significantly inhibited (because of decreased affinity for the stainless steel), suggesting that one or more charged lysine, tyrosine, or histidine residues are required for high-affinity fimbrial binding to stainless steel (Table 3). These observations, which suggest both a hydrophobic effect-mediated and a polar effect-mediated interaction, are consistent with what has been observed for numerous protein interactions.

Bacterial fimbriae should be expected to play a significant role in

Table 2. Effect of Salt and Methanol Concentrations on
P. aeruginosa PAK Fimbrial Binding to Stainless Steel
as Determined by a Solid-Phase Immunoassay

Concn		Relative	K_m
Salt (mM)	Methanol (%)	binding (%)	(µg/ml)
0			106
1			61.9
5			2.88
10			0.87
5	0	100	ND[a]
5	1	104.5	ND
5	5	96.5	ND
5	10	87.9	ND
5	20	89.2	ND

[a] ND, Not done.

mediating attachment to solid surfaces, and they may have evolved to mediate attachment not only to viable cells but also to the particular surfaces associated with the environmental niche that the organism fills. Clearly, *P. aeruginosa* fimbriae have at least two separate domains for mediating attachment, one for epithelial cells (59) and one for solid surfaces. It is likely that other organisms have evolved similar strategies, and the roles of various fimbriae in mediating attachment to a variety of surfaces should be further examined. The mechanism of attachment to solid surfaces will likely be found to involve polar as well as hydrophobic interactions.

Role of Fimbriae in Bacterial Adhesion to Cell Surfaces

Bacterial fimbriae mediate attachment to a variety of cells, including plant roots (53, 69), fungal cells (68), and diverse mammalian cells (5, 104). The initial tissue specificity of bacterial colonization noted by Gibbons (51) appears to be determined in large part by fimbrial receptor

Table 3. Effect of pH on the Affinity of
P. aeruginosa PAK Fimbriae for Stainless Steel

pH	K_m of fimbrial binding (µg/ml)
6.5	2.78
7.2	2.88
8.0	5.13
9.4	30.75

specificity and receptor distribution in tissues (8). The significance of the adhesion of a pathogen to an epithelial surface as the initial stage of an infection is now established (139), and vaccines directed against fimbriae are effective in preventing colonization and homologous infection (113, 128, 135).

There is considerable diversity in both the fimbriae and the receptor moieties that the fimbriae recognize, but many fimbriae have a lectinlike activity in that they demonstrate marked specificity for carbohydrates (74, 96, 130). Significantly, the recognition of D-mannose residues by E. coli type 1 fimbriae can be readily inhibited by a number of hydrophobic aromatic hydrocarbons (38). This finding indicates that there is a significant energetic contribution to the binding to the sugar from a hydrophobic effect. The high degree of specificity of the type 1 fimbriae for D-mannose residues (41, 42) also indicates that there is a high degree of complementation between the fimbrial adhesin (the type 1 fimbrial adhesin is a separate protein that is laterally assembled into the fimbrium [54, 70, 91]) and the sugar in order for the carbohydrate to be bound (likely through H bonding, van der Waals interactions, and a solvent-dependent contribution). The type 1 fimbrial lectin-binding domain likely consists of a fairly large hydrophobic cleft (130) that allows for both a hydrophobic and a polar contribution to the interaction with the appropriate receptors.

For at least one variant of the E. coli type 1 fimbriae, the mannose residues of human and animal mucins can serve as appropriate receptors (124, 125). Sajjan and Forstner (124, 125) also found that there is a significant hydrophobic contribution to the interaction of the type 1 fimbriae and mucin in addition to the specific interaction with the carbohydrate residues. These authors have suggested that the hydrophobic effect stabilizes the interaction of the fimbriae with the carbohydrate residues and that the hydrophobic effect-mediated interaction may actually involve interaction of the fimbriae with the mucin link protein. These studies are somewhat complicated by the facts that type 1 fimbrial adhesin is a separate entity from the pilin structural subunit and the interaction of the adhesin with the pilin subunit is poorly understood. Studies with P. aeruginosa whole cells and with purified P. aeruginosa PAK fimbriae indicate there is a relatively low affinity interaction with human mucins. The K_ms for PAK fimbrial binding to purified human sputal or small intestinal mucin is 24 or 22 μg/ml, respectively, and is of approximately the same affinity as PAK fimbrial binding to bovine serum albumin (K_m = 26 μg/ml) or polystyrene (K_m = 37 μg/ml) (unpublished observations from this laboratory). Bovine serum albumin does not inhibit P. aeruginosa whole-cell binding to human buccal epithelial cells or the binding of purified PAK fimbriae to human buccal epithelial cells (unpublished observations from this laboratory; J. Forstner, personal communication). These results indicate that the PAK fimbrial interaction with mucin is primarily a low-affinity hydrophobic effect.

While many fimbrial receptors are, or at least contain, carbohydrate components, there are a few examples of receptors that are proteinaceous. The *A. viscosus* type 1 fimbriae mediate attachment to saliva-coated hydroxylapatite (21) and more specifically to the proline-rich proteins of saliva that bind to the hydroxylapatite (20). The receptor moiety for the fimbriae is a conformation-dependent domain of the proteins that serves as a receptor only when the proteins are bound to a surface (75). The *E. coli* Dr fimbriae also recognize a protein receptor, the Dr blood antigen on the surface of erythrocytes (66, 97–100). The Dr fimbrial adhesin is apparently the pilin structural protein, and the receptor appears to include a tyrosine residue of the Dr blood antigen (98).

The molecular details of fimbrial adhesion to epithelial cells and the basis of the receptor specificity have not been well established because of the complexity of working with rather poorly defined polymers and very complex epithelial cells or epithelial surfaces. The *P. aeruginosa* fimbrial adhesin system offers the potential to alleviate some of these inherent difficulties and thus more accurately assess the molecular details of bacterial adhesion.

The *P. aeruginosa* Fimbrial Adhesin

P. aeruginosa adhesion to human respiratory epithelial cells is mediated by at least two distinct adhesins, the alginate capsular component (29) and fimbriae (30, 88, 118, 147). *P. aeruginosa* fimbriae bind specifically to human buccal epithelial cells and to the cilia of human ciliated tracheal epithelial cells (30, 59). The fimbriae bind to the same locations as do whole bacteria (48, 59); in fact, there is competitive inhibition between purified fimbriae and whole cells for receptor sites (30, 59). The fimbriae bind specifically to five separate buccal epithelial cell membrane glycopeptides of different sizes, and their receptor activity requires the presence of carbohydrate (Fig. 6; 27). Equilibrium analysis of whole-cell binding to buccal epithelial cells indicated that the fimbrial adhesin had a much higher affinity for epithelial cells than did the capsular adhesin (88). Competition assays have been used to establish that the adhesin with the higher affinity constant for a receptor will occupy the available receptor sites in a proportion which reflects the difference in affinities of the adhesins for their receptors (60). The *P. aeruginosa* fimbrial adhesin is thus the most significant for the initial colonization of a mucosal surface (for the initiation of an infection), and the interaction of

Figure 6. Identification of human buccal epithelial cell membrane glycoproteins that have receptor activity for *P. aeruginosa* PAK fimbriae. Lanes contained fimbriae in the following amounts per milliliter: 1, 200 μg; 2, 150 μg; 3, 100 μg; 4, 50 μg; 5, 0 μg (control); 6, 0 μg preincubated with a 100-μg/ml concentration of Fab fragments of monoclonal antibody PK99H (control); 7, 150 μg preincubated with a 100-μg/ml concentration of Fab fragments of monoclonal antibody PK99H (positive antibody control for specificity determination); 8, 0 μg preincubated with a 100-μg/ml concentration of Fab fragments of monoclonal antibody PK41C (control); 9, 150 μg preincubated with a 100-μg/ml concentration of Fab fragments of monoclonal antibody PK41C (negative control to verify the specificity of fimbrial receptor activity). Lane 10, Amido black-stained nitrocellulose blot of solubilized buccal epithelial cells; lane S, molecular size markers (positions indicated in kilodaltons on the right). (Reprinted from Doig et al. [27].)

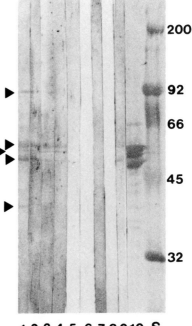

1 2 3 4 5 6 7 8 9 10 S

purified fimbriae with human respiratory epithelial cells accurately reflects the binding of whole cells and risk of acquisition of infection (139).

The *P. aeruginosa* PAK fimbria does not appear to have a separate adhesin, although a complete genetic analysis of the fimbrial operon has not yet been completed (28, 59, 101). It is clear that the PAK pilin structural protein does contain a domain that binds to human respiratory epithelial cells in a specific manner and competitively inhibits intact fimbrial binding to buccal epithelial cells (59). A synthetic peptide containing the PAK epithelial cell-binding domain binds to cilia on ciliated tracheal epithelial cells, as do intact fimbriae and whole cells (48, 59). Furthermore, the synthetic peptide binds to the same buccal epithelial cell surface glycopeptides as do intact fimbriae (27, 59). Thus, if there is a separate fimbria-associated adhesin, there must be substantial homology with the C-terminal region of the PAK pilin. However, we have no evidence of any such protein, and monoclonal antibodies PK99H and PK34C (antibodies that recognize epitopes specific for the C-terminal end of PAK pilin [28]), which inhibit both fimbrial and whole-cell binding to epithelial cells, bind only to the tip of the fimbriae (Fig. 4).

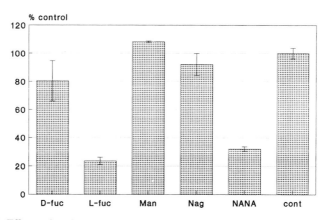

Figure 7. Effects of various monosaccharides on *P. aeruginosa* PAK fimbrial binding to human buccal epithelial cells (note that the D-fucose used in this study contained ~10% L-fucose). D-fuc, D-Fucose; L-fuc, L-fucose; Man, mannose; Nag, *N*-acetylglucosamine; NANA, *N*-acetylneuraminic acid; cont, control. (Reprinted from Doig et al. [27].)

The *P. aeruginosa* fimbrial binding domain appears to recognize a carbohydrate component, since both L-fucose and *N*-acetylneuraminic acid inhibit whole-cell binding to buccal epithelial cells (88). Recently, we have established that L-fucose (and not D-fucose) inhibits the binding of *P. aeruginosa* PAK fimbriae to buccal epithelial cells (Fig. 7; 27). Analyses of enzymatic digestion of buccal epithelial cells and immobilized buccal epithelial cell membrane proteins with fucosidase and neuraminidase have confirmed these results. The specificity of synthetic peptides containing the PAK fimbrial epithelial cell-binding domain for specific carbohydrates is now being investigated to unambiguously confirm that the epithelial cell-binding domain of *P. aeruginosa* PAK fimbriae has a lectinlike activity.

Several *P. aeruginosa* pilin proteins have now been sequenced (107, 108, 112), and in several cases the affinity constants for fimbrial binding to buccal epithelial cells have also been determined. Once the epithelial cell-binding domain has been more accurately determined through the use of synthetic peptide analogs, analysis of the effects of specific amino acid substitutions on receptor affinity and specificity should be possible. These studies should also enable determination of the particular molecular interactions involved in fimbria-mediated adhesion to epithelial cells. Isolation and characterization of the fimbrial receptor moiety will also undoubtedly enhance these studies.

Hypothesis Concerning the Molecular Basis of Fimbrial Binding to Epithelial Cell Surface Receptors

Examination of four separate classes of protein interactions reveals that the molecular strategy for protein-mediated binding is fairly consistent and appears to follow some general principles. The binding domains of fimbriae or fimbria-associated adhesins should be distinct, with respect to amino acid composition, from both the interior and the general surface of the fimbriae. It is likely that the binding of fimbrial adhesins to their appropriate receptor moieties will be found to follow the same principles that govern protein-mediated interactions.

One should therefore expect the binding domains of the adhesins to interact with their receptors by means of polar interactions such as ionic pairing (salt bridge formation), H bonding, dipole-dipole interactions, hydrophobic interactions due to van der Waals interactions, and the solvent effect due to the exclusion of water from the receptor-adhesin interaction site. The hydrophobic effect would be expected to contribute substantially to stabilizing and determining the equilibrium point of the interaction of the adhesin and receptor.

The initial interaction of the adhesin with its receptor could be initiated or facilitated either by hydrophobic interactions or by electrostatic interactions, resulting in a low-affinity binding of the adhesin to a receptor that may introduce a conformational change that facilitates a higher-affinity interaction. It is likely that the secondary events which would shortly follow and enhance the initial low-affinity interaction would then involve complementary structural pairing involving a full range of molecular interactions (i.e., both polar and hydrophobically mediated) and would generate a highly specific interaction (the net result being a highly specific high-affinity interaction). The binding domain of the adhesin would be expected to contain histidine, arginine, asparagine, tyrosine, tryptophan, and possibly negatively charged amino acid residues in a somewhat hydrophobic surface cleft. The C-terminal region of the *P. aeruginosa* PAK pilin protein satisfies a number of these predictions, and it will be interesting to discover whether these inferences hold.

Binding of the adhesin to a receptor should therefore be influenced by both the salt concentration and the pH of the environment. Changing the surface tension of the environment should have less effect on binding than altering the salt concentration because of the differential effect on protein hydration. The specificity of the interactions should be determined by the complementary nature of the receptor and the adhesin, i.e., the ability of the receptor to interact both electrostatically and hydrophobically with the binding domain of the adhesin. One would anticipate that most adhesins bind to a number of different epithelial cell surface components with differing affinities (which would reflect the degree their complementation to the binding domain of the adhesin). It is tempting to postulate that the initial interaction is generally hydrophobic in nature so that the greatest numbers of receptors can be bound. Thus, the fimbrial adhesins could mediate attachment to a variety of receptors from a variety of epithelial cell types from a variety of species. This is a more useful type of adhesin from a microbial ecology point of view.

CONCLUSIONS

Bacterial fimbriae are structural components that mediate the adhesion of a wide range of bacterial species to a variety of surfaces in their environment. The pilin structural proteins that constitute the bulk of fimbriae are hydrophobic, and consequently most fimbriae are quite hydrophobic. The hydrophobic amino acid residues of the pilin structural proteins play a central role in establishing a stable structural framework for the protein and in controlling the assembly of pilin monomers into intact fimbriae. In addition to the highly specific assembly process, there is clearly a substantial hydrophobic effect associated with the adhesive function of fimbriae. It is significant that the high affinity and specificity of these processes result from a mosaic of polar and hydrophobic interactions in a defined three-dimensional pattern. The role of the hydrophobic effect in fimbrial structure and function will be clarified when high-resolution structures of pilin and intact fimbriae become available.

Acknowledgments. The technical assistance of Richard Sherbourne, Kevin Ens, Marie Kaplan, Liane Kim, Peter Doig, and K. K. Lee is gratefully acknowledged.

This work has been generously supported by the Canadian Cystic Fibrosis Foundation and by the Natural Sciences and Engineering Research Council of Canada.

LITERATURE CITED

1. **Abraham, S. N., J. D. Goguen, D. Sun, P. Klemm, and E. H. Beachey.** 1987. Identification of two ancillary subunits of *Escherichia coli* type 1 fimbriae by using antibodies against synthetic oligopeptides of *fim* gene products. *J. Bacteriol.* **169:**5530–5536.
2. **Absolom, D. R., F. V. Lamberti, Z. Policova, W. Zingg, C. J. van Oss, and A. W. Neumann.** 1983. Surface thermodynamics of bacterial adhesion. *Appl. Environ. Microbiol.* **46:**90–97.
3. **Absolom, D. R., W. Zingg, C. Thomson, Z. Policova, C. J. van Oss, and A. W. Neumann.** 1985. Erythrocyte adhesion to polymer surfaces. *J. Colloid Interface Sci.* **104:**51–59.
4. **Akahane, K., and H. Umeyama.** 1986. A new method for calculating hydrophobic interaction energy in the biological system. *Chem. Pharm. Bull.* **34:**3492–3495.

5. **Beachey, E. H.** 1981. Bacterial adherence: adhesin-receptor interactions mediating the attachment of bacteria to mucosal surfaces. *J. Infect. Dis.* **143**:325–345.

6. **Beard, M. K. M., J. S. Mattick, L. J. Moore, M. R. Mott, C. F. Marrs, and J. R. Egerton.** 1990. Morphogenetic expression of *Moraxella bovis* fimbriae (pili) in *Pseudomonas aeruginosa. J. Bacteriol.* **172**:2601–2607.

7. **Bhattacharyya, L., and C. F. Brewer.** 1988. Lectin-carbohydrate interactions: studies of the nature of hydrogen bonding between D-galactose and certain D-galactose-specific lectins, and between D-mannose and concanavalin A. *Eur. J. Biochem.* **176**:207–212.

8. **Bloch, C. A., and P. E. Orndorff.** 1990. Impaired colonization by and full invasiveness of *Escherichia coli* K1 bearing a site-directed mutation in the type 1 pilin gene. *Infect. Immun.* **58**:275–278.

9. **Bradley, D. E.** 1972. A study of pili on *Pseudomonas aeruginosa. Genet. Res.* **19**:39–51.

10. **Bradley, D. E.** 1974. The adsorption of *Pseudomonas aeruginosa* pilus-dependent bacteriophages to a host mutant with nonretractile pili. *Virology* **58**:149–163.

11. **Bradley, D. E., and T. L. Pitt.** 1974. Pilus-dependence of four *Pseudomonas aeruginosa* bacteriophages with non-contractile tails. *J. Gen. Virol.* **23**:1–15.

12. **Bradley, D. L.** 1980. A function of *Pseudomonas aeruginosa* PAO polar pili: twitching motility. *Can. J. Microbiol.* **26**:146–154.

13. **Brinton, C. C., Jr.** 1959. Non-flagellar appendages of bacteria. *Nature* (London) **183**:782–786.

14. **Brinton, C. C., Jr.** 1965. The structure, function, synthesis and genetic control of bacterial pili and a molecular model for DNA and RNA transport in gram negative bacteria. *Trans. N.Y. Acad. Sci.* **27**:1003–1054.

15. **Brinton, C. C., Jr., J. Bryan, J. A. Dillon, N. Guerina, L. J. Jacobson, A. Labik, S. Lee, A. Levine, S. Lim, J. McMichael, S. Polen, K. Rogers, A. C.-C. To, and S. C.-M. To.** 1978. Uses of pili in gonorrhea control: role of bacterial pili in disease, purification and properties of gonococcal pili, and progress in the development of a gonococcal pilus vaccine for gonorrhea, p. 155–178. *In* G. F. Brooks, E. C. Gotschlich, K. K. Holmes, W. S. Sawyer, and F. E. Young (ed.), *Immunobiology of Neisseria gonorrhoeae.* American Society for Microbiology, Washington, D.C.

16. **Brooks, C. L., M. Karplus, and B. M. Pettitt.** 1988. Proteins: a theoretical perspective of dynamics, structure and thermodynamics. *Adv. Chem. Phys.* **71**:23–31.

17. **Burke, D. A., and A. T. R. Axon.** 1988. Hydrophobic adhesion of *E. coli* in ulcerative colitis. *Gut* **29**:41–43.

18. **Busscher, H. J., and A. H. Weerkamp.** 1987. Specific and non-specific interactions in bacterial adhesion to solid substrata. *FEMS Microbiol. Rev.* **46**:165–173.

19. **Caron, J., and J. R. Scott.** 1990. A *rns*-like regulatory gene for colonization factor antigen I (CFA/A) that controls expression of CFA/I pilin. *Infect. Immun.* **58**:874–878.

20. **Clark, W. B., J. E. Beem, W. E. Nesbitt, J. O. Cisar, C. C. Tseng, and M. J. Levine.** 1989. Pellicle receptors for *Actinomyces viscosus* type 1 fimbriae in vitro. *Infect. Immun.* **57**:3003–3008.

21. **Clark, W. B., T. T. Wheeler, and J. O. Cisar.** 1984. Specific inhibition of adsorption of *Actinomyces viscosus* T14V to saliva-treated hydroxyapatite by antibody against type 1 fimbriae. *Infect. Immun.* **43**:497–501.

22. **Costerton, J. W., K.-J. Cheng, G. G. Geesey, T. I. Ladd, J. C. Nickel, M. Dasgupta, and T. J. Marrie.** 1987. Bacterial biofilms in nature and disease. *Annu. Rev. Microbiol.* **41**:435–464.

23. **Costerton, J. W., R. T. Irvin, and K.-J. Cheng.** 1981. The bacterial glycocalyx in nature and disease. *Annu. Rev. Microbiol.* **35:**299–324.

24. **Dalrymple, B., and J. S. Mattick.** 1987. An analysis of the organization and evolution of the type 4 fimbrial (MePhe) subunit proteins. *J. Mol. Evol.* **25:**261–269.

25. **Davies, D. R., and H. Metzger.** 1983. Structural basis of antibody function. *Annu. Rev. Immunol.* **1:**87–117.

26. **de Ree, J. M., P. Schwillens, and J. F. van den Bosch.** 1987. Monoclonal antibodies raised against Pap fimbriae recognize minor component(s) involved in receptor binding. *Microb. Pathog.* **2:**113–121.

27. **Doig, P., W. Paranchych, P. A. Sastry, and R. T. Irvin.** 1989. Human buccal epithelial cell receptors of *Pseudomonas aeruginosa*: identification of polypeptides with pilus binding activity. *Can. J. Microbiol.* **35:**1141–1145.

28. **Doig, P., P. A. Sastry, R. S. Hodges, K. K. Lee, W. Paranchych, and R. T. Irvin.** 1990. Inhibition of pilus-mediated adhesion of *Pseudomonas aeruginosa* to human buccal epithelial cells by monoclonal antibodies directed against pili. *Infect. Immun.* **58:**124–130.

29. **Doig, P., N. R. Smith, T. Todd, and R. T. Irvin.** 1987. Characterization of the binding of *Pseudomonas aeruginosa* alginate to human epithelial cells. *Infect. Immun.* **55:**1517–1522.

30. **Doig, P., T. Todd, P. A. Sastry, K. K. Lee, R. S. Hodges, W. Paranchych, and R. T. Irvin.** 1988. Role of pili in the adhesion of *Pseudomonas aeruginosa* to human respiratory epithelial cells. *Infect. Immun.* **56:**1641–1646.

31. **Dorman, C. J., and C. F. Higgins.** 1987. Fimbrial phase variation in *Escherichia coli*: dependence on integration host factor and homologies with other site-specific recombinases. *J. Bacteriol.* **169:**3840–3843.

32. **Duguid, J. P.** 1959. Fimbriae and adhesive properties in *Klebsiella* strains. *J. Gen. Microbiol.* **21:**217–286.

33. **Duguid, J. P., I. W. Smith, G. Dempster, and P. N. Edmunds.** 1955. Non-flagellar filamentous appendages ("fimbriae") and haemagglutinating activity in *Bacterium coli*. *J. Pathol. Bacteriol.* **70:**335–358.

34. **Elleman, T. C., P. A. Hoyne, D. J. Stewart, N. M. McKern, and J. E. Peterson.** 1986. Expression of pili from *Bacteroides nodosus* in *Pseudomonas aeruginosa*. *J. Bacteriol.* **168:**574–580.

35. **Elleman, T. C., and J. E. Peterson.** 1987. Expression of multiple types of N-methyl Phe pili in *Pseudomonas aeruginosa*. *Mol. Microbiol.* **1:**377–380.

36. **Evans, D. G., and D. J. Evans, Jr.** 1978. New surface associated heat-labile colonization factor antigen (CFA/II) produced by enterotoxigenic *Escherichia coli* of serogroups O6 and O8. *Infect. Immun.* **21:**638–647.

37. **Every, D., and T. M. Skerman.** 1982. Protection of sheep against experimental footrot by vaccination with pili purified from *Bacteroides nodosus*. *N.Z. Vet. J.* **30:**156–158.

38. **Falkowski, W., M. Edwards, and A. J. Schaeffer.** 1986. Inhibitory effect of substituted aromatic hydrocarbons on adherence of *Escherichia coli* to human epithelial cells. *Infect. Immun.* **52:**863–866.

39. **Fenno, J. C., D. J. LeBlanc, and P. Fives-Taylor.** 1989. Nucleotide sequence analysis of a type 1 fimbrial gene of *Streptococcus sanguis* FW213. *Infect. Immun.* **57:**3527–3533.

40. **Finlay, B. B., B. L. Pasloske, and W. Paranchych.** 1986. Expression of the *Pseudomonas aeruginosa* PAK pilin gene in *Escherichia coli*. *J. Bacteriol.* **165:**625–630.

41. **Firon, N., I. Ofek, and N. Sharon.** 1982. Interaction of mannose-containing oligosac-

charides with the fimbrial lectin of *Escherichia coli*. *Biochem. Biophys. Res. Commun.* **105**:1426–1432.

42. **Firon, N., I. Ofek, and N. Sharon.** 1983. Carbohydrate specificity of the surface lectins of *Escherichia coli*, *Klebsiella pneumoniae*, and *Salmonella typhimurium*. *Carbohydr. Res.* **120**:235–249.

43. **Fletcher, M.** 1977. Significance of marine bacterial attachment to solid surfaces, p. 407–410. *In* D. Schlessinger (ed.), *Microbiology—1977*. American Society for Microbiology, Washington, D.C.

44. **Fletcher, M., and G. D. Floodgate.** 1973. An electron-microscopic demonstration of an acidic polysaccharide involved in the adhesion of a marine bacterium to solid surfaces. *J. Gen. Microbiol.* **74**:325–334.

45. **Fletcher, M., and J. H. Pringle.** 1985. The effect of surface free energy and medium surface tension on bacterial attachment to solid surfaces. *J. Colloid Interface Sci.* **104**:5–14.

46. **Folkhard, W., K. R. Leonard, S. Malsey, D. A. Marvin, J. Dubochet, A. Engel, M. Achtman, and R. Helmuth.** 1979. X-ray diffraction and electron microscopic studies on the structure of bacterial F-pili. *J. Mol. Biol.* **130**:145–160.

47. **Folkhard, W., D. A. Marvin, T. H. Watts, and W. Paranchych.** 1981. Structure of polar pili from *Pseudomonas aeruginosa* strains K and O. *J. Mol. Biol.* **149**:79–93.

48. **Franklin, A. L., T. Todd, G. Gurman, D. Black, P. M. Mankinen-Irvin, and R. T. Irvin.** 1987. Adherence of *Pseudomonas aeruginosa* to cilia of human tracheal epithelial cells. *Infect. Immun.* **55**:1523–1525.

49. **Frost, L. S., M. Carpenter, and W. Paranchych.** 1978. N-methylphenylalanine at the N-terminus of pilin isolated from *Pseudomonas aeruginosa* K. *Nature* (London) **271**:87–89.

50. **Fulks, K. A., C. F. Marrs, S. P. Stevens, and M. R. Green.** 1990. Sequence analysis of the inversion region containing the pilin genes of *Moraxella bovis*. *J. Bacteriol.* **172**:310–316.

51. **Gibbons, R. J.** 1977. Adherence of bacteria to host tissue, p. 395–406. *In* D. Schlessinger (ed.), *Microbiology—1977*. American Society for Microbiology, Washington, D.C.

52. **Goochee, C. F., R. T. Hatch, and T. W. Cadman.** 1987. Some observations on the role of type 1 fimbriae in *Escherichia coli* autoflocculation. *Biotechnol. Bioeng.* **29**:1024–1034.

53. **Haahtela, K. E. Tarkka, and T. K. Korhonen.** 1985. Type 1 fimbria-mediated adhesion of enteric bacteria to grass roots. *Appl. Environ. Microbiol.* **49**:1182–1185.

54. **Hanson, M. S., J. Hempel, and C. C. Brinton, Jr.** 1988. Purification of the *Escherichia coli* type 1 pilin and minor pilus proteins and partial characterization of the adhesin protein. *J. Bacteriol.* **170**:3350–3358.

55. **Heckels, J. E., and M. Virji.** 1985. Antigenic variation of gonococcal pili: immunochemical and biological studies, p. 288–296. *In* G. K. Schoolnik (ed.), *The Pathogenic Neisseriae*. American Society for Microbiology, Washington, D.C.

56. **Henrichsen, J.** 1983. Twitching motility. *Annu. Rev. Microbiol.* **37**:81–93.

57. **Honda, T., K. Kasemsuksakul, T. Oguchi, M. Kohda, and T. Miwatani.** 1988. Production and partial characterization of pili on non-O1 *Vibrio cholerae*. *J. Infect. Dis.* **157**:217–218.

58. **Houwink, A. L., and W. van Iterson.** 1950. Electron microscopical observations on bacterial cytology. II. A study on flagellation. *Biochim. Biophys. Acta* **5**:10–44.

59. **Irvin, R. T., P. Doig, K. K. Lee, P. A. Sastry, W. Paranchych, T. Todd, and R. S. Hodges.** 1989. Characterization of the *Pseudomonas aeruginosa* pilus adhesin: confir-

mation that the pilin structural protein subunit contains a human epithelial cell binding domain. *Infect. Immun.* **57**:3720–3726.

60. **Irvin, R. T., P. C. Doig, P. A. Sastry, B. Heller, and W. Paranchych.** 1989. Competition for bacterial receptor sites on respiratory epithelial cells by *Pseudomonas aeruginosa* strains of heterologous pilus type: usefulness of kinetic parameters in predicting the outcome. *Microbiol. Ecol. Health Dis.* **3**:39–47.

61. **Irvin, R. T., M. To, and J. W. Costerton.** 1984. Mechanism of adhesion of *Alysiella bovis* to glass surfaces. *J. Bacteriol.* **160**:569–576.

62. **Janin, J., S. Miller, and C. Chothia.** 1988. Surface, subunit interface and interior of oligomeric proteins. *J. Mol. Biol.* **204**:155–164.

63. **Jann, K., G. Schmidt, E. Blumenstock, and K. Vosbeck.** 1981. *Escherichia coli* adhesion to *Saccharomyces cerevisiae* and mammalian cells: role of piliation and surface hydrophobicity. *Infect. Immun.* **32**:484–489.

64. **Johnson, K., and S. Lory.** 1987. Characterization of *Pseudomonas aeruginosa* mutants with altered piliation. *J. Bacteriol.* **169**:5663–5667.

65. **Johnson, K., M. L. Parker, and S. Lory.** 1986. Nucleotide sequence and transcriptional initiation site of two *Pseudomonas aeruginosa* pilin genes. *J. Biol. Chem.* **261**:15703–15708.

66. **Kist, M. L., I. E. Salit, and T. Hofman.** 1990. Purification and characterization of the Dr hemagglutinin expressed by uropathogenic *Escherichia coli* strains. *Infect. Immun.* **58**:695–702.

67. **Klemm, P., and G. Christiansen.** 1987. Three *fim* genes required for the regulation of length and mediation of adhesion of *Escherichia coli* type 1 fimbriae. *Mol. Gen. Genet.* **208**:439–445.

68. **Korhonen, T. K.** 1979. Yeast cell agglutination by purified enterobacterial pili. *FEMS Microbiol. Lett.* **6**:421–425.

69. **Korhonen, T. K., E. Tarka, H. Ranta, and K. Haahtela.** 1983. Type 3 fimbriae of *Klebsiella* sp.: molecular characterization and role in bacterial adhesion to plant roots. *J. Bacteriol.* **155**:860–865.

70. **Krogfelt, K. A., H. Bergmans, and P. Klemm.** 1990. Direct evidence that the FimH protein is the mannose-specific adhesin of *Escherichia coli* type 1 fimbriae. *Infect. Immun.* **58**:1995–1998.

71. **Kuntz, I. D., and W. Kauzmann.** 1974. Hydration of proteins and polypeptides. *Adv. Protein Chem.* **28**:239–345.

72. **Kuntz, I. D., and A. Zipp.** 1977. Water in biological systems. *N. Engl. J. Med.* **297**:262–266.

73. **Lee, K. K., P. Doig, R. T. Irvin, W. Paranchych, and R. S. Hodges.** 1989. Mapping the surface regions of *Pseudomonas aeruginosa* PAK pilin: the importance of the C-terminal region for adherence to human buccal epithelial cells. *Mol. Microbiol.* **3**:1493–1499.

74. **Leffler, H., and C. Svanborg-Edén.** 1980. Chemical identification of a glycosphingolipid receptor for *Escherichia coli* attaching to human urinary tract epithelial cells and agglutinating human erythrocytes. *FEMS Microbiol. Lett.* **7**:127–134.

75. **Leung, K.-P., W. E. Nesbitt, W. Fischlschweiger, D. I. Hay, and W. B. Clark.** 1990. Binding of colloidal gold-labeled salivary proline-rich proteins to *Actinomyces viscosus* type 1 fimbriae. *Infect. Immun.* **58**:1987–1991.

76. **Levine, M. M., P. Ristaino, G. Marley, C. Smyth, S. Knutton, E. Boedeker, R. Black, C. Young, M. L. Clements, C. Cheney, and R. Patnaik.** 1984. Coli surface antigens 1 and 3 of colonization factor antigen II-positive enterotoxigenic *Escherichia coli*:

morphology, purification, and immune responses in humans. *Infect. Immun.* **44**:409–420.

77. **Lindahl, M., A. Faris, T. Wadström, and S. A. Hjertén.** 1981. A new test based on 'salting out' that measures relative surface hydrophobicity of bacterial cells. *Biochim. Biophys. Acta* **677**:471–476.

78. **Lindberg, F., B. Lund, L. Johansson, and S. Normark.** 1987. Localization of the receptor-binding protein adhesin at the tip of the bacterial pilus. *Nature* (London) **328**:84–87.

79. **Lowe, M. A., S. C. Holt, and B. I. Eisenstein.** 1987. Immunoelectron microscopic analysis of elongation of type 1 fimbriae in *Escherichia coli. J. Bacteriol.* **169**:157–163.

80. **Lund, B., F. Lindberg, B.-I. Marklund, and S. Normark.** 1987. The PapG protein is the α-D-galactopyranosyl-(1-4)-β-D-galactopyranose-binding adhesin of uropathogenic *Escherichia coli. Proc. Natl. Acad. Sci. USA* **84**:5898–5902.

81. **Lund, B., F. Lindberg, and S. Normark.** 1988. Structure and antigenic properties of the tip-located P pilus proteins of uropathogenic *Escherichia coli. J. Bacteriol.* **170**:1887–1894.

82. **Manoil, C., and J. P. Rosenbusch.** 1982. Conjugation-deficient mutants of *Escherichia coli* distinguish classes of functions of the outer membrane protein OmpA protein. *Mol. Gen. Genet.* **187**:148–156.

83. **Marshall, K. C.** 1973. Mechanism of adhesion of marine bacteria to surfaces, p. 625–632. *In* R. F. Acker, B. F. Brown, J. R. DePalma, and W. P. Iverson (ed.), *Proceedings of the Third International Congress on Marine Corrosion and Fouling.* Northwestern University Press, Evanston, Ill.

84. **Marshall, K. C., R. Stout, and R. Mitchell.** 1971. Mechanism of the initial events in the sorption of marine bacteria to surfaces. *J. Gen. Microbiol.* **68**:337–348.

85. **Marvin, D. A., and W. Folkhard.** 1986. Structure of F-pili: reassessment of the symmetry. *J. Mol. Biol.* **191**:299–300.

86. **Mattick, J. S., M. M. Bills, B. J. Anderson, B. Dalrymple, M. R. Mott, and J. R. Egerton.** 1987. Morphogenetic expression of *Bacteroides nodosus* fimbriae in *Pseudomonas aeruginosa. J. Bacteriol.* **169**:33–41.

87. **McConnell, M. M., P. Mullany, and B. Rowe.** 1987. A comparison of the surface hydrophobicity of enterotoxigenic *Escherichia coli* of human origin producing different adhesion factors. *FEMS Microbiol. Lett.* **42**:59–62.

88. **McEachran, D. W., and R. T. Irvin.** 1985. Adhesion of *Pseudomonas aeruginosa* to human buccal epithelial cells: evidence for two classes of receptors. *Can. J. Microbiol.* **31**:563–569.

89. **McEldowney, S., and M. Fletcher.** 1986. Effect of growth conditions and surface characteristics of aquatic bacteria on their attachment to solid surfaces. *J. Gen. Microbiol.* **132**:513–523.

90. **McEldowney, S., and M. Fletcher.** 1987. Adhesion of bacteria from mixed cell suspension to solid surfaces. *Arch. Microbiol.* **148**:57–62.

91. **Minion, F. C., S. N. Abraham, E. H. Beachey, and J. D. Goguen.** 1986. The genetic determinant of adhesive function in type 1 fimbriae of *Escherichia coli* is distinct from the gene encoding the fimbrial subunit. *J. Bacteriol.* **165**:1033–1036.

92. **Mirelman, D., and I. Ofek.** 1986. Introduction to microbial lectins and agglutinins, p. 1–19. *In* D. Mirelman (ed.), *Microbial Lectins and Agglutinins: Properties and Biological Activity.* John Wiley & Sons, Inc., New York.

93. **Mooi, F. R., and F. K. De Graaf.** 1985. Molecular biology of fimbriae of enterotoxigenic *Escherichia coli. Curr. Top. Microbiol. Immunol.* **118**:119–138.

94. **Morona, R., C. Kramer, and U. Henning.** 1985. Bacteriophage receptor area of outer membrane protein OmpA of *Escherichia coli* K-12. *J. Bacteriol.* **164**:539–543.
95. **Nesbitt, W. E., R. J. Doyle, and K. G. Taylor.** 1982. Hydrophobic interactions and the adherence of *Streptococcus sanguis* to hydroxylapatite. *Infect. Immun.* **38**:637–644.
96. **Normark, S., M. Baga, M. Goransson, F. P. Lindberg, B. Lund, M. Norgren, and B.-E. Uhlin.** 1986. Genetics and biogenesis of *Escherichia coli* adhesins, p. 113–143. *In* D. Mirelman (ed.), *Microbial Lectins and Agglutinins: Properties and Biological Activity.* John Wiley & Sons, Inc., New York.
97. **Nowicki, B., J. P. Barrish, T. Korhonen, R. A. Hull, and S. I. Hull.** 1987. Molecular cloning of the *Escherichia coli* O75X adhesin. *Infect. Immun.* **55**:3168–3173.
98. **Nowicki, B., A. Labigne, S. Moseley, R. Hull, S. Hull, and J. Moulds.** 1990. The Dr hemagglutinin, afimbrial adhesins AFA-I and AFA-III, and F1845 fimbriae of uropathogenic and diarrhea-associated *Escherichia coli* belong to a family of hemagglutinins with Dr receptor recognition. *Infect. Immun.* **58**:279–281.
99. **Nowicki, B., J. Moulds, R. Hull, and S. Hull.** 1988. A hemagglutinin of uropathogenic *Escherichia coli* recognizes the Dr blood group antigen. *Infect. Immun.* **56**:1057–1060.
100. **Nowicki, B., C. Svanborg-Edén, R. Hull, and S. Hull.** 1989. Molecular analysis and epidemiology of the Dr hemagglutinin of uropathogenic *Escherichia coli.* *Infect. Immun.* **57**:446–451.
101. **Nunn, D., S. Bergman, and S. Lory.** 1990. Products of three accessory genes, *pilB*, *pilC*, and *pilD*, are required for biogenesis of *Pseudomonas aeruginosa* pili. *J. Bacteriol.* **172**:2911–2919.
102. **Old, D. C., I. Corneil, L. F. Gibson, A. D. Thomson, and J. P. Duguid.** 1968. Fimbriation, pellicle formation and the amount of growth of salmonellas in broth. *J. Gen. Microbiol.* **50**:1–16.
103. **Ooi, T., M. Oobatake, G. Nemethy, and H. A. Scheraga.** 1987. Accessible surface areas as a measure of the thermodynamic parameters of hydration of peptides. *Proc. Natl. Acad. Sci. USA* **84**:3086–3090.
104. **Orndorff, P. E.** 1987. Genetic study of piliation in *Escherichia coli*: implications for understanding the microbe-host interactions at the molecular level. *Pathol. Immunopathol. Res.* **6**:82–92.
105. **Ottow, J. C. G.** 1975. Ecology, physiology, and genetics of fimbriae and pili. *Annu. Rev. Microbiol.* **29**:79–108.
106. **Padlan, E. A.** 1990. On the nature of antibody combining sites: unusual structural features that may confer on these sites an enhanced capacity for binding ligands. *Proteins Struct. Funct. Genet.* **7**:112–124.
107. **Paranchych, W., and L. S. Frost.** 1988. The physiology and biochemistry of pili. *Adv. Microb. Physiol.* **29**:53–114.
108. **Paranchych, W., P. A. Sastry, L. S. Frost, M. Carpenter, G. D. Armstrong, and T. H. Watts.** 1979. Biochemical studies on pili isolated from *Pseudomonas aeruginosa* strain PAO. *Can. J. Microbiol.* **25**:1175–1181.
109. **Parge, H. E., S. L. Bernstein, C. D. Deal, D. E. McRee, D. Christensen, E. D. Getzoff, and J. A. Tainer.** 1990. Biochemical purification and crystallographic characterization of the fiber-forming protein pilin from *Neisseria gonorrhoeae. J. Biol. Chem.* **265**:2278–2285.
110. **Pasloske, B. L., B. B. Finlay, and W. Paranchych.** 1985. Cloning and sequencing of *Pseudomonas aeruginosa* PAK pilin gene. *FEBS Lett.* **183**:408–412.
111. **Pasloske, B. L., and W. Paranchych.** 1988. The expression of mutant pilins in *Pseudomonas aeruginosa*: fifth position glutamate affects pilin methylation. *Mol. Microbiol.* **2**:489–495.

112. **Pasloske, B. L., P. A. Sastry, B. B. Finlay, and W. Paranchych.** 1988. Two unusual pilin sequences from different isolates of *Pseudomonas aeruginosa. J. Bacteriol.* **170:**3738–3741.

113. **Pecha, B., D. Low, and P. O'Hanley.** 1989. Gal-gal pili vaccines prevent pyelonephritis by piliated *Escherichia coli* in a murine model: single-component gal-gal pili vaccines prevent pyelonephritis by homologous and heterologous piliated *E. coli* strains. *J. Clin. Invest.* **83:**2102–2108.

114. **Perry, A. C. F., C. A. Hart, I. J. Nicolson, J. E. Heckels, and J. R. Saunders.** 1987. Inter-strain homology of pilin gene sequences in *Neisseria meningitidis* isolates that express markedly different antigenic pilus types. *J. Gen. Microbiol.* **133:**1409–1418.

115. **Pugh, G. W., D. E. Gughes, and G. D. Booth.** 1977. Experimentally induced infectious bovine keratonconjunctivitis: effectiveness of a pilus vaccine against exposure to homologous strains of *Moraxella bovis. Am. J. Vet. Res.* **38:**1519–1522.

116. **Quiocho, F. A.** 1986. Carbohydrate-binding proteins: tertiary structures and protein-sugar interactions. *Annu. Rev. Biochem.* **55:**287–315.

117. **Ramphal, R., and G. B. Pier.** 1985. Role of *Pseudomonas aeruginosa* mucoid exopolysaccharide in adherence to tracheal cells. *Infect. Immun.* **47:**1–4.

118. **Ramphal, R., J. C. Sadoff, M. Pyle, and J. D. Silipigni.** 1984. Role of pili in the adherence of *Pseudomonas aeruginosa* to injured tracheal epithelium. *Infect. Immun.* **44:**38–40.

119. **Read, R. J., and M. N. G. James.** 1986. Introduction to the protein inhibitors: X-ray crystallography, p. 301–336. *In* A. J. Barret and G. Salvesen (ed.), *Proteinase Inhibitors.* Elsevier Science Publishers BV, The Hague, The Netherlands.

120. **Richards, W. G., P. M. King, and C. A. Reynolds.** 1989. Solvation effects. *Protein Eng.* **2:**319–327.

121. **Rosenberg, M.** 1981. Bacterial adherence to polystyrene: a replica method of screening for bacterial hydrophobicity. *Appl. Environ. Microbiol.* **42:**375–377.

122. **Rothbard, J. B., R. Fernadez, L. Wang, N. N. H. Teng, and G. K. Schoolnik.** 1985. Antibodies to peptides corresponding to a conserved sequence of gonococcal pilins block bacterial adhesion. *Proc. Natl. Acad. Sci. USA* **82:**915–919.

123. **Rutter, P. R., and B. Vincent.** 1980. The adhesion of microorganisms to surfaces: physico-chemical aspects, p. 79–92. *In* R. C. W. Berkeley, J. M. Lynch, J. Melling, R. R. Rutter, and B. Vincent (ed.), *Microbial Adhesion to Surfaces.* Ellis Horwood Ltd., West Sussex, England.

124. **Sajjan, S. U., and J. F. Forstner.** 1990. Characteristics of binding of *Escherichia coli* serotype O157:H7 strain CL-49 to purified intestinal mucin. *Infect. Immun.* **58:**860–867.

125. **Sajjan, S. U., and J. F. Forstner.** 1990. Role of the putative "link" glycopeptide of intestinal mucin in binding of piliated *Escherichia coli* serotype O157:H7 strain CL-49. *Infect. Immun.* **58:**868–873.

126. **Sastry, P. A., B. B. Finlay, B. L. Pasloske, W. Paranchych, J. R. Pearlstone, and L. B. Smillie.** 1985. Comparative studies on the amino acid and nucleotide sequences of pilin derived from *Pseudomonas aeruginosa* PAK and PAO. *J. Bacteriol.* **164:**571–577.

127. **Sastry, P. A., J. R. Pearlstone, L. B. Smillie, and W. Paranchych.** 1985. Studies on the primary structure and antigenic determinants of pilin isolated from *Pseudomonas aeruginosa* K. *Can. J. Biochem. Cell Biol.* **63:**284–291.

128. **Schmidt, M. A., and P. O'Hanley.** 1988. Synthetic peptides corresponding to protective epitopes of *Escherichia coli* digalactoside-binding pilin prevent infection in a murine pyelonephritis model. *Proc. Natl. Acad. Sci. USA* **85:**1247–1251.

129. **Schoolnik, G. K., J. B. Rothbard, and E. C. Gotschlich.** 1986. Structure-function

analysis of gonococcal pili, p. 145–168. *In* D. Mirelman (ed.), *Microbial Lectins and Agglutinins: Properties and Biological Activity*. John Wiley & Sons, Inc., New York.

130. **Sharon, N., and I. Ofek.** 1986. Mannose specific bacterial surface lectins, p. 55–81. *In* D. Mirelman (ed.), *Microbial Lectins and Agglutinins: Properties and Biological Activity*. John Wiley & Sons, Inc., New York.

131. **Smyth, C. J., P. Jonsson, E. Olsson, O. Soderlind, J. Rosengren, S. Hjertén, and T. Wadström.** 1978. Difference in hydrophobic surface characteristics of porcine enteropathogenic *Escherichia coli* with or without K88 antigen as revealed by hydrophobic interaction chromatography. *Infect. Immun.* **22:**462–472.

132. **Steven, A. C., M. E. Bisher, B. L. Trus, D. Thomas, J. M. Zhang, and J. L. Cowell.** 1986. Helical structure of *Bordetella pertussis* fimbriae. *J. Bacteriol.* **167:**968–974.

133. **Strom, M. S., and S. Lory.** 1986. Cloning and expression of the pilin gene of *Pseudomonas aeruginosa* PAK in *Escherichia coli*. *J. Bacteriol.* **165:**367–372.

134. **Strom, M. S., and S. Lory.** 1987. Mapping of export signals of *Pseudomonas aeruginosa* pilin with alkaline phosphatase fusions. *J. Bacteriol.* **169:**3187–3188.

135. **Svennerholm, A.-M., C. Wenneras, J. Holmgren, M. M. McConnell, and B. Rowe.** 1990. Roles of different coli surface antigens of colonization factor antigen II in colonization by and protective immunogenicity of enterotoxigenic *Escherichia coli* in rabbits. *Infect. Immun.* **58:**341–346.

136. **Tanford, C.** 1973. *The Hydrophobic Effect: Formation of Micelles and Biological Membranes*. John Wiley & Sons, Inc., New York.

137. **Tanford, C.** 1978. The hydrophobic effect and the organization of living matter. *Science* **200:**1012–1018.

138. **Tinsley, C. R., and J. E. Heckels.** 1986. Variation in the expression of pili and outer membrane protein by *Neisseria meningitidis* during the course of meningococcal infection. *J. Gen. Microbiol.* **132:**2483–2490.

139. **Todd, T., A. L. Franklin, G. Gurman, P. M. Mankinen-Irvin, and R. T. Irvin.** 1989. Augmented bacterial adherence to ciliated tracheal epithelial cells in an intensive care unit population. *Am. Rev. Respir. Dis.* **140:**1585–1589.

140. **Totten, P. A., J. C. Lara, and S. Lory.** 1990. The *rpoN* gene product of *Pseudomonas aeruginosa* is required for expression of diverse genes, including the flagellin gene. *J. Bacteriol.* **172:**389–396.

141. **van Verseveld, H. W., P. Bakker, T. van der Woude, C. Terleth, and F. K. De Graaf.** 1985. Production of fimbrial adhesins K99 and F41 by enterotoxigenic *Escherichia coli* as a function of growth-rate domain. *Infect. Immun.* **49:**159–163.

142. **Wadström, T., A. Faris, J. Freer, D. Habte, D. Hallberg, and Å. Ljungh.** 1980. Hydrophobic surface properties of enterotoxigenic *E. coli* (ETEC) with different colonization factors (CFA/I, CFA/II, K88 and K99) and attachment to intestinal epithelial cells. *Scand. J. Infect. Dis. Suppl.* **24:**148–153.

143. **Watts, T. H., C. M. Kay, and W. Paranchych.** 1983. Spectral properties of three quaternary arrangements of *Pseudomonas* pilin. *Biochemistry* **22:**3640–3646.

144. **Watts, T. H., P. A. Sastry, R. S. Hodges, and W. Paranchych.** 1983. Mapping of antigenic determinants of *Pseudomonas aeruginosa* PAK polar pili. *Infect. Immun.* **42:**113–121.

145. **Watts, T. H., D. G. Scraba, and W. Paranchych.** 1982. Formation of 9-nm filaments from pilin monomers obtained by octyl-glucoside dissociation of *Pseudomonas aeruginosa* pili. *J. Bacteriol.* **151:**1508–1513.

146. **Watts, T. H., E. A. Worobec, and W. Paranchych.** 1982. Identification of pilin pools in the membranes of *Pseudomonas aeruginosa*. *J. Bacteriol.* **152:**687–691.

147. **Woods, D. E., D. C. Straus, W. G. Johanson, Jr., V. K. Berry, and J. A. Bass.** 1980.

Role of pili in the adherence of *Pseudomonas aeruginosa* to mammalian buccal epithelial cells. *Infect. Immun.* **29**:1146–1151.

148. **Worobec, E. A., L. S. Frost, P. Pieroni, G. D. Armstrong, R. S. Hodges, J. M. R. Parker, B. B. Finlay, and W. Paranchych.** 1986. Location of the antigenic determinants of conjugative F-like pili. *J. Bacteriol.* **167**:660–665.

149. **Yoshimura, F., K. Takahashi, Y. Nodasaka, and T. Suzuki.** 1984. Purification and characterization of a novel type of fimbriae from the oral anaerobe *Bacteroides gingivalis. J. Bacteriol.* **160**:949–957.

150. **Zhou, N. E., C. T. Mant, and R. S. Hodges.** 1990. Effect of preferred binding domains on peptide retention behavior in reversed-phase chromatography: amphipathic α-helices. *Peptide Res.* **3**:8–20.

Microbial Cell Surface Hydrophobicity
Edited by R. J. Doyle and M. Rosenberg
© 1990 American Society for Microbiology, Washington, DC 20005

Chapter 6

Adhesion of Bacteria to Plant Cells: Role of Specific Interactions versus Hydrophobicity

Gerrit Smit and Gary Stacey

A large variety of microorganisms live on the surfaces of plants, but most have only minor effects on plant growth. Accordingly, the adhesion mechanism utilized by these microbes has not been well studied. In contrast, symbiotic and pathogenic interactions have a strong influence on the growth of plants and consequently have been intensively examined. Of the many potential systems for study, only two, the *Rhizobium/ Bradyrhizobium*-legume symbiosis and *Agrobacterium* infection, provide the majority of the literature on the role of attachment in plant-bacterium interaction. In the cases of *Rhizobium* and *Agrobacterium* species, bacterial attachment is thought to be a prerequisite for infection and virulence, respectively, whereas in other cases, e.g., pathogenic *Pseudomonas* strains, attachment seems to be avoided during pathogenesis (38, 103, 146, 165, 183). This chapter focuses on attachment of bacteria to plant cell surfaces, and more specifically on attachment of members of the family *Rhizobiaceae* (i.e., *Rhizobium*, *Bradyrhizobium*, and *Agrobacterium* species).

Adhesion of bacteria to plant roots can be assigned to two general mechanistic classes: (i) nonspecific adhesion, in which no specific adhesins and receptors are known and which most likely is based solely on physicochemical interactions, and (ii) specific adhesion, in which adhes-

Gerrit Smit • Department of Plant Molecular Biology, Leiden University, Nonnensteeg 3, 2311 VJ Leiden, The Netherlands.   **Gary Stacey** • Center for Legume Research, Department of Microbiology, and Graduate Program of Ecology, University of Tennessee, Knoxville, Tennessee 37996.

ins localized on the bacterial cell surface recognize receptors on the plant cell surface. However, the latter type of adhesion does not exclude the possible importance of ionic and hydrophobic interactions between adhesin and receptor. Specific interactions can subsequently be divided into host-specific adhesion and non-host-specific adhesion. Host specificity as found in a number of plant-bacterium interactions, e.g., the *Rhizobium*-legume symbiosis (see below), implies that only those bacteria capable of establishing a stable association will adhere to the surface of a particular plant host species. Information on the role of nonspecific adhesion mechanisms in microbe-plant interactions is limited, and the literature clearly reflects the inclination of investigators to search for specific adhesion mechanisms. Here we present an overview of current knowledge about the mechanism of attachment of bacteria to plant cells.

ATTACHMENT IN THE *RHIZOBIUM*-LEGUME SYMBIOSIS

Infection of Legumes by Rhizobia

More than a century ago, it was discovered that nodules found on the roots of leguminous plants were formed after infection by soil bacteria. In the root nodules, the bacteria fix atmospheric nitrogen, thereby enabling the host plant to grow in a nitrogen-limited environment (10, 58). The symbiosis between *Rhizobium* and legumes is a complex interaction involving the following steps (131, 175):

1. Excretion of flavonoid compounds by the host plant.
2. Root colonization (motility and chemotaxis).
3. Attachment to root hairs.
4. Production of root hair deformation factors by the rhizobia and induction of nodule meristems upon induction of the rhizobia by flavonoids.
5. Root hair curling (shepherd crook) leading to entrapment of rhizobia.
6. Infection thread formation and growth of the infection thread into the cortex.
7. Outgrowth of nodules and infection of nodule cells by rhizobia released from the infection thread; formation of symbiosome.
8. Conversion of the rhizobia into bacteroids; nitrogen fixation.
9. Senescense of the nodules, followed by a release of rhizobia into the rhizosphere.

This interaction is a host plant-specific process (e.g., *Rhizobium leguminosarum* bv. *viciae* nodulates pea, lentil, and common vetch but not clover or bean, whereas *R. leguminosarum* bv. *trifolii* nodulates only clover). In the so-called fast-growing rhizobia (e.g., *R. leguminosarum* and *R. meliloti*), many essential nodulation (*nod*) genes, including the genes encoding host specificity, are located on a large plasmid, the Sym (for symbiosis) plasmid (11, 40, 62, 69, 81, 98, 116, 133, 142). In addition, a number of *nod* genes have been located on the chromosome or on other plasmids (14, 23, 44, 46, 50, 87, 88, 97, 139). In the slow-growing rhizobia (e.g., *Bradyrhizobium japonicum*), the *nod* genes are apparently located on the chromosome (60, 154, 157). Compounds secreted by the plant and identified as flavonoids (6, 121, 127, 180, 185) are involved in transcriptional activation of the *nod* genes mediated by a positive regulatory protein encoded by the *nodD* gene (65, 113, 131, 184). It appears that the biochemical products of some of the *nod* genes may in turn act as signals to the plant root cells (5, 45, 56, 166, 184). At least one of these factors is excreted by the rhizobia and causes root hair deformation. Direct contact of the rhizobia with the plant root hair, in addition to production of this factor, results in specific root hair deformations in which the angle of curvature might exceed 360°, the so-called shepherd crooks (20, 41, 134, 166, 182, 184). After curling of the hair, the entrapped rhizobia infect the root hair by forming an infection thread. Growth of this tubular structure, which contains the invading rhizobia, is explained as being a redirection of growth of the root hair tip (165; R. Bakhuizen, Ph.D. thesis, Leiden University, Leiden, The Netherlands, 1988). Before or during infection thread formation, rhizobial factors likely act as signals that initiate formation of the root nodule meristem (8, 9, 21, 90, 91). The infection thread grows in the direction of the meristem, and subsequently the young nodule cells are infected (25). Bacteria are released from the infection thread and take the form of a quasi-organelle, the symbiosome (135–137), enclosed by a membrane of plant origin, the symbiosome membrane (76, 77, 128, 129, 136, 137). After bacteria are released from the infection thread, they differentiate morphologically into bacteroids, in which nitrogen fixation occurs (175). During the infection process, a number of nodule-specific plant gene products (nodulins) are expressed (47, 109, 110). The biochemical functions of only a few of these plant nodulins are currently known (114).

Role of Bacterial Attachment in Nodulation

Attachment of rhizobia to the root hair surface (Fig. 1) has been proposed to be an essential element in two features of the *Rhizobium*-legume symbiosis: (i) marked curling of root hairs and (ii) determination of host plant specificity (e.g., reference 130). It is generally assumed that

Figure 1. Scanning electron micrograph of attached *B. japonicum* bacteria to root hair cells (courtesy of E. Roth).

attachment precedes marked root hair curling in the infection process. A computer simulation study of root hair curling demonstrated that attachment of rhizobia is a prerequisite for induction of marked curling (165). However, conclusive experimental evidence is lacking.

Heterologous rhizobial strains usually fail to form infection threads, indicating that host specificity is expressed at a step preceding the outgrowth of the infection thread in the root hair (20, 89, 153). Bohlool and Schmidt (12) and Dazzo and Hubbell (29) hypothesized that host specificity is a consequence of specific binding of the homologous rhizobial partner to the root hair cell surface of the host plant. The molecular basis for the recognition mechanism was proposed to be binding of a host plant-lectin to the bacterial surface. This hypothesis was tested by examining the attachment ability of nodulating or nonnodulating rhizobial strains in the absence or presence of lectin-haptenic monosaccharides (30). Initial binding of heterologous (nonnodulating) rhizobia to clover root hair tips was found to be approximately 10 times lower than that of the nodulating rhizobia (*R. leguminosarum* bv. *trifolii*). The initial bacterial attachment to clover root hairs appeared to be a rapid event and could be prevented by the addition of clover lectin haptens (e.g., 2-deoxyglucose). Moreover, lectin-mediated attachment to clover root hairs appeared to be dependent on the presence of the Sym plasmid (26,

28, 30, 31, 188). A low level of non-host-specific attachment was observed; however, this type of attachment occurred less rapidly and could not be prevented by addition of lectin haptens. This nonspecific attachment was proposed to be mediated by rhizobial fibrils (31, 34, 115). Host-specific attachment appeared to be dependent on the growth phase of the bacteria, with optimal plant attachment and lectin binding at the late exponential phase of growth in batch culture (27, 32, 111). Lectin receptor sites were demonstrated in the extracellular capsular polysaccharides (EPS) and in the O-antigenic part of the lipopolysaccharide (LPS) of the rhizobial cell wall (1, 64, 71, 72, 111, 123, 167, 181). Data using other *Rhizobium*-legume combinations supporting the lectin recognition hypothesis were reported by Stacey et al. (155) and Kato et al. (73).

In contrast to Dazzo et al. (26, 30) and Zurkowski (188), several authors were unable to demonstrate specific lectin-mediated binding of homologous rhizobia to legume root hairs (4, 18, 19, 21, 22, 108, 124, 148, 149, 152, 171, 172, 174). Heterologous rhizobia appeared to adhere to host and non-host plant surfaces as well as did homologous rhizobia, whereas absence of the Sym plasmid did not affect rhizobial attachment behavior (106, 148). Moreover, if attachment is a prerequisite for marked root hair curling, the lectin recognition hypothesis does not explain the observed induction of root hair curling by heterologous rhizobia (33, 179).

Vesper et al. (171, 172, 174) described fimbriae of *B. japonicum* and implicated these structures in attachment of bacteria to soybean roots. The fimbria-mediated attachment was found to be "firm," indicating that the bacteria were not removed from their target cells after vortexing of the roots for severai minutes. In these studies, a correlation was found between the number of fimbriated cells in culture, adherence to hydrophobic surfaces (e.g., plastic), and the attachment to soybean roots. These results pointed to the involvement of fimbriae in a non-host-specific attachment process. Since attachment to plastic was positively correlated with the ability of the rhizobia to attach to soybean roots, hydrophobic interactions are likely to be involved. In support of a role of fimbriae in attachment, it was found that fimbria-negative mutants had reduced attachment ability and that addition of antibodies against purified fimbriae inhibited attachment. Since fimbria-negative mutants nodulated normally (173), this type of attachment is apparently not essential for nodulation.

It was recently shown that bacterial growth conditions, and especially nutrient limitations, strongly affect the way in which rhizobia attach to plant root hairs. Using several kinds of growth limitations, Smit et al. (148–151) and Kijne et al. (78, 79) were able to identify molecules involved in attachment of *R. leguminosarum* bv. *viciae* to pea root hair tips. In these studies, attachment of rhizobia to the root hairs was studied, since

Table 1. Influence of Various Treatments of *R. leguminosarum*
bv. *viciae* 248 Cells on Attachment to Pea Root Hair Tips[a]

Pretreatment	% Attachment in class[b]:			
	1	2	3	4
C-limited cells				
None	3	23	14	60
Cellulase[c]	21	77	2	0
Rhicadhesin[d]	57	21	13	9
LPS[e]	6	28	16	50
Mn^{2+}-limited cells				
None	2	23	15	60
Cellulase[c]	4	55	9	32
Rhicadhesin[d]	33	40	10	17
Ca^{2+}-limited cells				
None	64	33	0	3
Rhicadhesin[d]	9	23	18	50

[a] Bacteria were cultivated in TY medium (C-limited cells), YM medium (Mn^{2+}-limited cells), or TY medium with low (0.35 mM) CaCl$_2$ (Ca^{2+}-limited cells) and harvested at an A_{620} of 0.70, 0.85, or 0.70, respectively. (Adapted from references 79, 149, 150, and 151.)
[b] Class 1, No attached bacteria; class 2, few attached bacteria; class 3, apical portion of the root hair covered with bacteria; class 4, many attached bacteria forming a caplike structure on top of the root hair.
[c] Bacteria were harvested and suspended in 100 mM sodium phosphate buffer (pH 5.0) supplemented with 1 mg of purified cellulase per ml. After 2 h of incubation at 28°C, the bacteria were harvested by centrifugation, washed, and suspended in 25 mM phosphate buffer (pH 7.5) for incubation with the pea roots in the attachment assay.
[d] Roots were incubated with 20 ng of rhicadhesin per ml in 25 mM phosphate buffer (pH 7.5), washed, and subsequently incubated with the bacteria in the attachment assay.
[e] LPS, purified from *R. leguminosarum* bv. *viciae*, was added during the attachment assay in a final concentration of 250 µg/ml.

these cells are the site of infection initiation. Attachment in these studies was microscopically quantified and ranked as class 1 (no attached bacteria), class 2 (few attached bacteria), class 3 (the apical portion of the root hair covered with bacteria), or class 4 (many attached bacteria forming a caplike aggregate on top of the root hair) (Table 1). In this attachment assay, the roots were washed before determination of attachment, indicating that only irreversible attachment was measured. Three components were identified as important in attachment of rhizobia to pea root hair tips: (i) a proteinaceous bacterial adhesin, called rhicadhesin (see below), (ii) bacterial cellulose fibrils, and (iii) pea lectin. Attachment was found to be a two-step process in which bacteria first adhere to the plant through rhicadhesin and subsequently form aggregates on root hair tips through cellulose fibril- and plant lectin-mediated bacterium-bacterium interactions (Fig. 2).

Carbon limitation results in a non-host-specific attachment of *R. leguminosarum* in which neither host plant lectins nor Sym plasmid-

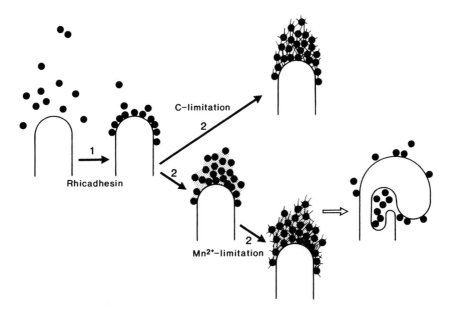

Figure 2. Model for attachment of rhizobial cells to pea root hair tips. Step 1 attachment is mediated by rhicadhesin and leads to the attachment of single rhizobial cells to the surface of the root hair tip (151). The mechanism of step 2 attachment depends on the growth conditions of the rhizobia and results in the formation of aggregates of bacteria on the tip of the root hair. In the case of carbon-limited bacteria, rhizobial cellulose fibrils are involved in the second step of attachment (149), whereas in the case of manganese-limited rhizobia, host plant lectins are also involved. In the latter case, the aggregates are formed within a shorter period of time. Lectin-mediated accumulation is correlated with optimal infection, as indicated by the open arrow (79).

localized *nod* genes are involved (148, 149). Cellulose fibrils were found to mediate aggregation (cap formation) of carbon-limited rhizobia at the root hair tips (149; Table 1). Tn5-induced fibril-overproducing mutants showed greatly increased cap-forming ability, whereas fibril-negative mutants lost this ability completely. Aggregation in liquid medium was greatly increased in the case of overexpression of cellulose fibrils and abolished in the case of the fibril-negative mutants (149). It was found that the fibril mutants were affected in hydrophobicity (G. Smit and J. W. Kijne, unpublished data) as measured by contact angle determination (169, 170). Fibril-negative mutants showed a significantly lower hydrophobicity than did the parent strain, whereas the fibril-overproducing mutants showed a higher hydrophobicity. These data suggest that the aggregation of rhizobia on top of the root hairs (cap formation) and in liquid medium is based on hydrophobic interactions between the rhizobia. This finding coincides with the observation that rhizobia cultivated from a surface pellicle are highly fibrillated and show a strong cap-forming ability (148). However, in the same study it was reported that NaCl and other salts inhibited cap formation, a result which makes it unlikely that cap formation is based solely on hydrophobic interactions, since physicochemical interactions should be enhanced under high-ionic-strength conditions (138). Therefore, it is likely that electrostatic interactions are also involved in cap formation.

Both fibril-overproducing and fibril-negative mutants showed normal nodulation properties, indicating that neither cellulose fibrils nor fibril-mediated cap formation is a prerequisite for successful nodulation under the conditions used.

Manganese limitation appeared to result in a different attachment mechanism (78, 79). In contrast to the attachment of carbon-limited cells, accumulation of manganese-limited rhizobia (cap formation) is already in full progress after 10 min of incubation, is partially resistant to sodium chloride, is partially resistant to pretreatment of the rhizobia with purified cellulase (Table 1), and is significantly delayed in the presence of a strong pea lectin haptenic monosaccharide, 3-O-methyl-D-glucose. The pea lectin hapten did not inhibit direct binding of manganese-limited bacteria to the root hair surface. The subsequent accumulation of manganese-limited rhizobia at the root hair tip is apparently accelerated by pea lectin molecules and inhibited by the hapten. Manganese-limited cells were found to be more infective than C-limited cells or exponentially growing cells, indicating a possible involvement of lectin in infection thread formation. The hypothesis that lectin is essential for infection thread formation is contrary to the earlier proposal that lectin is involved primarily in bacterium-root hair adhesion. However, this hypothesis is consistent with the above-mentioned role of lectin in cap formation and with the work of Halverson and Stacey (54, 55) and, more recently, Díaz et al. (37). Halverson and Stacey (55) showed that a nodulation-defective mutant of *B. japonicum*, capable of normal root hair attachment and curling, could be restored to normal nodulation by pretreatment with soybean lectin. The implication here is that some step in the infection process subsequent to root hair curling is affected by lectin. Díaz et al. (37) investigated the possible role of pea lectin in the determination of host-specific nodulation. *R. leguminosarum* bv. *viciae*, the normal symbiont of pea, can attach to clover root hairs and induce root hair curling, but infection threads are not formed (148, 179). Díaz et al. (37) constructed transgenic clover plants expressing the pea lectin gene and found that these plants could be nodulated by *R. leguminosarum* bv. *viciae*. Again, these results clearly point to the involvement of lectin in a nodulation step subsequent to root hair curling, with the formation of the infection thread being most likely.

A role for lectin in plant-bacterium adhesion has been difficult to establish because of the conflicting results published from different laboratories. The current information showing that bacterial growth conditions have a large influence on bacterium-plant adhesion offers an explanation for these conflicting reports. In addition, it is well established that culture age affects the expression of lectin receptors on the rhizobial cell surface (27, 111, 112). Therefore, a role of lectin in adhesion can be shown but only under specific growth conditions, such as with manganese-limited cells but not with C-limited cells as mentioned above.

It would be interesting to learn whether nutrient limitation is a prerequisite for attachment and nodulation under field conditions. If so, the symbiosis might be viewed as a survival mechanism by which the bacteria can avoid nutrient limitation. In view of the composition of plant root exudates (e.g., references 15 and 96), carbon is probably not a growth-limiting nutrient in the rhizosphere. Thus, lectin-enhanced accumulation might be a normal feature of the *Rhizobium*-legume association under natural conditions.

Calcium limitation is the only limitation that was found not to lead to optimal attachment of rhizobia to pea root hairs. Growth of rhizobia, as well as agrobacteria (see below), under low-Ca^{2+} conditions strongly reduced the initial direct attachment of rhizobia to pea root hair surfaces, suggesting the presence of a Ca^{2+}-dependent adhesin (149–152; Fig. 2 and Table 1). Although growth under low-Ca^{2+} conditions affects a number of cell surface characteristics (e.g., loss of the O-antigenic part of the LPS and flagella), these characteristics do not cause the attachment-minus (Att$^-$) phenotype (150; Table 1). The apparent lack of involvement of LPS in rhizobial attachment is interesting since this component affects hydrophobicity and the net charge of the cell (35). In addition, LPS has repeatedly been proposed as a receptor for plant lectins. With regard to the role of LPS as a lectin receptor, it should be noted that these studies were performed with carbon-limited cells, a condition under which an involvement of lectin in attachment cannot be shown (see above). Thus, these results do not exclude a role of LPS in lectin-enhanced accumulation as found under manganese-limited growth conditions.

Although no role for physicochemical adhesion mechanisms was shown in these studies, it is remarkable that the lack of binding of Ca^{2+}-limited cells to pea roots is correlated with the inability of these cells to bind to glass at the air-liquid interface (148–150).

A Ca^{2+}-dependent adhesin was isolated from the surface of *R. leguminosarum* bv. *viciae* and has been purified on the basis of its ability to inhibit attachment of both carbon- and manganese-limited bacteria to the surface of pea root hairs (151; Table 1). This protein, called rhicadhesin, has the following properties: (i) it is sensitive to boiling and protease treatment; (ii) it has an isoelectric point of 5.1 and a molecular weight of 14,000; (iii) it is a Ca^{2+}-binding protein with approximately five Ca^{2+} binding sites per monomer; (iv) Ca^{2+} is required for anchoring it to the bacterium, not to the plant; and (v) it is released in the growth medium when the rhizobia are grown under low-Ca^{2+} conditions (99, 150; G. Smit, Ph.D. thesis, Leiden University, Leiden, The Netherlands, 1988). Rhicadhesin was found to be common among all genera of the family *Rhizobiaceae*, including *Agrobacterium* (see below), and also appears to be present in *Azotomonas* spp. Adhesin activity was not found associated with bacteria representing a number of other genera. Its expression was found to be independent of the presence or absence of a Sym (in rhizobia) or Ti (tumor-inducing; in agrobacteria) plasmid.

Rhicadhesin-mediated binding of *Rhizobiaceae* cells to root surfaces was found not only with pea but with all tested plants, including monocotyledonous species. Therefore, it seems that this mechanism is common to members of the *Rhizobiaceae* and is not host specific. Attachment can be efficiently blocked after preincubation of the roots with very low amounts of rhicadhesin (151; Table 1), suggesting that binding occurs at specific sites on the root hairs. Therefore, it is likely that rhicadhesin-mediated attachment is based on specific interactions with a common plant surface component. The identity of the receptor site remains to be determined.

From the available data, it can be concluded that homologous as well as heterologous rhizobia are able to attach to host plant root hairs, the presence of a Sym plasmid is not a prerequisite for attachment, more than one type of binding occurs in certain *Rhizobium*-plant root hair associations (although some types of binding, e.g., fimbria-mediated attachment, [173], are not a prerequisite for nodulation), and attachment of homologous infective rhizobia might be enhanced or otherwise influenced by the compatible interaction with the root hair cell. In the latter case, host plant lectins might be involved in the attachment process.

ATTACHMENT OF PLANT-PATHOGENIC BACTERIA TO THEIR TARGET CELLS

Tumor formation by Agrobacteria

Agrobacterium tumefaciens causes tumors, called crown galls, on many dicotyledonous plants. The bacterial genes involved in tumorigenicity are present on a large plasmid, the Ti plasmid (63, 168, 186).

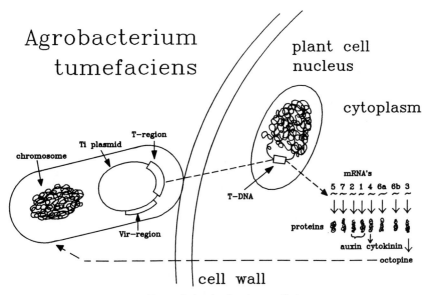

Figure 3. Tumor induction by *A. tumefaciens*.

Agrobacteria transfer a portion of the Ti plasmid, the T-DNA, into the host plant chromosome via a mechanism resembling bacterial conjugation (Fig. 3) (24, 67, 119, 120, 162). The *vir* region of the Ti plasmid is essential for the transfer of T-DNA but is itself not transferred (61, 80). As with the previously discussed *nod* genes, expression of the *vir* genes is specifically induced by plant phenolic compounds (13), identified as acetosyringone and α-hydroxyacetosyringone by Stachel et al. (158). T-DNA contains a number of genes that are expressed in the transformed plant cells, some of which encode the production of the plant hormones indole acetic acid and cytokinin and of unusual amino acid derivatives (opines) (2, 7, 17, 160, 163). These latter compounds can be utilized as sole carbon and nitrogen source by agrobacteria. Although the molecular basis of T-DNA transfer is not yet fully understood, the binding of the bacteria to the host plant cells appears to be an essential step in tumor formation (38, 39, 51, 84, 92, 102, 103).

Role of Bacterial Attachment in Tumor Formation

In contrast to the *Rhizobium* infection process, *A. tumefaciens* bacteria induce tumors from a wound site and do not penetrate the host plant cells. Attachment of *Agrobacterium* cells has been studied in two ways. In some studies, tumor formation or number has been used as an indirect assay for attachment. This method relies on the assumption that factors which reduce tumor formation do so by competitively interfering with binding of tumorigenic agrobacteria. Compounds of competing bacteria are added to wound sites together with tumorigenic *Agrobacterium* cells, and any inhibition of tumor formation is measured (92–95, 125, 177). These studies appeared to indicate that the bacterial adhesin involved in binding to the host cell is the LPS fraction of the agrobacterial cell wall (177).

A second method used in attachment studies is based on direct counting of attached agrobacteria to individual plant cells (84, 101, 105, 117). These studies have been interpreted as revealing that attachment of agrobacteria is a two-step process in which at least two factors are involved. The first step in attachment is direct binding of bacteria to the plant cell surface. The nature of the bacterial component involved in this step is proposed to be LPS as well as proteins (102, 103, 105). However, recently it was shown that the first step in attachment of agrobacteria is also rhicadhesin mediated (151; Smit, Ph.D. thesis, 1988). Consistent with these observations, Krens et al. (84) reported that divalent cations were essential for *Agrobacterium* attachment. In these studies, agrobacteria were grown under low-Ca^{2+} conditions, a condition under which members of the *Rhizobiaceae* produce rhicadhesin but excrete it into the medium. The presence of divalent cations retains rhicadhesin on the bacterial surface and restores attachment ability (99, 150, 151; Smit, Ph.D. thesis, 1988; Table 1).

The second step in the attachment process of agrobacteria to plant cells involved accumulation of bacteria at the attachment site. This step was found to be due to bacterium-bacterium interactions, mediated by cellulose fibrils, which are synthesized by the agrobacteria in response to plant factors (101, 104). These fibrils apparently anchor the bacteria tightly to the plant cells. *A. tumefaciens* mutants defective in cellulose fibril synthesis were found to be virulent, demonstrating that these structures are not essential for tumorigenicity (101). As should be obvious, the two-step model for *Agrobacterium* attachment closely resembles that previously described for rhizobia. However, in the case of rhizobia, there is no indication that plant factors induce the synthesis of cellulose fibrils (149).

Transposon-induced avirulent mutants of *A. tumefaciens* were isolated that appeared to be affected in their ability to attach to most plant cells (23, 38, 39, 102, 103, 161), although they apparently were able to attach to some plant cells, e.g., *Solanum tuberosum* (57). In all cases, the mutation was found to be chromosomally located. Among these chromosomal virulence (*chv*) mutants, mutants in the *chvA* or *chvB* gene have been the most intensively studied. Mutations in these genes result in a pleiotropic phenotype; these bacteria are nonmotile since they lack flagella, show an altered bacteriophage sensitivity, do not conjugate efficiently, and are affected in cyclic β-1,2 glucan production (16, 38, 39, 126). Other mutants lacking flagella and with an altered phage sensitivity were found to retain virulence and attachment ability (16). Since no mutants were found with normal virulence but affected in the synthesis of β-1,2 glucan, it was suggested that this component is essential for virulence and attachment (44, 126). This hypothesis was supported by the finding that avirulent *exoC* mutants, although pleiotropic, are also defective in cyclic β-1,2 glucan production (22). The *chvB* gene encodes a 235-kilodalton membrane protein involved in glucan synthesis (187). Mutations in the *chvA* gene apparently affect the ability to excrete the β-1,2 glucan (66). Genes comparable to *chvA* and *chvB* have been identified in *Rhizobium meliloti* (*ndvA* and *ndvB*, respectively) and shown to be essential for nodulation (44, 50). Miller et al. (107) suggested that the cyclic β-1,2 glucan was essential for maintaining the osmotic pressure of the periplasmic space in agrobacteria. Given these data and the pleiotropic phenotype of *chvA* and *chvB* mutations, we favor a model in which modification of the periplasm by affecting β-1,2 glucan synthesis disrupts normal function such as transport or protein insertion into the membrane. Consistent with this model, it was recently demonstrated that *chvB* mutants could be restored in attachment ability as well as virulence after treatment with rhicadhesin, purified from either *Rhizobium* or *Agrobacterium* wild-type cells (99; Smit, Ph.D. thesis, 1988). Therefore, the lack of rhicadhesin on the cell surface appears to be the primary cause of the Att$^-$ virulence-minus (Vir$^-$) phenotype of these mutants.

In addition to the above-described *chvA* and *chvB* mutants, Matthysse (103) isolated attachment-negative mutants of *A. tumefaciens* and showed them to be affected in the synthesis of cell surface proteins. However, no evidence was presented that the proteins absent in these mutants are directly involved in binding of the bacteria to the host plant cells. Unlike the *chvB* or *chvA* mutants described above, these mutants were not affected in motility or β-1,2 glucan synthesis. In addition, they were not affected in fimbriae, cellulose fibrils, or hydrophobicity. *chvC* mutants isolated by Thomashow et al. (161) as well as the *exoC* mutants described by Cangelosi et al. (22) also appeared to be affected in attachment, which is most likely due to their inability to synthesize cellulose fibrils. However, it is unlikely that this latter trait accounts for the Vir$^-$ phenotype of the *chvC* mutants, since other mutants defective in cellulose fibril synthesis were shown to be virulent (101).

At present, there is confusion in the literature over a possible role of the Ti plasmid in attachment to the plant surface. A number of studies showed Ti plasmid-cured agrobacteria to have reduced attachment ability (105, 177), whereas others could not demonstrate any role of the Ti plasmid in attachment (84, 100, 117, 149, 151). Since all point mutations affecting attachment ability have been mapped to the chromosome, and since the presence of rhicadhesin is independent of the Ti plasmid, it is likely that attachment is primarily chromosomally encoded.

Attachment of Pathogenic *Pseudomonas* spp. to Plant Cells

P. solanacearum

The wilt-inducing pathogen *Pseudomonas solanacearum* exists in the form of numerous strains with widely diverse host ranges. In contrast to *A. tumefaciens* infection, in which attachment is an essential step for virulence, phytopathogenic *P. solanacearum* bacteria appear to avoid attachment in a compatible interaction (i.e., resulting in disease). Attachment of incompatible strains to tobacco mesophyll cells leads to a rapid hypersensitive response (HR) and a drastic reduction in bacterial growth. It has been suggested that attachment of *P. solanacearum* cells to the plant surface triggers a recognition event leading to a plant defense response. Most work on *P. solanacearum* attachment has been focused on the natural shift of the virulent strain K60 to the avirulent mutant B1. The wild-type strain K60 produces much EPS (EPS$^+$) and does not induce an HR when inoculated on tobacco plants, whereas strain B1 is EPS$^-$ and induces an HR. These strains also differ in a number of other cell surface characteristics such as motility and production of indole acetic acid, fimbriae, and the O-polysaccharide portion of LPS (74, 75, 122, 159, 178). B1 cells were found to produce more fimbriae than did the wild-type strain, and purified fimbriae were able to agglutinate tobacco cell walls as well as polystyrene spheres. This agglutination could be prevented by low amounts of EPS obtained from the wild-type strain K60. The aggregation of polystyrene spheres suggests that hydrophobic interactions are involved in this process, although the inhibition of aggregation by EPS is difficult to explain solely by this mechanism. The inhibition of cell wall aggregation by EPS is significant, since it correlates with the differences of the two strains to adhere to tobacco cells.

Sequeira and Graham (147) reported that B1 cells, in contrast to K60 cells, could be agglutinated by a hydroxyproline-rich glycoprotein (HPRG), which was found to be present ubiquitously in the cell walls of potato. This HPRG has similarities to potato lectin but was found not to be identical (85, 86). Because of the highly basic nature of the cell wall HPRG, the observation that several other polycations could agglutinate

B1 cells, and the finding that certain polycations, as well as high salt concentration, inhibited agglutination of B1 cells by HPRG, it seems likely that this agglutination process is based on ionic interactions.

Since agglutination of B1 cells and aggregation of their LPS could be prevented by EPS, it was suggested that LPS was the bacterial receptor for the plant cell wall HPRG and that EPS specifically competed for binding sites, thus inhibiting agglutination of the virulent strains. In the attachment of *P. solanacearum*, EPS is able to inhibit both fimbria-mediated agglutination and LPS-HPRG binding. Since the wild-type strain produces copious amounts of EPS, it is possible that EPS prevents the attachment of these cells to tobacco mesophyll cells and subsequently prevents an HR. However, induction of an HR is not determined solely by adhesion of the pseudomonads to their target cells, since other recognition events are also involved (59, 183).

Although conclusive data are not available, Sequeira and co-workers (42, 43, 145) suggested that binding of B1 bacteria to plant cells is a two-step process in which ionic forces, presumably between bacterial LPS and plant HPRG, are involved in an initial reversible binding, followed by a fimbria-mediated second step leading to firm attachment. The second step is most likely dependent on hydrophobic interactions. Again, this model is generally reminiscent of those proposed to explain *Rhizobium* and *Agrobacterium* attachment.

P. syringae pv. phaseolica

The phytopathogenic bacterium *Pseudomonas syringae* pv. *phaseolicola* causes halo blight of bean. Initiation of infection apparently depends on the ability of the cells to adhere to the target cell surface. Fimbriated cells were found to adhere specifically to stomata cells, whereas nonfimbriated mutants showed no preference for stomata but adhered in lower numbers throughout the plant cell surface (132). These results suggest the presence of a specific receptor for fimbriae on the stomata or, alternatively, that the thick stomatal wax layer adsorbs the fimbriae, which are hydrophobic. The former explanation seems more convincing, since the total aerial surface of plants is covered with cuticular wax, which is most likely identical in structure on epidermal cells and stomatal cells. The fact that attachment of this bacterium to bean leaf surfaces is insensitive to low or high salt concentrations might point to an involvement of hydrophobic interaction rather than ionic interaction. Nothing is known about the plant receptor in this interaction. Fimbria-negative mutants of *P. syringae* pv. *phaseolicola* were found to be virulent when injected into the leaf, indicating that this type of attachment is not essential for virulence.

Table 2. Classes of Bacterial Mutants Affected in Attachment
or Physicochemical Properties

Organism	Phenotype	Gene cluster, genes, or gene loci	Role	Reference(s)
Agrobacterium spp.	Vir⁻ Att⁻	*chvA*	β-1,2 Glucan modification, excretion	38, 39, 66
	Vir⁻ Att⁻	*chvB*	β-1,2 Glucan production, rhicadhesin production	38, 39, 99, 126
	Vir⁻ Att⁻	*chvC*	Cellulose fibril production	161
	Vir⁻ Att⁻	*exoC*	Involved in production of EPS	22
	Vir⁻ Att⁻	*att*	Membrane polypeptide production	102, 103
	Vir⁺ Att⁻	*cel*	Cellulose fibril synthesis	101
Rhizobium spp.	Nod⁺ Att⁻		Cellulose fibril synthesis	149
	Nod⁺ Att⁺		LPS production, cell surface characteristics	35, 150
Bradyrhizobium spp.	Nod⁺ Att⁻		Fimbria production	171–174
	Nod⁻ Att⁻		LPS production	156
Pseudomonas solanacearum	Vir⁻ Att⁺		Fimbria production, EPS production	75, 122, 159
P. putida	Att⁺		LPS production, cell surface characteristics	36
P. syringae pv. *phaseolica*	Vir⁺ Att⁻		Fimbria production	132

However, virulence was found to be strongly reduced under more natural conditions, such as washing of the leaf after surface inoculation (132).

Table 2 summarizes mutant bacterial strains that are altered in attachment or physicochemical properties.

PLANT GROWTH-PROMOTING BACTERIA

Recently, scientific attention has been focused on bacteria that can enhance the growth of plants. The plant growth-promoting ability of these bacteria has been variously attributed to the fixation of atmospheric nitrogen (e.g., *Klebsiella* and *Enterobacter* species), the production of iron-chelating siderophores (fluorescent *Pseudomonas* species), or the synthesis of plant growth-promoting substances (118, 143). Unfortunately, in most cases very little research has been directed to understanding the mechanism by which these bacteria adhere to plant surfaces. For example, *Azospirillum* strains enter into an associative symbiosis with the roots of cereals and forage grasses and possess the ability to invade the root systems of these plants. Inoculation with azospirilla has been shown in certain instances to have a significant effect on plant growth (118, 141). However, it is unknown whether attachment of azospirilla to plant root cells is a prerequisite for plant growth promotion. Recently, positive hybridization to *Azospirillum* genomic DNA has been shown by using the *Agrobacterium chvA* and *chvB* genes as probes (176). However, the importance of these results to the biology of *Azospirillum*-plant interaction is unclear.

Attachment of Noninfective Bacterial Species to Plant Roots

Examples of noninfective interactions between bacteria and plants are the association of *Klebsiella* spp. and *Enterobacter agglomerans* with the roots of grasses (cereals) and root colonization by fluorescent *Pseudomonas* spp. (e.g., with corn and potato).

Attachment of Fluorescent *Pseudomonas* spp. to Plant Roots

Fluorescent *Pseudomonas* spp. have the potential to act as biocontrol agents in agriculture (144); the mechanism of action has been attributed to the ability of these bacteria to excrete siderophores, iron-chelating compounds, and antibiotics, the latter of which reduce the growth of deleterious microorganisms. Pseudomonads are found to be strong colonizers of plant roots, and attachment is thought to be an important step in the colonization process. Firm binding of *Pseudomonas* cells to radish, bean, and potato roots has been reported, but the molecular mechanism of this process is poorly understood (3, 36, 68). James et al. (68) and De Weger et al. (36) studied the role of cell surface characteristics (e.g., hydrophobicity) in attachment of these bacteria to radish and potato roots as well as in attachment to various abiotic surfaces. It was found that alterations in either hydrophobicity or cell surface charge did not affect the attachment ability of these bacteria. These results indicate that differences in physicochemical cell surface properties do not play a major role in firm attachment of these bacteria to plant roots. However, in these studies only firm adhesion was measured; therefore, a role for physicochemical interactions cannot be excluded.

Anderson et al. (3) reported that *Pseudomonas putida* bacteria were agglutinated by a bean root surface glycoprotein. Mutants of *P. putida* with decreased agglutinability showed a reduced binding and colonization ability, indicating that this type of specific interaction could play a role in attachment to root surfaces.

Adhesion of *Klebsiella* and *Enterobacter* spp. to Grass Roots

Korhonen et al. (83) and Haahtela et al. (52, 53) have examined the in vitro adhesion of associative nitrogen-fixing *Klebsiella* and *Enterobacter* spp. to grass roots. Attachment was found to be dependent on inoculum density, incubation time, temperature, pH, ionic strength, and bacterial growth phase. Type 1 and type 3 fimbriae were isolated from the bacteria and examined for their possible involvement in attachment to roots. Both types were found to be involved in attachment on the basis of the observation that purified fimbriae bound to grass roots in vitro, and receptor analogs (e.g., mannose or α-methyl-D-mannoside in the case of type 1 fimbriae) inhibited root binding of *Klebsiella* and *Enterobacter* cells. It should be noted that this inhibition was not complete, an observation which might be explained by the high hydrophobicity of these proteins (82). Binding that is not inhibited by mannoside might reflect unspecific hydrophobic binding. Fab antibody fragments against fimbriae inhibited bacterial attachment to root surfaces (52, 83). Type 3 fimbriae were found to be more efficient than type 1 fimbriae in promoting adhesion (52, 83). It remains to be determined whether this type of attachment is essential for the intimate association of these bacteria with grass roots.

Attachment of *Azospirillum* spp. to Grass Roots

Attachment of *Azospirillum brasiliense* Sp7 to root hairs of pearl millet can be viewed as favoring the colonization of the root surface. In this way, these bacteria may occupy niches that take advantage of root exudate. Attached azospirilla were associated with granular material on root hairs and fibrillar material on undifferentiated epidermal cells. Attachment to the root hairs was significantly lower in the presence of 5 mM potassium nitrate, whereas attachment to undifferentiated epidermal cells was unaffected. Root exudate from pearl millet was found to contain nondialyzable and protease-sensitive compounds that bound to azospirilla and promoted attachment to millet root hairs but not to undifferentiated epidermal cells (164). These results indicate that more than one mechanism of attachment exists, a result consistent with those reported by Gafny et al. (49) for attachment of strain Cd to corn roots. Interestingly, it was recently reported that *A. brasiliense* Sp7 as well as a number of other *Azospirillum* strains showed DNA homology with *Agrobacterium* chromosomal virulence (*chv*) genes (176). However, rhicadhesin-mediated attachment of *A. brasiliense* Sp7 could not be shown (151), indicating that these results might not reflect homology in the attachment mechanism of this bacterium with species of members of the *Rhizobiaceae*.

Figure 4. Simplified common model for attachment of bacteria to plant cell surfaces. Step 1 attachment is mediated by a specific attachment mechanism and appears to be essential for establishment of the bacterium-plant interaction. Step 2 attachment is mediated by bacterial fibrillous appendages, plant compounds (i.e., plant lectins), or both. Physicochemical interactions are most evidently shown in step 2 attachment. It should be noted that this two-step attachment model does not necessarily imply that in all cases step 1 attachment should precede step 2 attachment.

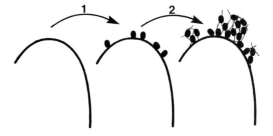

FINAL REMARKS

As reviewed in this chapter, bacteria use different strategies to attach to plant cells. This variation might reflect differences in the type of interaction as well as differences in the surfaces of the target cells. With respect to the latter, it is surprising that *A. tumefaciens* attachment to wound sites appears to be mechanistically similar to *Rhizobium* attachment to root hairs. Another striking point is that attachment of bacteria to plant cells generally seems to be a two-step process (Fig. 4).

In cases in which the two-step mechanism has been extensively studied (i.e., rhizobia and agrobacteria), the first step involves a specific, protein-mediated, reversible adhesion to the plant surface. Members of the *Rhizobiaceae* synthesize a Ca^{2+}-dependent adhesin, rhicadhesin. However, in the case of *P. solanacearum*, a protein has also been postulated to be involved in the first, reversible step of adhesion. In this case, the protein is a plant-produced hydroxyproline-rich agglutinin.

Fibrillous appendages of bacteria appear to be involved in the second step of attachment, leading to bacterial aggregation and anchoring of bacteria to the plant cell surface. Recent data indicate that in most (if not all) cases, the fibril-mediated attachment step is not essential for establishing the interaction between the bacterium and the plant. This appears to be true for both cellulose fibrils and proteinaceous fimbriae. In contrast, fimbriae are apparently essential for bacterium-animal cell adhesion (as reviewed in references 48 and 70). Why then is the second, fibril-mediated step prevalent in microbe-plant interactions?

In some cases, there are indications that extracellular fibrils enable a tight anchoring of the bacteria, which might be necessary under certain conditions (e.g., flushing of attachment sites by water) (101, 132). Bacterial appendages such as fimbriae and cellulose fibrils affect the physicochemical properties of the cell surface and therefore may have general effects on the cell environment. For instance, fimbria-mediated adhesion, which is reported in several plant-bacterium associations, likely involves hydrophobic interactions. Various studies have shown that adhesion to plant cells is positively correlated with adhesion to abiotic surfaces (36, 52, 148, 150, 171, 172). Thus, although data on the role of physicochemical interactions do not support a major role in attachment of bacteria to plant cells and probably are not of prime importance in plant-bacterium associations, these interactions might be important for the establishment of an environment beneficial to the bacteria (e.g., bacterium-bacterium aggregation).

It is also important to note that in most attachment studies, aspecific (reversible) attachment may have been largely overlooked, since the plant tissue was vigorously washed before attachment was examined. This fact makes it difficult to determine whether reversible, aspecific attachment is mediated by physicochemical interactions in some cases. It is conceivable that such aspecific attachment is important for bacterial colonization of plant tissue before specific interaction. Such colonization might be very important under field conditions, e.g., as a determinant of competition between various microorganisms, but may be overlooked under laboratory conditions, in which attachment is usually studied under axenic conditions.

Since so few bacterial-plant associations have been examined with regard to the precise mechanism of bacterial adhesion to the plant surface, it is difficult to judge whether the foregoing statements should be viewed as generalizations or merely indicative of the microscopic view now available. This conundrum should be carefully considered, since historical analysis of the literature clearly shows the tendency of investigators to prejudice their research approaches according to the dogma prevalent at the time. We hope that this review will not contribute to the dogma but instead lead to new approaches.

Finally, although some plant components (e.g., lectins) have been studied, much more research is needed to examine the importance of plant surface components in bacterium-plant adhesion. Special effort should be placed on identifying the plant surface receptors that interact with bacterial cell surface components. The literature in this area is encouraging that a better knowledge of the mechanisms of bacterium-plant adhesion could lead to methods for improving plant productivity.

Acknowledgments. The work referenced from our laboratory was supported by Public Health Service grants GM 33494-04 and GM 40183-01 from the National Institutes of Health and grant 62-600-1636 from the U.S. Department of Agriculture.

LITERATURE CITED

1. Abe, M., J. E. Sherwood, R. I. Hollingsworth, and F. B. Dazzo. 1984. Stimulation of clover root hair infection by lectin-binding oligosaccharides from the capsular and extracellular polysaccharides of *Rhizobium trifolii*. *J. Bacteriol.* **160**:517–520.

2. Akiyoshi, D. E., R. O. Morris, R. Hinz, B. S. Mischke, T. Kosuge, D. J. Garfinkel, M. P. Gordon, and E. W. Nester. 1983. Cytokinin-auxin balance in crown gall tumors is regulated by specific loci in the T-DNA. *Proc. Natl. Acad. Sci. USA* **80**:407–411.

3. Anderson, A. J., P. Habibzadegah-Tari, and C. S. Tepper. 1988. Molecular studies on the role of root surface agglutinin in adherence and colonization by *Pseudomonas putida*. *Appl. Environ. Microbiol.* **54**:375–380.

4. Badenoch-Jones, J., D. J. Flanders, and B. G. Rolfe. 1985. Association of *Rhizobium* strains with roots of *Trifolium repens*. *Appl. Environ. Microbiol.* **49**:1511–1520.

5. Banfalvi, Z., and A. Kondorosi. 1989. Production of root hair deformation factors by *Rhizobium meliloti* nodulation genes in *Escherichia coli*: HsnD (*NodH*) is involved in the plant host-specific modification of the NodABC factor. *Plant Mol. Biol.* **13**:1–12.

6. Banfalvi, Z., A. Nieuwkoop, M. Schell, L. Besl, and G. Stacey. 1988. Regulation of *nod* gene expression in *Bradyrhizobium japonicum*. *Mol. Gen. Genet.* **214**:420–424.

7. Barry, G. F., S. G. Rogers, R. T. Fraley, and L. Brand. 1984. Identification of a cloned cytokinin biosynthetic gene. *Proc. Natl. Acad. Sci. USA* **81**:4776–4780.

8. Bauer, W. D. 1981. Infection of legumes by rhizobia. *Annu. Rev. Plant Physiol.* **32**:407–449.

9. Bauer, W. D., T. V. Bhuvaneswari, H. E. Calvert, I. J. Law, N. S. A. Malik, and S. J. Vesper. 1985. Recognition and infection by slow-growing rhizobia, p. 247–253. *In* H. J. Evans, P. J. Bottomly, and W. E. Newton (ed.), *Nitrogen Fixation Progress*. Martinus Nijhoff Publishers B.V., Amsterdam.

10. Beyerinck, M. W. 1888. Die Bacterien der Papilionaccen-Knöllchen. *Bot. Zeitung* **46–50**:725–804.

11. Beynon, J. L., J. E. Beringer, and A. W. B. Johnston. 1980. Plasmids and host-range in *Rhizobium leguminosarum* and *Rhizobium phaseoli*. *J. Gen. Microbiol.* **120**:421–429.

12. Bohlool, B. B., and E. L. Schmidt. 1974. Lectins: a possible basis for specificity in the *Rhizobium*-legume root nodule symbiosis. *Science* **185**:269–271.

13. Bolton, G. W., E. W. Nester, and M. P. Gordon. 1986. Plant phenolic compounds induce expression of the *Agrobacterium tumefaciens* loci needed for virulence. *Science* **232**:983–985.

14. Borthakur, D., C. E. Barber, J. W. Lamb, M. J. Daniels, J. A. Downie, and A. W. B. Johnston. 1986. A mutation that blocks exopolysaccharide synthesis prevents nodulation of peas by *Rhizobium leguminosarum* but not of beans by *R. phaseoli* and is corrected by cloned DNA from *Rhizobium* or the phytopathogen *Xanthomonas*. *Mol. Gen. Genet.* **203**:320–323.

15. Boulter, D., J. J. Jeremy, and M. Wilding. 1966. Amino acids liberated into the culture medium by pea seedling roots. *Plant Soil* **24**:121–127.

16. Bradley, D. E., C. J. Douglas, and J. Peschon. 1984. Flagella-specific bacteriophages of *Agrobacterium tumefaciens*: demonstration of virulence of nonmotile mutants. *Can. J. Microbiol.* **30**:676–681.

17. Braun, A. C. 1958. A physiological basis for autonomous growth of crown gall tumor cell. *Proc. Natl. Acad. Sci. USA* **44**:344–349.

18. Caetano Anolles, G., and G. Favelukes. 1986. Quantitation of adsorption of rhizobia in low numbers to small legume roots. *Appl. Environ. Microbiol.* **52**:371–376.

19. Caetano Anolles, G., and G. Favelukes. 1986. Host-symbiont specificity expressed

during early adsorption of *Rhizobium meliloti* to the root surface of alfalfa. *Appl. Environ. Microbiol.* **52**:377–382.

20. **Callaham, D. A., and J. G. Torrey.** 1981. The structural basis for infection of root hairs of *Trifolium repens* by *Rhizobium*. *Can. J. Bot.* **62**:1647–1664.

21. **Calvert, H. E., M. Pence, N. S. A. Malik, and W. D. Bauer.** 1984. Anatomical analysis of the development and distribution of *Rhizobium* infections in soybean roots. *Can. J. Bot.* **62**:2735–2744.

22. **Cangelosi, G. A., L. Hung, V. Puvanesarajah, G. Stacey, D. A. Ozga, J. A. Leigh, and E. W. Nester.** 1987. Common loci for *Agrobacterium tumefaciens* and *Rhizobium meliloti* exopolysaccharide synthesis and their roles in plant interactions. *J. Bacteriol.* **169**:2086–2091.

23. **Carlson, R. W., S. Kalembasa, D. Turowski, P. Pachori, and K. D. Noel.** 1987. Characterization of the lipopolysaccharide from a *Rhizobium phaseoli* mutant that is defective in infection thread development. *J. Bacteriol.* **169**:4923–4928.

24. **Chilton, M.-D., M. H. Drummond, D. J. Merlo, D. Sciaky, A. L. Montoyo, M. P. Gordon, and E. W. Nester.** 1977. Stable incorporation of plasmid DNA into higher plant cells: the molecular basis of crown gall tumorigenesis. *Cell* **11**:263–271.

25. **Dart, P.** 1974. The infection process, p. 381–429. *In* A. Quispel (ed.), *The Biology of Nitrogen Fixation*. Elsevier/North-Holland Publishing Co., Amsterdam.

26. **Dazzo, F. B.** 1981. Bacterial attachment as related to cellular recognition in the *Rhizobium*-legume symbiosis. *J. Supramol. Struct. Cell. Biochem.* **16**:29–41.

27. **Dazzo, F. B., and W. J. Brill.** 1979. Bacterial polysaccharide which binds *Rhizobium trifolii* to clover root hairs. *J. Bacteriol.* **137**:1362–1373.

28. **Dazzo, F. B., R. I. Hollingsworth, J. E. Sherwood, M. Abe, E. M. Hrabak, A. E. Gardiol, H. S. Pankratz, K. B. Smith, and H. Yang.** 1985. Recognition and infection of clover root hairs by *Rhizobium trifolii*, p. 239–245. *In* H. J. Evans, P. J. Bottomly, and W. E. Newton (ed.), *Nitrogen Fixation Research Progress*. Martinus Nijhoff Publishers B.V., Dordrecht, The Netherlands.

29. **Dazzo, F. B., and D. H. Hubbell.** 1975. Cross-reactive antigens and lectins as determinants of symbiotic specificity in the *Rhizobium*-clover association. *Appl. Microbiol.* **30**:1018–1033.

30. **Dazzo, F. B., C. A. Napoli, and D. Hubbell.** 1976. Adsorption of bacteria to roots as related to host specificity in the *Rhizobium*-clover symbiosis. *Appl. Environ. Microbiol.* **32**:166–171.

31. **Dazzo, F. B., G. L. Truchet, J. E. Sherwood, E. M. Hrabak, M. Abe, and S. H. Pankratz.** 1984. Specific phases of root hair attachment in the *Rhizobium trifolii* clover symbiosis. *Appl. Environ. Microbiol.* **48**:1140–1150.

32. **Dazzo, F. B., M. R. Urbano, and W. J. Brill.** 1979. Transient appearance of lectin receptors on *Rhizobium trifolii*. *Curr. Microbiol.* **2**:15–20.

33. **Debellé, F., S. B. Sharma, C. Rosenberg, J. Vasse, F. Mailet, G. Truchet, and J. Dénairié.** 1986. Respective roles of common and specific *Rhizobium meliloti nod* genes in the control of lucerne infection, p. 17–28. *In* B. Lugtenberg (ed.), *Recognition in Microbe-Plant Symbiotic and Pathogenic Interactions*. NATO ASI Series vol. H4. Springer-Verlag KG, Berlin.

34. **Deinema, M. H., and L. P. T. M. Zevenhuizen.** 1971. Formation of cellulose fibrils by gram-negative bacteria and their role in bacterial flocculation. *Arch. Microbiol.* **78**:42–57.

35. **De Maagd, R. A., A. S. Rao, I. H. M. Mulders, L. Goossen-de Roo, M. van Loosdrecht, C. A. Wijffelman, and B. J. J. Lugtenberg.** 1989. Isolation and characterization of mutants of *Rhizobium leguminosarum* biovar *viciae* strain 248 with altered lipopoly-

saccharides: possible role of surface charge or hydrophobicity in bacterial release from the infection thread. *J. Bacteriol.* **171**:1143–1150.

36. **De Weger, L. A., M. C. M. van Loosdrecht, H. E. Klaassen, and B. J. J. Lugtenberg.** 1989. Mutational changes in physicochemical cell surface properties of plant-growth-stimulating *Pseudomonas* spp. do not influence the attachment properties of the cells. *J. Bacteriol.* **171**:2756–2761.

37. **Díaz, C. L., L. S. Melchers, P. J. J. Hooykaas, B. J. J. Lugtenberg, and J. W. Kijne.** 1989. Root lectin as a determinant of host specificity in the *Rhizobium*-legume symbiosis. *Nature* (London) **338**:579–581.

38. **Douglas, C., W. Halperin, and E. W. Nester.** 1982. *Agrobacterium tumefaciens* affected in attachment to plant cells. *J. Bacteriol.* **152**:1265–1275.

39. **Douglas, C. J., R. J. Staneloni, R. A. Rubin, and E. W. Nester.** 1985. Identification and genetic analysis of an *Agrobacterium tumefaciens* chromosomal virulence region. *J. Bacteriol.* **161**:850–860.

40. **Downie, J. A., G. Hombrecher, Q.-S. Ma, C. D. Knight, B. Wells, and A. W. B. Johnston.** 1983. Cloned nodulation genes of *Rhizobium leguminosarum* determine host-range specificity. *Mol. Gen. Genet.* **190**:359–365.

41. **Downie, J. A., C. D. Knight, A. W. B. Johnston, and L. Rossen.** 1985. Identification of genes and gene products involved in the nodulation of peas by *Rhizobium leguminosarum*. *Mol. Gen. Genet.* **198**:255–262.

42. **Duvick, J. P., and L. Sequeira.** 1984. Interaction of *Pseudomonas solanacearum* lipopolysaccharide and extracellular polysaccharide with agglutinin from potato tubers. *Appl. Environ. Microbiol.* **48**:192–198.

43. **Duvick, J. P., and L. Sequeira.** 1984. Interaction of *Pseudomonas solanacearum* with suspension-cultured tobacco cells and tobacco leaf cell walls in vitro. *Appl. Environ. Microbiol.* **48**:199–205.

44. **Dylan, T., L. Ielpi, S. Stanfield, L. Kashyap, C. Douglas, M. Yanofsky, E. Nester, D. R. Helinski, and G. Ditta.** 1986. *Rhizobium meliloti* genes required for nodule development are related to chromosomal virulence genes in *Agrobacterium tumefaciens*. *Proc. Natl. Acad. Sci. USA* **83**:4403–4407.

45. **Faucher, C., S. Camut, J. Dénarié, and G. Truchet.** 1989. The *nodH* and *nodQ* host range genes of *Rhizobium meliloti* behave as avirulence genes in *R. leguminosarum* bv. *viciae* and determine changes in the production of plant-specific extracellular signals. *Mol. Plant Microbe Interact.* **2**:291–300.

46. **Finan, T. M., A. M. Hirsch, J. A. Leigh, E. Johanson, G. A. Kuldau, S. Deegan, G. C. Walker, and E. R. Signer.** 1985. Symbiotic mutants of *Rhizobium meliloti* that uncouple plant from bacterial differentiation. *Cell* **40**:869–877.

47. **Franssen, H. J., J.-P. Nap, T. Gloudemans, W. Stiekema, H. van Dam, F. Govers, J. Louwerse, A. van Kammen, and T. Bisseling.** 1987. Characterization of cDNA for nodulin-75 of soybean: a gene product involved in early stages of root nodule development. *Proc. Natl. Acad. Sci. USA* **84**:4495–4499.

48. **Gaastra, W., and F. K. de Graaf.** 1982. Host-specific fimbrial adhesins on noninvasive enterotoxigenic *Escherichia coli* strains. *Microbiol. Rev.* **46**:129–161.

49. **Gafny, R., Y. Okon, and Y. Kapulnik.** 1981. Adsorption of *Azospirillum brasilense* to corn roots. *Soil Biol. Biochem.* **18**:69–75.

50. **Geremia, R. A., S. Cavaignac, A. Zorrequieta, N. Toro, J. Olivares, and R. A. Ugalde.** 1987. A *Rhizobium meliloti* mutant that forms ineffective pseudonodules in alfalfa produces exopolysaccharide but fails to form β-(1→2)glucan. *J. Bacteriol.* **169**:880–884.

51. **Glogowski, W., and A. G. Galsky.** 1978. *Agrobacterium tumefaciens* site attachment as

a necessary prerequisite for crown gall tumor formation on potato discs. *Plant Physiol.* **61**:1031–1033.

52. **Haahtela, K., E. Tarkka, and T. K. Korhonen.** 1985. Type 1 fimbria-mediated adhesion of enteric bacteria to grass roots. *Appl. Environ. Microbiol.* **49**:1182–1185.

53. **Haahtela, K., T. Laakso, and T. K. Korhonen.** 1986. Associative nitrogen fixation by *Klebsiella* spp.: adhesion sites and inoculum effects on grass roots. *Appl. Environ. Microbiol.* **52**:1074–1079.

54. **Halverson, L. J., and G. Stacey.** 1985. Host recognition in the *Rhizobium*-legume symbiosis: evidence for the involvement of lectin in nodulation. *Plant Physiol.* **77**:621–625.

55. **Halverson, L. J., and G. Stacey.** 1986. Effect of lectin on nodulation by wild-type *Bradyrhizobium japonicum* and a nodulation-deficient mutant. *Appl. Environ. Microbiol.* **51**:753–760.

56. **Halverson, L. J., and G. Stacey.** 1986. Signal exchange in plant-microbe interactions. *Microbiol. Rev.* **50**:193–225.

57. **Hawes, M. C., and S. G. Pueppke.** 1989. Variation in binding and virulence of *Agrobacterium tumefaciens* chromosomal virulence mutants on different plant species. *Plant Physiol.* **91**:113–118.

58. **Hellriegel, H., and H. Wilfarth.** 1888. *Untersüchungen über die Stickstoffnahrung der Gramineën und Leguminosen*, p. 1–234. Beilageheft Z. Ver. Rubenzucker Industrie, Deutschen Reichs.

59. **Hendrick, C. A., and L. Sequeira.** 1984. Lipopolysaccharide-defective mutants of the wilt pathogen *Pseudomonas solanacearum*. *Appl. Environ. Microbiol.* **48**:94–101.

60. **Hennecke, H., H. M. Fischer, S. Ebeling, M. Gubler, B. Thöny, M. Göttfert, J. Lamb, M. Hahn, T. Ramseier, B. Regensburger, A. Alvarez-Morales, and D. Studer.** 1987. *Nif*, *fix* and *nod* gene clusters in *Bradyrhizobium japonicum*, and *nifA*-mediated control of symbiotic nitrogen fixation, p. 191–196. *In* D. P. S. Verma and N. Brisson (ed.), *Molecular Genetics of Plant-Microbe Interactions*. Martinus Nijhoff Publishers B.V., Dordrecht, The Netherlands.

61. **Hille, J., I. Klaasen, and R. Schilperoort.** 1982. Construction and application of R-prime plasmid from *Agrobacterium tumefaciens* for complementation of vir genes. *Plasmid* **7**:107–118.

62. **Hirsch, P. R., M. Van Montagu, A. W. B. Johnston, N. J. Brewin, and J. Schell.** 1980. Physical identification of bacteriocinogenic, nodulation and other plasmids in strains of *Rhizobium leguminosarum*. *J. Gen. Microbiol.* **120**:403–412.

63. **Hooykaas, P. J. J., P. M. Klapwijk, M. P. Nuti, R. A. Schilperoort, and A. Rorsch.** 1977. Transfer of the *Agrobacterium tumefaciens* Ti plasmid to avirulent agrobacteria and to *Rhizobium ex planta*. *J. Gen. Microbiol.* **98**:477–484.

64. **Hrabak, E. M., M. R. Urbano, and F. B. Dazzo.** 1981. Growth-phase-dependent immunodeterminants of *Rhizobium trifolii* lipopolysaccharide which bind trifoliin A, a white clover lectin. *J. Bacteriol.* **148**:697–711.

65. **Innes, R. W., P. L. Keumpel, J. Plazinski, H. Canter-Cremers, B. G. Rolfe, and M. A. Djordjevic.** 1985. Plant factors induce expression of nodulation and host-range genes in *Rhizobium trifolium*. *Mol. Gen. Genet.* **201**:426–432.

66. **Inon de Iannino, N., and R. A. Ugalde.** 1989. Biochemical characterization of avirulent *Agrobacterium tumefaciens chvA* mutants: synthesis and excretion of β-(1-2)glucan. *J. Bacteriol.* **171**:2842–2849.

67. **Inze, D., A. Follin, M. Van Lijsebettens, C. Simeons, C. Genetello, M. Van Montagu, and J. Schell.** 1984. Genetic analysis of the individual T-DNA genes of *Agrobacterium*

tumefaciens: further evidence that two genes are involved in indole-3-acetic acid synthesis. *Mol. Gen. Genet.* **194**:265–274.

68. James, D. W., Jr., T. V. Suslow, and K. E. Steinback. 1985. Relationship between rapid, firm adhesion and long-term colonization of roots by bacteria. *Appl. Environ. Microbiol.* **50**:392–397.

69. Johnston, A. W. B., J. L. Beynon, A. V. Buchanan-Wollaston, S. M. Setchell, P. R. Hirsch, and J. E. Beringer. 1978. High frequency transfer of nodulating ability between strains and species of *Rhizobium*. *Nature* (London) **276**:635–636.

70. Jones, G. W., and R. E. Isaacson. 1983. Proteinaceous bacterial adhesins and their receptors. *Crit. Rev. Microbiol.* **10**:229–260.

71. Kamberger, W. 1979. Role of surface polysaccharides in the *Rhizobium*-pea symbiosis. *FEMS Microbiol. Lett.* **6**:361–365.

72. Kamberger, W. 1979. An Ouchterlony double diffusion study on the interactions between legume lectins and rhizobial cell surface antigens. *Arch. Microbiol.* **121**:83–90.

73. Kato, G., Y. Maruyama, and M. Nakamura. 1980. Role of bacterial polysaccharides in the adsorption process of the *Rhizobium*-pea symbiosis. *Agric. Biol. Chem.* **44**:2843–2855.

74. Kelman, A., and E. B. Cowling. 1965. Cellulase of *Pseudomonas solanacearum* in relation to pathogenesis. *Phytopathology* **55**:148–155.

75. Kelman, A., and J. Hruschka. 1973. The role of motility and aerotaxis in the selective increase of avirulent bacteria in still broth of *Pseudomonas solanacearum*. *J. Gen. Microbiol.* **76**:177–188.

76. Kijne, J. W. 1975. The fine structure of pea root nodules. I. Vacuolar changes after endocytotic host cell infection by *Rhizobium leguminosarum*. *Physiol. Plant Pathol.* **5**:75–79.

77. Kijne, J. W. 1975. The fine structure of pea root nodules. II. Senescence and disintegration of the bacteroid tissue. *Physiol. Plant Pathol.* **7**:17–21.

78. Kijne, J. W., G. Smit, C. L. Díaz, and B. J. J. Lugtenberg. 1986. Attachment of *Rhizobium leguminosarum* to pea root hair tips, p. 101–111. *In* B. Lugtenberg (ed.), *Recognition in Microbe-Plant Symbiotic and Pathogenic Interactions*. NATO ASI Series vol. H4. Springer-Verlag KG, Berlin.

79. Kijne, J. W., G. Smit, C. L. Díaz, and B. J. J. Lugtenberg. 1988. Lectin-enhanced accumulation of manganese-limited *Rhizobium leguminosarum* cells on pea root hair tips. *J. Bacteriol.* **170**:2994–3000.

80. Klee, H. J., F. F. White, V. N. Iyer, M. P. Gordon, and E. W. Nester. 1983. Mutational analysis of the virulence region of an *Agrobacterium tumefaciens* Ti plasmid. *J. Bacteriol.* **153**:878–883.

81. Kondorosi, E., Z. Banfalvi, and A. Kondorosi. 1984. Physical and genetic analysis of a symbiotic region of *Rhizobium meliloti*: identification of nodulation genes. *Mol. Gen. Genet.* **193**:445–452.

82. Korhonen, T. K., E. Nurmiako, H. Ranta, and C. Svanborg-Eden. 1980. New method for isolation of immunological pure pili from *Escherichia coli*. *Infect. Immun.* **27**:569–575.

83. Korhonen, T. K., E. Tarkka, H. Ranta, and H. Haahtela. 1983. Type 3 fimbriae of *Klebsiella* spp.: molecular characterization and role in bacterial adhesion to plant roots. *J. Bacteriol.* **155**:860–865.

84. Krens, F. A., L. Molendijk, G. J. Wullems, and R. A. Schilperoort. 1985. The role of bacterial attachment in the transformation of cell-wall-regenerating tobacco protoplasts by *Agrobacterium tumefaciens*. *Planta* **166**:300–308.

85. Leach, J. E., M. A. Cantrell, and L. Sequeira. 1982. A hydroxyproline-rich bacterial

agglutinin from potato: its localization by immunofluorescense. *Physiol. Plant Pathol.* **21**:319–325.

86. **Leach, J. E., M. A. Cantrell, and L. Sequeira.** 1982. Hydroxyproline-rich bacterial agglutinin from potato: extraction, purification, and characterization. *Plant Physiol.* **70**:1353–1358.

87. **Leigh, J. A., J. W. Reed, J. F. Hanks, A. M. Hirsch, and G. C. Walker.** 1987. *Rhizobium meliloti* mutants that fail to succinylate their calcofluor-binding exopolysaccharide are defective in nodule-invasion. *Cell* **51**:579–587.

88. **Leigh, J. A., E. R. Signer, and G. C. Walker.** 1985. Exopolysaccharide-deficient mutants of *Rhizobium meliloti* that form ineffective nodules. *Proc. Natl. Acad. Sci. USA* **82**:6231–6235.

89. **Li, D., and D. H. Hubbell.** 1969. Infection thread formation as a basis of nodulation specificity in *Rhizobium trifolii-Trifolium fragerum* associations. *Can. J. Microbiol.* **15**:1133–1136.

90. **Libbenga, K. R., and R. J. Bogers.** 1974. Root nodule morphogenesis, p. 430–472. *In* A. Quispel (ed.), *The Biology of Nitrogen Fixation*. Elsevier/North-Holland Publishing Co., Amsterdam.

91. **Libbenga, K. R., F. Van Iren, R. J. Bogers, and M. F. Schraag-Lamers.** 1973. The role of hormones and gradients in the initiation of cortex proliferation and nodule formation in *Pisum sativum* L. *Planta* **114**:29–39.

92. **Lippincott, B. B., and J. A. Lippincott.** 1969. Bacterial attachment to a specific wound site as an essential stage in tumor initiation by *Agrobacterium tumefaciens. J. Bacteriol.* **97**:620–628.

93. **Lippincott, B. B., M. H. Whatley, and J. A. Lippincott.** 1979. Tumor induction of *Agrobacterium* involves attachment of bacterium to a site on the host plant cell wall. *Plant Physiol.* **59**:388–390.

94. **Lippincott, J. A., and B. B. Lippincott.** 1975. The genus *Agrobacterium* and plant tumorigenesis. *Annu. Rev. Microbiol.* **29**:377–405.

95. **Lippincott, J. A., and B. B. Lippincott.** 1980. Microbial adherence in plants, p. 375–398. *In* E. H. Beachey (ed.), *Microbial Adherence*, series B, vol. 6. Chapman & Hall, Ltd., London.

96. **Lipton, D. S., R. W. Blanchar, and D. G. Blevins.** 1987. Citrate, malate, and succinate concentrations in exudates from P-sufficient and P-stressed *Medicago sativa* L. seedlings. *Plant Physiol.* **85**:315–317.

97. **Long, S., J. W. Reed, J. Himawan, and G. C. Walker.** 1988. Genetic analysis of a cluster of genes required for synthesis of the calcofluor-binding exopolysaccharide of *Rhizobium meliloti. J. Bacteriol.* **170**:4239–4248.

98. **Long, S. R., W. J. Buikema, and F. B. Ausubel.** 1982. Cloning of *Rhizobium meliloti* nodulation genes by direct complementation of nod⁻ mutants. *Nature* (London) **298**:485–489.

99. **Lugtenberg, B. J. J., G. Smit, C. L. Díaz, and J. W. Kijne.** 1989. Attachment of *Rhizobium leguminosarum* cells to pea root hair tips seems not to be sufficient to generate signals for symbiosis, p. 129–136. *In* B. J. J. Lugtenberg (ed.), *Molecular Signals in Microbe-Plant Symbiotic and Pathogenic Systems*. NATO ASI Series vol. H36. Springer-Verlag KG, Berlin.

100. **Marton, J., G. J. Wullems, L. Molendijk, and R. A. Schilperoort.** 1979. In vitro transformation of cultured cells from *Nicotiana tabacum* by *Agrobacterium tumefaciens. Nature* (London) **277**:129–131.

101. **Matthysse, A. G.** 1983. Role of bacterial cellulose fibrils in *Agrobacterium tumefaciens* infections. *J. Bacteriol.* **154**:906–915.

102. **Matthysse, A. G.** 1986. Attachment of *Agrobacterium* to plant host cells, p. 219–227. *In* B. Lugtenberg (ed.), *Recognition in Microbe-Plant Symbiotic and Pathogenic Interactions*. NATO ASI Series vol. H4. Springer-Verlag KG, Berlin.

103. **Matthysse, A. G.** 1987. Characterization of nonattaching mutants of *Agrobacterium tumefaciens*. *J. Bacteriol.* **169**:313–323.

104. **Matthysse, A. G., K. V. Holmes, and R. H. G. Gurlitz.** 1981. Elaboration of cellulose fibrils by *Agrobacterium tumefaciens* during attachment to carrot cells. *J. Bacteriol.* **145**:583–595.

105. **Matthysse, A. G., P. M. Wyman, and K. V. Holmes.** 1978. Plasmid-dependent attachment of *Agrobacterium tumefaciens* to plant tissue culture cells. *Infect. Immun.* **22**:516–522.

106. **Menzel, G., H. Uhig, and G. Weischael.** 1972. Settling of rhizobia and other soil bacteria on the roots of some legumes and non-legumes. *Zentralbl. Bakteriol. Parasitenkd. Infektionskr. Hyg.* **127**:348–358.

107. **Miller, K. J., E. P. Kennedy, and V. N. Reinhold.** 1986. Osmotic adaptation by gram-negative bacteria: possible role for periplasmic oligosaccharides. *Science* **231**: 48–51.

108. **Mills, K. M., and W. D. Bauer.** 1985. *Rhizobium* attachment to clover roots. *J. Cell Sci. Suppl.* **2**:333–345.

109. **Morrison, N. A., T. Bisseling, and D. P. S. Verma.** 1988. Development and differentiation of the root nodule: involvement of plant and bacterial genes, p. 405–425. *In* L. W. Browder (ed.), *Developmental Biology*, vol. 5. Plenum Publishing Corp., New York.

110. **Morrison, N., and D. P. S. Verma.** 1987. A block in the endocytosis of *Rhizobium* allows cellular differentiation in nodules but affects the expression of some peribacteroid membrane nodulins. *Plant Mol. Biol.* **9**:185–196.

111. **Mort, A. J., and W. D. Bauer.** 1980. Composition of the capsular and extracellular polysaccharides of *Rhizobium japonicum*: changes with culture age and correlation with binding of soybean seed lectin to the bacteria. *Plant Physiol.* **66**:158–163.

112. **Mort, A. J., and W. D. Bauer.** 1981. Structure of the capsular and extracellular polysaccharides of *Rhizobium japonicum* that bind soybean lectin. Application of two new methods for cleavage of polysaccharides into specific oligosaccharide fragments. *J. Biol. Chem.* **257**:1870–1875.

113. **Mulligan, J. T., and S. R. Long.** 1985. Induction of *Rhizobium meliloti nodC* expression by plant exudate requires *nodD*. *Proc. Natl. Acad. Sci. USA* **82**:6609–6613.

114. **Nap, J.-P., and T. Bisseling.** 1990. Nodulin function and nodulin gene regulation in root nodule development, p. 181–229. *In* P. M. Gresshoff (ed.), *Molecular Biology of Symbiotic Nitrogen Fixation*. CRC Press, Inc., Boca Raton, Fla.

115. **Napoli, C., F. Dazzo, and D. Hubbell.** 1975. Production of cellulose fibrils by *Rhizobium*. *Appl. Microbiol.* **30**:123–131.

116. **Nuti, M. P., A. A. Lepidi, R. H. Prakash, R. A. Schilperoort, and F. C. Cannon.** 1979. Evidence for nitrogen fixation (*nif*) genes on indigous *Rhizobium* plasmids. *Nature* (London) **282**:533–535.

117. **Ohyama, K., L. E. Pelcher, and A. Schaefer.** 1979. In vitro binding of *Agrobacterium tumefaciens* to plant cells from suspension culture. *Plant Physiol.* **63**:382–387.

118. **Okon, Y.** 1985. *Azospirillum* as a potential inoculant for agriculture. *Trends Biotechnol.* **3**:223–228.

119. **Ooms, G., A. Bakker, L. Molendijk, G. J. Wullems, M. P. Gordon, E. W. Nester, and R. A. Schilperoort.** 1982. T-DNA organization in homogenous and heterogenous octopine-type crown gall tissues of *Nicotiana tabacum*. *Cell* **30**:589–597.

120. **Ooms, G., P. J. J. Hooykaas, G. Moolenaar, and R. A. Schilperoort.** 1981. Crown gall

plant tumors of abnormal octopine Ti plasmids: analysis of T-DNA functions. *Gene* **14**:33–50.

121. **Peters, N. K., J. W. Frost, and S. R. Long.** 1986. A plant flavone, luteolin, induces expression of *Rhizobium meliloti* nodulation genes. *Science* **233**:977–980.

122. **Phelps, R. H., and L. Sequiera.** 1967. Synthesis of indoleacetic acid by a cell-free system from virulent and avirulent strains of *Pseudomonas solanacearum. Phytopathology* **57**:1182–1190.

123. **Planque, K., and J. W. Kijne.** 1977. Binding of pea lectins to a glycan type polysaccharide in the cell walls of *Rhizobium leguminosarum. FEBS Lett.* **73**:64–66.

124. **Pueppke, S. G.** 1984. Adsorption of slow- and fast-growing rhizobia to soybean and cowpea roots. *Plant Physiol.* **75**:924–928.

125. **Pueppke, S. G., and U. K. Benny.** 1983. *Agrobacterium* tumorigenesis in potato: effect of added *Agrobacterium* lipopolysaccharides and the degree of methylation of added plant galacturans. *Physiol. Plant Pathol.* **23**:439–446.

126. **Puvanesarajah, V., F. M. Schell, G. Stacey, C. J. Douglas, and E. W. Nester.** 1985. Role for 2-linked-β-glucan in the virulence of *Agrobacterium tumefaciens. J. Bacteriol.* **164**:102–106.

127. **Redmond, J. W., M. Batley, M. A. Djordjevic, R. W. Innes, P. L. Keumpel, and B. G. Rolfe.** 1986. Flavones induce the expression of the nodulation genes in *Rhizobium. Nature* (London) **323**:632–635.

128. **Robertson, J. G., P. Lyttleton, S. Bullivant, and G. E. Grayson.** 1978. Membranes in lupin root nodules. I. The role of Golgi bodies in biogenesis of infection threads and peribacteroid membranes. *J. Cell Sci.* **30**:129–149.

129. **Robertson, J. G., P. Lyttleton, and C. E. Pankhurst.** 1981. Preinfection and infection processes in the legume-*Rhizobium* symbiosis, p. 280–291. *In* A. H. Gibson and W. E. Newton (ed.), *Current Perspectives in Nitrogen Fixation.* Australian Academy of Science, Canberra.

130. **Rolfe, B. G., M. A. Djordjevic, K. F. Scott, J. E. Hughes, J. Badenoch-Jones, P. M. Gresshoff, Y. Cen, W. F. Dudman, W. Zurkowski, and J. Shine.** 1981. Analysis of nodule forming ability of fast-growing *Rhizobium* strains, p. 142–145. *In* A. H. Gibson and W. E. Newton (ed.), *Current Perspectives in Nitrogen Fixation.* Australian Academy of Science, Canberra.

131. **Rolfe, B. G., R. W. Innes, P. R. Schofield, J. M. Watson, C. L. Sargent, P. L. Keumpel, J. Plazinski, H. Canter-Cremers, and M. A. Djordjevic.** 1985. Plant-secreted factors induce the expression of *R. trifolii* nodulation and host-range genes, p. 79–85. *In* H. J. Evans, P. J. Bottomly, and W. E. Newton (ed.), *Nitrogen Fixation Progress.* Martinus Nijhoff Publishers B.V., Amsterdam.

132. **Romanschuk, M., and D. H. Bamford.** 1986. The causal agent of halo blight in bean, *Pseudomonas syringae* pv. *phaseolicola*, attaches to stomata via its pili. *Microb. Pathog.* **1**:139–148.

133. **Rosenberg, C., P. Boistard, J. Dénarié, and F. Casse-Delbart.** 1981. Genes controlling early and late functions in symbiosis are located on a megaplasmid in *R. meliloti. Mol. Gen. Genet.* **184**:326–333.

134. **Rossen, L., A. W. B. Johnston, and J. A. Downie.** 1984. DNA sequences of the *Rhizobium leguminosarum* nodulation genes *nodAB* and *C* required for root hair curling. *Nucleic Acids Res.* **12**:9497–9508.

135. **Roth, E., K. Jeon, and G. Stacey.** 1988. Homology in endosymbiotic systems: the term "symbiosome," p. 220–225. *In* R. Palacios and D. P. S. Verma (ed.), *The Molecular Genetics of Plant-Microbe Interactions.* American Phytopathological Society Press, St. Paul, Minn.

136. Roth, L. E., and G. Stacey. 1989. Bacterium release into host cells of nitrogen-fixing soybean nodules: the symbiosome membrane comes from three sources. *Eur. J. Cell Biol.* **49**:13–23.

137. Roth, L. E., and G. Stacey. 1989. Cytoplasmic membrane systems involved in bacterium release into soybean nodule cells as studies with two *Bradyrhizobium japonicum* mutant strains. *Eur. J. Cell Biol.* **49**:24–32.

138. Rutter, P. R., and B. Vincent. 1984. Physicochemical interactions of the substratum, microorganisms, and the fluid phase, p. 21–38. *In* K. C. Marshall (ed.), *Microbial Adhesion and Aggregation.* Springer-Verlag KG, Berlin.

139. Sanders, R. E., R. W. Carlson, and P. Albersheim. 1978. A *Rhizobium* mutant incapable of nodulation and normal polysaccharide secretion. *Nature* (London) **271**:240–242.

140. Sanderson, K. E., and P. E. Hartman. 1978. Linkage map of *Salmonella typhimurium*, edition 5. *Microbiol. Rev.* **42**:471–519.

141. Sarig, S., Y. Kapulnik, and Y. Okon. 1986. Effect of *Azospirillum brasiliense* on nitrogen fixation and growth of several winter legumes. *Plant Soil* **90**:335–342.

142. Schofield, P. R., R. W. Ridge, B. G. Rolfe, J. Shine, and J. Watson. 1984. Host-specific nodulation is encoded on a 14 kb DNA fragment in *Rhizobium trifolii. Plant Mol. Biol.* **3**:3–11.

143. Schroth, M. N., and J. G. Hancock. 1981. Selected topics in biological control. *Annu. Rev. Microbiol.* **35**:453–476.

144. Schroth, M. N., and J. G. Hancock. 1982. Disease-suppressive soil and root-colonizing bacteria. *Science* **216**:1376–1381.

145. Sequeira, L. 1984. Recognition systems in plant-pathogen interactions. *Biol. Chem.* **51**:281–286.

146. Sequeira, L., G. Gaard, and G. A. De Soeten. 1977. Interaction of bacteria and host cell walls: its relation to mechanisms of induced resistance. *Physiol. Plant Pathol.* **10**:43–50.

147. Sequeira, L., and T. L. Graham. 1977. Agglutination of avirulent strains of *Pseudomonas solanacearum* by potato lectin. *Physiol. Plant Pathol.* **11**:43–54.

148. Smit, G., J. W. Kijne, and B. J. J. Lugtenberg. 1986. Correlation between extracellular fibrils and attachment of *Rhizobium leguminosarum* to pea root hair tips. *J. Bacteriol.* **168**:821–827.

149. Smit, G., J. W. Kijne, and B. J. J. Lugtenberg. 1987. Both cellulose fibrils and a Ca^{2+}-dependent adhesin are involved in the attachment of *Rhizobium leguminosarum* to pea root hair tips. *J. Bacteriol.* **169**:4294–4301.

150. Smit, G., J. W. Kijne, and B. J. J. Lugtenberg. 1989. Roles of flagella, lipopolysaccharide, and a Ca^{2+}-dependent cell surface protein in attachment of *Rhizobium leguminosarum* biovar *viciae* to pea root hair tips. *J. Bacteriol.* **171**:569–572.

151. Smit, G., T. J. J. Logman, M. E. T. I. Boerrigter, J. W. Kijne, and B. J. J. Lugtenberg. 1989. Purification and partial characterization of the Ca^{2+}-dependent adhesin from *Rhizobium leguminosarum* biovar *viciae*, which mediates the first step in attachment of *Rhizobiaceae* cells to plant root hair tips. *J. Bacteriol.* **171**:4054–4062.

152. Smit, G., A. A. van der Baan, J. W. Kijne, and B. J. J. Lugtenberg. 1986. The attachment mechanism of *Rhizobium. Antonie van Leeuwenhoek J. Microbiol. Serol.* **52**:362–363.

153. Spaink, H. P., C. A. Wijffelman, E. Pees, R. J. H. Okker, and B. J. J. Lugtenberg. 1987. *Rhizobium* nodulation gene *nodD* as a determinant of host specificity. *Nature* (London) **328**:337–340.

154. Stacey, G., L. J. Halverson, T. Nieuwkoop, Z. Banfalvi, M. G. Schell, D. Gerhold, N.

Deshmane, J. S. So, and K. M. Sirotkin. 1986. Nodulation of soybean: *Bradyrhizobium japonicum* physiology and genetics, p. 87–100. *In* B. Lugtenberg (ed.), *Recognition in Microbe-Plant Symbiotic and Pathogenic Interactions*. NATO ASI Series vol. H4. Springer-Verlag KG, Berlin.

155. **Stacey, G., A. S. Paau, and W. J. Brill.** 1980. Host recognition in the *Rhizobium*-soybean symbiosis. *Plant Physiol.* **66:**609–614.

156. **Stacey, G., L. A. Pocratsky, and V. Puvanesarajah.** 1984. Bacteriophage that can distinguish between wild-type *Rhizobium japonicum* and a non-nodulating mutant. *Appl. Environ. Microbiol.* **48:**68–72.

157. **Stacey, G., M. G. Schell, and N. Deshmane.** 1989. Determinants of host specificity in the *Bradyrhizobium japonicum*-soybean symbiosis, p. 395–400. *In* B. J. J. Lugtenberg (ed.), *Molecular Signals in Microbe-Plant Symbiotic and Pathogenic Systems*. NATO ASI Series vol. H36. Springer-Verlag KG, Berlin.

158. **Stachel, S. E., E. Messens, M. Van Montagu, and P. Zambryski.** 1985. Identification of the plant signal molecules produced by wounded plant cells that activate T-DNA transfer in *Agrobacterium tumefaciens*. *Nature* (London) **318:**624–629.

159. **Stemmer, W. P. C., and L. Sequeira.** 1985. Possible role of fimbriae in attachment of plant-associated bacteria to plant cell walls, p. 199–201. *In* A. A. Szaley and R. P. Legocky (ed.), *Proceedings of the 2nd International Symposium on the Molecular Genetics of Bacteria-Plant Interactions*. Media Services, Cornell University, Ithaca, N.Y.

160. **Thomashow, M. F., S. Hughly, W. G. Buchholz, and L. S. Thomashow.** 1986. Molecular basis for the auxin-independent phenotype of crown gall tumor tissues. *Science* **23:**616–618.

161. **Thomashow, M. F., J. E. Karlinsey, J. R. Marks, and R. E. Hurlbert.** 1987. Identification of a new virulence locus in *Agrobacterium tumefaciens* that affects polysaccharide composition and plant cell attachment. *J. Bacteriol.* **169:**3209–3216.

162. **Thomashow, M. F., R. Nutter, A. L. Montoya, M. P. Gordon, and E. W. Nester.** 1980. Integration and organization of Ti-plasmid sequences in crown gall tumors. *Cell* **19:**729–739.

163. **Thomashow, L. S., S. Reevers, and M. F. Tomashow.** 1984. Crown gall oncogenesis: evidence that a T-DNA gene from the *Agrobacterium* Ti plasmid pTIA6 encodes an enzyme that catalyzes synthesis of indole acetic acid. *Proc. Natl. Acad. Sci. USA* **81:**5071–5075.

164. **Umali-Garcia, M., D. H. Hubbell, M. H. Gaskins, and F. B. Dazzo.** 1980. Association of *Azospirillum* with grass roots. *Appl. Environ. Microbiol.* **39:**219–226.

165. **Van Batenburg, F. D. H., R. Jonker, and J. W. Kijne.** 1986. *Rhizobium* induces marked root hair curling by redirection of tip growth, a computer simulation. *Physiol. Plant.* **66:**476–480.

166. **Van Brussel, A. A. N., S. A. J. Zaat, H. C. J. Canter Cremers, C. A. Wijffelman, E. Pees, T. Tak, and B. J. J. Lugtenberg.** 1986. Role of plant root exudate and Sym plasmid-located nodulation genes in the synthesis by *Rhizobium leguminosarum* of Tsr factor, which causes thick and short roots on common vetch. *J. Bacteriol.* **165:**517–522.

167. **Van der Schaal, I. A. M., J. W. Kijne, C. L. Díaz, and F. van Iren.** 1983. Pea lectin binding by *Rhizobium*, p. 531–538. *In* T. C. Bøg-Hansen and G. A. Spengler (ed.), *Lectins*, vol. 3. Walter de Gruyter, Berlin.

168. **Van Larebeke, N., G. Engler, M. Holsters, S. Van den Elsacker, I. Zaenen, R. A. Schilperoort, and J. Schell.** 1974. Large plasmids in *Agrobacterium tumefaciens* essential for crown gall-inducing ability. *Nature* (London) **242:**171–172.

169. Van Loosdrecht, M. C. M., J. Lyklema, W. Norde, G. Schraa, and A. J. B. Zehnder. 1987. The role of bacterial cell wall hydrophobicity in adhesion. *Appl. Environ. Microbiol.* **53**:1893–1897.

170. Van Loosdrecht, M. C. M., J. Lyklema, W. Norde, G. Schraa, and A. J. B. Zehnder. 1987. Electrophoretic mobility and hydrophobicity as a measure to predict the initial steps of bacterial adhesion. *Appl. Environ. Microbiol.* **53**:1898–1901.

171. Vesper, S. J., and W. D. Bauer. 1985. Characterization of *Rhizobium* attachment to soybean roots. *Symbiosis* **1**:139–162.

172. Vesper, S. J., and W. D. Bauer. 1986. Role of pili (fimbriae) in attachment of *Bradyrhizobium japonicum* to soybean roots. *Appl. Environ. Microbiol.* **52**:134–141.

173. Vesper, S. J., and T. V. Bhuvaneswari. 1988. Nodulation of soybean roots by an isolate of *Bradyrhizobium japonicum* with reduced firm attachment capability. *Arch. Microbiol.* **150**:15–19.

174. Vesper, S. J., N. S. A. Malik, and W. D. Bauer. 1987. Transposon mutants of *Bradyrhizobium japonicum* altered in attachment to host roots. *Appl. Environ. Microbiol.* **53**:1959–1961.

175. Vincent, J. M. 1980. Factors controlling the legume-*Rhizobium* symbiosis, p. 103–129. *In* W. E. Newton and W. H. Orme-Johnston (ed.), *Nitrogen Fixation*, vol. 2. University Park Press, Baltimore.

176. Waelkens F., M. Maris, C. Verreth, J. VanderLeyden, and A. Van Gool. 1987. *Azospirillum* DNA shows homology with *Agrobacterium* chromosomal virulence genes. *FEMS Microbiol. Lett.* **43**:241–246.

177. Whatley, M. H., J. S. Bodwin, B. B. Lippincott, and J. A. Lippincott. 1976. Role for *Agrobacterium* cell envelope lipopolysaccharide in infection site attachment. *Infect. Immun.* **13**:1080–1083.

178. Whatley, M. H., N. Hunter, M. A. Cantrell, C. A. Hendrick, L. Sequeira, and K. Keegstra. 1980. Lipopolysaccharide composition of the wilt pathogen, *Pseudomonas solanacearum.* Correlation with the hypersensitive response in tobacco. *Plant Physiol.* **65**:557–559.

179. Wijffelman, C. A., E. Pees, A. A. N. van Brussel, R. J. H. Okker, and B. J. J. Lugtenberg. 1985. Genetic and functional analysis of the nodulation region of the *Rhizobium leguminosarum* Sym plasmid pRL1JI. *Arch. Microbiol.* **143**:225–232.

180. Wijffelman, C., B. Zaat, H. Spaink, I. Mulders, T. van Brussel, R. Okker, E. Pees, R. de Maagd, and B. Lugtenberg. 1986. Induction of *Rhizobium nod* genes by flavonoids: differential adaptation of promotor, *nodD* gene and inducers for various cross-inoculation groups, p. 123–135. *In* B. Lugtenberg (ed.), *Recognition in Microbe-Plant Symbiotic and Pathogenic Interactions.* NATO ASI Series vol. H4. Springer-Verlag KG, Berlin.

181. Wolpert, J. S., and P. Albersheim. 1976. Host symbiont interactions. I. The lectins of legumes interact with the O-antigen containing lipopolysaccharides of their symbiont rhizobia. *Biochem. Biophys. Res. Commun.* **70**:729–737.

182. Yao, P. Y., and J. M. Vincent. 1976. Factors responsible for the curling and branching of clover root hairs by *Rhizobium. Plant Soil* **45**:1–16.

183. Young, D. H., and L. Sequeira. 1986. Binding of *Pseudomonas solanacearum* fimbriae to tobacco leaf cell walls and its inhibition by bacterial extracellular polysaccharides. *Physiol. Mol. Plant Pathol.* **28**:393–402.

184. Zaat, S. A. J., A. A. N. van Brussel, T. Tak, E. Pees, and B. J. J. Lugtenberg. 1987. Flavonoids induce *Rhizobium leguminosarum* to produce *nodDABC* gene-related factors that cause thick, short roots and root hair responses on common vetch. *J. Bacteriol.* **169**:3388–3391.

185. **Zaat, S. A. J., C. A. Wijffelman, H. P. Spaink, A. A. N. van Brussel, R. J. H. Okker, and B. J. J. Lugtenberg.** 1987. Induction of the *nodA* promotor of *Rhizobium* Sym plasmid pRL1JI by plant flavones and flavanones. *J. Bacteriol.* **169:**198–204.
186. **Zaenen, I., N. Van Larebeke, H. Teuchy, M. Van Montagu, and J. Schell.** 1974. Supercoiled circular DNA in crown-gall inducing *Agrobacterium* strains. *J. Mol. Biol.* **86:**109–127.
187. **Zorrequieta, A., M. E. Tolmasky, and R. J. Staneloni.** 1985. The enzyme synthesis of β1-2 glucans. *Arch. Biochem. Biophys.* **238:**368–372.
188. **Zurkowski, W.** 1980. Specific adsorption of bacteria to clover root hairs, related to the presence of plasmid pWZ2 in cells of *Rhizobium trifolii. Microbios* **27:**27–32.

Microbial Cell Surface Hydrophobicity
Edited by R. J. Doyle and M. Rosenberg
© 1990 American Society for Microbiology, Washington, DC 20005

Chapter 7

Hydrophobicity in the Aquatic Environment

Yeshaya Bar-Or

The title given to this chapter might seem to be self-contradictory, since cell surface hydrophobicity (CSH) implies the avoidance of water and rejection of cells possessing this property out of aquatic environments. Yet many different microorganisms inhabiting solid-water and air-water interfaces possess hydrophobic characteristics and at the same time require the aqueous milieu for growth and development. Hydrophobic interactions are important for adhesion to these surfaces (11, 16, 48, 68, 77) and enable the cells to benefit from advantages that might be derived from the association with surfaces while immersed in water.

In examining the biology of microbial CSH, several questions arise: What are the advantages of life at aqueous-nonaqueous interfaces? Are hydrophobic interactions of cells with surfaces reversible? If so, under which conditions and by which mechanisms? Are there unique cell-wall components common to all hydrophobic microorganisms? What is the nature of the hydrophobins (68) conferring CSH to the cells? Is there any interplay between the hydrophobic effect and other cell-to-surface relationships such as electrostatic or lectin-sugar interactions? Is colony hydrophobicity dependent on the hydrophobicity of individual cells?

In addressing these questions, one should bear in mind that CSH is often measured on whole cell suspensions; the results therefore reflect an average of the cell population and the cell surface properties, ignoring variations within the population or at the cell surface. Such variations

Yeshaya Bar-Or • Division of Water Resources, Ministry of the Environment, P.O. Box 6234, 91061 Jerusalem, Israel.

may be the very mechanisms through which microbial communities are able to adapt to environmental changes.

HYDROPHOBICITY AND ADHESION TO INTERFACES

The upper 1-cm-thick microlayer of large bodies of water is usually populated much more densely with microorganisms than is the bulk water (7, 27, 74). The same is true for solid-water interfaces, where microbial mats or films are found adhering to the benthos of shallow streams characterized by extreme conditions of alkalinity, redox potential, etc. (15), in the proximity of deep ocean hydrothermal vents (37), or attached to suspended inorganic particles (43). Copious microbial growth, termed biofouling or biofilm, is a regular feature of objects submerged in or continually washed by water (13). This characteristic indicates that interfaces may provide a better environment for survival and growth than does the bulk water phase as a result of such advantages as accumulation of organic nutrient molecules at interfaces (17), protection against predators (62), inactivation of toxic substances and degrading enzymes by the sorbing surface (10, 46), and enhancement of metabolic activity (19, 23, 82) that may be triggered by changes in cell wall permeability caused by the adsorption process. Phototrophic microorganisms require light for growth, and this necessitates close proximity to the water surface, especially under turbid conditions prevailing in many rivers and lakes. Similarly, adhesion to interfaces may be of particular value when the sorbing surface can be metabolized, as is the case with hydrocarbon- or cellulose-degrading microorganisms.

Accumulation at a surface may be due to various attractive forces, such as phototaxis (O. Lieberman, Ph.D. thesis, The Hebrew University, Jerusalem, Israel, 1988), chemotaxis (80), or charge-charge interactions (31), but hydrophobic interactions are probably the most important mechanism for adhesion in aquatic ecosystems (16, 42, 77).

CSH is, of course, of significance only in conjunction with water. This is best illustrated by comparing cyanobacterial mats colonizing flooded versus arid sands. The cyanobacterial species inhabiting the sandy basins of the Tel Aviv wastewater reclamation project are highly hydrophobic, and this feature apparently facilitates their initial attachment to the flooded benthos (40). Later production of slime may further contribute to consolidation of the cyanobacterial "rafts" with underlying sand particles. On the other hand, the cyanobacterial mats covering extensive areas of sandy soil in the arid southern part of Israel are quite hydrophilic, and their mechanisms of attachment to the soil seem to be by forming three-dimensional networks that envelop and trap soil particles and by production of copious amounts of adhesive slimes (T. Yisraeli and Y. Bar-Or, unpublished data). Interestingly, the CSH of a cyanobacterial culture isolated from a shallow depression, where water could accumulate and stand for extended periods of time, was significantly higher (45% according to the microbial adhesion to hydrocarbons [MATH] test [67]) than in other cultures isolated from truly arid areas (15% or less). Thus, quite different mechanisms have evolved to allow the attachment of aquatic and terrestrial populations to similar media, depending on and reflecting unique ecological conditions.

Interfaces of Water and Air (Neuston)

The major mechanism for the transfer of microbial cells to the upper layer in open oceans is rising bubbles (6, 7, 12). Hydrophobic cells, as well as gas bubbles, are surrounded by a layer of water molecules that is more rigidly organized than bulk water. When such a cell approaches and then adheres to a bubble, there is a decrease in the total number of water molecules enveloping them, with a resultant increase in entropy. There is therefore a thermodynamic advantage to hydrophobic cells over hydrophilic ones in being adsorbed to bubbles. This advantage is reflected in the enrichment of hydrophobic cells at the interface of water and air; and a higher proportion of hydrophobic cells is found in surface samples than in the bulk water (16).

The outermost part of natural surface microlayers consists of hydrophobic substances, including various lipids, proteins, and polysaccharides (50). Norkrans, Kjelleberg, and others have shown that the surface film formed by these molecules strongly influences bacterial accumulation. The density of *Serratia* cells was found to be 10 times higher in surface layers than in the bulk phase, but when a lipid layer was spread, the enrichment factor increased up to 100 (54). Bacteria exhibiting high enrichment factors tended to remain bound to the surface, whereas those species which remained in the bulk phase were constantly moving between the two strata (30).

Experiments designed to test the ability of bacteria to utilize surface films layered on water have shown that this ability depends on several factors, including compactness of the film, electrostatic interactions with charged fatty acids, and hydrophobicity of the cells (45, 54). In situ measurements of metabolic activity revealed no significant difference between surface-bound and bulk water cells (30), but this observation might be attributed to exchange of less adhering cells, which scavenge surface-localized nutrients, between the phases. Studies on scavenging of fatty acids from a liquid-solid interface revealed that a hydrophobic strain of *Serratia marcescens* was much more efficient than a hydrophilic mutant (42). It was theorized that adhesion to surfaces (via hydrophobic interactions) allows for better interaction with and utilization of sorbed nutrients.

There are some disadvantages to life at the surface microlayer, too. This habitat is exposed to greater and more rapid fluctuations in temperature, salinity, and dissolved gases than is subsurface water, and neustonic organisms are subject to higher intensities of visible light and UV radiation (28). Photooxidative death was indeed recorded in cyanobacterial populations exposed to sunlight at the upper water layer (1, 84). As a result, the bacterio- and phytoneuston is characterized by a reduced species diversity and less stability than in planktonic populations (28, 84).

Interfaces of Water and Solid and of Immiscible Liquids

As already mentioned, hydrophobic interactions are of major importance in the firm adhesion of many diverse microorganisms to water-solid interfaces (11, 77). The nonselective nature of CSH enables aquatic cells to attach to such diverse surfaces as sediments (20), ship hulls (57, 83), and various plastics (48, 49, 61). CSH also facilitates adhesion of hydrocarbon-utilizing bacteria (69) and yeasts (39) to interfaces of water and hydrocarbon.

Highly specific relationships between various aquatic microorganisms and biotic surfaces can also be due to hydrophobic interactions, as was shown for the symbiotic attachment of *Anabaena* cells to the *Azolla* fern (20). Similarly, the initial adhesion of *Aeromonas salmonicida* to salmonids, which is a prerequisite for infection and formation of skin lesions in the fish, is perhaps made possible by the CSH of the pathogen (56). CSH may also be important for uptake by and survival in macrophages within which the bacteria proliferate (16a, 56).

Although CSH may be sufficient for initial adhesion to take place, it is certainly not sufficient by itself to ensure growth and proliferation. This fact has been shown by the studies of Rosenberg and colleagues, who tested the ability of *Streptococcus pyogenes* and *Acinetobacter calcoaceticus* to adhere to human epithelial cells and to hexadecane (67, 69). Both organisms adhered to the two substrates, although *S. pyogenes* is exclusively associated with animate tissues, and *A. calcoaceticus* is never found in such ecosystems and was originally isolated from seawater as a hydrocarbon-degrading bacterium. It is quite obvious that the types of nutrients available, metabolic pathways of the attached cells, and resistance to various biological, physical, and chemical stresses ultimately determine the success of colonizing a surface. Therefore, it seems that CSH-dependent colonization of interfaces is of potential benefit to the cell but is certainly not without risk.

Indiscriminate accumulation at solid surfaces, which is the typical result of nonspecific adhesion, may even be harmful under certain environmental and physiological conditions. Sloughing of biofilms off trickling filter media was attributed to lack of oxygen and other nutrients (33). Furthermore, the metabolic activity of surface-associated cells is sometimes similar to or even lower than that of cells sampled from bulk water (8, 22, 30). The dynamic equilibrium maintained between attached and nonattached cells may make it difficult to meaningfully assess utilization of adsorbed and dissolved nutrients by adsorbed and free-living cells.

It is therefore of interest to examine the degree to which CSH can be modified during the life cycle of the cells and under various growth conditions.

CSH: A VARIABLE PROPERTY

CSH is a genetically inherited property, and mutants lacking it could be isolated on the basis of nonadhesion to hydrophobic media (81). CSH can, however, be modified by environmental and physiological conditions. McEldowny and Fletcher (48) measured CSH and attachment to hydrophobic and hydrophilic surfaces in several bacterial freshwater isolates grown under various nutrient regimens. They found that whereas a *Flexibacter* sp. was not significantly influenced by changes in medium composition, a *Chromobacterium* sp. did exhibit differences in attachment which depended on the carbon source and the growth phase. The growth temperature can also influence the degree of hydrophobicity (35).

The effect of medium components was similarly shown in cells of *Corynebacterium glutamicum*, which were markedly more hydrophobic when grown in phosphate-rich media than were phosphate-depleted cells (9). This finding was attributed to the synthesis of lipoteichoic acids by phosphate-enriched cells. Lipoteichoic acids were also demonstrated to be the major components responsible for CSH of group A streptococci (51; see also reference 14).

In the euryhaline bacterium *Halomonas elongata*, CSH increased when the cells were grown in high concentrations of NaCl and approached the stationary growth phase (29). This bacterium produces more charged phospholipids when grown in high NaCl concentrations (79), which may explain the increased hydrophobicity of the cell surface.

The synthesis of a hydrophobic capsule could be reversed in a *Pseudomonas* sp., isolated from sewage sludge, according to growth conditions. Cells growing in dilute medium formed a capsule, grew in clumps, and were hydrophobic. In rich medium, the culture was composed of single cells that were hydrophilic and lacked a capsule (75). Apparently, the capsule plays a role in CSH.

CSH was found in several species to increase during starvation (44) and was correlated with increased "irreversible" binding to glass surfaces (18). This greater adhesiveness may be a kind of a survival mechanism, since it makes surface-bound nutrients more efficiently utilized. CSH is, however, often markedly reduced in the presence of metabolic inhibitors (2, 58), enabling the cells to migrate to new sources of nutrients.

Surface characteristics can be modified by masking of the cell surface with an additional layer of amphiphilic molecules. The hydrophobic region of these substances may be bound to the hydrophobic cell surface, whereas the hydrophilic part is turned outward and confers relative hydrophilicity to the cell. This phenomenon is illustrated in the following examples.

Early-stationary-phase cells of *A. calcoaceticus* are hydrophobic and are thus able to attach to hydrocarbon droplets, which they can metabolize (69; E. Rosenberg, personal communication). Emulsan, an anionic heteropolysaccharide to which fatty acids are linked (5, 85), then accumulates as a minicapsule on the cell wall. This accumulation leads to a decrease in CSH, which allows for detachment from the hydrocarbon droplets just as growth proceeds and depletes the medium of nutrients (63, 64). During the late stationary phase, which represents growth under unfavorable conditions, the capsular polymer is released to the medium and the cells become hydrophobic once again (60), retaining their ability to reattach to fresh surfaces.

A similar process has been found in the benthic cyanobacterium *Phormidium* J-1. Filaments from stationary-phase cultures exhibit reduced CSH (20), a feature that has been attributed to the production of emulcyan, a protein-polysaccharide complex possessing emulsifying and surface-active properties (21; Y. Elkana and N. Hershkowitz, unpublished data). In filaments from early-stationary-phase cultures, there is no observable accumulation of an additional defined wall layer, but after batchwise incubation for an extended period (3 weeks or more), an extensive slime layer is formed (3) that may be composed, at least partially, of emulcyan. Interestingly, the decrease in CSH is accompanied by reduced electrostatic binding to clay particles (4); therefore, emulcyan-slime production modulates both hydrophobic and electrostatic interactions with the benthos in a coordinated fashion.

It has been hypothesized that microorganisms from near-shore environments are more adherent under optimal conditions, whereas open-ocean species tend to adhere when starved (30). Both strategies promote survival, because detachment of benthic shallow water cells allows their transport to adjacent, nutritionally favorable niches, whereas attachment of deep-sea bacteria to solids, such as suspended particles, avails the cells of the nutrients adsorbed at the solid-water interface (41). This hypothesis does not apparently take into account the decreased CSH of freshwater cyanobacteria that have been incubated with chloramphenicol (2).

In conclusion, CSH is not a fixed, rigidly preserved property of the cell envelope but rather is continuously modified to suit changing environmental circumstances and physiological conditions.

HYDROPHOBIC COMPONENTS AND STRUCTURES OF THE CELL WALL

CSH is obviously governed by the chemical composition and physical organization of the outer layer(s) of the cell wall. Contrary to what might be expected, lipids per se do not necessarily play a major role as hydrophobins.

Proteins as Hydrophobic Components of the Cell Wall

Several lines of evidence indicate the involvement of surface proteins in CSH.

(i) The CSH of many microorganisms is lost upon superficial treatment with proteases (2, 58). In the benthic, hydrophobic cyanobacterium *Phormidium* J-1, the cells possessed a double-layered minicapsule that was largely destroyed by protease treatment, with a simultaneous sharp decrease in CSH (2). In this case, proteolytic digestion was effective only after pretreatment with the detergent sodium dodecyl sulfate, which liberated large amounts of proteins and carbohydrates without affecting CSH by itself. These results led to the construction of a model (Fig. 1) describing individual hydrophobic protein molecules as uniformly dispersed in the minicapsule and surrounded by hydrophilic proteins and carbohydrates that protect them against hydrolytic enzymes (2).

(ii) Inhibition of protein synthesis in *Vibrio proteolytica* was accompanied by a marked decrease in CSH, and incubation with the sulfhydryl-blocking agent *P*-chloromercuribenzoate had the same effect (J. H. Paul, presented at the Conference on Microbial Adhesion and Corrosion in the Marine Environment, La Jolla, Calif., 1983).

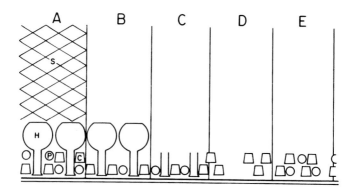

Figure 1. Schematic representation of the structure and composition of the *Phormidium* J-1 external layer. S, Slime; P, nonhydrophobic proteins; H, hydrophobic proteins; C, carbohydrates. (A) Mature cells having a slime layer. The underlying external layers are composed of carbohydrates, hydrophobic proteins, and nonhydrophobic proteins. (B) After sodium dodecyl sulfate treatment, removing much of the carbohydrates and nonhydrophobic proteins along with the slime. (C) After a subsequent pronase treatment, which digests the hydrophobic proteins, rendering the cell surface hydrophilic. (D) After chloramphenicol treatment, which interrupts protein synthesis. Cell surface proteins are turned over, making the cell surface more hydrophilic. (E) After Omnimixer treatment, which shears the cell surface and uproots the loosely bound hydrophobic proteins. (Reprinted from reference 2 with permission.)

(iii) Wild-type strains of the fish pathogen *Aeromonas salmonicida*, possessing the A protein (which contains a high proportion of hydrophobic and uncharged amino acids [76]) were markedly more hydrophobic than A-protein-negative mutants (56). Interestingly, the A-protein layer (A layer) was found to cover most of the cell surface; therefore, hydrophobicity may also be due to masking of the hydrophilic O side chains of the lipopolysaccharide molecules. The greater adhesiveness of the hydrophobic strains may be related to pathogenesis by allowing attachment to the fish skin; a similar hypothesis has been proposed for pathogenic strains of group A streptococci (55) and *Serratia marcescens* (66).

(iv) *Candida tropicalis*, a hydrocarbon-utilizing yeast, exhibited a lower affinity to hydrocarbons after proteolytic digestion (39). In this case, a protein-bound polysaccharide-fatty acid complex was isolated (38) and was postulated to be associated with binding of hydrocarbon droplets as the first step in their internalization. Ultrastructural studies revealed a hairy fringe around cells grown in the presence of a hydrocarbon, which was much less visible in cells treated with protease (38).

Nonprotein Hydrophobic Components of the Cell Wall

CSH is not universally associated with surface proteins, as is shown in the following cases.

(i) An adhesive *Rhodococcus* sp., isolated from pond water, produced a rhamnose-rich polysaccharide that was found both in the growth medium and on the cell surface and is thought to be involved in hydrophobic adhesiveness of the cells (53).

(ii) Smooth strains of *Salmonella typhimurium* and other gram-negative bacteria, carrying intact lipopolysaccharide molecules, appear to be more hydrophilic than the corresponding rough mutants (31), in which sequential loss of the O side chain and core oligosaccharide monomers lead to increased CSH.

(iii) Hydrophobicity in *Serratia marcescens* was long considered to be associated with prodigiosin, a surface-localized pyrrol-containing red pigment (32). However, more recent findings show that even nonpigmented mutants retain their CSH (65, 66).

Hydrophobicity-conferring molecules can be of more than one type in the same cell: in *A. calcoaceticus* RAG-1, hydrophobicity was related to both fimbriae and another cell wall component, which conferred high hydrophobicity and the ability to grow on hydrocarbons even in the absence of fimbriae (59).

X-ray photoelectron spectroscopy, which can provide information on the elemental composition of the outermost layer (2 to 5 nm) of microorganisms, was used to find correlations between surface chemistry and hydrophobicity (52). Oxygen-bound carbon was found to be inversely correlated with CSH, whereas hydrogen-bound carbon frequency on the cell surface was directly correlated with hydrophobicity in gram-negative and gram-positive bacteria. Hydrocarbon (C-H) groups represent nonpolar regions with reduced tendency to interact with water. These groups can be part of apolar moieties in proteins, lipoproteins, lipids, and lipopolysaccharides. However, in the bacteria tested by this method, CSH was inversely proportional to the N/P ratio, indicating that proteins did not play a role in CSH. In yeasts, however, N/P ratio, not C-H/C ratio, was the factor of importance in determining CSH.

Distribution and Orientation of Hydrophobic Components on the Cell Wall

The localization and distribution of the hydrophobic components along the cell envelope determine the degree of association with interfaces. Hydrophobic molecules can affect overall cell surface properties only when forming an extensive cover or patches over the cell envelope.

When hydrophilic and hydrophobic groups are mixed together on the cell surface, hydrophobic interactions may be neutralized because the local structuring of water molecules is dominated by their interactions with the hydrophilic groups (36). It can therefore be expected either that cells forming hydrophobic interactions will be hydrophobic all along the cell or that their hydrophobic components will be concentrated in patches or poles. Geitler (26) described various microalgae that inhabit surface layers of quiescent water bodies and divided them into three categories: (i) epineustonic, extending part of the cell into the surface film while the rest is suspended in air; (ii) neustonic, a state in which the whole organism is immersed in the interface microlayer; and (iii) hyponeustonic, a state in which a stalk anchors the cell to the interface while the cell itself is suspended in the underlying water phase. This diversity may represent differences in distribution of surface hydrophobicity.

In the now classical work of Marshall and Cruickshank (47), the perpendicular orientation of *Flexibacter* sp. and *Hyphomicrobium* sp. toward interfaces of water and oil, water and solid, and water and air was shown experimentally to indicate localization of CSH at the poles, which are consequently repelled from the aqueous phase. On the other hand, in the cyanobacterium *Phormidium* J-1, the full length of the filaments or cells becomes adsorbed to interfaces of water and hydrocarbon (20), indicating an evenly distributed CSH.

In a study on the natural microbiota colonizing the skin of healthy fish (73), it was found that lawns of 12 of 13 bacterial isolates repelled water droplets, but only 7 were hydrophobic according to the hexadecane-water partitioning test (MATH). The authors suggested that this discrepancy between colonial and cellular hydrophobicity be explained by a (theoretical) production of slime by the bacterial lawn which is washed off during suspension in the hexadecane-water mixture. However, another explanation might be that hydrophobicity is limited to only small regions in the cell surface which become oriented away from the (solidified) aqueous medium, that is, toward the air. When the cells are in suspension, this orientation is lost or perhaps the hydrophobic sites become evenly distributed such that macroscopic CSH is lost.

CSH-related proteins and polysaccharides may serve merely as a bridge connecting associated lipids or fatty acids to the cell wall (38). Alternatively, relatively hydrophobic regions of protein (or even rhamnose-rich, hydroxyl-poor polysaccharide [53]) molecules can be oriented so as to be exposed at the surface while the hydrophilic portions of the molecules anchor them to the cell wall. Such amphiphilic molecules, possessing both hydrophobic and hydrophilic regions, may be of profound significance for the ability of attached cells to migrate through aquatic media and colonize new surfaces. This would be quite impossible if the cell surface were continuously and totally hydrophobic.

INTERPLAY OF HYDROPHOBICITY AND SURFACE CHARGE

Since most surfaces immersed in aquatic environments are charged, it is unlikely that hydrophobic interactions are the only factor in determining the degree of adhesion of microorganisms, which themselves are charged to varying degrees. The relative importance of the hydrophobic and electrostatic attractive and repulsive interactions of bacteria with surfaces should therefore be discerned when one assesses their contribution to promoting adhesion.

In a study by van Loosdrecht and colleagues of 23 bacterial species (not all from aquatic habitats), it was concluded that hydrophobic cells adhered to polystyrene to a greater extent than did relatively hydrophilic ones, irrespective of surface charge (measured as electrophoretic mobility) (78). However, electrokinetic potential became more important in the less hydrophobic cells. When results of several experiments (77, 78) were evaluated together, it became evident that the more hydrophobic bacteria also possessed high negative charge (78). Certain species even showed a simultaneous increase in both CSH and surface charge upon starvation (44). Charged groups were calculated to occupy only a minor fraction of the total surface area (no more than 8%) (78); therefore, a concomitant increase in the hydrophobic surface area is possible on theoretical grounds as well. It has been argued (77) that a combination of high hydrophobicity and low surface charge would lead to such strong adhesion that cells possessing these properties would escape standard microbiological isolation procedures and would therefore be unknown to us. More to the point, this combination of properties would prevent spreading and colonization, and therefore such cells would be automatically annihilated.

Hydrophobicity was found in some bacteria to increase with increasing dilution rate of the growth medium, whereas electrophoretic mobility did not change markedly (77). Electrolytes, detergents, and pH had a highly variable effect on attachment to and detachment from hydrophobic petri dish polystyrene and the less hydrophobic tissue culture dish polystyrene (TCD) (49). This finding indicates that a given bacterium can maintain either predominantly hydrophobic or electrostatic interactions with surfaces or both, depending on the physicochemical properties of the cell, the sorbing surface, and the suspension liquid.

In *V. proteolytica*, there was a preferred orientation of the cell with

respect to the adsorbing surface (57). Triton X-100, a surfactant, completely inhibited attachment of *V. proteolytica* to polystyrene but had
almost no effect on attachment to a hydrophilic TCD. On the other hand,
mannose partially inhibited attachment to the TCD but had no effect on
attachment to polystyrene. Pronase and other proteolytic enzymes
strongly decreased adhesion to polystyrene but not to hydrophilic surfaces (glass and TCD). This finding again indicates that attachment of *V.
proteolytica* to hydrophobic surfaces is aided by mechanisms different
from those used for attachment to hydrophilic surfaces (57).

The relative contribution of hydrophobic and electrostatic interactions between cells and surfaces depends not only on cell surface
properties but also on the substratum characteristics (24). Hydrophobic
interactions are created by the release of structured water molecules,
enveloping two hydrophobic surfaces, to the bulk phase when these
surfaces approach and contact each other (70). This fact implies that when
the sorbing surface is polar, the hydrophobic interaction is neutralized (at
least in part) because the ensuing thermodynamic gain is smaller. When
the CSH is also low, there will be an even smaller chance for the
formation of hydrophobic interactions with hydrophilic, polar surfaces.
Accordingly, although even cells isolated from the bulk water phase
usually exhibit some degree of hydrophobicity (27), they do not do so to
an extent sufficient to allow stable adsorption to interfaces. This characteristic was demonstrated by measuring the attachment of several bacterial species, isolated from a river, to various surfaces (61). Maximal
attachment to any given surface was species specific, but there was an
overall preference for hydrophobic surfaces (polyvinyl chloride and
polystyrene) over hydrophilic ones (borosilicate glass and polyethylene
terephthalate).

COMMUNAL VERSUS INDIVIDUAL HYDROPHOBICITY

Measurements of the contact angle formed when an air bubble or a
water drop approaches a bacterial lawn (25) and of the direction of
spreading of a drop of water placed at the border between a bacterial lawn
and a nonbiological hydrophobic or hydrophilic medium (72) are both
methods that yield information on the surface hydrophobicity of a mass of
cells combined or grown as a continuum. The MATH (67) and hydrophobic interaction chromatography (HIC) methods measure the hydrophobic
interaction of individual cells, suspended in water, with hydrophobic solid
or liquid media. However, when the MATH and HIC methods are used,
results are usually presented as if the values obtained refer to the whole

cell population. Thus, a culture may be reported to be 70% hydrophobic. This is misleading, since when 70% of the cell population adhere to a hydrophobic medium, then these cells by definition are 100% hydrophobic while the remaining 30% are completely hydrophilic in terms of the test in use. In other words, results of contact angle and direction-of-spreading measurements reflect the hydrophobic properties shared by the cells, whereas MATH and HIC distinguish between hydrophobic and hydrophilic subpopulations. This means that whenever CSH is less than 100% according to MATH (as is usually the case), some of the cells of "hydrophobic populations" would not adhere to a surface and remain in the bulk water. This, in turn, may explain why different techniques for measuring CSH often yield quite different results (25, 73).

When CSH is protein dependent, incubation in the presence of protein synthesis inhibitors often leads to a much reduced hydrophobicity (2, 34, 58). Continuous protein synthesis is apparently required to maintain CSH, and it might well be that switching on and off of hydrophobicity-related protein synthesis facilitates the existence of a hydrophilic subpopulation and consequently the exchange of cells between the interface and the bulk phase. Such processes may have many ecological implications: they enable the cells to scavenge nutrients accumulated at the interface without having to attach permanently; survival of at least part of the population is secured even if the attached or unattached subpopulation is eliminated; and mobile cells, ready for dispersal and colonization, are always on hand.

The collective hydrophobicity of bacterial lawns or films may have profound importance in changing the macroscopic properties of the adsorbing surface. Microbial conditioning, a prerequisite for macrofouling of ship hulls (83), may contribute to attachment of larval invertebrates not only through lectin-sugar interactions (R. Mitchell, presented at the Conference on Microbial Adhesion and Corrosion in the Marine Environment, La Jolla, Calif., 1983) but also by creating a surface with an altered hydrophobicity. Likewise, the slippery nature of fish skin (73) may be due to hydrophobic slimes of microbial origin that reduce drag forces and thus assist in motion through water.

CONCLUDING REMARKS

It is hard to form a unified concept of the ways by which CSH influences microbial life in aquatic environments. As shown above, hydrophobicity may be related to different surface components that may or may not be involved in metabolic activities of the cells. CSH may either decrease or increase under starvation conditions and may be tuned to specific needs under specific conditions by changes in surface charge. The differences in expression and regulation of CSH reflect the diversity of ecological niches supporting growth and the capacity of microorganisms for adaptation to inherently different and ever-changing physicochemical parameters. Although a great deal has been learned, vital information is still distressingly lacking in several areas. (i) Little is known about the relevance of laboratory methods for measuring CSH, using hydrocarbons or derivatized Sepharose, to actual interactions with much less hydrophobic objects submerged in water. (ii) Compared with current knowledge on structure-function relationships in hydrophobic membrane-proteins (71), little is known about the ultrastructural organization and stereochemistry of hydrophobic outermost components of microbial cell walls. (iii) Little is known about the biochemical and structural mechanisms through which such hydrophobic components are modified under natural conditions.

The integration of studies on phenomena of microbial ecology related to hydrophobicity and adhesion with research on the molecular level has gained momentum during the past 5 years and holds promise for further characterization of the basic processes underlying biological interaction with interfaces.

Acknowledgment. This chapter is dedicated to the memory of the late professor Moshe Shilo, a great scientist and beloved teacher and colleague.

LITERATURE CITED

1. **Abeliovich, A., and M. Shilo.** 1972. Photooxidative death in blue-green algae. *J. Bacteriol.* **111:**682–689.

2. **Bar-Or, Y., M. Kessel, and M. Shilo.** 1985. Modulation of cell surface hydrophobicity in the benthic cyanobacterium *Phormidium* J-1. *Arch. Microbiol.* **142:**21–27.

3. **Bar-Or, Y., M. Kessel, and M. Shilo.** 1989. Mechanisms for release of the benthic cyanobacterium *Phormidium* strain J-1 to the water column, p. 214–218. *In* Y. Cohen and E. Rosenberg (ed.), *Microbial Mats: Physiological Ecology of Benthic Microbial Communities.* American Society for Microbiology, Washington, D.C.

4. **Bar-Or, Y., and M. Shilo.** 1988. The role of cell-bound flocculants in coflocculation of benthic cyanobacteria with clay particles. *FEMS Microbiol. Ecol.* **53:**169–174.

5. **Belsky, I., D. L. Gutnick and E. Rosenberg.** 1979. Emulsifier of *Arthrobacter* RAG-1: determination of emulsifier-bound fatty acids. *FEBS Lett.* **101:**175–178.

6. **Bezdek, H. F., and A. F. Carlucci.** 1972. Surface concentration of marine bacteria. *Limnol. Oceanogr.* **17:**566–569.

7. **Blanchard, D. C., and L. D. Syzdek.** 1970. Mechanism for the water-to-air transfer and accumulation of bacteria. *Science* **170:**626–628.

8. **Bright, J. J., and M. Fletcher.** 1983. Amino acid assimilation and respiration by attached and free-living populations of a marine *Pseudomonas* sp. *Microb. Ecol.* **9:**215–226.

9. **Buchs, J., N. Mozes, C. Wandrey, and P. G. Rouxhet.** 1988. Cell adsorption control by culture conditions. *Appl. Microbiol. Biotechnol.* **29:**119–128.

10. **Burton, G. A., D. Gunnison, and G. R. Lanza.** 1987. Survival of pathogenic bacteria in various freshwater sediments. *Appl. Environ. Microbiol.* **53:**633–638.

11. **Busscher, H. J., and A. H. Weerkamp.** 1987. Specific and non-specific interactions in bacterial adhesion to solid substrata. *FEMS Microbiol. Rev.* **46:**165–173.

12. **Carlucci, A. F., and H. F. Bezdek.** 1972. On the effectiveness of a bubble for scavenging bacteria from sea water. *J. Geophys. Res.* **77:**6608–6610.

13. **Characklis, W. G., and K. E. Cooksey.** 1983. Biofilms and microbial fouling. *Adv. Appl. Microbiol.* **29:**93–138.

14. **Christensen, G. D., W. A. Simpson, and E. M. Beachey.** 1985. Adhesion of bacteria to animal tissues: complex mechanisms, p. 279–305. *In* D. Savage and M. Fletcher (ed.), *Bacterial Adhesion.* Plenum Publishing Corp., New York.

15. **Cohen, Y., and E. Rosenberg (ed.).** 1989. *Microbial Mats: Physiological Ecology of Benthic Microbial Communities.* American Society for Microbiology, Washington, D.C.

16. **Dahlback, B., M. Hermansson, S. Kjelleberg, and B. Norkrans.** 1981. The hydrophobicity of bacteria—an important factor in their initial adhesion at the air-water interface. *Arch. Microbiol.* **128:**267–270.

16a. **Daly, J. G., and R. M. W. Stevenson.** 1987. Hydrophobic and haemagglutinating properties of *Renibacterium salmoninarum. J. Gen. Microbiol.* **133:**3575–3580.

17. **Davies, J. T., and E. K. Rideal.** 1963. *Interfacial Phenomena.* Academic Press, Inc. (London), Ltd., London.

18. **Dawson, M. P., B. Humphrey, and R. C. Marshall.** 1981. Adhesion: a tactic in the survival strategy of a marine vibrio during starvation. *Curr. Microbiol.* **6:**195–201.

19. **Diab, S., and M. Shilo.** 1988. Effect of adhesion to particles on the survival and activity of *Nitrosomonas* sp. and *Nitrobacter* sp. *Arch. Microbiol.* **150:**387–393.

20. **Fattom, A., and M. Shilo.** 1984. Hydrophobicity as an adhesion mechanism of benthic cyanobacteria. *Appl. Environ. Microbiol.* **47:**135–143.

21. **Fattom, A., and M. Shilo.** 1985. Production of emulcyan by *Phormidium* J-1: its activity and function. *FEMS Microbiol. Ecol.* **31:**3–9.

22. **Fletcher, M.** 1979. A microautoradiographic study of the activity of attached and free-living bacteria. *Arch. Microbiol.* **122:**271–274.

23. **Fletcher, M.** 1986. Measurement of glucose utilization by *Pseudomonas fluorescens* that are free-living and that are attached to surfaces. *Appl. Environ. Microbiol.* **52:**672–676.

24. **Fletcher, M., and G. I. Loeb.** 1979. Influence of substratum characteristics on the attachment of a marine pseudomonad to solid surfaces. *Appl. Environ. Microbiol.* **37:**67–72.

25. **Fletcher, M., and K. C. Marshall.** 1982. Bubble contact angle method for evaluating substratum interfacial characteristics and its relevance to bacterial attachments. *Appl. Environ. Microbiol.* **44:**189–192.

26. **Geitler, L.** 1942. Zur Kenntnis der Bewohner des oberflachenhautchens einheimischer Gewasser. *Biol. Gen.* **16:**450–475.

27. **Guerin, W. F.** 1989. Phenanthrene degradation by estuarine surface microlayer and bulk water microbial populations. *Microb. Ecol.* **17:**89–104.

28. **Hardy, J. T.** 1973. Phytoneuston ecology of a temperate marine lagoon. *Limnol. Oceanogr.* **18:**525–533.

29. **Hart, D. J., and R. H. Vreeland.** 1988. Changes in the hydrophobic-hydrophilic cell surface character of *Halomonas elongata* in response to NaCl. *J. Bacteriol.* **170:**132–135.

30. **Hermansson, M., and B. Dahlback.** 1983. Bacterial activity at the air/water interface. *Microb. Ecol.* **9:**317–328.

31. Hermansson, M., S. Kjelleberg, T. K. Korhonen, and T. A. Stenstrom. 1982. Hydrophobic and electrostatic characterization of surface structures of bacteria and its relationship to adhesion to an air-water interface. *Arch. Microbiol.* **131**:308–312.

32. Hermansson, M., S. Kjelleberg, and B. Norkrans. 1979. Interaction of pigmented wild type and pigmentless mutant of *Serratis marcescens* with a lipid surface film. *FEMS Microbiol. Lett.* **6**:129–132.

33. Howell, J. A., and B. Atkinson. 1976. Sloughing of microbial film in trickling filters. *Water Res.* **10**:307–315.

34. Humphrey, B., S. Kjelleberg, and K. C. Marshall. 1983. Responses of marine bacteria under starvation conditions at a solid-water interface. *Appl. Environ. Microbiol.* **45**:43–47.

35. Ishiguro, E. E., W. W. Kay, T. Ainsworth, J. B. Chamberlain, R. A. Austen, J. T. Buckley, and T. J. Trust. 1981. Loss of virulence during culture of *Aeromonas salmonicida* at high temperatures. *J. Bacteriol.* **148**:333–340.

36. Israelachvili, J., and R. Pashley. 1982. The hydrophobic interaction is long range, decaying exponentially with distance. *Nature* (London) **300**:341–342.

37. Jannasch, H. W. 1984. Microbes in the oceanic environment, p. 97–122. *In* D. P. Kelly and N. G. Carr (ed.), *The Microbe*, part II. *Prokaryotes and Eukaryotes*. Cambridge University Press, Cambridge.

38. Kappeli, O., and A. Fiechter. 1977. Component from the cell surface of the hydrocarbon-utilizing yeast *Candida tropicalis* with possible relation to hydrocarbon transport. *J. Bacteriol.* **131**:917–921.

39. Kappeli, O., P. Walther, M. Mueller, and A. Fiechter. 1984. Structure of the cell surface of the yeast *Candida tropicalis* and its relation to hydrocarbon transport. *Arch. Microbiol.* **138**:279–282.

40. Katznelson, R. 1989. Clogging of groundwater recharge basins by cyanobacterial mats. *FEMS Microbiol. Ecol.* **62**:231–242.

41. Kefford, B., B. A. Humphrey, and K. C. Marshall. 1986. Adhesion: a possible survival strategy for leptospires under starvation conditions. *Curr. Microbiol.* **13**:247–256.

42. Kefford, B., S. Kjelleberg, and K. C. Marshall. 1982. Bacterial scavenging: utilization of fatty acids localized at a solid-liquid interface. *Arch. Microbiol.* **133**:257–260.

43. Kirchman, D., and R. Mitchell. 1982. Contribution of particle-bound bacteria to total microheterotrophic activity in five ponds and two marshes. *Appl. Environ. Microbiol.* **43**:200–209.

44. Kjelleberg, S., and M. Hermansson. 1984. Starvation-induced effects on bacterial surface characteristics. *Appl. Environ. Microbiol.* **48**:497–503.

45. Kjelleberg, S., and T. A. Stenstrom. 1980. Lipid surface films: interaction of bacteria with free fatty acids and phospholipids at the air/water interface. *J. Gen. Microbiol.* **116**:417–423.

46. Lechavalier, M. W., C. D. Cawthon, and R. G. Lee. 1988. Factors promoting survival of bacteria in chlorinated water supplies. *Appl. Environ. Microbiol.* **54**:649–654.

47. Marshall, K. C., and R. H. Cruickshank. 1973. Cell surface hydrophobicity and the orientation of certain bacteria at interfaces. *Arch. Mikrobiol.* **91**:29–40.

48. McEldowney, S., and M. Fletcher. 1986. Effect of growth conditions and surface characteristics of aquatic bacteria on their attachment to solid surfaces. *J. Gen. Microbiol.* **132**:513–523.

49. McEldowney, S., and M. Fletcher. 1986. Variability of the influence of physicochemical factors affecting bacterial adhesion to polystyrene substrata. *Appl. Environ. Microbiol.* **52**:460–465.

50. McIntyre, F. 1974. The top millimeter of the ocean. *Sci. Am.* **May**:62–67.

51. **Miörner, H., G. Hansson, and G. Kronvall.** 1983. Lipoteichoic acid is the major cell wall component responsible for surface hydrophobicity of group A streptococci. *Infect. Immun.* **39:**336–343.

52. **Mozes, N., A. J. Leonard, and P. G. Rouxhet.** 1988. On the relations between the elemental surface composition of yeasts and bacteria and their charge and hydrophobicity. *Biochim. Biophys. Acta* **945:**324–334.

53. **Neu, T. R., and K. Poralla.** 1988. An amphiphilic polysaccharide from an adhesive *Rhodococcus* strain. *FEMS Microbiol. Lett.* **49:**389–392.

54. **Norkrans, B., and F. Sorensson.** 1977. On the marine lipid surface microlayer: bacterial accumulation in model systems. *Bot. Mar.* **20:**473–478.

55. **Ofek, I., E. Whitnack, and E. H. Beachey.** 1983. Hydrophobic interactions of group A streptococci with hexadecane droplets. *J. Bacteriol.* **154:**139–145.

56. **Parker, N. D., and C. B. Munn.** 1984. Increased cell surface hydrophobicity associated with possession of an additional surface protein by *Aeromonas salmonicida*. *FEMS Microbiol. Lett.* **21:**233–237.

57. **Paul, J. H., and W. H. Jeffrey.** 1985. The effect of surfactants on the attachment of marine and estuarine bacteria to surfaces. *Can. J. Microbiol.* **31:**224–228.

58. **Paul, J. H., and W. H. Jeffrey.** 1985. Evidence for separate adhesion mechanisms for hydrophilic and hydrophobic surfaces in *Vibrio proteolytica*. *Appl. Environ. Microbiol.* **50:**431–437.

59. **Pines, O., and D. Gutnick.** 1984. Alternate hydrophobic sites on the cell surface of *Acinetobacter calcoaceticus* RAG-1. *FEMS Microbiol. Lett.* **22:**307–311.

60. **Pines, O., and D. Gutnick.** 1986. Role for emulsan in growth of *Acinetobacter calcoaceticus* RAG-1 on crude oil. *Appl. Environ. Microbiol.* **51:**661–663.

61. **Pringle, J. H., and M. Fletcher.** 1983. Influence of substratum wettability on attachment of freshwater bacteria to solid surfaces. *Appl. Environ. Microbiol.* **45:**811–817.

62. **Roper, M. M., and K. C. Marshall.** 1974. Modification of the interaction between *Escherichia coli* and bacteriophage in saline sediment. *Microb. Ecol.* **1:**1–13.

63. **Rosenberg, E., A. Gottlieb, and M. Rosenberg.** 1983. Inhibition of bacterial adherence to hydrocarbons and to epithelial cells by emulsan. *Infect. Immun.* **39:**1024–1028.

64. **Rosenberg, E., N. Kaplan, O. Pines, M. Rosenberg, and D. Gutnick.** 1983. Capsular polysaccharides interfere with adherence of *Acinetobacter calcoaceticus* to hydrocarbon. *FEMS Microbiol. Lett.* **17:**157–160.

65. **Rosenberg, M.** 1984. Isolation of pigmented and nonpigmented mutants of *Serratia marcescens* with reduced cell surface hydrophobicity. *J. Bacteriol.* **160:**480–482.

66. **Rosenberg, M., Y. Blumberger, H. Judes, R. Bar-Nes, E. Rubinstein, and Y. Mazor.** 1986. Cell surface hydrophobicity of pigmented and nonpigmented clinical *Serratia marcescens* strains. *Infect. Immun.* **51:**932–935.

67. **Rosenberg, M., D. L. Gutnick, and E. Rosenberg.** 1980. Adherence of bacteria to hydrocarbons: a simple method for measuring cell-surface hydrophobicity. *FEMS Microbiol. Lett.* **9:**29–33.

68. **Rosenberg, M., and S. Kjelleberg.** 1986. Hydrophobic interactions: role in bacterial adhesion. *Adv. Microb. Ecol.* **9:**353–393.

69. **Rosenberg, M., A. Perry, E. A. Bayer, D. L. Gutnick, E. Rosenberg, and I. Ofek.** 1981. Adherence of *Acinetobacter calcoaceticus* RAG-1 to human epithelial cells and to hexadecane. *Infect. Immun.* **33:**29–33.

70. **Rossky, P. J., and H. L. Friedman.** 1980. Benzene-benzene interactions in aqueous solution. *J. Phys. Chem.* **84:**587–589.

71. **Rothfield, L. I. (ed.).** 1971. *Structure and Function of Biological Membranes.* Academic Press, Inc., New York.

72. **Sar, N.** 1987. Direction of spreading (DOS): a simple method for measuring the hydrophobicity of bacterial lawns. *J. Microbiol. Methods* **6:**211–219.

73. **Sar, N., and E. Rosenberg.** 1987. Fish skin bacteria: colonial and cellular hydrophobicity. *Microb. Ecol.* **13:**193–202.

74. **Sieburth, J. M., P. J. Johnson, K. M. Burney, D. M. Lavoie, K. R. Hinga, D. A. Caron, F. W. French III, P. W. Johnson, and R. G. Davis.** 1976. Dissolved organic matter and heterotrophic neuston in the surface microlayers of the North Atlantic. *Science* **194:**1415–1418.

75. **Singh, K. K., and W. S. Vincent.** 1987. Clumping characteristics and hydrophobic behavior of an isolated bacterial strain from sewage sludge. *Appl. Microbiol. Biotechnol.* **25:**396–398.

76. **Trust, T. J., W. W. Kay, and E. E. Ishiguro.** 1983. Cell surface hydrophobicity and macrophage association of *Aeromonas salmonicida*. *Curr. Microbiol.* **9:**315–318.

77. **van Loosdrecht, M. C. M., J. Lyklema, W. Norde, G. Schraa, and A. J. B. Zehnder.** 1987. The role of bacterial cell wall hydrophobicity in adhesion. *Appl. Environ. Microbiol.* **53:**1893–1897.

78. **van Loosdrecht, M. C. M., J. Lyklema, W. Norde, G. Schraa, and A. J. B. Zehnder.** 1987. Electrophoretic mobility and hydrophobicity as a measure to predict the initial steps of bacterial adhesion. *Appl. Environ. Microbiol.* **53:**1898–1901.

79. **Vreeland, R. H., R. Anderson, and R. G. E. Murray.** 1984. Cell wall and phospholipid composition and their contribution to the salt tolerance of *Halomonas elongata*. *J. Bacteriol.* **160:**879–883.

80. **Wiegel, J., and M. Dykstra.** 1984. *Clostridium thermocellum* adhesion and sporulation while adhered to cellulose and hemicellulose. *Appl. Microbiol. Biotechnol.* **20:**59–65.

81. **Wolkin, R. H., and J. L. Pate.** 1985. Selection for non-adherent or nonhydrophobic mutants co-selects for nonspreading mutants of *Cytophaga johnsonae* and other gliding bacteria. *J. Gen. Microbiol.* **131:**737–750.

82. **ZoBell, C. E.** 1943. The effect of solid surfaces upon bacterial activity. *J. Bacteriol.* **46:**39–56.

83. **ZoBell, C. E., and E. C. Allen.** 1935. The significance of marine bacteria in the fouling of submerged surfaces. *J. Bacteriol.* **29:**239–251.

84. **Zohary, T.** 1985. Hyperscums of the cyanobacterium *Microcystis aeruginosa* in a hypertrophic lake. *J. Plankton Res.* **7:**399–409.

85. **Zuckerberg, A., A. Diver, Z. Peeri, D. L. Gutnick, and E. Rosenberg.** 1979. Emulsifier of *Arthrobacter* RAG-1: chemical and physical properties. *Appl. Environ. Microbiol.* **37:**414–420.

Microbial Cell Surface Hydrophobicity
Edited by R. J. Doyle and M. Rosenberg
© 1990 American Society for Microbiology, Washington, DC 20005

Chapter 8

Changes in Bacterial Surface Hydrophobicity during Morphogenesis and Differentiation

Eugene Rosenberg and Nechemia Sar

The ability of bacterial cells to alter in size and shape in response to changing environmental conditions is widespread in the procaryotic world. Many bacteria undergo subtle changes in size and shape in relation to nutrient availability. For example, bacteria often become smaller and more spherical when starved (6, 34). In some cases, this morphogenesis is dramatic and leads to functionally differentiated cells (12), such as cyanobacterial heterocysts, endospores of gram-positive bacteria, myxospores of myxococci, exospores of streptomycetes, and cysts of azotobacters. In other cases, morphogenesis is associated with spatial differentiation, such as streptomycetes substratum filaments and aerial exospores. In this chapter, we will attempt to bring together some of the scattered data on the surface hydrophobicity of bacteria undergoing morphogenesis. First, we will briefly discuss techniques that distinguish between cell and colonial surface hydrophobicity.

MEASUREMENTS OF BACTERIAL HYDROPHOBICITY: DISTINCTION BETWEEN COLONY AND CELL SURFACE HYDROPHOBICITY

In nature, bacterial growth and development generally occur in the form of colonies, swarms, or clumps rather than as individual dispersed

Eugene Rosenberg • Department of Microbiology, Tel Aviv University, Ramat Aviv, Israel 69978.   Nechemia Sar • H. Steinitz Marine Biology Laboratory, P.O. Box 469, Eilat, Israel 88103.

cells. The physiological properties of these multicellular aggregates are not necessarily the sum of the individual cells composing them. This is particularly true with regard to surface properties. It is therefore important to distinguish between cell surface hydrophobicity and colonial hydrophobicity.

Bacterial cell surface hydrophobicity is generally measured by hydrophobic interaction chromatography, e.g., octyl- or phenyl-Sepharose (19, 21, 58), and microbial adhesion to liquid hydrocarbons (MATH test), e.g., hexadecane (45, 46). In these methods (44), the dispersed bacteria are free to rotate and even undergo structural changes in order to orient themselves at the hydrocarbon-water interface at a minimum free energy. For example, Marshall and Cruickshank (31) demonstrated that certain bacteria are hydrophobic only at one end of the cell and will adhere by polar attachment. The observation that many bacteria concentrate at hydrocarbon-water interfaces rather than remain in the aqueous phase or penetrate into the hydrocarbon indicates that these cells have both hydrophilic and hydrophobic surface regions.

The relative hydrophobicity of colonies or surface layers of bacteria can be estimated from the contact angle of a drop of water on the given surface (1, 5, 37). For technical reasons, these measurements are difficult to perform and interpret. Somewhat more reproducible data can be obtained by collecting the bacteria on micropore filters before making the contact angle measurements (30). However, this procedure of preparing the bacterial layers disrupts any geometry that may have existed in the growing colony and washes off readily soluble surface components.

The direction-of-spreading (DOS) method for measuring colonial hydrophobicity (51, 53, 54) is rapid, requires no special equipment, and is noninvasive. This simple technique consists of introducing a droplet of water at the border between two surfaces and recording the direction of spreading. By definition, the water drop moves away from the more hydrophobic one. An illustration of this method is shown in Fig. 1. The NS-27 bacterial lawn was more hydrophobic than the agar surface (Fig. 1A, plate i) and the glass slide (Fig. 1A, plate ii) and had a hydrophobicity similar to that of polystyrene (since the water drop spread equally in both directions) (Fig. 1A, plate iii). The NS-24 bacterial lawn (Fig. 1B) was more hydrophobic than the agar and glass surfaces but less hydrophobic than polystyrene. Using agar, glass, and polystyrene as reference surfaces, a scoring system of 1 to 10 was used to evaluate the relative colonial hydrophobicity of bacteria (Fig. 2). By this scoring system, strains NS-27 and NS-24 had values of 9 and 7, respectively.

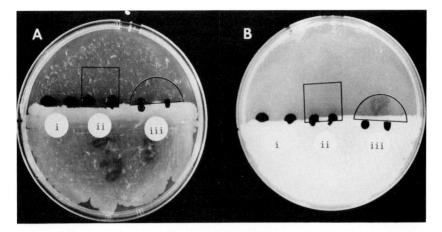

Figure 1. Measurement of the colonial hydrophobicity of strains NS-27 (A) and NS-24 (B) by the DOS method. After the bacterial lawns were allowed to develop for 48 h, two 5-μl water drops were introduced at each of the borders between the bacterial growth (lower part) and one of the following surfaces: 2% agar (i), a cover glass (ii), and polystyrene (iii).

EXAMPLES OF CHANGES IN SURFACE HYDROPHOBICITY DURING PROCARYOTIC DEVELOPMENT

Bacillus Endospores

One of the best-studied systems of procaryotic cellular differentiation is the formation of endospores in members of the family *Bacillaceae*. It has been known for a long time that these endospores are extremely hydrophobic relative to the vegetative cells. In fact, two of the earliest procedures for separating endospores from vegetative cells, foam flotation (4, 17, 47) and polymer two-phase systems (50), are based on the high cell surface hydrophobicity of the endospores. Visualization of the high cell surface hydrophobicity of endospores is shown in Fig. 3.

Surprisingly, few experimental studies have been directed toward understanding the chemical basis of *Bacillus* spore hydrophobicity or the timing of the transition from a hydrophilic vegetative cell to a hydrophobic spore (stage of endosporulation) and back to the vegetative cell (germination-outgrowth). Doyle and co-workers (11) reported that all *Bacillus* spores tested (*Bacillus anthracis*, *B. cereus*, *B. mycoides*, *B. subtilis*, and *B. thuringiensis*) were hydrophobic compared with the corresponding vegetative cells, as measured by adhesion to hexadecane and hydrophobic chromatography. The binding to octyl-Sepharose was enhanced by salts and inhibited by detergents. Treatment of *B. subtilis* 168 and *B. thuringiensis* 4040 with hot water, guanidine hydrochoride, or

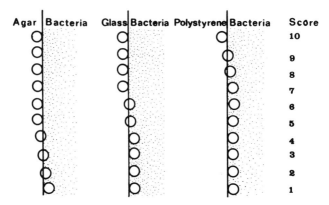

Figure 2. Scoring system for measuring the surface hydrophobicity of microbial lawns by the DOS method. A small drop of water was placed at the interface between a bacterial lawn and agar, a bacterial lawn and a glass cover slip, and a bacterial lawn and a polystyrene cover slip. For an example, see Fig. 1. The direction of spread of the drops determines the score, which is measured from 1 (the least hydrophobic bacteria) to 10 (the most hydrophobic bacteria).

1% sodium dodecyl sulfate increased significantly the hydrophobicity of the spores. The authors suggest that these treatments exposed hydrophobic residues on the coat proteins. Alternatively, these treatments could have extracted proteins that were blocking some of the hydrophobic sites (which need not be proteins). In support of the latter concept, it was reported that lysozyme treatment of *B. subtilis* spores also enhanced hydrophobicity, presumably by removing peptidoglycan-containing components. In several other bacteria, it is known that exopolysaccharides inhibit adhesion (42, 49).

    In addition to their ability to form resistant endospores, members of the genus *Bacillus* are characterized by their elaboration of peptide antibiotics. One group of peptides, consisting of gramicidins, tyrocidins, and polymyxins, modify membrane structure and function (24). Gramicidins are composed of a linear array of hydrophobic amino acids in alternating D and L configurations, tyrocidins are cyclic peptides containing hydrophobic amino acids with a few free amino groups, and polymyxins are composed of a cyclic peptide that is connected to a hydrophobic branched-chain fatty acid. These molecules are synthesized by specific synthesis without the direct participation of mRNA. It has been suggested that gramicidins and tyrocidins play a role in the early stages of endosporulation by regulating transport to and from the prespore and in controlling the direction of membrane synthesis (18). Polymyxins are

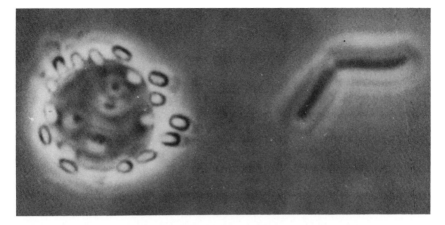

Figure 3. Adhesion of *B. cereus* spores to hexadecane. When a mixture of spores and vegetative cells of *B. cereus* were mixed with the hydrocarbon, spores tended to adhere to the hydrocarbon droplets, whereas vegetative cells remained in the aqueous phase.

formed late in the sporulation process, bind to subsurface layers, and may be involved in the uptake of the dipicolinic acid-calcium chelate.

Recent experiments have demonstrated that the hydrophobicity of *Bacillus brevis* spores is due to the presence of gramicidin S (GS) on the spore surface (41). The antibiotic GS [cyclo-(Val^1-Orn^2-Leu^3-D-Phe^4-Pro^5)$_2$] (Fig. 4) is synthesized during late exponential growth and early stationary phase. Essentially all of the synthesized antibiotics are bound to spores when they are released from the sporangium (32, 39). Mutants lacking one of the GS synthetase activities fail to produce measurable amounts of GS but still form resistant spores (10, 28). However, the GS$^-$ mutants have an extended outgrowth period during germination (29, 38).

Spores of the wild-type *B. brevis* adhered avidly to hexadecane, whereas the GS$^-$ mutant demonstrated no adhesion (Table 1). Addition of

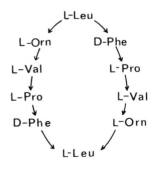

Figure 4. Structure of the antibiotic gramicidin S.

Table 1. Effect of GS on the Hydrophobicity
of *B. brevis*[a]

Conditions[b]	Adhesion to hexadecane (%)[c]
Vegetative cells	0
WT spores..	93
GS⁻ mutant spores	0
+ 2 μg of GS/ml..............................	40
+ 4 μg of GS/ml..............................	85
Germinating WT spores	
2 h	80
3 h	2

[a] Data adapted from Rosenberg et al. (41).
[b] The strains used were *B. brevis* Nagano (wild type [WT]) and its GS⁻ mutant, *B. brevis* B1-7.
[c] Measured by the method of Rosenberg et al. (46), using 0.2 ml of hexadecane and 2 ml of the cell suspension (MATH).

GS (which binds to spores) increased the hydrophobicity of the mutant spores in a dose-dependent manner. During germination, the wild-type spores lost their hydrophobicity after the germination-outgrowth stage (0.5 to 1.5 h) and coincident with the beginning of vegetative growth (2 to 3 h). Addition of GS to mutant spores delayed germination and outgrowth by 0.5 and 8 h with 0.5 and 1.0 μg of GS, respectively. The mutant spores, containing the added GS, lost their hydrophobicity at the completion of the delayed germination-outgrowth period. The data indicate that GS is responsible for the hydrophobicity of *B. brevis* spores and for delaying late stages in germination.

By analogy with the detailed molecular analysis of how GS binds to nucleic acids and forms phase transition complexes (26), it was suggested that positively charged ornithine residues of GS (8 Å [0.8 nm] apart) bind the negatively charged surface of the spore. The side chains of the remaining hydrophobic amino acids would then be oriented outward, imparting a high cell surface hydrophobicity on *B. brevis* spores.

Decreased Hydrophobicity during Myxosporulation

Myxococcus xanthus is a gram-negative, rod-shaped bacterium that has a complex life cycle, involving cell-to-cell interactions at all stages of growth and development (22, 40). Cells are able to glide slowly over solid surfaces (by a still unknown mechanism), forming feeding swarms. The most common substrates for the growth of *M. xanthus* are proteins and peptides. When grown on protein, the cells export proteinases and express a cell density-dependent growth phenomenon (43): the more cells, the more extracellular proteinases, the higher the concentration of permeable peptides and amino acids, and the faster the growth rate. When the swarms of cells become starved, they aggregate, forming raised mounds of 10^5 to 10^6 cells, referred to as fruiting bodies. The cells then undergo morphogenesis to form spherical, environmentally resistant resting cells called myxospores. The conversion of vegetative cells to myxospores can also be artificially induced in liquid growth medium by addition of glycerol (13), thereby bypassing the formation of fruiting bodies.

Kupfer and Zusman (27) determined changes in cell surface hydrophobicity during myxosporulation in fruiting bodies and in liquid growth medium. In both cases, myxosporulation was accompanied by a hydrophilic shift. Before induction of development, the vegetative cells adhered to hexadecane (50 to 60%) and xylene (80 to 90%). Neither the glycerol-induced nor fruiting body myxospores adhered significantly to hexadecane or xylene. During fruiting body formation, the hydrophilic shift took place 12 to 36 h after development began. This corresponds to the time of cellular aggregation and mound formation but precedes sporulation by several hours. With glycerol-induced spores, the hydrophilic shift occurred at 60 to 120 min, corresponding to the time of morphogenesis of long rods into ovoids. Aggregation-defective, sporulation-proficient mutants behaved like wild-type cells with regard to the hydrophilic shift during development. Mutants blocked in sporulation, however, failed to undergo this shift.

The transition from a hydrophobic cell to a hydrophilic myxospore appears to involve the noncovalent blocking of hydrophobic sites on the cell surface. When fruiting body spores were purified on sucrose gradients, they showed the same hydrophobicity as vegetative cells. Thus, the hydrophilic shift was probably due to the formation of exopolysaccharides that were loosely bound to the myxospore. Since glycerol-induced myxospores retained their hydrophilic properties, even after purification by sucrose gradient centrifugation, the blocking groups may be more tightly bound when produced in shake flask cultures.

Spatial Differentiation in Streptomycetes

Members of the genus *Streptomyces* are gram-positive bacteria characterized by the formation of a branching substrate mycelium that is occasionally interrupted by a cross wall (7, 14). Young colonies consist entirely of such substrate mycelia. When the cells are starved of nutrients, aerial hyphae develop directly from the uppermost substrate mycelia. At the tips of the aerial mycelium, the cells differentiate into chains of exospores. Subsequently, new aerial hyphae grow over the initial sporu-

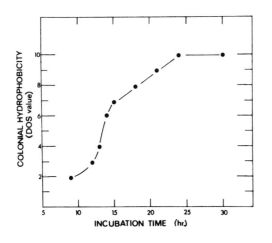

Figure 5. Changes in colonial hydrophobicity during the development of *S. clavuligerus*. A spore suspension of *S. clavuligerus* was spread evenly onto sectors of agar medium and incubated at 30°C. Colonial hydrophobicity was measured by the DOS method.

lation zone and again sporulate. Such waves of sporulation continue several times. Eventually, the colony consists mostly of aerial hyphae and exospores, most of the substrate mycelium having lysed. One of the differences between substrate mycelia and aerial mycelia-exospores is that the latter are extremely hydrophobic whereas the former are not. This was demonstrated qualitatively many years ago by the fact that aerial mycelia, but not substrate mycelia, stain with Sudan black (23).

The development with time of hydrophobic aerial mycelia and exospores of *Streptomyces clavuligerus* is summarized in Fig. 5. Nine hours after incubation, the spores had germinated and given rise to a visible vegetative mycelium. The DOS value (Fig. 2) of this substrate mycelium was 2, indicating that mycelium surface growth was less hydrophobic than growth on agar, glass, and polystyrene. By 18 h, the white aerial mycelium was prominent and the DOS value had reached 8 (more hydrophobic than agar and glass but less hydrophobic than polystyrene). At 24 h, the aerial mycelium had developed further and contained mature spores. At this stage, the DOS value was 10, more hydrophobic than agar, glass, and polystyrene.

It has been suggested that the hydrophobicity of aerial mycelia and exospores is due to a sheath which encloses them (20, 48). The sheath is lacking in vegetative mycelia. Spores become hydrophilic after removal of the sheath or after treatment with benzene, xylene, or ethyl alcohol. The particular chemical components of the sheath responsible for its hydrophobic nature are unknown. It is interesting that at least some antibiotics appear to concentrate in the sheath.

Differentiation and Hydrophobicity of Marine Vibrios

Members of the genus *Vibrio* are gram-negative bacteria that are found mostly in the marine environment. Many of the marine vibrios have two cell forms: the swimming cell and the swarming cell. When grown in a liquid environment, they appear as short motile rods propelled by a single polar flagellum (the swimming cell). When propagated on a solid surface, the cells elongate and produce hundreds of unsheathed lateral flagella, which function to translocate the bacteria over the surface (the swarming cells) (57, 62). It has recently been demonstrated that the polar flagellum acts as a dyanometer to regulate swimming cell differentiation. Conditions that restrict polar flagellum movement (i.e., solid surfaces, high viscosity, and antipolar flagellum antibodies) all induce swarmer cell differentiation (2, 33). Several hours after the onset of differentiation, the small colony starts to rapidly expand over the agar (swarming). Expansion of the colony over the agar surface is controlled by chemotaxis, as demonstrated by the finding that specific chemotaxis mutants failed to swarm (52). During the swarming over the agar surface, several distinct zones of bacterial growth appear (Fig. 6). In addition to the differences in macroscopic appearance, each zone contains cells of characteristic morphology. The center of the colony is composed of irregular cells, including bent rods, club-shaped bacteria, and polarly flagellated ovoids. The outermost zone is composed of typical swarming cells, long rods containing many lateral flagella in addition to the polar flagellum (54).

We have studied the properties of the swarming colony of *Vibrio* sp. strain NS-30. Strain NS-30 was isolated from the skin of a fast-swimming barracuda (53). It was proposed that bacterial films may improve the hydrodynamic properties of certain fast-swimming fish. The colonial and cellular hydrophobicities of swarming colonies of NS-30 were measured by the DOS and MATH methods, respectively. Each of the distinct growth zones had a characteristic colonial hydrophobicity (Fig. 7).

The spreading (diameter) of 5-μl water drops placed directly on the bacterial surface increased from the center to the periphery, indicating qualitatively that surface hydrophobicity was highest at the center and lowest at the periphery of the colony. The higher surface hydrophobicity at the center of the colony (zone 1) than at the periphery (zone 4) was also demonstrated by the fact that the water drop at the center spread toward the glass, whereas at the periphery it spread away from the glass. More accurate measurements of colonial hydrophobicity were obtained by the DOS method (Table 2). DOS values of 9 (similar to the value for polystyrene), 8 (slightly less hydrophobic than polystyrene but more

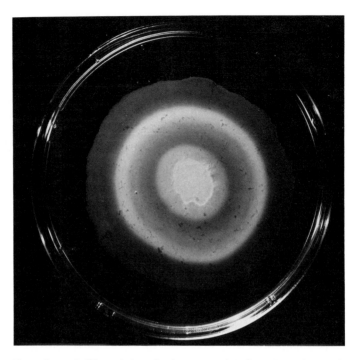

Figure 6. Swarming and differentiation of *Vibrio* sp. strain NS-30. Bacteria were inoculated onto the center of the plate and allowed to incubate at 27°C for 20 h. The photograph was taken with proper illumination to show the distinct concentric growth rings.

hydrophobic than glass), 5 (similar to the value for glass), and 1 (less hydrophobic than 2% agar) were recorded for zones 1 to 4, respectively.

Bacteria removed from any of the zones in the colony and suspended in buffer had low cellular hydrophobicity (MATH test). The fact that inner zones showed high surface hydrophobicity, although cells suspended from these zones failed to adhere to hexadecane, suggested that extracellular materials, presumably amphipathic in nature, were responsible for the high surface hydrophobicities. These materials remained in the supernatant fluid when the cell suspensions were centrifuged before the MATH test. Evidence in support of this hypothesis came from an examination of the hydrocarbon-in-water emulsifying activities of the supernatant fluids derived from the centrifuged cells from each growth zone. The order of increasing emulsifying activity was identical with the order of increasing colonial hydrophobicity values: emulsifying activity decreased from the center (zone 1) toward the colony edge (zone 4) (Fig. 8).

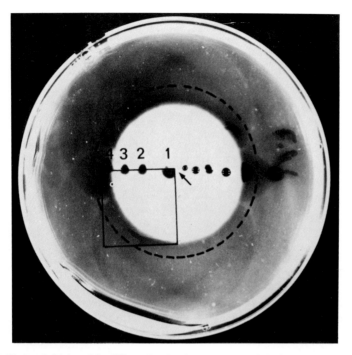

Figure 7. Hydrophobicity of the differentiated *Vibrio* sp. strain NS-30 colony. After a square cover glass was placed over part of the colony, 5-μl drops of water were placed at intervals across the colony. On the portion of the colony containing the cover slip, the droplets were placed at the glass-bacterium interface. The diameter of the drops (right side) and direction of movement of the drops (toward or away from the glass) indicate the relative hydrophobicity of the bacterial lawn at the indicated position. The broken line corresponds to the translucent outer boundary of the colony, as seen in Fig. 6.

Scanning electron microscopy of the colony of NS-30 grown for 2 h on SLB agar by the method of Shapiro (55) showed that the entire surface of the colony except for zone 4 was covered with a continuous surface layer. The scanning electron microscopy analysis did not resolve the thickness and structure of the surface layer.

The composition of the colony changed with time. Twenty-four hours after inoculation, the differentiated colony covered the entire plate. From 24 to 48 h, the colony appeared less and less differentiated. After 48 h, the whole plate was covered by bacterial growth that had the traits of zone 2: short rods, no cell adhesion to hexadecane, colonial hydrophobicity DOS value of 8, and presence of extracellular emulsifying activity. The ability of strain NS-30 to cover large areas in a short time (24 h to cover a diameter of 9 cm) and its high colonial hydrophobicity indicate the

Table 2. Changes in Surface Hydrophobicity of *E. coli* during Intraperiplasmic Development of *Bdellovibrio bacteriovorus*[a]

Time (min)[b]	Hydrophobicity (%)[c]
1	5
15	50
30	40
60	55
90	60
120	60

[a] Data taken from Cover and Rittenberg (8).
[b] A synchronous culture of *B. bacteriovorus* (6×10^{10} cells per ml) was grown on radioactively labeled *E. coli* (3×10^{10} cells per ml).
[c] Percentage of cells bound to octyl-Sepharose.

potential of such bacteria to increase the water repellency of the fish skin, thereby potentially decreasing frictional drag.

The chemical nature and mechanism of regulation of secretion of the extracellular emulsifying materials are not known. It will be interesting to find out whether morphogenesis and production of the emulsifier of NS-30 are regulated by a common mechanism. It is possible that the same emulsifying agent is produced in all of the growth zones and that the different hydrophobicity values reflect the local concentration of the common material. An alternative explanation is that different emulsifying agents are produced in different locations. Differential gene activity

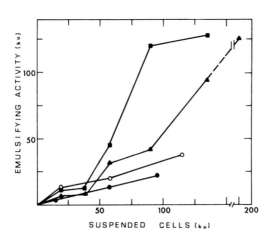

Figure 8. Extracellular emulsifying activity during differentiation of *Vibrio* sp. strain NS-30. Bacteria were allowed to develop on plates as illustrated in Fig. 6 and 7 for 20 h. Cells were then suspended from each of the four growth zones (as shown in Fig. 7), and their turbidities were measured. After removal of cells by centrifugation, the clear supernatant fluid was assayed for hydrocarbon-in-water emulsifying activity. The abscissa indicates the turbidity of the cell suspension before centrifugation. Symbols: ■, zone 1; ▲, zone 2; ○, zone 3; ●, zone 4.

across the colony is likely to exist because formation of the various cell types described above requires changes in the activities of many genes. The finding that an extracellular mechanism can be responsible for the surface properties of a colony demonstrates a new mechanism by which bacteria may interact and alter the local environment. Theoretically, these changes in colonial hydrophobicity can be achieved at a low price. Even a monomolecular layer covering the colony surface could be responsible for large changes in the surface properties of the colony.

Bdelloplast Formation

Bdellovibrios are predators that grow and multiply in the intraperiplasm of their bacterial prey (63). Shortly after a bdellovibrio attaches to and penetrates into its prey bacterium (e.g., *Escherichia coli*), the prey cell undergoes morphogenesis to a spherical cell, referred to as a bdelloplast. The penetration and bdelloplast formation steps require the action of bdellovibrio enzymes (59, 60). During the intraperiplasmic growth of the bdellovibrio, the substrate cell becomes more hydrophobic (Table 2). The initial high hydrophobicity (at 15 min) is probably due to the highly hydrophobic bdellovibrios which have not yet penetrated the prey bacteria and tether the substrate cell to the octyl-Sepharose. The evidence for this view is that mild shearing to remove attached but unpenetrated bdellovibrios caused a reduction in hydrophobicity. At later stages (60 to 120 min), shearing had little effect on the hydrophobicity measurement.

Changes in cell hydrophobicity associated with the morphogenesis to bdelloplasts are due to changes in the substrate cell lipopolysaccharide resulting from bdellovibrio enzymatic activity. At least part of this change is due to solubilization of lipopolysaccharide amino sugars during attack (35, 36, 61). In addition, the cell wall of the prey bacterium is deacetylated and then acylated with fatty acids. It has been suggested that these reactions are responsible for the morphogenesis to the spherical bdelloplast and its resistance to lysozyme and osmotic shock (61).

GENERALIZATIONS

The five examples presented above were chosen because the morphogenesis events are dramatic and because at least some experimental data are available on changes in surface hydrophobicity during the morphogenesis. Changes in hydrophobicity associated with the life cycle of cyanobacteria have been discussed by Shilo (56). These examples represent only a small fraction of the morphogenesis and hydrophobicity changes that occur among procaryotes. Many bacterial species become smaller and more spherical and hydrophobic as a result of starvation. Starvation-induced effects on bacterial size and surface characteristics have been particularly well studied in marine bacteria (see, for examples, references 9 and 25).

In an attempt to derive some generalizations with regard to bacterial morphogenesis and hydrophobicity, we have summarized the data in Table 3. Both increases and decreases in bacterial hydrophobicity occur during morphogenesis. The decreases all occur in systems in which the altered cells regain their activity by passing through an aqueous system rather than by aerial dispersion. For example, myxospores inside fruiting bodies are believed to germinate and begin their feeding and swarming activity after being immersed in water. All aquatic adherent cyanobacteria have a hydrophobic cell surface. In order to spread and colonize new suitable surfaces, the cells must detach and disperse through the water column. The hydrophobic spores of bacilli and streptomycetes, on the other hand, spread by aerial dispersion. It has been suggested that aerial spores may play an important role in evolution (3).

There seems to be no common mechanism by which the cells change in hydrophobicity. In some cases, there appears to be de novo synthesis of hydrophobic outer envelopes (endospores of bacilli and the outer sheath of streptomycete exospores). In other cases, an existing cell envelope is modified to make it more hydrophobic (bdelloplasts and some marine bacteria). Another important mechanism for changing surface hydrophobicity appears to be the noncovalent blocking of sites on the cell surface. Old cells of the filamentous cyanobacterium *Phormidium* J-1 become hydrophilic as a result of the masking of surface hydrophobic sites by a high-molecular-weight polysaccharide, referred to as emulcyan (15, 16, 56). During myxosporulation, a similar hydrophilic shift occurs as a result of loosely bound polymers that block hydrophobic sites. In *B. brevis* spores, the binding of polypeptide antibiotic GS to the surface is responsible for the hydrophobic transition. A phase transition complex is formed, involving positively charged ornithine residues binding to the negatively charged spore surface, with the remaining hydrophobic groups of the antibiotic oriented outward.

In conclusion, changes in cell surface hydrophobicity often occur during procaryotic morphogenesis. The underlying mechanisms and natural roles of these hydrophobic transitions, however, appear to differ from species to species. Thus, bacterial morphogenesis and the accompanying changes in surface hydrophobicity are further examples of the diversity, rather than unity, of the procaryotic world.

Table 3. Hydrophobic Changes during Morphogenesis of Bacteria

Morphogenetic event	Hydrophobic transition	Inducing event	Postulated mechanism of shift in hydrophobicity	Possible function
Endosporulation	Increase	Nutritional downshift	Formation of outer coat; binding of hydrophobic peptides	Aerial dispersion; binding of hydrophobic antibiotics
Myxosporulation	Decrease	Developmental program	Noncovalent blocking of hydrophobic sites	Fruiting body formation?
Streptomycete differentiation	Aerial mycelium and spore increase	Developmental program	Formation of outer sheath	Aerial dispersion of exospores; binding of hydrophobic antibiotics
Vibrio colonial differentiation	Decrease	Swarming	Escape of swarmer cell from extracellular amphipathic emulsifier	Unknown
Bdelloplast formation	Increase	Bdellovibrio penetration and enzymatic activity	Alteration of lipopolysaccharide and peptidoglycan	Formation and stabilization of bdelloplast
Cyanobacterial aging	Decrease	Starvation that blocks synthesis of hydrophobic proteins	Blocking of hydrophobic sites by emulcyan	Detachment and dispersal
Marine bacteria	Generally increase	Starvation	Changes in outer surface; fibril formation	Adhesion and survival tactic

LITERATURE CITED

1. **Absolom, D. R., F. V. Lamberti, Z. Policova, W. Zingg, C. J. van Oss, and A. W. Neumann.** 1983. Surface thermodynamics of bacterial adhesion. *Appl. Environ. Microbiol.* **46**:90–97.
2. **Beias, R., M. Simon, and M. Silverman.** 1986. Regulation of lateral flagella gene transcription in *Vibrio parahaemolyticus*. *J. Bacteriol.* **167**:210–218.
3. **Bisset, K. A.** 1950. Evolution in bacteria and the significance of the bacterial spore. *Nature* (London) **166**:431–432.
4. **Boyles, W. A., and R. E. Lincoln.** 1958. Separation and concentration of bacterial spores and vegetative cells by foam flotation. *Appl. Microbiol.* **6**:327–334.
5. **Busscher, H. J., A. H. Weerkamp, H. C. van der Mei, A. W. J. van Pelt, H. P. de Jong, and J. Arends.** 1984. Measurement of the surface free energy of bacterial cell surfaces and its relevance for adhesion. *Appl. Environ. Microbiol.* **48**:980–983.
6. **Casida, L. E., Jr.** 1977. Small cells in pure cultures of *Agromyces ramosus* and in natural soil. *Can. J. Microbiol.* **23**:214–216.
7. **Chater, K. F., and D. A. Hopwood.** 1973. Differentiation in actinomycetes. *Symp. Soc. Gen. Microbiol.* **23**:143–160.
8. **Cover, W. H., and S. C. Rittenberg.** 1981. Change in the surface hydrophobicity of substrate cells during bdelloplast formation by *Bdellovibrio bacteriovorus* 109J. *J. Bacteriol.* **157**:391–397.
9. **Dawson, M. P., B. Humphrey, and K. C. Marshall.** 1981. Adhesion: a tactic in the survival strategy of marine vibrio during starvation. *Curr. Microbiol.* **6**:195–198.
10. **Demain, A. L., and J. M. Piret.** 1979. Relationship between antibiotic biosynthesis and sporulation, p. 183–188. *In* M. Luckner and K. Schreiber (ed.), *Regulation of Secondary Product and Plant Hormone Metabolism.* Pergamon Press, New York.
11. **Doyle, R. J., F. Nedjat-Haiem, and J. S. Singh.** 1984. Hydrophobic characteristics of *Bacillus* spores. *Curr. Microbiol.* **10**:329–332.
12. **Dworkin, M.** 1985. *Developmental Biology of the Bacteria.* The Benjamin-Cummings Publishing Co. Inc., Menlo Park, Calif.
13. **Dworkin, M., and S. M. Gibson.** 1964. A system for studying microbial morphogenesis: rapid formation of microcysts in *Myxococcus xanthus*. *Science* **146**:243–244.
14. **Ensign, J. C.** 1978. Formation, properties and germination of actinomycete spores. *Annu. Rev. Microbiol.* **32**:185–219.
15. **Fattom, A., and M. Shilo.** 1984. Hydrophobicity as an adhesion mechanism of benthic cyanobacteria. *Appl. Environ. Microbiol.* **47**:135–143.
16. **Fattom, A., and M. Shilo.** 1985. Production of emulcyan by *Phormidium* J-1: its activity and function. *FEMS Microbiol. Ecol.* **31**:3–9.
17. **Gaudin, A. M., A. L. Mular, R. F., and O'Conor.** 1960. Separation of microorganisms by flotation. *Appl. Microbiol.* **8**:91–97.
18. **Gerhardt, P., R. Scherrer, and S. H. Black.** 1972. Molecular sieving by dormant spore structure, p. 68–74. *In* H. D. Halvorson, R. Hanson, and L. L. Campbell (ed.), *Spores V.* American Society for Microbiology, Washington, D.C.
19. **Hermansson, M., S. Kjelleberg, T. K. Korhonen, and T. A. Stenstrom.** 1982. Hydrophobic and electrostatic characterization of surface structures of bacteria and its relationship to adhesion at an air-water surface. *Arch. Microbiol.* **131**:308–312.
20. **Higgins, M. L., and L. K. Silvey.** 1966. Slide culture observations of two freshwater actinomycetes. *Trans. Am. Microsc. Soc.* **85**:390–398.
21. **Hjertén, S., J. Rosengren, and S. Pahlman.** 1974. Hydrophobic interaction chromatog-

raphy. The synthesis and the use of some alkyl and aryl derivatives of agarose. *J. Chromatogr.* **101**:281–288.

22. **Kaiser, D., C. Manoil, and M. Dowrkin.** 1979. Myxobacteria: cell interactions, genetics, and development. *Annu. Rev. Microbiol.* **33**:595–639.

23. **Kalakoutskii, L. V., and N. S. Agre.** 1976. Comparative aspects of development and differentiation in actinomycetes. *Bacteriol. Rev.* **40**:469–524.

24. **Katz, E., and A. L. Demain.** 1977. The peptide antibiotics of *Bacillus*: chemistry, biogenesis, and possible functions. *Bacteriol. Rev.* **41**:449–474.

25. **Kjelleberg, S., and M. Hermansson.** 1984. Starvation-induced effects on bacterial surface characteristics. *Appl. Environ. Microbiol.* **48**:497–503.

26. **Krauss, E. M., and I. S. Chan.** 1984. Complexation and phase transfer of nucleic acids by gramicidin S. *Biochemistry* **23**:73–77.

27. **Kupfer, D., and D. R. Zusman.** 1984. Changes in cell surface hydrophobicity of *Myxococcus xanthus* are correlated with sporulation-regulated events in the developmental program. *J. Bacteriol.* **159**:776–779.

28. **Lazaridis, I., M. Frangou-Lazaridis, F. C. MacCuish, S. Nandi, and B. Seddon.** 1980. Gramicidin S content and germination and outgrowth of *Bacillus brevis* spores. *FEMS Microbiol. Lett.* **7**:229–232.

29. **Marahiel, M., W. Danders, M. Krause, and H. Kleinkauf.** 1979. Biological role of gramicidin S in spore functions. Studies on gramicidin S-negative mutants of *Bacillus brevis* ATCC 9999. *Eur. J. Biochem.* **99**:49–55.

30. **Mark, C. M. van L., L. Johannes, N. Willem, S. Gosse, and A. J. B. Zehnder.** 1987. The role of bacterial cell wall hydrophobicity in adhesion. *Appl. Environ. Microbiol.* **53**:1893–1897.

31. **Marshall, K. C., and R. H. Cruickshank.** 1973. Cell surface hydrophobicity and the orientation of certain bacteria at interfaces. *Arch. Mikrobiol.* **91**:29–40.

32. **Matteo, C. C., M. Glade, A. Tanaka, J. M. Piret, and A. L. Demain.** 1975. Microbiological studies on the formation of gramicidin S synthetases. *Biotechnol. Bioeng.* **17**:129–142.

33. **McCarter, I., H. Marcia, and M. Silverman.** 1989. Flagellar dyanometer control of swarmer cell differentiation of *Vibrio parahaemolyticus*. *Cell* **54**:345–351.

34. **Morita, R. Y.** 1982. Starvation-survival of heterotrophs in the marine environment. *Adv. Microb. Ecol.* **6**:171–198.

35. **Nelson, D. R., and S. C. Rittenberg.** 1981. Incorporation of substrate cell lipid A components into the lipopolysaccharide of intraperiplasmically grown *Bdellovibrio bacteriovorus*. *J. Bacteriol.* **147**:860–868.

36. **Nelson, D. R., and S. C. Rittenberg.** 1981. Partial characterization of lipid A of intraperiplasmically grown *Bdellovibrio bacteriovorus*. *J. Bacteriol.* **147**:869–874.

37. **Neumann, A. W., D. R. Absolom, D. W. Francis, and C. J. van Oss.** 1980. Conversion tables of contact angles to surface tensions. *Sep. Purif. Methods* **9**:69–163.

38. **Piret, J. M., and A. L. Demain.** 1982. Germination initiation and outgrowth of spores of *Bacillus brevis* strain Nagano and its gramicidin S-negative mutant. *Arch. Microbiol.* **133**:38–43.

39. **Piret, J. M., and A. L. Demain.** 1983. Sporulation and spore properties of *Bacillus brevis* and its gramicidin S-negative mutants. *J. Gen. Microbiol.* **129**:1309–1319.

40. **Rosenberg, E. (ed.).** 1984. *Myxobacteria: Development and Cell Interactions.* Springer-Verlag, New York.

41. **Rosenberg, E., D. R. Brown, and A. L. Demain.** 1985. The influence of gramicidin S on hydrophobicity of germinating *Bacillus brevis* spores. *Arch. Microbiol.* **142**:51–54.

42. **Rosenberg, E., N. Kaplan, O. Pines, M. Rosenberg, and D. Gutnick.** 1983. Capsular polysaccharides interfere with adherence of *Acinetobacter calcoaceticus* to hydrocarbons. *FEMS Microbiol. Lett.* **17:**157–160.

43. **Rosenberg, E., K. H. Keller, and M. Dworkin.** 1977. Cell density-dependent growth of *Myxococcus xanthus* on casein. *J. Bacteriol.* **129:**770–777.

44. **Rosenberg, M., and S. Kjelleberg.** 1986. Hydrophobic interactions: role in bacterial adhesion. *Adv. Microb. Ecol.* **9:**353–393.

45. **Rosenberg, M., and E. Rosenberg.** 1981. Role of adherence in growth of *Acinetobacter calcoaceticus* on hexadecane. *J. Bacteriol.* **148:**51–57.

46. **Rosenberg, M., E. Rosenberg, and D. Gutnick.** 1980. Bacterial adherence to hydrocarbons, p. 541–542. *In* R. C. W. Berkeley, J. M. Lynch, J. Melling, P. R. Rutter, and B. Vincent (ed.), *Microbial Adhesion to Surfaces*. Society of Chemical Industry, London.

47. **Rubin, A. J., E. A. Cassel, O. Henderson, J. D. Johnson, and J. C. Lamb III.** 1966. Microflotation: new low gas-flow rate foam separation technique for bacteria and algae. *Biotechnol. Bioeng.* **8:**135–151.

48. **Ruddick, S. M., and S. T. Williams.** 1972. Studies on the ecology of actinomycetes in soil. V. Some factors influencing the dispersal and adsorption of spores in soil. *Soil Biol. Biochem.* **4:**93–103.

49. **Runnels, P. L., and H. W. Moon.** 1984. Capsule reduces adherence of enterotoxigenic *Escherichia coli* to isolated intestinal epithelial cells of pigs. *Infect. Immun.* **45:**737–740.

50. **Sacks, L. E., and G. Alderton.** 1961. Behavior of bacterial spores in aqueous polymer two-phase systems. *J. Bacteriol.* **82:**331–341.

51. **Sar, N.** 1987. Direction of spreading (DOS): a simple method for measuring the hydrophobicity of bacterial lawns. *J. Microbiol. Methods* **6:**211–219.

52. **Sar, N., L. McCarter, M. Simon, and M. Silverman.** 1990. Chemotactic control of the two flagellar systems of *Vibrio parahaemolyticus*. *J. Bacteriol.* **172:**334–341.

53. **Sar, N., and E. Rosenberg.** 1987. Fish skin bacteria: colonial and cellular hydrophobicity. *Microb. Ecol.* **13:**193–202.

54. **Sar, N., and E. Rosenberg.** 1989. Colonial differentiation and hydrophobicity of a marine *Vibrio* sp. *Curr. Microbiol.* **18:**331–334.

55. **Shapiro, J.** 1985. Scanning electron microscope study of *Pseudomonas putida*. *J. Bacteriol.* **164:**1171–1181.

56. **Shilo, M.** 1989. The unique characteristics of benthic cyanobacterium, p. 207–218. *In* Y. Cohen and E. Rosenberg (ed.), *Microbial Mats: Physiological Ecology of Benthic Microbial Communities*. American Society for Microbiology, Washington, D.C.

57. **Shinoda, S., and K. Okamoto.** 1977. Formation and function of *Vibrio parahaemolyticus* lateral flagella. *J. Bacteriol.* **129:**1266–1271.

58. **Smyth, C. J., P. Jonsson, E. Olsson, O. Soderlind, J. Rosengren, S. Hjertén, and T. Wadström.** 1978. Differences in hydrophobic surface characteristics of porcine enteropathogenic *Escherichia coli* with or without K88 antigen as revealed by hydrophobic interaction chromatography. *Infect. Immun.* **22:**462–472.

59. **Thomashow, M. F., and S. C. Rittenberg.** 1978. Intraperiplasmic growth of *Bdellovibrio bacteriovorus* 109J: solubilization of *Escherichia coli* peptidoglycan. *J. Bacteriol.* **135:**998–1007.

60. **Thomashow, M. F., and S. C. Rittenberg.** 1978. Intraperiplasmic growth of *Bdellovibrio bacteriovorus* 109J: N-deacetylation of *Escherichia coli* peptidoglycan amino sugars. *J. Bacteriol.* **135:**1008–1014.

61. **Thomashow, M. F., and S. C. Rittenberg.** 1978. Intraperiplasmic growth of *Bdellovibrio*

bacteriovorus 109J: attachment of long-chain fatty acids to *Escherichia coli* peptidogly-can. *J. Bacteriol.* **135:**1015–1024.

62. **Ulizur, S., and M. Kessel.** 1973. Giant flagella bundles of *Vibrio alginolyticus* (NCMB 1803). *Arch. Microbiol.* **94:**331–339.

63. **Varon, M., and M. Shilo.** 1968. Interaction of *Bdellovibrio bacteriovorus* and host bacteria. I. Kinetic studies of attachment and invasion of *Escherichia coli* B by *Bdellovibrio bacteriovorus. J. Bacteriol.* **95:**744–753.

Microbial Cell Surface Hydrophobicity
Edited by R. J. Doyle and M. Rosenberg
© 1990 American Society for Microbiology, Washington, DC 20005

Chapter 9

Cell Surface Hydrophobicity of Medically Important Fungi, Especially *Candida* Species

Kevin C. Hazen

Interest in surface hydrophobicity of fungi has increased dramatically during the past 5 years, likely because of the development of simple and inexpensive methods for assessing fungal cell surface hydrophobicity (CSH) and the substantial evidence in bacterial systems that CSH contributes to various physiologic and pathogenic activities. Two genera, *Candida* and *Saccharomyces*, have been the predominant foci of fungal CSH studies. Concern about CSH of *Saccharomyces* species is primarily due to the use of these organisms in the fermentation industry. Information about surface interactions involved in flocculation and nutrient uptake would provide insight into conditions needed to optimize fermentation production. Interest in CSH of *Candida* species, particularly *Candida albicans*, has developed because these organisms are frequent agents of life-threatening opportunistic disease in immunocompromised patients, especially those suffering hematogenous cancers. Knowledge about surface interactions involved in adhesion to host and prosthetic surfaces, proliferation in tissues, immunological responses, and antifungal susceptibility is needed before comprehensive and effective prophylactic treatment for candidiasis can be realized. *C. albicans* and *Saccharomyces cerevisiae* are also particularly favorable models for CSH expression studies because of the extensive base of genetic and biochemical information available for these organisms.

Kevin C. Hazen • Department of Pathology, Box 214, University of Virginia Health Sciences Center, Charlottesville, Virginia 22908.

Despite the growing interest in fungal hydrophobicity, little is known about the physiological and biochemical basis of CSH expression. Most attention has been given to determining whether hydrophobic interactions are involved in attachment to various plastics, in attachment to host cells, and in coadhesion. These studies have shown that hydrophobic interactions contribute to fungal attachment to plastics but that fungal growth conditions, assay conditions, and the type of plastic can influence the degree to which the hydrophobic interactions participate. Similarly, hydrophobic interactions involved in fungal attachment to host tissues are determined by multiple factors, not the least of which are fungal isolate and target tissue.

This chapter is designed to acquaint the reader with the developments in CSH research of *Candida* species and other medically important fungi. Occasional reference to observations made on fungal plant pathogens and other nonmedically important fungi will be given. Hydrophobicity of *S. cerevisiae* and other fungi involved in the fermentation industry are discussed elsewhere in this volume. The methods used to assess fungal CSH will be presented first. Here, emphasis will be placed on the problems and advantages of the various methods for studying fungi and not on the general methodology and chemical basis of the methods. This section will be followed by a review of the observations concerning CSH in fungal activities, especially physiology and pathogenicity. The biochemical basis of CSH will then be discussed in the context of the preceding section.

METHODS FOR ASSESSING CSH OF FUNGI

In 1924, Mudd and Mudd (82, 83) devised a method to explore the behavior of bacteria at surface films of varying surface tension. The method involved partitioning of bacteria between an aqueous phase and an oil phase and therefore also indicated whether a bacterial population contained hydrophobic cells. Since then, a variety of methods for evaluating CSH of bacteria have been developed (see Rosenberg and Doyle, this volume). However, the number of methods that have been applied to fungi is limited, in part because fewer investigators are interested in fungal CSH an in part because certain methods cannot be directly transferred to fungal systems. As a consequence, less is known about CSH of fungi as a group than is known about bacterial CSH. Recent demonstrations showing that CSH plays a significant role in pathogenesis by *C. albicans* should stimulate greater interest in fungal CSH.

Inherent Problems Associated with Measuring Fungal CSH

Fungi are achlorophyllous organisms that possess a cell wall throughout their life cycle. Many fungi are monomorphic; their vegetative phase is either a mold or a yeast. Some fungi are dimorphic; that is, they can exhibit two vegetative forms (for example, yeast and mold or spherule and mold). The systemic fungal pathogens of humans exhibit dimorphism. Generally, as saprobes they grow in the mold form but as pathogens they produce the yeast (e.g., *Histoplasma capsulatum*) or spherule (e.g., *Coccidioides immitis*) form. *C. albicans*, on the other hand, grows in the yeast form when it is a commensal of the human host but exhibits hyphal and pseudohyphal production during parasitism (19, 95, 98).

Size and shape of yeast forms vary from one species to another and can vary within a particular isolate simply as a result of growth temperature (40). Hyphae vary in length (ranging from less than 20 μm to greater than 10 mm) and in width from species to species. Hyphal growth occurs apically, resulting in a gradient of cellular ages along a hypha, with the oldest cells possibly being dead.

Reproductive propagules of fungi are extremely variable in size, shape, topography, texture, color, and surface composition. The asexual reproductive spores of many of the medically important fungi are conidia. Depending on the organism, conidia may be released singly, in clumps, or in chains. From the point of view of initiation of pathogenesis of fungal diseases, introduction of a fungus into an appropriate host site, such as the lung, is usually due to exposure to conidia. Whether the conidia can establish an infection depends on their ability to attach to host tissues and avoid the host immune response.

Measuring CSH of fungi begs the question of deciding which fungal form to study. Unlike bacterial populations, in which the cell size and shape of the organisms are relatively invariant, fungi offer a multiplicity of forms and structures. Assessing the CSH of each form produced by a fungus is appealing because it would provide useful information about the organism's surface biology throughout its life cycle. Until recently, this assessment had not been possible because the available CSH measurement assays were insufficiently amenable to be used for each fungal morphology. With only a few exceptions, fungal hydrophobicity measurements have been performed on yeast cells.

Another problem associated with CSH measurements of fungi concerns the cell wall. This structure is complex and contains numerous types of macromolecules. It is composed primarily of polysaccharides (chitin, cellulose, chitosan, mannans, or glucans) but also contains proteins and a small amount of lipid (for reviews, see references 27, 28,

33, and 95). The external surface, the site of the cell wall that is evaluated in many hydrophobicity assays, can be smooth, rough, or covered with microfibrils. Yeast cell walls have a birth scar and one or more bud scars. A single budding cell can display different topographies simultaneously.

An important question that must be considered when one performs a hydrophobicity assay is, What does an apparently positively hydrophobic cell indicate? The answer is a function of the assay. Except for the hydrophobic microsphere attachment assay (HMA) (36, 41), none of the fungal CSH measurement methods can provide information on the distribution of hydrophobic sites along a fungal cell. It is not possible to determine whether the hydrophobic sites are uniformly arranged or in patches along a cell. Thus, for instance, in the microbial adhesion to hydrocarbons assay (MATH), the hydrophobic bud of a cell may cause the cell to enter the hydrocarbon phase and make the cell appear hydrophobic when, in fact, the mother cell is hydrophilic. Whether this constraint is an important issue for an investigator depends on the purpose for which CSH is measured.

Five methods have been used to assess fungal CSH: contact angle measurements; MATH; HMA; hydrophobic interaction chromatography (HIC); and the salt aggregation test (SAT). The general procedures and results obtained by various groups of investigators are presented in Table 1. Few groups have attempted to compare CSH methods to determine whether a particular method is reliable. Given that the various groups conduct each assay differently, hydrophobicity levels obtained by one group for a particular strain can not be unconditionally compared with the levels obtained by another group for the same strain even if the same growth conditions are used. The strengths and weaknesses of each method are as follows.

Contact angle measurements

In the contact angle measurements method, the angle of contact formed by a drop of "sensing liquid" (e.g., distilled water or α-bromonaphthalene) on a lawn of yeast cells is determined. The yeast cell lawn must first be partially dried before consecutive droplets of sensing liquid form reproducible contact angles. Droplet contact angles formed on a lawn of yeast cells are the aggregate result of the cumulative contribution of negatively and positively hydrophobic sites of each cell. Unlike the other hydrophobicity measurement methods, contact angle measurements do not provide information on the percentage of cells within a yeast population that is hydrophobic. However, they do provide a direct means to determine the surface tension of a cell population and an indirect method to assess a population's surface free energy (1, 32, 85). These parameters are useful for evaluating the effect of the yeast cell population in a liquid and the ability of the cells to adhere to various surfaces.

Contact angle measurements have been limited to yeast cells, although spherical to ovate conidia are likely amenable to the method. A potential problem would be to obtain homogeneous populations of conidia, a task that is difficult with many fungi. Because conidia are generally larger than bacterial cells, a conidial cell layer is likely much rougher than a layer formed by bacterial cells. Hyphal cells are not good candidates for contact angle measurements because the requisite smooth lawn cannot be formed and, as mentioned earlier, they vary in age and size.

It is not clear what effect partially drying a yeast cell lawn has on the surface composition of each cell. It is reasonable to assume that the partially dehydrated cell surface does not exhibit hydrophobicity to the same extent as a hydrated surface because of changes in macromolecular folding. This difference could account for the disagreement in results obtained by contact angle measurements and aqueous-hydrocarbon biphasic assay when the two methods are used to determine relative hydrophobic strengths of different *Candida* species (Table 2; 60, 77).

MATH

The basic MATH procedure involves evaluating the change in absorbance or optical density of a cell population suspended in an aqueous buffer as a result of the cell population being exposed to a hydrocarbon. The strengths of the MATH lie in its simple equipment requirements, low cost, ease of performance, and objective result. These advantages have made this method popular for both fungal and bacterial studies since its description by Rosenberg et al. (101). The parameters that must be considered for running the assay have been described elsewhere (100). Two factors that need to be emphasized are the use of acid-cleaned glassware and the use of reagent-grade buffers and hydrocarbons. Lack of observance of these two factors can lead to spurious results.

Several problems with the MATH limit its usefulness for mycological investigations. Hyphae and germ tubes cannot be expected to give reliable results, since these can occupy all of the available area at the water-hydrocarbon interface with a relatively low number of cells. In addition, absorbance values for a hyphal or germ tube population before and after exposure to hydrocarbon may not involve equivalent populations. The cell subpopulation associated with the hydrocarbon may differ

Table 1. Summary of Hydrophobicity Methods Applied to Medically Important Yeasts[a]

General method	Specific steps	Comments	Reference(s)
Contact angle measurements	Performed at 20°C Air-dried cell layer Water used as sensing liquid	Various *Candida* species	77, 78, 87
	Used contact angle meter Dried cells for 3 h Bromonaphthalene used as sensing liquid Projected image of liquid bubble to determine angle	*C. albicans*	60, 61
HMA	Various microsphere diameters and buffers tested Standard assay involved 10^6 cells and 4.5 × 10^8 microspheres (0.845-μm diameter)/ml Conducted at room temperature	*C. albicans* Buffer composition and pH affects apparent CSH level	35–37, 41, 59
SAT Improved SAT	10^8 cells/ml of saline Mix 20 μl of cells with 20 μl of ammonium sulfate solution Add methylene blue in improved SAT	*C. albicans* Some isolates autoaggregate Different results obtained with SAT and improved SAT	70
HIC	$2 × 10^9$ cells Made column in Pasteur pipette	*C. albicans*	94
	Mix 20 ml of cells with 1.0 ml of HIC-Sepharose Pour onto sieve and elute cells	*Cryptococcus neoformans*	62

MATH	Mix cell suspension in basal medium with 20% n-tetradecane in basal medium	Candida intermedia, Candida tropicalis, Saccharomyces cerevisiae First report of MATH with yeasts	41, 43
	Cell density (A_{600} = 0.400) 1.2 ml of cell suspension in PUM buffer 0.3 ml of hydrocarbon Vortex 3 min 12- by 75-mm tubes	C. albicans C. tropicalis	77, 87
	Cell density (A_{660} = 0.5) 3 ml of cells in PBS 0–500 µl of hydrocarbon Mix 60 s 10-mm-diameter tubes	Various Candida species	
	10^7 cells/ml 2.5 ml of cell suspension in PUM buffer or D5.45 0.5 ml of hexadecane 16-mm-diameter tubes Preincubate 10 min, vortex 2 min	C. albicans, C. rugosa, hyphal cells of C. albicans	56
	10^8 cells/ml 2.5 ml of cell suspension in PUM buffer 0.5 ml of hexadecane 12-mm-diameter tubes Preincubate 10 min, vortex 2 min	C. albicans, white-opaque variants	53

(Continued on next page)

Table 1—*Continued*

General method	Specific steps	Comments	Reference(s)
	"Method of Rosenberg et al, 1980" (101) Cell density (A_{400} = 1.4–1.6) 1.2 ml of cell suspension in PUM buffer 0.2 ml of hexadecane (not stated in ref. 101) 10-mm tubes Preincubate 10 min, vortex 2 min, 30°C	One isolate of *C. albicans* High yeast cell numbers	73
	Cell density (A_{660} = 0.5) 2.0 ml of cells suspension in PUM buffer 0.5 ml of hexadecane 12- by 105-mm tubes Mix 60 s	Various *Candida* species and iso-lates	80
	Cell density (A_{540} = 1.0) Volume of cell suspension not stated 10-mm-diameter tubes Various volumes of xylene Vortex 60 s	One isolate of *C. albicans*	49
	5×10^6 cells/ml 1.5 ml of cell suspension, HBSS 1.0 ml of various hydrocarbons Equilibrate at 37°C Read A_{530} of aqueous phase 12- by 75-mm tubes Vortex 2 min	Multiple isolates of different *Candida* species	60

[a] The specific procedure for each method is described in the text. PUM, Phosphate-urea-magnesium; HBSS, Hanks balanced salts solution.

morphologically from the cell subpopulation residing in the aqueous phase, thus affecting the cell geometry involved in causing light absorbance. Each of these subpopulations will differ from the original population suspended in the aqueous phase before mixing with the hydrocarbon. Another problem with obtaining hydrophobicity values for a population of germ tubes or hyphae is that these structures tend to clump, which would grossly affect the absorbance value. This problem could be exacerbated by the hydrocarbon forming a coat over the cells (50). It may also be necessary to consider the physical effect of buoyancy on the apparent hydrophobicity of the hyphal cell population.

The MATH has proven to be an effective and useful method for assessing CSH of yeast cells. The resultant CSH value represents the percentage of cells that possess hydrophobic surface areas of sufficient nonpolarity and abundance that the cells partition at the hydrocarbon phase. Several factors inherent to a yeast cell population, such as cell size and density, could influence its apparent CSH. Large, heavy cells could separate from the water-hydrocarbon interface, making the cell population appear hydrophilic. The release of surface macromolecules, particularly ones with surfactant activity (59), into the aqueous phase or during mixing could alter the surface tension of the interface and cause lower CSH values.

The MATH was performed differently by the various investigative groups cited in Table 1. Differences in the size of the glass test tubes, volume of hydrocarbon, volume of yeast suspension, composition of the aqueous buffer, and yeast cell concentration are apparent. Each of these factors can influence the apparent CSH of the yeast cell population (100). A high cell concentration could overload the system, resulting in a false low CSH value (100). It is noteworthy that in some cases, adhesion of *C. albicans* to hexadecane is very low at low initial cell densities, but avid adhesion occurs at high initial cell concentrations (M. Barki, S. Goldberg, R. J. Doyle, and M. Rosenberg, unpublished observations). This observation suggests that positive cooperativity is operative in adhesion of *C. albicans* to hexadecane. When the volume of hydrocarbon is below a threshold level, the percentage of cells that appear hydrophobic decreases (Fig. 1). Similar results were obtained by Minagi et al. (77). We have recently determined that the effect of decreasing the hydrocarbon volume is most dramatic on the CSH level of cells having surfaces of low hydrophobic avidity (K. C. Hazen and W. G. LeMelle, unpublished observations). These observations make clear that investigators who plan to use the MATH should first determine the optimum concentration of cells (that is, the highest concentration that results in the highest percent change in absorbance after mixing with hydrocarbon but is not the result

Table 2. Comparison of Different CSH Assessment Methods for *Candida* Yeast Cells[a]

Data set	CSH method Contact angle measurements[b]	MATH	HIC	Growth conditions	Comments	Reference
1	C. tropicalis C. krusei C. parapsilosis C. glabrata C. albicans	C. tropicalis C. krusei C. parapsilosis C. glabrata C. albicans		Yeast nitrogen base + 250 mM glucose, 37°C, late exponential phase	Laboratory strains	87
2	C. pseudotropicalis C. albicans (J) C. krusei (1) C. tropicalis (2) = C. lipolytica (8661) C. albicans (1755) C. glabrata (2) C. krusei (2) C. tropicalis (1) C. parapsilosis (2) C. parapsilosis (1)	C. lipolytica (8661)[b] C. tropicalis (2) = C. pseudotropicalis (2) C. parapsilosis (2) C. albicans (2) C. glabrata (2) C. krusei (1) C. parapsilosis (1) C. albicans (1755) C. tropicalis (1) C. krusei (2)		Sabouraud dextrose broth, 26°C, 21 h		60
3	C. tropicalis C. krusei C. glabrata C. parapsilosis C. albicans C. stellatoidea	C. krusei C. glabrata C. tropicalis C. parapsilosis C. stellatoidea C. albicans		Glucose-yeast extract-peptone, 37°C, OD$_{660}$ = 2.0	Growth phase not stated	77

| 4 | C. albicans
LGH1095-23
LGH870-23
LGH490-23
LGH490-37
LGH870-37 | C. albicans
LGH870-23
LGH1095-23
LGH490-23
LGH870-37
LGH490-37 =
LGH1095-37 | Sabouraud dextrose broth 23°C, 37°C, 24 h | Values for LGH870-23 and LGH1095-23 not statistically different | 41 |
| 5 | C. albicans (glucose)
C. albicans (hexadecane) | C. albicans (glucose) = C. albicans (hexadecane) | Yeast nitrogen base + glucose or hexadecane 26°C, 20–120 h | Single C. albicans isolate | 59 |

[a] Numbers and letters in parentheses represent my strain designations. OD_{660}, Optical density at 660 nm.
[b] Species are listed in descending order of CHS level.

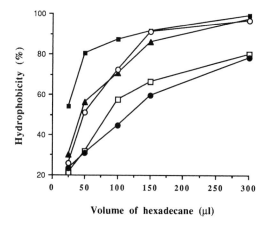

Figure 1. Effect of hydrocarbon volume on the apparent CSH level of five isolates of *C. albicans* in the aqueous-hydrocarbon biphasic assay. Symbols: ■, LGH870; ▲, LGH1095; □, UMC9385; ●, 9938; ○, CK4. Yeasts were suspended in 1.2 ml of phosphate-urea-magnesium buffer.

of positive cooperativity, which is a separate phenomenon) and the volume of hydrocarbon that does not cause false low levels of CSH but is not so great that poor mixing occurs.

HIC

In the HIC method, a known number of yeast cells is passed over a hydrophobic affinity column composed of octyl- or phenyl-Sepharose, and the number of cells retained on the column after several washes is considered indicative of the hydrophobic level of the cell population. HIC has been used for assessing CSH of medically important fungi by only two groups (62, 94). Why so few groups have used the method is unclear. However, our experience suggests that one problem is that *C. albicans* yeast cells may be mechanically trapped by the Sepharose gel, leading to false high CSH levels. This problem apparently was also observed by others (54, 62), who suggested that simply mixing the cells with the gel and then washing the gel provides a more reliable estimate of the hydrophobicity level of a yeast cell population. Care must be taken not to let the Sepharose gel stack during washing; otherwise, mechanical trapping of cells will occur. Another precaution is to ensure that yeast cell attachment to the gel is not due to interactions with the Sepharose resin (20). Parallel experiments in which unmodified Sepharose is used should be run. HIC is entirely unsuitable for germinated cells or hyphae.

SAT

Originally devised by Lindahl et al. (69) for assessing CSH of *Escherichia coli*, the SAT method involves exposing a fixed density of cells to increasing concentrations of ammonium salts. Lower concentrations of ammonium salt are needed to cause aggregation of highly hydrophobic cells. The SAT has been applied to medically important yeasts by Macura (70), who used not only the SAT as described by Lindahl et al. (69) but also an "improved" SAT (developed by Rozgonyi et al. [105]) in which methylene blue is added to the cell-ammonium salt mix. The SAT is easy to perform, rapid, and inexpensive but appears to have limited applicability to, at least, *Candida* yeast cells. One reason is that these yeast cells tend to aggregate at relatively low concentrations of ammonium salts (unpublished observation), making it difficult to distinguish strains of varying CSH levels. The basis for the high sensitivity of these yeast cells to ammonium salts has not been explored.

HMA

In 1984, Lachica and Zink (64) reported that CSH of *Yersinia enterocolitica* strains could be evaluated by assessing the ability of the cells to agglutinate latex particles. The particle diameter that they found most useful was 5.70 μm. The use of such large particles renders visualization of specific sites of attachment difficult. The recently developed HMA method involves mixing low-sulfate (i.e., hydrophobic), surface-unmodified, polystyrene microspheres with yeast cells and determining the percentage of cells with three or more attached microspheres (36, 37, 41). This percentage represents the population hydrophobic level. The use of three or more attached microspheres as an indicator of CSH was based on correlative studies using the MATH (41). To date, the HMA has been used almost exclusively by my laboratory. It has proven to be an extremely powerful method for studying hydrophobicity of fungi. It is easy to perform, inexpensive, and rapid.

The HMA has several advantages over other CSH measurement methods. It is the only method that can be applied to hyphae and germ tubes under conditions identical to those used to assess yeast cells. Instead of enumerating how many microspheres are attached to the fungal structures, the degree of hydrophobicity is determined semiquantitatively (37). By this method, it has been possible to show that germ tubes produced by *C. albicans* are hydrophobic (37, 41). A significant advantage with the HMA versus other CSH measurement methods is that the distribution of hydrophobic sites on a cell can be determined (Fig. 2). This is possible by using microspheres with diameters smaller than those of the fungal cell to be evaluated. Whether a bud and its mother cell differ in surface hydrophobicity status is possible to discern. A third advantage is that the cells and microspheres are suspended and mixed in an aqueous medium which contains no chemical reagents that could alter the apparent

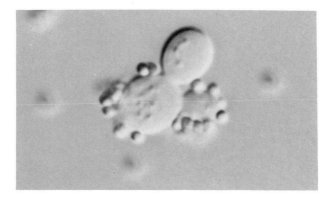

Figure 2. Appearance of a hydrophobic yeast cell as detected by the HMA. A positive cell (mother and daughter spheres are considered one cell) is one with three or more attached microspheres. This cell demonstrates one of the advantages of the HMA. That is, the distribution of hydrophobic sites along a cell surface can be determined. Note that one of the daughter spheres appears hydrophilic.

surface hydrophobicity of a cell. The same conditions can be used to assess CSH of mammalian cells which may be the sites of adhesion for a pathogenic fungus (36). A fourth and potentially powerful advantage to the HMA is that an investigator can modify the criteria for designating a cell as hydrophobic such that different degrees of hydrophobicity are demonstrated. For example, instead of specifying that a cell composed of mother cell and bud have three or more attached microspheres to be considered, an investigator could define a hydrophobic cell as one in which mother and daughter cells each have one or more attached microspheres. Such information could then be used to distinguish different hydrophobic subpopulations. The availability of microspheres of various surface properties, chemical composition, color, and size is another potentially useful advantage to the microsphere assay, since these would allow an investigator to test different surface attributes of various fungal populations.

There are several cautions that one must heed when performing the standard HMA, all of which can lead spuriously low CSH levels. The use of surfactant-free, low-sulfate microspheres is imperative. Microspheres with a high surface density of negatively or positively charged groups have too little hydrophobic surface area ("parking area") to interact with fungal cells by predominantly hydrophobic interactions (110). If surfactant is present, it could alter the CSH properties of the fungal cell. We use microspheres supplied by Serva Fine Biochemicals (Westbury, N.Y.), but other sources include Interfacial Dynamics Corp. (Portland, Ore.), Polysciences (Warrington, Pa.), Duke Scientific Corp. (Palo Alto, Calif.), and Seragen Diagnostics (Indianapolis, Ind.). Carboxy, carboxylate, and mixed-function latex microspheres are not suitable for detecting CSH. Use of hydrophobic latex microspheres possessing positively charged groups (e.g., amidine latex) will result in spuriously high CSH levels as a result of polar interactions with the negatively charged surface of fungal cells.

The size of the microspheres can be varied by the investigator. However, the microspheres should be individually discernible by bright-field microscopy. In general, as the radius of the particle increases, its hydrophobic strength decreases (1, 6). The use of microspheres of uniform size is necessary to ensure that different levels of microsphere hydrophobicity are not involved, allowing for easier interpretation of the data. The maximum size of microspheres is also limited by their colloidal properties. In general, microspheres larger than 1 μm in diameter do not act as colloids (110). Larger microspheres could easily sediment out of suspension during the assay or not mix well with the fungal cells. Microspheres that begin to aggregate in the stock suspension should be replaced with fresher suspensions.

All glassware that comes in contact with cells and microspheres should first be scrupulously cleaned. This is also true with the aqueous-hydrocarbon biphasic assay (97). We have found that washing glassware with detergents specifically designed for tissue culture glassware (e.g., Contrad 70; Polysciences) twice followed by overnight soaking in 1% HCl and seven rinses with deionized water works well. Heating glassware at 170°C for 4 h also helps increase the hydrophilicity of the glass surface, preventing hydrophobicity-mediated cell attachment to the sides of the glass tubes during the assay.

The microsphere concentration can affect the apparent CSH level of a cell population (Fig. 3). The effect appears to be related to the avidity of the hydrophobic surface of each cell (42). From this observation, we have recently developed a modification of the microsphere assay based on enzyme binding theory to discriminate yeast populations that appear otherwise similarly hydrophobic in which a hydrophobic avidity constant (K_H) of the hydrophobic cell surface is determined (42).

One condition which must be met when one compares two yeast strains for CSH expression by the HMA is that the cells are similar in their sphere-to-cell ratios. That is, the number of spherical units (buds and mother cell) within a contiguous cellular unit should be approximately the same. *C. albicans* changes its ratio of spherical units per contiguous cell unit during growth (40). By stationary phase (24 h in Sabouraud dextrose

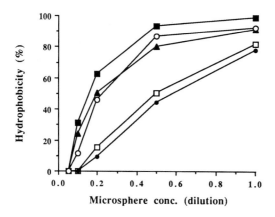

Figure 3. Effect of varying the microsphere concentration on the apparent CSH levels of five isolates of *C. albicans*. Symbols: ■, LGH870; ▲, LGH1095; □, UMC9385; ●, 9938; ○, CK4. Yeast cells were kept at a constant concentration and tested against various concentrations of polystyrene microspheres. The standard concentration (final concentration, 4.5×10^8 microspheres per ml) is indicated as 1.0.

broth [SDB]), the ratio is typically between 1.3 and 2.0. Exponential-phase cells may have a ratio of 5.0. The standard HMA defines a hydrophobic cell (contiguous unit) as one with three or more attached microspheres. The more spherical units per contiguous unit, the more potential hydrophobic sites available for attachment to the hydrophobic microspheres. A cell population having a high ratio of spherical units per contiguous unit, such as an exponential-phase cell population of *C. albicans* could appear more hydrophobic than a cell population in which the sphere-to-cell ratio is low (stationary-phase population).

Comparison of Yeast CSH Methods

Few research groups have attempted to compare the effectiveness of two or more CSH methods to determine CSH levels of yeast populations, although several such comparisons have been performed for bacterial cells (see, for example, references 24, 81, and 122). When such comparisons have been performed for yeast populations, they have involved the MATH. On the basis of their overall ranking order, the MATH and contact angle measurements provide similar results (Table 2). This is particularly evident from the first set of data in Table 2. For that study, the investigators used various *Candida* species that differed significantly from each other with respect to both contact angles and absorption to the hydrocarbon phase. When these differences were not so great, the ranking orders obtained by the MATH and contact angle measurements did not agree exactly (data sets 2 and 3 in Table 2). These results indicate that contact angle measurements and the MATH provide similar results with respect to ranking orders but cannot be relied on to discriminate identically between yeast cell populations that differ slightly in CSH.

Only two studies have compared the HMA and another CSH assessment method. In one study (data set 4 in Table 2), excellent agreement was obtained when the HMA was compared with the MATH, but here again the *C. albicans* strains differed significantly from each other in the MATH. The HMA did not provide results similar to those of the MATH and contact angle measurements in a second study (data set 5 in Table 2). However, in this study only two strains of *C. albicans*, one of which produced abundant surfactant material, were used. Also, the size homogeneity and surface characteristics of the microspheres were not stated. Further comparative studies need to be performed with the HMA and other CSH methods. The development of a kinetic method to assess cell surface hydrophobic avidity provides more definitive information about the hydrophobicity status of a yeast cell population than any of the present yeast cell CSH assays (43).

GENERAL CHARACTERISTICS OF FUNGAL HYDROPHOBICITY

CSH levels of various medically important fungi have been determined by several groups of investigators, but studies concerning the intrinsic and extrinsic factors that influence CSH expression have been few. Most of the latter studies have been limited to *C. albicans*. They have revealed, similar to findings for bacterial systems (102), that many intrinsic and extrinsic factors determine whether a cell population will appear hydrophobic. These same factors will also influence what macromolecules are exposed on the surface of a fungal cell (114). From these observations, it is clear that the cell wall surface properties and composition of fungi are consequences of multiple factors. Each of these factors must be considered when one attempts to interpret data concerning surface-dependent activities such as adhesion.

Growth Conditions

Temperature

The temperature at which yeast cells of *C. albicans* isolates are grown can influence their relative CSH levels (35–37, 40, 43, 55). *C. albicans* isolates generally appear more hydrophobic when grown to stationary phase at room temperature (23 to 25°C) than at 37°C. *C. albicans* isolates that have similar hydrophobicity levels at these two growth temperatures have been observed (38, 43). The expression of CSH by *C. albicans* is not an all-or-nothing event. An entire yeast cell surface is not necessarily hydrophobic or hydrophilic but may have patches of surface hydrophobicity (unpublished observations). Also, yeast cell pop-

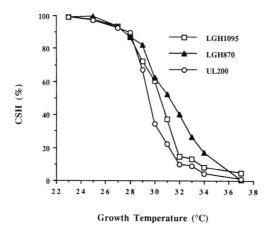

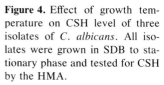

Figure 4. Effect of growth temperature on CSH level of three isolates of *C. albicans*. All isolates were grown in SDB to stationary phase and tested for CSH by the HMA.

Growth Temperature (°C)

ulations can be moderately hydrophobic. The CSH level of a stationary-phase yeast cell population is directly related to growth temperature (B. W. Hazen and K. C. Hazen, unpublished results). We have determined that with increasing growth temperature beginning at 23°C and ending at 37°C, the CSH level decreases (Fig. 4). This phenomenon was seen with all four isolates tested. This study also revealed that there is an approximately 4-degree range at which the CSH level changes most dramatically for each isolate. In general, highest CSH levels were obtained when the growth temperature was less than 26°C and lowest levels were obtained when the temperature was greater than 33°C. The physiologic basis for the 4-degree range has not been investigated.

Growth temperature was found to influence the CSH levels of other *Candida* species besides *C. albicans* (*C. glabrata*, *C. tropicalis*, and *C. stellatoidea*), but it did not affect the levels obtained with the monomorphic yeasts *S. cerevisiae* and *Cryptococcus neoformans* (37, 43). The distribution of hydrophobic sites on *S. cerevisiae* was, however, affected by growth temperature (Fig. 5). Hydrophobic polystyrene microspheres were distributed over the entire surface of cells when the yeasts were grown at 23°C but were polarly dispersed when the cells were grown at 37°C (37).

Growth medium

Not surprisingly, growth medium can influence the relative hydrophobicity level of a *C. albicans* cell population (43, 55, 69). When yeast cells are grown at 37°C in media that are conducive to germ tube formation, the CSH level is higher than when the cells are grown in a

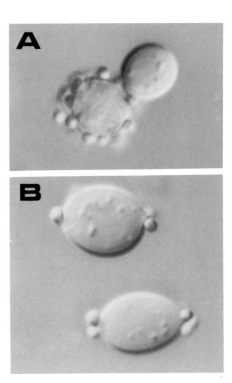

Figure 5. Effect of growth temperature on the distribution of hydrophobic sites on *S. cerevisiae*. When cells are grown at 23°C, the hydrophobic sites are relatively random (A), but the sites become polarly distributed when the cells are grown at 37°C (B). (Adapted from reference 40 with permission from the New York Botanical Garden.)

complex medium such as SDB or glucose-yeast extract-peptone broth (43, 55). Elevating the concentration of glucose or galactose in SDB to 250 to 500 mM results in higher levels of CSH than those obtained in SDB containing lower concentrations of the sugars (K. C. Hazen and S. Murthy, unpublished observations). Klotz and Penn (61) were able to obtain a highly hydrophobic variant of a *C. albicans* isolate by repeated exposure to growth medium containing hexadecane.

Kennedy and Sandin (55) investigated the effect of growth temperature and 11 growth media on the CSH level of one isolate of *C. albicans*. In this study, the MATH was used to assess CSH. Except for one medium (LBC) in which the cells were grown at 25°C, CSH levels of the yeast cell populations with all of the media were below 14%. This result likely is isolate dependent or is an artifact of the MATH conditions (too low a volume of hydrocarbon for the diameter of the test tubes; see discussion above).

The specific medium components that cause high levels of CSH have not been investigated, nor have medium pH, viscosity, and ionic potential.

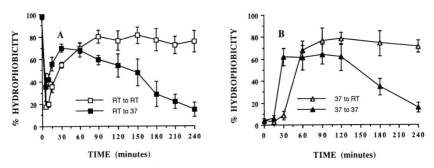

Figure 6. Effect of early growth phase and temperature on apparent CSH levels of *C. albicans* LGH1095. Yeasts cells were grown to stationary phase in SDB at either room temperature (23 to 25°C) (A) or 37°C (B) and then diluted into fresh SDB at either room temperature or 37°C. CSH levels were monitored with the HMA during lag and early exponential phases. These results demonstrate that yeast cells undergo a rapid (within 60 min) but transient change in CSH status during early growth. The bars represent the standard error of the mean. (Adapted from reference 35 with permission.)

Growth phase

Expression of CSH by *C. albicans* yeast cells is dependent on the cell growth phase (Fig. 6). Upon release into fresh medium at 23°C, hydrophobic cells become hydrophilic with 30 min but increase in CSH level as they approach stationary phase (35, 43). Upon release into fresh medium, hydrophilic cells become hydrophobic within 15 min (35). As these cells enter exponential phase, they become moderately hydrophobic. By stationary phase, they are once again hydrophilic. Moderate levels of hydrophobicity at exponential phase are also seen for cells that were highly hydrophobic before release into fresh medium (35, 43).

Solid versus liquid media

Little information is available about the influence of medium form on CSH levels of medically important fungi. Kennedy and Sandin (55) investigated whether CSH levels of a single isolate of *C. albicans* were affected by growing the yeast cells in broths or on agar plates of Sabouraud dextrose medium. No significant difference in CSH levels was noted regardless of whether the cells were grown at 25 or 37°C. It is likely that yeast cells of most isolates of *Candida* species obtained by growth on solid media will express highly variable CSH levels because of the nutritional and growth-phase dynamics of colonial formation. As noted above, growth phase influences the apparent CSH level of yeast cells. In a colony, yeast cells are at various phases of growth. Cells located at the center of the colony are also exposed to concentrations of nutrients and microenvironmental conditions different from those of cells located contiguous to the agar surface or at the top or edges of the colony. We have found that such variability in CSH levels does occur when cells are obtained from colonies on solid media (K. C. Hazen and B. W. Hazen, unpublished observations).

CSH Levels of Medically Important Yeasts

The clinical significance of fungal CSH in candidiasis is unknown. No attempts to relate CSH expression of the more medically significant *Candida* species to their prevalence have been made. However, CSH levels of one to three isolates of these species after growth under specific conditions have been assessed (Table 2). No clear agreement among the studies is evident. *C. albicans*, the most frequent agent of nearly all forms of candidiasis (89), has been variously ranked as one of the most hydrophobic species to one of the least (compare data sets 1 and 2 in Table 2).

The reasons for the disagreement between studies are manifold. Most important, the various groups of investigators used different growth conditions to prepare cells for CSH measurement. The effects of growth temperature, medium, and phase have been discussed. Inherent differences among strains of each *Candida* species should also be expected. Such differences were demonstrated by Klotz et al. (60). This point is further exemplified by the report of Miyake et al. (80) in which clinical and laboratory strains of various *Candida* species exhibited different CSH levels. The ranking order for the species was also different between clinical and laboratory strains. Investigations in which the relative order of CSH potency for the medially important *Candida* species is to be determined should be restricted to the freshest isolates available. Repeated subculturing of an isolate could result in loss of hydrophobic expression. The investigations should also include multiple isolates of each species. The use of a single isolate or just a few isolates will not provide statistically significant information about the *Candida* species.

It may be noteworthy that *C. tropicalis* ranked as one of the most hydrophobic *Candida* species regardless of growth temperature in all the studies (Table 2). *C. tropicalis* has been reported to be the most virulent species in patients suffering acute leukemia (124–126). The results from the CSH studies suggest the possibility that CSH expression contributes to *C. tropicalis* virulence. Demonstration of whether this speculation is correct awaits further work.

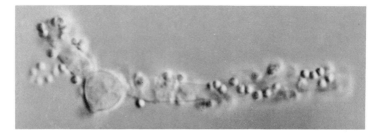

Figure 7. Distribution of hydrophobic microspheres on germ tubes of *C. albicans*. Germ tubes are highly hydrophobic. However, the mother cell, as seen here, may not be hydrophobic.

Dynamic Expression of CSH

Surface antigen expression by *C. albicans* has been shown to undergo dynamic changes in vitro during growth and morphogenesis (7–10, 12, 118, 119). Altered surface antigen expression has also been found to occur during pathogenesis of *C. albicans* (91–93). It is not surprising then to find that CSH expression also undergoes dynamic changes during growth. As mentioned above, exponential-phase cells of *C. albicans* grown in SDB express moderate levels of hydrophobicity but stationary-phase cells are either highly hydrophobic or hydrophilic, depending on growth temperature. Such surface changes are also seen with lag-phase cells; they are initially hydrophilic but increase in CSH before evidence of growth occurs (35).

The distribution of hydrophobic sites on a cell also appears to change during growth. When hydrophilic *C. albicans* yeast cells are placed in media conducive to germination at 37°C, they become hydrophobic within 30 to 45 min (35). Hydrophobic germ tubes are produced subsequently by the hydrophobic mother cells. Once the germ tubes are produced, the mother cell may no longer express surface hydrophobicity (Fig. 7). The reasons for this are not clear. However, hyphae, which are the result of further extension of germ tubes and septum formation, may show a CSH gradient (unpublished observations). The cells closest to the apex are most hydrophobic, whereas more distal cells are less so. The older cells may no longer express hydrophobicity. These observations suggest that cell aging can cause changes in surface hydrophobicity levels. Age differences may account for nonuniform distribution of hydrophobic sites among the hyphae of a mycelium (37).

CSH IN FUNGAL PHYSIOLOGY

Growth and Nutrition

All fungi obtain nutrients for growth by absorption. They produce a wide range of exoenzymes, cell wall-associated enzymes, cell membrane enzymes, and specialized macromolecules (e.g., chelators and surfactants) devoted to breaking down large substrates to absorbable size, capturing nutrients, transporting nutrients to the cell cytoplasm, or sensing environmental nutritional signals (31). How surface hydrophobicity is involved in these growth and nutritional activities is not known. However, several observations suggest that surface hydrophobicity expression is nutritionally beneficial to fungi.

Hydrophobic interactions could be involved in attachment of fungi to solid substrata that contain useful nutrients (e.g., attachment of hydrophobic conidia to the waxy surface of plants prior to production of appressoria and haustoria [44]). Thus, the fungus would be optimally situated to obtain nutrients directly from the surface. On the other hand, hydrophobicity-mediated attachment by starved cells to a solid substratum may serve as a means to keep a fungus at a particular location until a fresh nutrient influx occurs (71, 72). This may be the situation for the aquatic fungi. Similar to the case for bacteria, the value of surface hydrophobicity-mediated attachment to solid substrata is dependent on the fungus (29).

Hydrophobic surface materials may serve as surfactants or bioemulsifiers that help make hydrocarbon-containing substrata more readily accessible as nutrient sources. Many of the fungal surfactants and bioemulsifiers are extracellular materials, such as liposan produced by the non-medically important *Candida* species *Candida lipolytica* (13, 14), sophorolipids released by *Torulopsis* (*Candida*) *bombicola* (16, 48), and other glycolipids (45).

Medically important *Candida* species also produce surfactants that may be either cell wall associated or released into the extracellular milieu (50–52, 59, 61). The cell wall-associated surfactants make the cells more hydrophobic. By growing *C. tropicalis* in the presence of alkanes, Kaeppeli and Fiechter (51) induced the production of a polysaccharide-fatty acid complex on the yeast cell surface. Further study revealed that the polysaccharide was mannan and the fatty acids comprised 4% of the total dry weight of the complex (52). The inducible mannan-fatty acid complex appears to be involved in hydrocarbon uptake (50, 79). Cells producing the complex bind better than glucose-grown cells to the interface formed between hexadecane and water, indicating the cells are more hydrophobic (50, 51). Miura et al. (79) have also demonstrated that hydrophobic *C. tropicalis* and *Candida intermedia* isolates were more capable than hydrophilic isolates of assimilating hydrocarbons.

Utilizing a procedure similar to that used by Kaeppeli and Fiechter (50), Klotz and Penn (61) induced surfactant production by *C. albicans* in a medium containing hexadecane as the sole carbon source. The surfactant material was released into the medium and was also cell associated, where it caused the cell surface to be more hydrophobic (58, 59). The cell-associated surfactant appears to contribute to the ability of *C. albicans* to bind to epithelial cells (61) and to inert substrata (59).

Morphogenesis

How surface hydrophobicity expression participates in yeast cell formation, germ tube and hypha production, and conidiogenesis for those fungal species that exhibit morphology-dependent CSH is not understood. It is clear, however, that CSH expression accompanies morphogenesis. For example, the hyphae of *Aspergillus fumigatus* are weakly hydrophobic but the conidia are highly hydrophobic (37), indicating that conidiogenesis involves CSH expression. Likewise, germ tube formation by hydrophilic yeast cells of *C. albicans* is preceded by and coincident with CSH expression (35). The evidence with *C. albicans* strongly suggests that CSH expression is required for successful germ tube formation. Inhibition of germination by morphogenic autoregulatory substance (39) resulted in reduction of CSH expression (35). A hydrophobic variant of *C. albicans*, that is, one that is hydrophobic when grown at either 23 or 37°C, produces elongated cells and germ tubes under conditions that induce only yeast forms of the parent strain (Hazen and Hazen, unpublished observations). A hydrophilic variant also becomes markedly hydrophobic before germination is evident. In the case of *C. albicans*, CSH expression may contribute to germination by plasticizing the wall, thus making the wall more pliable for tubular evagination. The increased pliability of the wall is needed to overcome the thermodynamic pressures to form a spherical cell.

Spore Dispersion

Hydrophobic fungal spores are more aerobuoyant than hydrophilic spores. Hydrophobic spores can disperse over wide geographic distances and thus help ensure that suitable hosts will be found for growth and sexual reproduction. In contact with an aqueous medium, hydrophobic spores can enter the water-air interface and form a surface film that will allow the spores to be distributed by streaming to more nutritionally rich environments (4, 5, 123). Relative to the medically important *Aspergillus* species, aero- and hydrobuoyancy due to conidial surface hydrophobicity probably contributes to the high incidence of nosocomial aspergillosis in immunosuppressed patients.

FUNGAL SURFACE HYDROPHOBICITY IN THE MYCOSES

During the initiation of a mycosis, the cell wall is the only fungal structure that comes in contact with the host. The wall serves at least two functions in successful parasitism. First, it is the site involved in attaching the fungus to host tissue. In this context, the cell wall must allow sufficiently close contact between the fungus and host target site so that adhesion can occur. The cell wall must also function to protect the fungus from preexisting host-specific (immune) and nonspecific resistance mechanisms. Expression of CSH appears to accomplish both functions for *C. albicans*, *Aspergillus* species, and *Rhizopus* species. It is likely that CSH expression by infectious forms of other parasitic fungi should act identically, but further work is needed to confirm this. For medically important *Candida* species other than *C. albicans*, CSH does participate in adhesion to inert materials, such as those used in prosthetic devices (60, 78).

Attachment and Adhesion to Inert Surfaces

For the purpose of this review, attachment and adhesion will be used interchangeably. Colonization by *Candida* species of plastic prosthetic devices and catheters can serve as the nidus for subsequent local or disseminated infection. Denture stomatitis due to proliferation of colonized *C. albicans* from the surface of acrylic dentures is a frequent complaint of the elderly (89). Hospitalized patients with indwelling catheters run the risk of developing candidemia and possibly candida septicemia (104). Knowledge about fungal mechanisms involved in adhesion to plastics and resins could provide insight into how to prevent colonization. However, attachment of microorganisms to inert surfaces is a complex phenomenon (34, 110). Evidence accumulated over the past decade indicates that hydrophobic interactions can play a role in attachment of *Candida* species to acrylic and other plastics but that the degree of involvement depends on the surface characteristics of the plastic and the fungus. This point is demonstrated by the study of Minagi et al. (78). The relationship of surface free energy (which is directly influenced by surface hydrophobicity) to adhesion of single isolates of *C. albicans* and *C. tropicalis* to various denture base resins that vary in surface hydrophobicity as measured by contact angles was investigated. The isolate of *C. tropicalis* was highly hydrophobic (contact angle = 118.6°), but the *C.*

albicans isolate was hydrophilic (contact angle = 51.1°). Their results demonstrate that hydrophobic interactions are involved in attachment of a highly hydrophobic yeast isolate to resins. With decreasing surface hydrophobicity of the resins, yeast adhesion decreased. On the other hand, adhesion of the hydrophilic yeast isolate to the resins was generally related to the degree of hydrophilicity of the resin.

Acrylic

The contact angle data presented in Klotz et al. (60) indicate that polymethylmethacrylate or acrylic resin has an approximate water contact angle of 60°. In contrast, Teflon, which is highly hydrophobic, has a water contact angle of ca. 115°, untreated hydrophobic polystyrene has a contact angle of ca. 90°, and heat-treated, hydrophilic Pyrex glass has a contact angle of 10°. This finding indicates that acrylic resin is only moderately hydrophobic. It must be noted that Miyake et al. (80) obtained a water contact angle of 79.5° for acrylic. They did not determine contact angles for the other plastics. The difference between the results is probably due to the type of acrylic resin used and how the contact angle measurement was performed. One prediction from these results, however, is that fungal adhesion to acrylic could be mediated by a combination of hydrophobic and hydrophilic (electrostatic) interactions. Which mechanism is operative or predominant depends on the hydrophobic cell wall status of the fungus.

Environmental conditions leading to attachment of *Candida* species to acrylic as well as the fungal growth conditions that stimulate acrylic adhesiveness are summarized in Table 3. Unfortunately, only two studies have investigated the relationship between CSH and adhesion to acrylic (59, 80). Each demonstrates that adhesion to acrylic by *Candida* yeast cells is enhanced by higher levels of CSH. The study by Miyake et al. (80) used correlative statistics to obtain this conclusion. However, the data are based on CSH levels of various *Candida* species and not on a wide range of isolates of a single species. Klotz (59) looked at a single isolate of *C. albicans* but used only two cell populations that differed in CSH.

In the studies in which *C. albicans* was grown at 37°C and in complex media (glucose-yeast extract-peptone, Trypticase soy broth, and SDB), attachment to acrylic was frequently poor (Table 3). This result is not surprising, since the conditions have been shown to support hydrophilic cell wall expression (see above). When cells were grown at 37°C in yeast nitrogen base or mycological peptone with elevated levels of glucose, galactose, or sucrose, increased attachment to acrylic occurred. We have observed that elevated levels of galactose or glucose in SDB and yeast nitrogen base induce higher levels of CSH in *C. albicans* (K. C. Hazen et al., unpublished data). Taken together, these results suggest that CSH is involved in adhesion of *C. albicans* to acrylic. However, definitive studies are needed to confirm this point and to determine whether the same is also true for other *Candida* species. If possible, individual strains of *C. albicans* should be grown under conditions in which only one growth parameter is modified (e.g., temperature) but different levels of CSH result.

The assay conditions, such as incubation temperature, incubation period, and cell density, can also influence the apparent adhesion level (35, 41, 54, 113). Hydrophobic interactions are entropically dependent, indicating that they are stronger with increasing temperature (46, 47, 90). If hydrophobic interactions are involved in adhesion, this should be evident by a decreased level of adhesion when the incubation temperature is lowered. However, the effect of temperature on adhesion may not be clear if one does not also consider the incubation period. *Candida* species have been shown to undergo significant surface changes within minutes after exposure to new environmental conditions (7, 9, 35). Contact with a solid surface can also trigger various fungal surface changes that could increase the strength of binding to an inert surface (63, 126).

Plastics

Similar to the observations obtained with acrylic, the contribution of CSH to adhesion of *Candida* species to plastic depends on the degree of hydrophobicity of the fungal cell and the plastic. Conflicting results have been obtained concerning the relationship between CSH and adhesion to polystyrene (Table 3). These conflicts can be explained by considering the relative degree of hydrophobicity of the interacting solids. As described for the HMA, polystyrenes can differ in their relative degree of surface hydrophobicity according to how they are made and whether they have been surface modified. When adhesion of *Candida* species to polystyrene was investigated, different polystyrenes were used (53, 54, 60). Bacteriological petri dish polystyrene is hydrophobic, but it possesses some degree of electronegativity (60). This form of polystyrene was used by Rosenberg (99) for developing a rapid screening assay for bacterial CSH and has been shown to correlate well with the aqueous-hydrocarbon biphasic assay when applied to various gram-negative and -positive bacteria (24). However, microtiter plate polystyrene (particularly plates containing fewer than 96 wells each) may not be hydrophobic since it is typically tissue culture treated. That is, the polystyrene of these plates is modified so that it is highly hydrophilic to allow for good mammalian cell

Table 3. Summary of Studies Evaluating the Relationship between *C. albicans* CSH and Attachment to Inert Surfaces[a]

Inert surface	CSH assay	Attachment assay and method of quantitation	Fungal growth conditions	General results	Comments	Reference
Acrylic[b]	MATH	37°C, 60 min Direct enumeration	Glucose-yeast extract-peptone, 37°C, mid-exponential phase	Good correlation between CSH and adhesion.	Used various *Candida* species.	80
Acrylic, PVC	MATH and contact angle measurements	26°C, 24 h Direct enumeration	Yeast nitrogen base + glucose or hexadecane, 26°C	More hydrophobic cells attached best to both inert surfaces.	Used one isolate of hydrophobic cells obtained by growth in hexadecane.	59
Acrylic	ND	37°C, 50 min Direct enumeration	Yeast nitrogen base, 28°C, exponential phase	*C. albicans* adhered best of all species tested.	Some variability among *C. albicans* strains.	112
	ND	37°C, 1 h Image analysis	Sabouraud dextrose broth, 37°C, early exponential phase	Roughened acrylic allowed best adhesion. Cell concentration dependence for adhesion.		113
	ND	35°C, drip cells onto surface for 30 min Colonial development	Brain heart infusion + sucrose, 35°C, "overnight"	No attachment of *C. albicans* to surface was detected.		97

Substrate		Method	Growth conditions	Results	Comments	Ref.
	ND	RT, 60 min Microscopic enumeration	Yeast nitrogen base + 50 mM glucose or 500 mg of galactose, 37°C, stationary phase (24 h)	Cells grown in galactose adhered best.	Used one isolate of *C. albicans*.	17
	ND	RT, 60 min Direct enumeration	Yeast nitrogen base + various sugar concentrations, 37°C, 24 h	Increasing sugar concentration caused increased attachment. Cells in H_2O adhered more than cells in PBS. Addition of Ca^{2+} increased adhesion.		74
	ND	RT, 60 min Direct enumeration	Mycological peptone ± sugar, 37°C, 18 h	Increased sucrose in growth medium led to increased adhesion.		106
	ND	RT, 60 min Direct enumeration	Mycological peptone ± 500 mM sugar, 37°C, 18 h	Sucrose and glucose caused greater adhesion than did other sugars. Coating acrylic with serum caused increased adhesion.		107
Chlorhexidine-treated acrylic vs untreated acrylic	ND	RT, 60 min Direct enumeration	Yeast nitrogen base + 500 mM sucrose, 37°C, 24 h	Chlorhexidine reduced adhesion.		76

(Continued on next page)

Table 3—*Continued*

Inert surface	CSH assay	Attachment assay and method of quantitation	Fungal growth conditions	General results	Comments	Reference
Modified glass	Contact angle measurements and MATH	37°C, 3 h Direct enumeration	Yeast nitrogen base + 250 mM glucose, 37°C, late exponential phase	Hydrophobic cells adhered to hydrophobic surfaces. Hydrophilic cells adhered to charged surfaces.		87
Polystyrene[b]	Contact angle measurements and MATH	26°C, 20 h Enumeration from photographs	Sabouraud dextrose broth, 26°C, growth phase not stated	Hydrophobicity was correlated with adhesion. Formaldehyde-treated cells had lower adhesion. Carbodiimide-treated cells gave more adhesion.		60
Modified polystyrene (hydrophilic)	MATH	37°C, 1.5 h Direct enumeration	Modified Lee medium, 24°C, stationary phase	Opaque phenotype appeared more hydrophobic but there was variability among species for adhesion.	Shape and size of cells may have affected results.	53
	MATH	25°C, 1 h, varying buffer conditions	Yeast nitrogen base + 50 mM glucose or 500 mM galactose, modified Lee medium, 25°C, 48 h	Adhesion level depended on strain and assay conditions. Hydrophobicity played minor role in adhesion.		54

PVC,[b] Teflon[b]	ND	37°C, 60 min	Not stated	Adhesion to Teflon less than to PVC. *C. tropicalis* adhered greater than *C. albicans*.	104
Polypropylene	MATH	37°C, 15–360 min	Trypticase soy broth, 37°C, "overnight"	Cells were not hydrophobic, and no attachment occurred.	49

[a] PVC, Polyvinyl chloride; ND, Not done; RT, room temperature; PBS, phosphate-buffered saline.
[b] As stated in the text, the order of surface hydrophobicity for the various inert surfaces is acrylic << polystyrene < Teflon < polyvinyl chloride (59).

binding. This form of polystyrene was apparently used by Kennedy et al. (53, 54) since their commercial source makes only tissue culture-treated polystyrene microtiter plates (information obtained from technical services, Costar, Cambridge, Mass.). The hydrophilicity of the microtiter plates and the use of variously hydrophobic yeast strains explains why these investigators found such variation in the environmental conditions that influence the attachment of the *C. albicans* isolates to the plastic. Contact angle measurements of the microtiter plate polystyrene would have provided useful information about the hydrophobicity of a polystyrene.

When adhesion of *C. albicans* to hydrophobic polystyrene was studied, increasing concentrations of NaCl resulted in enhanced binding (60). This result indicated that hydrophobic interactions were operative, since the increased NaCl results in "salting out" of hydrophobic moieties (84). Klotz et al. (60) further demonstrated that although electrostatic interactions do participate in yeast adhesion to bacteriological plastic, hydrophobic interactions predominate. This was shown by altering the surface charge density of the yeast cell surfaces with carbodiimide and formaldehyde. When two or more strains of different *Candida* species were investigated for their ability to bind to the polystyrene plastic, typically the most hydrophobic isolate was most adherent. This was not the case for *Candida krusei*. No explanation for this disparity was given. One possibility, however, is that the cells had undergone significant surface changes during the 20-h incubation period.

Glass

Pyrex brand borosilicate glass (Corning Glass Works, Corning, N.Y.) is hydrophilic (60). However, improperly washed and heated glass can have hydrophobic characteristics. Other forms of glass such as aluminosilicate (Corex; Corning) may differ from borosilicate in their hydrophobic status. Given these considerations, investigators who wish to study attachment to glass must first be aware of the hydrophobicity characteristic of the glass. This can easily be obtained by contact angle measurements (this same consideration is applicable as well when plastic is the substrate; see above). Nikawa et al. (87) studied the relationship of cell hydrophobicity of different *Candida* species and attachment to glass that was modified to have different levels of hydrophobicity or different surface charges. They found that hydrophobic yeast cells adhered best to hydrophobic glass. The yeast strain having low surface hydrophobicity but high ζ potential (due to surface electronegativity) adhered best to modified glass having a net positive surface. These and other results indicate that highly hydrophobic yeast cells which possess a low ζ potential bind to glass through hydrophobic interactions. Less hydrophobic cells bind through electrostatic interactions. While hydrophobic interactions can participate in attachment, the degree of adhesion is dependent primarily on the magnitude and charge of the glass surface and on the ζ potential of the yeast cells.

The influence of monosaccharides on hydrophobicity-mediated attachment of *C. albicans* to glass beads (PolyGlas; Polysciences) was studied by Reinhart et al. (94). No sugar decreased adhesion. All amino sugars appeared to enhance attachment. These same sugars, with the exception of glucosamine, appeared to have little effect on yeast attachment to a phenyl-Sepharose HIC column. Glucosamine caused approximately a 15% reduction in attachment (no standard deviation or data range was provided). These results suggest that CSH was not involved in attachment of the isolate of *C. albicans* to the glass beads and that various sugars can enhance the attachment. Regardless of the surface components involved in yeast cell expression of CSH, they do not appear to utilize carbohydrates for hydrophobic interactions. Unfortunately, these interpretations are not cogent. The CSH level of the control cell population and the relative degree of surface hydrophobicity of the glass beads were not stated. In addition, the validity of the HIC method used by the investigators is questionable (see above).

Attachment to Host Surfaces

The role of microbial CSH in adhesion of bacteria to human cells has been unequivocally established (reviewed in reference 102 and in this volume). However, the role of fungal hydrophobicity in attachment to host tissue has not been explored as extensively. Disagreement among investigators is evident (Table 4). In contrast to attachment to inert, relatively homogeneous surfaces, human cells present a vast menage of surface components, topographies, and surface component distributions. However, similar the case for to attachment to inert components, the participation of fungal surface hydrophobicity in adhesion should be evident only if the host target surface displays hydrophobic sites and the fungal cells are hydrophobic.

The problem of determining the relative contribution of hydrophobic interactions in adhesion to human tissue is complicated by the potential participation of multiple physicochemical adhesion mechanisms for any given set of fungal and host tissue cells (61, 111). This problem necessitates that the participation of CSH to adhesion of a fungus to host tissue be evaluated on an isolate-to-isolate basis (38). A recent study has

Table 4. Summary of Studies in Which the Relationship between CSH and Adherence to Host Tissues Was Assessed

Host surface[a]	CSH assay	Attachment assay	Fungal growth conditions[b]	General results	Comments	Reference
VEC	HIC	VEC: yeasts (1:50), 37°C, 90 min	25°C, 24 h, PPGB	No correlation between HIC and adhesion values.		94
BEC	SAT, improved SAT	BEC: yeasts (1:1,000)	37°C, 24 h, SDB	Hydrophobic strains more adherent.	No autoaggregation of yeast cells occurred in adhesion assay.	70
BEC	MATH	BEC: yeasts (1:100), 37°C, 60 min	White-opaque cells, 24°C, 24 h, MLBCB	Opaque cells more hydrophobic, also coadhered. Opaque cells differed from white cells in size and shape. Hydrophobic cells adhered less.	MATH involved initial yeast concentration. Only three isolates used.	53
BEC	MATH	BEC: yeasts (1:100) 37°C, 60 min	25°C, 37°C, 24 h, various growth media	When CSH value was high, adhesion to BEC was high.	Only one isolate tested. CSH levels were low except for one condition.	55
HIE	MATH, contact angle measurements	Monolayer assay, radiometric	SDB, YNB + glucose, YNB + hexadecane	Hydrophobic cells were more adherent.	Only one isolate tested. Made hydrophobic by growth in hexadecane.	61
IMHP	MATH	Monolayer assay, 37°C, 60 min	37°C, 20 h, nutrient 37°C, 4–20 h, TSB	Low CSH, low adhesion.	One *C. albicans* isolate tested.	73
HeLa, HIE, BEC	HMA	Monolayer assay (HeLa, HIE), 37°C, 15 min BEC: yeasts (1:20), 37°C, 60 min	23°C, 37°C, 24 h, SDB, YNBG, GYEP	High correlation between CSH and adhesion for each isolate.	Demonstrates isolate dependence for CSH and adhesion. Multiple isolates tested.	38

[a] VEC, Vaginal epithelial cells; BEC, buccal epithelial cells; HIE, human intestinal epithelial cells; IMHP, human embryonic lung fibroblasts.
[b] PPGB, Phytone-peptone glucose broth; SDB, Sabouraud dextrose broth; MLBCB, modified Lee-Buckley-Campbell broth; YNB, yeast nitrogen base; TSB, Trypticase soy broth; YNBG, yeast nitrogen base plus glucose; GYEP, glucose-yeast extract-peptone.

demonstrated that when CSH and adhesion levels of various *C. albicans* isolates are grouped together, poor correlation between hydrophobicity and adhesion is obtained (38). When each isolate's CSH level and corresponding adhesion levels are analyzed, good correlation is obtained. This one factor, i.e., grouping values for different isolates, probably explains the discrepancies seen in studies by various investigative groups. However, other factors that may also contribute include differences in the CSH assessment methods, tissue targets, and adhesion assay protocols used.

The human pathogenic fungus investigated most comprehensively for the role of CSH in adhesion is *C. albicans* (Table 4). Its ability to adhere to buccal epithelial cells (BECs) and vaginal epithelial cells (VECs) is prerequisite for establishment of oral and vaginal thrush. A frequent problem associated with fungal adhesion to these cells is yeast cell coadhesion (38, 53, 54). The participation of CSH in coadhesion is likely since "hydrophobic bond breaking agents" (30), such as fetuin and methylumbelliferone, result in disaggregation of the yeast cell clumps (Hazen and Murthy, unpublished observations). The role of CSH to BECs and VECs is confused when coadhesion occurs. Macura (70) reported that coadhesion did not occur in her investigation of adhesion of more than 100 isolates of *C. albicans* to BECs. In this study, hydrophobic yeast strains adhered more abundantly than hydrophilic strains to BECs. No correlation statistical analysis between hydrophobicity and adhesion levels was performed. Reinhart et al. (94) suggested that CSH is not involved in adhesion of *C. albicans* to VECs. These investigators did not state whether coadhesion occurred.

Mucosal membranes covered by mucin-rich mucus are less susceptible to adhesion by *C. albicans* (11, 53), probably because of inhibition of surface hydrophobic interactions (61). The surfactant property of mucin possibly lowers the surface tension at the interface between an epithelial cell surface and the surrounding liquid, causing the absorption of yeast cells to be inhibited (59).

Adhesion of *Candida* species to epithelial cells other than VECs and BECs appears to involve at least two mechanisms, adhesin-receptor and hydrophobic interactions (25, 26, 103). The adhesin appears to be a mannoprotein (18, 75). The fungal hydrophobic attachment molecules have not been discerned (see below). Direct evidence that hydrophobic interactions are involved in attachment of *C. albicans* to epithelial cells has come from several observations (38, 61, 73). Using a single strain of *C. albicans*, Klotz and Penn (61) demonstrated that the hydrophobic form, which was induced by growth in hexadecane, adhered more than the hydrophilic form to human intestinal epithelial cell monolayers.

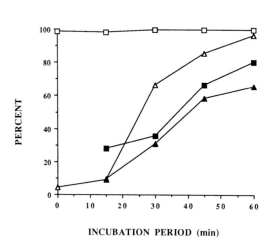

Figure 8. Effect of incubation time on *C. albicans* LGH1095 CSH levels (solid symbols) and adhesion (open symbols) to HeLa cell monolayers at 37°C. Yeast cells were grown at either 23°C (squares) or 37°C (triangles). These results demonstrate that within 30 min the CSH level of initially hydrophilic cells changes, thus making it difficult to distinguish the effect of CSH status on adhesion. However, at 15 min, the CSH levels of the yeast cells have not changed. At this time point, it is clear that hydrophobic yeast cells are more adherent than hydrophilic cells to HeLa cells. (Adapted from reference 38 with permission.)

INCUBATION PERIOD (min)

Martin et al. (73) also used a single strain of *C. albicans*. This strain was poorly hydrophobic and poorly adherent to lung fibroblasts.

In a comprehensive study, the effect of tissue target, incubation time, growth medium, and yeast isolate variability on the relationship between CSH and adhesion was investigated (38). The results demonstrate that CSH is involved in adhesion of *C. albicans* to various types of human epithelial cells. The influence of CSH on adhesion was not medium dependent. The study also revealed several important aspects of *C. albicans* surface hydrophobicity-adhesion dynamics. First, if the incubation period during which yeast cells are exposed to epithelial cells is lengthy, then equivocal results may occur. Hydrophilic yeast cells were shown to undergo CSH status changes within 30 min during incubation. Initially hydrophilic cells were substantially hydrophobic by 60 min. No change was evident at 15 min. Initially hydrophobic cells remain hydrophobic throughout a 60-min incubation period (Fig. 8). Adhesion to HeLa cell monolayers is significantly greater by hydrophobic yeast cells than by hydrophilic yeast cells at 15 min, but the difference is not evident for most isolates by 60 min. Thus, the only point at which the effect of CSH on adhesion can be assessed is early in the assay incubation period.

Another important observation was that the contribution of CSH to adhesion is isolate dependent. When correlation statistical analysis of CSH and adhesion levels for various isolates of *C. albicans* is performed on the aggregate data, CSH is found to correlate weakly with adhesion, particularly for yeast cell populations that were initially hydrophobic (38). However, when the analyses are restricted to a given isolate that is allowed to vary in CSH, it is clear that hydrophobicity is involved. These results show the drawback of attempting to relate a single surface attribute to adhesion for a group of yeast isolates that display other surface adhesion attributes to different degrees.

A third significant observation was that once cells reach a certain threshold hydrophobicity level, the contribution of CSH to adhesion does not continue to increase. Cells, therefore, need only show a certain level of CSH to obtain the maximum participation of surface hydrophobicity in adhesion. Other adhesion mechanisms then determine the level of adhesion that is possible. The threshold level of CSH is likely to be isolate dependent.

Several observations suggest that surface hydrophobicity participates in adhesion of germ tubes to epithelial cells. Germ tubes adhere more abundantly than yeast cells to BECs and VECs (57, 104, 108, 117). Germ tubes are also highly hydrophobic (35, 41). Germ tubes are also better suited than yeast cells to penetrate the electrostatic double layer surrounding an epithelial cell target, thus allowing hydrophobic interactions to contribute to adhesion (reviewed by K. C. Hazen, *in* M. J. Kennedy, ed., *Fungal Adhesion and Aggregation*, in press). Definitive studies concerning germ tube hydrophobicity in adhesion are needed.

Adhesion of hydrophobic and hydrophilic yeast cells of *C. albicans* to nonepithelial mouse tissues ex vivo appears to differ (K. C. Hazen, D. L. Brawner, M. A. Jutila, and J. E. Cutler, submitted for publication). Whereas hydrophilic cells tended to bind to regions within spleen, kidney, and lymph node that were richly populated with phagocytic cells, hydrophobic yeast cells bound throughout the tissues. These results suggest that after clearance from the bloodstream, hydrophobic cells are more likely than hydrophilic cells to seed and colonize various organs of an infected host.

Adhesion of *C. albicans* to fibronectin, which is found in fibrin matrices and associated with surfaces of many tissues, is also likely mediated, in part, by hydrophobic interactions. Fibronectin is a highly hydrophobic protein (2). Adhesion to fibrin matrices is greater by hydrophobic than hydrophilic yeast cells (unpublished observations). This evidence, in view of the results obtained with the ex vivo experiments discussed above, suggests that hydrophobic interactions may be a significant factor in the development of *Candida* endocarditis (109).

CSH and Host Defenses

Nonspecific and specific host responses, especially cell-mediated immunity (CMI), play critical roles in preventing the development of mycoses. When nonspecific defense mechanisms have been bridged, dysfunction of CMI leads to greater levels of disease severity for the systemic mycoses. A poor CMI response may be due to host defects or stimulated by the infecting fungus (96, 98). The role of fungal CSH in circumventing the CMI response is not known, but recent studies reveal that CSH can contribute to the ability of *Rhizopus*, *Aspergillus*, and *Candida* species to avoid phagocytic killing.

Rhizopus and *Aspergillus* conidia

The genera *Rhizopus* and *Aspergillus* contain species that can cause rapidly destructive fatal disease. Inhalation is one of the primary routes of entry to a host. Once the organism is in the lungs, it must evade phagocytosis. Resting conidia enter the alveoli and become active, whereupon the conidia swell (swollen conidia) and germinate. Hyphae then invade through lung tissue and enter the bloodstream. In the case of aspergilloma, asexual reproductive structures complete with conidiogenous cells and attached conidia may be produced (98).

The resting conidia of *A. fumigatus* and *Rhizopus oryzae* are hydrophobic (15, 42, 66). Swollen conidia are hydrophilic. These two types of conidia thus present an excellent natural system with which to study the role of CSH in evasion of phagocytosis. Using this system, Diamond and his colleagues have demonstrated that hydrophobic conidia are killed less effectively than hydrophilic conidia by human polymorphonucleated neutrophils (PMN) (for a review, see reference 21). Levitz et al. (65, 66, 68) demonstrated that hydrophilic conidia of both species are more susceptible to PMN oxidative products (e.g., hypoiodous and hypochlorous acids) and to rabbit PMN cationic peptides. In addition, the resting conidia of *A. fumigatus* stimulated less production than did swollen conidia of various PMN microbicidal factors. Macrophages were also less able to kill the hydrophobic conidia than were the hydrophilic conidia (68). These results suggest that hydrophobic conidia somehow cause poor killing responses by PMN and macrophages.

C. albicans

Murine PMN also appear to be less effective at killing *C. albicans* hydrophobic yeasts in the absence of normal serum (3). The hydrophobic yeast cells appear able to germinate and escape destruction by PMN. Whether poor candidicidal activity by PMN is due to the same mechanisms as obtain for *Rhizopus* and *Aspergillus* species remains to be determined. However, the addition of normal mouse serum to the PMN-yeast cell mixture results in equal levels of killing of hydrophobic and hydrophilic cells. This result suggests that opsonic factors may play a crucial role in stimulating killing of hydrophobic and hydrophilic cells.

Phagocytic killing of *C. albicans* hyphae has been addressed by Diamond (21). Although it was not stated, the hyphae used in this study were likely hydrophobic as judged by the results obtained by others (37, 41). Nonopsonized hyphae are killed less than opsonized hyphae (22, 23). This result appears to be due to the decreased ability of unopsonized hyphae to stimulate membrane depolarization and superoxide production (67, 128). Whether the hydrophobic surface of hyphae is directly involved in these events is not known.

BIOCHEMISTRY OF FUNGAL CSH

Little has been learned about the molecular basis of CSH of medically important fungi, in part perhaps because of the possible difficulty of purifying hydrophobic molecules which may stick to glassware. Production of presumably somewhat hydrophobic bioemulsifiers and surfactants by various fungi, including *C. albicans*, has been demonstrated (13, 14, 16, 45, 48, 58). These substances are either primarily carbohydrate and a small amount of protein or glycolipids. All of these substances have been shown to be released by cells. Whether they are also on the surface of cells and responsible for at least a transient appearance of CSH has not been shown.

As mentioned earlier, the conidia of *Aspergillus* species, particularly *Aspergillus niger*, are hydrophobic. Cole et al. (15) chemically analyzed the conidial cell wall of *A. niger* and determined that the outer conidial wall is made up of protein (ca. 60% by dry weight) and lipid (ca. 24%). Little carbohydrate was detected. The outer wall contains repeating structures called rodlets which appear to be covered by an amorphous layer. The authors suggest that this layer is responsible for the surface hydrophobicity of conidia but have not chemically characterized the layer. A similar layer, also covering the rodlets of the conidia, has been demonstrated in the dermatophyte *Trichophyton mentagrophytes* (127). This layer is composed of a complex of glycoproteins and lipids. Cell wall polysaccharides appear involved in exposure of CSH by *C. albicans*. Shimokawa and Nakayama (115, 116) have isolated a hydrophobic variant of *C. albicans* that displays a reduced level of cell wall mannan and mannan phosphorylation.

The cell wall macromolecules involved in attachment of *C. albicans* germ tubes to polystyrene plastic petri dishes have been investigated by Tronchin et al. (120). Initial attachment was presumably due to hydrophobic interactions. These investigators allowed the yeast cells to adhere to the plastic for 3 h before scraping the cells off and analyzing the adherent material. The incubation period was used to allow the yeast cells to germinate. Recent studies with the plant pathogen *Erysiphe graminis* have demonstrated that attachment to an inert surface may trigger the secretion of adherent materials that may be unrelated to surface hydrophobicity (86). The adherent molecules were removed from the plastic surface by dithiothreitol and analyzed. At least four mannoproteins were involved in attachment to the plastic. Whether these proteins are responsible for surface hydrophobicity of germ tubes is not clear.

Using HMA to assess CSH during analysis, we have been investigating the chemical components involved in hydrophobicity of *C. albicans* (Hazen et al., submitted). Enzymatic treatments revealed that CSH is due in part to proteins. The proteins appear to be covalently attached to the cell wall, since their removal was accomplished only by treatment with the β-1,3 glucanase, lyticase. Dithiothreitol, heat, and detergents failed to extract the hydrophobic proteins. Hydrophilic material, on the other hand, was released by dithiothreitol as well as by a sodium dodecyl sulfate-heat treatment, suggesting that this material is loosely associated with the cell wall. When the hydrophilic material was removed, the cell wall became hydrophobic, indicating that hydrophilic cell walls also possess the hydrophobic proteins.

The surface proteins of *C. albicans* were also radioiodinated in situ, extracted with lyticase, and subjected to sodium dodecyl sulfate-polyacrylamide gel electrophoresis analysis (Hazen et al., submitted). Depending on the isolate of *C. albicans*, at least four low-molecular-weight (<70-kilodalton) proteins were associated with hydrophobic cells and not hydrophilic cells. However, the hydrophilic cells possessed at least two high-molecular-weight proteins not present in the hydrophobic yeast cell extract. Preliminary evidence suggests that these high-molecular-weight proteins are highly glycosylated. How the high-molecular-weight proteins influence the exposure of the hydrophobic surface proteins is under investigation.

FUTURE DIRECTIONS IN FUNGAL CSH

The study of fungal CSH is in its infancy. There is much to learn. The results and insights described in this chapter form the basis for future investigations. It is clear that CSH is involved in a wide range of fungal activities, including pathogenesis and morphogenesis. The energetic demand to expose hydrophobic molecules in an aqueous environment must be high but must also be a thermodynamically reasonable trade-off to ensure fungal survival.

Further studies will be aided by the isolation and characterization of the molecules responsible for CSH. Comparison of these molecules from different fungi may help demonstrate the relatedness of the organisms and also provide information about the feasibility of developing vaccines that have various degrees of specificity. How CSH expression is regulated at the transcriptional and translational levels is also a fertile area for investigation. If the experimental results obtained by bacteriologists studying CSH are an indication, investigating fungal CSH will be a fascinating and enlightening endeavor.

Acknowledgments. I thank Beth W. Hazen for critically reviewing the manuscript.

Work cited from my laboratory was supported by Public Health Service grant AI 24925 from the National Institutes of Health.

LITERATURE CITED

1. **Absolom, D. R.** 1986. Measurement of surface properties of phagocytes, bacteria, and other particles. *Methods Enzymol.* **132:**16–95.
2. **Absolom, D. R., W. Zingg, and A. W. Neumann.** 1987. Protein adsorption to polymer particles: role of surface properties. *J. Biomed. Mater. Res.* **21:**161–171.
3. **Antley, P. P., and K. C. Hazen.** 1988. Role of yeast cell growth temperature on *Candida albicans* virulence in mice. *Infect. Immun.* **56:**2884–2890.
4. **Bandoni, R. J.** 1972. Terrestrial occurrence of some aquatic hyphomycetes. *Can. J. Bot.* **50:**2283–2288.
5. **Bandoni, R. J., and R. E. Koske.** 1974. Monolayers and microbial dispersal. *Science* **183:**1079–1081.
6. **Bangham, A. D., and B. A. Pethica.** 1960. The adhesiveness of cells and the nature of chemical groups at their surface. *Proc. R. Phys. Soc. Edinburgh* **28:**43–52.
7. **Brawner, D. L., and J. E. Cutler.** 1984. Variability in expression of a cell surface determinant on *Candida albicans* as evidenced by an agglutinating monoclonal antibody. *Infect. Immun.* **43:**966–972.
8. **Brawner, D. L., and J. E. Cutler.** 1985. Changes in surface topography of *Candida albicans* during morphogenesis. *Sabouraudia* **23:**389–393.
9. **Brawner, D. L., and J. E. Cutler.** 1986. Ultrastructural and biochemical studies of two dynamically expressed cell surface determinants on *Candida albicans*. *Infect. Immun.* **51:**327–336.
10. **Brawner, D. L., and J. E. Cutler.** 1987. Cell surface and intracellular expression of two *Candida albicans* antigens during in vitro and in vivo growth. *Microb. Pathog.* **2:**249–257.
11. **Butrus, S. I., and S. A. Klotz.** 1986. Blocking *Candida* adhesion to contact lenses. *Curr. Eye Res.* **5:**745–750.
12. **Chaffin, W. L., J. Skudlarek, and K. J. Morrow.** 1988. Variable expression of a surface determinant during proliferation of *Candida albicans*. *Infect. Immun.* **56:**302–309.

13. **Cirigliano, M. C., and G. M. Carman.** 1984. Isolation of a bioemulsifier from *Candida lipolytica. Appl. Environ. Microbiol.* **48**:747–750.

14. **Cirigliano, M. C., and G. M. Carman.** 1985. Purification and characterization of liposan, a bioemulsifier from *Candida lipolytica. Appl. Environ. Microbiol.* **50**:846–850.

15. **Cole, G. T., T. Sekiya, R. Kasai, T. Yokoyama, and Y. Nozawa.** 1979. Surface ultrastructure and chemical composition of the cell walls of conidial fungi. *Exp. Mycol.* **3**:132–156.

16. **Cooper, D. G., and D. A. Paddock.** 1984. Production of a biosurfactant from *Torulopsis bombicola. Appl. Environ. Microbiol.* **47**:173–176.

17. **Critchley, I. A., and L. J. Douglas.** 1985. Differential adhesion of pathogenic *Candida* species to epithelial and inert surfaces. *FEMS Microbiol. Lett.* **28**:199–203.

18. **Critchley, I. A., and L. J. Douglas.** 1987. Isolation and partial characterization of an adhesin from *Candida albicans. J. Gen. Microbiol.* **133**:629–636.

19. **Cutler, J. E., and K. C. Hazen.** 1983. Yeast/mold morphogenesis in *Mucor* and *Candida albicans*, p. 267–306. *In* J. W. Bennett and A. Ciegler (ed.), *Secondary Metabolism and Differentiation in Fungi.* Marcel Dekker, Inc., New York.

20. **Darnell, K. R., M. E. Hart, and F. R. Champlin.** 1987. Variability of cell surface hydrophobicity among *Pasteurella multocida* somatic serotype and *Actinobacillus lignieresii* strains. *J. Clin. Microbiol.* **25**:67–71.

21. **Diamond, R. D.** 1988. Fungal surfaces: effects of interactions with phagocytic cells. *Rev. Infect. Dis.* **10**:S428–S431.

22. **Diamond, R. D., and R. Krzesicki.** 1978. Mechanisms of attachment of neutrophils to *Candida albicans* pseudohyphae in the absence of serum and of subsequent damage to pseudohyphae by microbicidal processes of neutrophils *in vitro. J. Clin. Invest.* **61**:360–369.

23. **Diamond, R. D., R. Krzesicki, and W. Jao.** 1978. Damage to pseudohypal forms of *Candida albicans* by neutrophils in the absence of serum in vitro. *J. Clin. Invest.* **61**:349–359.

24. **Dillon, J. K., J. A. Fuerst, A. C. Hayward, and G. H. G. Davis.** 1986. A comparison of five methods for assaying bacterial hydrophobicity. *J. Microbiol. Methods* **6**:13–19.

25. **Douglas, L. J.** 1987. Adhesion to surfaces, p. 239–280. *In* A. H. Rose and J. S. Harrison (ed.), *The Yeasts*, 2nd ed., vol. 2. Academic Press, Inc., New York.

26. **Douglas, L. J.** 1987. Adhesion of *Candida* species to epithelial surfaces. *Crit. Rev. Microbiol.* **15**:27–43.

27. **Farkas, V.** 1979. Biosynthesis of cell walls of fungi. *Microbiol. Rev.* **43**:117–144.

28. **Fleet, G. H.** 1985. Composition and structure of yeast cell walls. *Curr. Top. Med. Mycol.* **1**:24–56.

29. **Fletcher, M.** 1984. Comparative physiology of attached and free-living bacteria, p. 223–232. *In* K. C. Marshall (ed.), *Microbial Adhesion and Aggregation.* Springer-Verlag, New York.

30. **Garber, N., N. Sharon, D. Shohet, J. S. Lam, and R. J. Doyle.** 1985. Contribution of hydrophobicity to hemagglutination reactions of *Pseudomonas aeruginosa. Infect. Immun.* **50**:336–337.

31. **Garraway, M. O., and R. C. Evans.** 1984. *Fungal Nutrition and Physiology.* John Wiley & Sons, Inc., New York.

32. **Gerson, D. F.** 1981. Methods in surface physics for immunology, p. 105–138. *In* I. Lefkovits and B. Pernis (ed.), *Immunological Methods*, vol. II. Academic Press, Inc., Orlando, Fla.

33. **Gooday, G. W., and A. P. J. Trinci.** 1980. Wall structure and biosynthesis in fungi, p.

207–251. *In* G. W. Gooday, D. Lloyd, and A. P. J. Trinci (ed.), *The Eucaryotic Microbial Cell.* Cambridge University Press, Cambridge.

34. **Gristina, A. G.** 1987. Biomaterial-centered infection: microbial adhesion versus tissue integration. *Science* **237**:1588–1595.

35. **Hazen, B. W., and K. C. Hazen.** 1988. Dynamic expression of cell surface hydrophobicity during initial yeast cell growth and before germ tube formation of *Candida albicans. Infect. Immun.* **56**:2521–2525.

36. **Hazen, B. W., and K. C. Hazen.** 1988. Modification and application of a simple, surfaced hydrophobicity detection method to immune cells. *J. Immunol. Methods* **107**:157–163.

37. **Hazen, B. W., R. E. Liebert, and K. C. Hazen.** 1988. Relationship of cell surface hydrophobicity to morphology of monomorphic and dimorphic fungi. *Mycologia* **80**:348–355.

38. **Hazen, K. C.** 1989. Participation of yeast cell surface hydrophobicity in adhesion of *Candida albicans* to human epithelial cells. *Infect. Immun.* **57**:1894–1900.

39. **Hazen, K. C., and J. E. Cutler.** 1983. Isolation and purification of morphogenic autoregulatory substance produced by *Candida albicans. J. Biochem.* **94**:777–783.

40. **Hazen, K. C., and B. W. Hazen.** 1987. Temperature-modulated physiological characteristics of *Candida albicans. Microbiol. Immunol.* **31**:497–508.

41. **Hazen, K. C., and B. W. Hazen.** 1987. A polystyrene microsphere assay for detecting surface hydrophobicity variations within *Candida albicans* populations. *J. Microbiol. Methods* **6**:289–299.

42. **Hazen, K. C., and W. G. LeMelle.** 1990. Improved assay for surface hydrophobic avidity of *Candida albicans* cells. *Appl. Environ. Microbiol.* **56**:1974–1976.

43. **Hazen, K. C., B. J. Plotkin, and D. M. Klimas.** 1986. Influence of growth conditions on cell surface hydrophobicity of *Candida albicans* and *Candida glabrata. Infect. Immun.* **54**:269–271.

44. **Hoch, H. C., and R. C. Staples.** 1987. Structural and chemical changes among the rust fungi during appressorium development. *Annu. Rev. Phytopathol.* **25**:231–247.

45. **Hommel, R., O. Stuwer, W. Stuber, D. Haferburg, and H.-P. Kleber.** 1987. Production of water-soluble surface active exolipids by *Torulopsis apicola. Appl. Microbiol. Biotechnol.* **26**:199–205.

46. **Israelachvili, J., and R. Pashley.** 1982. The hydrophobic interaction is long range, decaying exponentially with distance. *Nature* (London) **300**:341–342.

47. **Israelachvili, J. N., and P. M. McGuiggan.** 1988. Forces between surfaces in liquids. *Science* **241**:795–800.

48. **Ito, S., and S. Inoue.** 1982. Sophorolipids from *Torulopsis bombicola*: possible relation to alkane uptake. *Appl. Environ. Microbiol.* **43**:1278–1283.

49. **Jacques, M., T. J. Marrie, and J. W. Costerton.** 1986. *In vitro* quantitative adhesion of microorganisms to intrauterine contraceptive devices. *Curr. Microbiol.* **13**:133–137.

50. **Kaeppeli, O., and A. Fiechter.** 1976. The mode of interaction between the substrate and cell surface of the hydrocarbon-utilizing yeast *Candida tropicalis. Biotechnol. Bioeng.* **18**:967–974.

51. **Kaeppeli, O., and A. Fiechter.** 1977. Component from the cell surface of the hydrocarbon-utilizing yeast *Candida tropicalis* with possible relation to hydrocarbon transport. *J. Bacteriol.* **131**:917–921.

52. **Kaeppeli, O., M. Mueller, and A. Fiechter.** 1978. Chemical and structural alterations at the cell surface of *Candida tropicalis*, induced by hydrocarbon substrate. *J. Bacteriol.* **133**:952–958.

53. **Kennedy, M. J., A. L. Rogers, L. R. Hanselmen, D. R. Soll, and R. J. Yancey.** 1988.

Variation in adhesion and cell surface hydrophobicity in *Candida albicans* white and opaque phenotypes. *Mycopathologia* **102**:149–156.

54. **Kennedy, M. J., A. L. Rogers, and R. J. Yancey, Jr.** 1989. Environmental alteration and phenotypic regulation of *Candida albicans* adhesion to plastic. *Infect. Immun.* **57**:3876–3881.

55. **Kennedy, M. J., and R. L. Sandin.** 1988. Influence of growth conditions on *Candida albicans* adhesion, hydrophobicity, and cell wall ultrastructure. *J. Med. Vet. Mycol.* **26**:79–92.

56. **Kennedy, M. J., P. A. Volz, C. A. Edwards, and R. J. Yancey.** 1987. Mechanisms of association of *Candida albicans* with intestinal mucosa. *J. Med. Microbiol.* **24**: 333–341.

57. **Kimura, L. H., and N. N. Pearsall.** 1980. Relationship between germination of *Candida albicans* and increased adhesion to human buccal epithelial cells. *Infect. Immun.* **28**:464–468.

58. **Klotz, S. A.** 1988. A bioemulsifier produced by *Candida albicans* enhances yeast adhesion to intestinal cells. *J. Infect. Dis.* **158**:636–639.

59. **Klotz, S. A.** 1989. Surface-active properties of *Candida albicans*. *Appl. Environ. Microbiol.* **55**:2119–2122.

60. **Klotz, S. A., D. J. Drutz, and J. E. Zajic.** 1985. Factors governing adhesion of *Candida* species to plastic surfaces. *Infect. Immun.* **50**:97–101.

61. **Klotz, S. A., and R. L. Penn.** 1987. Multiple mechanisms may contribute to the adhesion of *Candida* yeasts to living cells. *Curr. Microbiol.* **16**:119–122.

62. **Kozel, T. R.** 1983. Dissociation of a hydrophobic surface from phagocytosis of encapsulated and nonencapsulated *Cryptococcus neoformans*. *Infect. Immun.* **39**: 1214–1219.

63. **Kunoh, H., N. Yamaoka, H. Yoshioka, and R. L. Nicholson.** 1988. Preparation of the infection court by *Erysiphe graminis*. I. Contact-mediated changes in morphology of the conidium surface. *Exp. Mycol.* **12**:325–335.

64. **Lachica, R. V., and D. L. Zink.** 1984. Determination of plasmid-associated hydrophobicity of *Yersinia enterocolitica* by a latex particle agglutination test. *J. Clin. Microbiol.* **19**:660–663.

65. **Levitz, S. M., and R. D. Diamond.** 1984. Killing of *Aspergillus fumigatus* spores and *Candida albicans* yeast phase by the iron-hydrogen peroxide-iodide cytotoxic system: comparison with the myeloperoxidase-hydrogen peroxide-halide system. *Infect. Immun.* **43**:1100–1102.

66. **Levitz, S. M., and R. D. Diamond.** 1985. Mechanisms of resistance of *Aspergillus fumigatus* conidia to killing by neutrophils in vitro. *J. Infect. Dis.* **152**:33–42.

67. **Levitz, S. M., C. A. Lyman, T. Murata, J. A. Sullivan, G. L. Mandell, and R. D. Diamond.** 1987. Cytosolic calcium changes in individual neutrophils stimulated by opsonized and unopsonized *Candida albicans* hyphae. *Infect. Immun.* **55**:2783–2788.

68. **Levitz, S. M., M. E. Selsted, T. Ganz, R. I. Lehrer, and R. D. Diamond.** 1986. In vitro killing of spores and hyphae of *Aspergillus fumigatus* and *Rhizopus oryzae* by rabbit neutrophil cationic peptides and bronchoalveolar macrophages. *J. Infect. Dis.* **154**: 483–489.

69. **Lindahl, M., A. Faris, T. Wadström, and S. Hjertén.** 1981. A new test based on 'salting-out' to measure relative surface hydrophobicity of bacterial cells. *Biochim. Biophys. Acta* **677**:471–476.

70. **Macura, A. B.** 1987. Hydrophobicity of *Candida albicans* related to their adhesion to mucosal epithelial cells. *Zentralbl. Bakteriol. Parasitenkd. Infektionskr. Hyg. Abt. 1 Reihe A* **266**:491–496.

71. **Marshall, K. C.** 1980. Microorganisms and interfaces. *BioScience* **30**:246–249.
72. **Marshall, K. C.** 1986. Adsorption and adhesion processes in microbial growth at interfaces. *Adv. Colloid Interface Sci.* **25**:59–86.
73. **Martin, D., L. G. Mathieu, J. Lecomte, and J. deRepentigny.** 1986. Adhesion of gram-positive and gram-negative bacterial strains to human lung fibroblasts in vitro. *Exp. Biol.* **45**:323–334.
74. **McCourtie, J., and L. J. Douglas.** 1981. Relationship between cell surface composition of *Candida albicans* and adhesion to acrylic after growth on different carbon sources. *Infect. Immun.* **32**:1234–1241.
75. **McCourtie, J., and L. J. Douglas.** 1984. Relationship between cell surface composition, adhesion, and virulence of *Candida albicans*. *Infect. Immun.* **45**:6–12.
76. **McCourtie, J., T. W. MacFarlane, and L. P. Samaranayake.** 1985. Effect of chlorhexidine gluconate on the adhesion of *Candida* species to denture acrylic. *J. Med. Microbiol.* **20**:97–104.
77. **Minagi, S., Y. Miyake, Y. Fujioka, H. Tsuru, and H. Suginaka.** 1986. Cell-surface hydrophobicity of *Candida* species as determined by the contact-angle and hydrocarbon-adhesion methods. *J. Gen. Microbiol.* **132**:1111–1115.
78. **Minagi, S., Y. Miyake, K. Inagaki, H. Tsuru, and H. Suginaka.** 1985. Hydrophobic interaction in *Candida albicans* and *Candida tropicalis* adhesion to various denture base resin materials. *Infect. Immun.* **47**:11–14.
79. **Miura, Y., M. Okazaki, S.-I. Hamada, S.-I. Murakawa, and R. Yugen.** 1977. Assimilation of liquid hydrocarbon by microorganisms. I. Mechanisms of hydrocarbon uptake. *Biotechnol. Bioeng.* **19**:701–714.
80. **Miyake, Y., Y. Fujita, S. Minagi, and H. Suginaka.** 1986. Surface hydrophobicity and adhesion of *Candida* to acrylic surfaces. *Microbios* **46**:7–14.
81. **Mozes, N., and P. G. Rouxhet.** 1987. Methods for measuring hydrophobicity of microorganisms. *J. Microbiol. Methods* **6**:99–112.
82. **Mudd, S., and E. B. H. Mudd.** 1924. The penetration of bacteria through capillary spaces. IV. A kinetic mechanism in interfaces. *J. Exp. Med.* **40**:633–645.
83. **Mudd, S., and E. B. H. Mudd.** 1924. Certain interfacial tension relations and the behavior of bacteria in films. *J. Exp. Med.* **40**:647–660.
84. **Nesbitt, W. E., R. J. Doyle, K. G. Taylor, R. H. Staat, and R. R. Arnold.** 1982. Positive cooperativity in the binding of *Streptococcus sanguis* to hydroxylapatite. *Infect. Immun.* **35**:157–165.
85. **Neumann, A. W., R. J. Good, C. J. Hope, and M. Sejpal.** 1974. An equation-of-state approach to determine surface tensions of low-energy solids from contact angles. *J. Colloid Interface Sci.* **49**:291–304.
86. **Nicholson, R. L., H. Yoshioka, N. Yamaoka, and H. Kunoh.** 1988. Preparation of the infection court by *Erysiphe graminis*. II. Release of esterase enzyme from conidia in response to a contact stimulus. *Exp. Mycol.* **12**:336–349.
87. **Nikawa, H., S. Sadamori, T. Hamada, N. Satou, and K. Okuda.** 1989. Non-specific adhesion of *Candida* species to surface-modified glass. *J. Med. Vet. Mycol.* **27**: 269–271.
88. **Odds, F. C.** 1985. Morphogenesis in *Candida albicans*. *Crit. Rev. Microbiol.* **12**:45–93.
89. **Odds, F. C.** 1988. *Candida and Candidosis*, 2nd ed. Bailliere Tindall, London.
90. **Pashley, R. M., P. M. McGuiggan, B. W. Ninham, and D. F. Evans.** 1985. Attractive forces between uncharged hydrophobic surfaces: direct measurements in aqueous solution. *Science* **229**:1088–1089.
91. **Poulain, D., V. Hopwood, and A. Vernes.** 1985. Antigenic variability of *Candida albicans*. *Crit. Rev. Microbiol.* **12**:223–270.

92. **Poulain, D., G. Tronchin, B. Lefebvre, and M. O. Husson.** 1982. Antigenic variability between *Candida albicans* blastospores isolated from healthy subjects and patients with *Candida* infection. *Sabouraudia* **20**:173–177.
93. **Poulain, D., G. Tronchin, A. Vernes, R. Popeye, and J. Biguet.** 1983. Antigenic variations of *Candida albicans* in vivo and in vitro—relationships between P antigens and serotypes. *Sabouraudia* **21**:99–112.
94. **Reinhart, H., G. Muller, and J. D. Sobel.** 1985. Specificity and mechanism of in vitro adhesion of *Candida albicans*. *Ann. Clin. Lab. Sci.* **15**:406–413.
95. **Reiss, E.** 1985. Cell wall composition, p. 57–101. *In* D. H. Howard (ed.), *Fungi Pathogenic for Humans and Animals*, part B. *Pathogenicity and Detection.* Marcel Dekker, Inc., New York.
96. **Reiss, E.** 1986. *Molecular Immunology of Mycotic and Actinomycotic Infections.* Elsevier Science Publishers, New York.
97. **Richards, S., and C. Russell.** 1987. The effect of sucrose on the colonization of acrylic by *Candida albicans* in pure and mixed culture in an artificial mouth. *J. Appl. Bacteriol.* **62**:421–427.
98. **Rippon, J. W.** 1988. *Medical Mycology. The Pathogenic Fungi and the Pathogenic Actinomycetes*, 3rd ed. The W. B. Saunders Co., Philadelphia.
99. **Rosenberg, M.** 1981. Bacterial adhesion to polystyrene: a replica method of screening for bacterial hydrophobicity. *Appl. Environ. Microbiol.* **42**:375–377.
100. **Rosenberg, M.** 1984. Bacterial adhesion to hydrocarbons: a useful technique for studying cell surface hydrophobicity. *FEMS Microbiol. Lett.* **22**:289–295.
101. **Rosenberg, M., D. Gutnick, and E. Rosenberg.** 1980. Adhesion of bacteria to hydrocarbons: a simple method for measuring cell-surface hydrophobicity. *FEMS Microbiol. Lett.* **9**:29–33.
102. **Rosenberg, M., and S. Kjelleberg.** 1986. Hydrophobic interactions: role in bacterial adhesion. *Adv. Microb. Ecol.* **9**:353–393.
103. **Rotrosen, D., R. A. Calderone, and J. E. Edwards, Jr.** 1986. Adhesion of *Candida* species to host tissues and plastic surfaces. *Rev. Infect. Dis.* **8**:73–85.
104. **Rotrosen, D., T. R. Gibson, and J. E. Edwards, Jr.** 1983. Adhesion of *Candida* species to intravenous catheters. *J. Infect. Dis.* **147**:594–595.
105. **Rozgonyi, F., K. R. Szitha, Å. Ljungh, S. B. Baloda, S. Hjertén, and T. Wadström.** 1985. Improvement of the salt aggregation test to study bacterial cell-surface hydrophobicity. *FEMS Microbiol. Lett.* **30**:131–138.
106. **Samaranayake, L. P., and T. W. MacFarlane.** 1980. An in-vitro study of the adhesion of *Candida albicans* to acrylic surfaces. *Arch. Oral Biol.* **25**:603–609.
107. **Samaranayake, L. P., J. McCourtie, and T. W. MacFarlane.** 1980. Factors affecting the in-vitro adhesion of *Candida albicans* to acrylic surfaces. *Arch. Oral Biol.* **25**:611–615.
108. **Sandin, R. L., A. L. Rogers, R. J. Patterson, and E. S. Beneke.** 1982. Evidence for mannose-mediated adhesion of *Candida albicans* to human buccal cells in vitro. *Infect. Immun.* **35**:79–85.
109. **Scheld, W. M., R. W. Strunk, G. Balian, and R. A. Calderone.** 1985. Microbial adhesion to fibronectin *in vitro* correlates with production of endocarditis in rabbits. *Proc. Soc. Exp. Biol. Med.* **180**:474–482.
110. **Seaman, G. V. F., and J. W. Goodwin.** 1986. Physicochemical factors in latex rapid agglutination tests. *Am. Clin. Prod. Rev.* **June**:25–31.
111. **Segal, E.** 1987. Pathogenesis of human mycoses: role of adhesion to host surfaces. *Microbiol. Sci.* **4**:344–347.
112. **Segal, E., O. Lehrman, and D. Dayan.** 1988. Adhesion in vitro of various *Candida* species to acrylic surfaces. *Oral Surg. Oral Med. Oral Pathol.* **66**:670–673.

113. **Shakespeare, A. P., and J. Verran.** 1988. The use of automated image analysis for rapid measurement of the in vitro attachment of *Candida albicans* to transparent acrylic. *Lett. Appl. Microbiol.* **6:**79–83.

114. **Shepherd, M. G.** 1987. Cell envelope of *Candida albicans*. *Crit. Rev. Microbiol.* **15:**7–25.

115. **Shimokawa, O., and H. Nakayama.** 1984. Isolation of a *Candida albicans* mutant with reduced content of cell wall mannan and deficient mannan phosphorylation. *Sabouraudia* **22:**315–322.

116. **Shimokawa, O., and H. Nakayama.** 1986. A *Candida albicans* rough-type mutant with increased cell surface hydrophobicity and a structural defect in the cell wall mannan. *J. Med. Vet. Mycol.* **24:**165–168.

117. **Sobel, J. D., P. G. Myers, D. Kaye, and M. E. Levisen.** 1981. Adhesion of *Candida albicans* to human vaginal and buccal epithelial cells. *J. Infect. Dis.* **143:**76–82.

118. **Sundstrom, P. M., and G. E. Kenny.** 1984. Characterization of antigens specific to the surface of germ tubes of *Candida albicans* by immunofluorescence. *Infect. Immun.* **43:**850–855.

119. **Sundstrom, P. M., and G. E. Kenny.** 1985. Enzymatic release of germ tube-specific antigens from cell walls of *Candida albicans*. *Infect. Immun.* **49:**609–614.

120. **Tronchin, G., J.-P. Bouchara, R. Robert, and J.-M. Senet.** 1988. Adhesion of *Candida albicans* germ tubes to plastic: ultrastructural and molecular studies of fibrillar adhesins. *Infect. Immun.* **56:**1987–1993.

121. **Tronchin, G., D. Poulain, and A. Vernes.** 1984. Cytochemical and ultrastructural studies of *Candida albicans*. *Arch. Microbiol.* **139:**221–224.

122. **van der Mei, H. C., A. H. Weerkamp, and H. J. Busscher.** 1987. A comparison of various methods to determine hydrophobic properties of streptococcal cell surfaces. *J. Microbiol. Methods* **6:**277–287.

123. **Webster, J., P. F. Sanders, and E. Descals.** 1978. Tetraradiate aquatic propagules in two species of *Entomophthora*. *Trans. Br. Mycol. Soc.* **70:**472–479.

124. **Wingard, J. R., J. D. Dick, W. G. Merz, G. R. Sandford, R. Saral, and W. H. Burns.** 1980. Pathogenicity of *Candida tropicalis* and *Candida albicans* after gastrointestinal inoculation in mice. *Infect. Immun.* **29:**808–813.

125. **Wingard, J. R., J. D. Dick, W. G. Merz, G. R. Sandford, R. Saral, and W. H. Burns.** 1982. Differences in virulence of clinical isolates of *Candida tropicalis* and *Candida albicans* in mice. *Infect. Immun.* **37:**833–836.

126. **Wingard, J. R., W. G. Merz, and R. Saral.** 1979. *Candida tropicalis*: a major pathogen in immunocompromised patients. *Ann. Intern. Med.* **91:**539–543.

127. **Wu-Yuan, C. D., and T. Hashimoto.** 1977. Architecture and chemistry of microconidial walls of *Trichophyton mentagrophytes*. *J. Bacteriol.* **129:**1584–1592.

128. **Wysong, D. R., C. A. Lyman, and R. D. Diamond.** 1989. Independence of neutrophil respiratory burst oxidant generation from the early cytosolic calcium response after stimulation of unopsonized *Candida albicans* hyphae. *Infect. Immun.* **57:**1499–1505.

Microbial Cell Surface Hydrophobicity
Edited by R. J. Doyle and M. Rosenberg
© 1990 American Society for Microbiology, Washington, DC 20005

Chapter 10

Significance of Hydrophobicity in the Adhesiveness of Pathogenic Gram-Negative Bacteria

R. Victor Lachica

Adhesiveness as a common attribute of pathogenic bacteria is well established (5, 23). This property is generally accepted as a requirement for pathogens to counter the host mechanisms that are designed to cleanse the surfaces physically of unwanted particles, including microorganisms. These mechanisms include peristalsis in the small intestine, micturition in the urinary tract, and the mucociliary escalator of the respiratory tract. After attachment, some microorganisms proliferate at and remain on the surface of the host, whereas others proceed to penetrate to deeper tissue or other locations.

Considerable effort has been directed at defining the role of the nonflagellar filamentous surface appendages, termed fimbriae or pili, as adhesins in the specific interaction with epithelial cell surface receptors (13, 60). In addition, hydrophobic interaction is believed to play a role in overcoming the repulsive forces between the bacterial cell and the host cell (5, 93). This chapter provides a brief survey of the latest developments in the adhesion capabilities of certain pathogenic gram-negative bacteria. Current knowledge about the role of cell surface hydrophobicity in the initial binding of these bacteria to host cells is evaluated.

R. Victor Lachica • U.S. Army Natick Research, Development and Engineering Center, Natick, Massachusetts 01760-5020.

ESCHERICHIA COLI

Members of the species *Escherichia coli* consist of innocuous strains that reside in the intestines of humans and animals as well as those that cause urinary tract infection and distinct syndromes of diarrheal disease. The diarrheagenic strains include (i) enterotoxigenic *E. coli* (ETEC), (ii) enteroinvasive *E. coli* (EIEC), (iii) enteropathogenic *E. coli* (EPEC), and more recently (iv) enterohemorrhagic *E. coli* (EHEC) (90). The uropathogenic and ETEC strains provide a clear correlation between the presence of a specific fimbrial component and the ability of bacteria harboring it to attach to a given host epithelium (13, 60).

ETEC

Strains belonging to the ETEC group are associated with diarrheal disease in pigs, calves, lambs, and humans. Two general requirements are involved for the onset of disease: (i) the ability to attach to and colonize the mucosal epithelium and (ii) the elaboration of heat-labile or heat-stable enterotoxins that produce the diarrheal response. Both properties are plasmid encoded; a plasmid may simultaneously carry genes for both properties (29).

K88 antigen was the first fimbrial adhesin to be described as a virulence factor (51). It is present on many ETEC isolates from piglets with severe diarrhea (13, 60) and mediates the specific attachment of the bacteria to the porcine intestinal epithelium. Other recognized adhesins include K99 (associated with diarrhea in calves, lambs, and piglets), 987p (associated with diarrhea in piglets), and F41 (associated with diarrhea in calves) (13, 60). Adhesins associated with diarrhea in humans include CFA/I, CFA/II, CFA/III, and CFA/IV (13, 37, 60, 62). CFA/II consists of three fimbrial antigens (CS1, CS2, and CS3). CFA/IV also consists of three antigens, two of which (CR4 and CR5) are fimbrial antigens; CS6 has no demonstrable fimbrial structure (62).

Smyth et al. (103) first reported the association of K88 antigen with high cell surface hydrophobicity. Subsequently, similar observations were made with strains possessing antigens K99 (12) and F41 (75). However, strains carrying 987p fimbriae expressed relatively low cell surface hydrophobicity (113).

Strains with CFA/I and CFA/II adhesins expressed higher surface hydrophobicity than did the animal ETEC strains (21, 69). The screening of human ETEC strains for cell surface hydrophobicity led to the recognition of the CFA/III fimbriae (36, 37). All the strains with CFA/IV adhesin were highly hydrophobic (73) except those possessing only the CS6 antigen. The latter strains were subsequently shown to be nonadherent to human small intestinal antigens and cultured human intestinal mucosa (62).

Uropathogenic *E. coli*

Uropathogenic *E. coli* resides harmlessly in the colon but is the important pathogen in urinary tract infections (57, 106). Attachment to the human urinary tract epithelium is mediated by a variety of adhesins termed P fimbriae (63, 106). Successful infection of the urinary tract definitely requires the presence of such adhesins (106), which are not affected by D-mannose and are correlated with mannose-resistant agglutination of human erythrocytes.

There is general agreement that most uropathogenic *E. coli* strains express cell surface hydrophobicity. However, there is much disagreement over which fimbriae (P or type 1) are associated with elevated surface hydrophobicity (20, 40, 44, 70, 79, 80). Sherman et al. (101) observed differences in the cell surface hydrophobicity and binding capability of a series of *E. coli* expressing type 1 fimbriae. There was a close correlation between increased cell surface hydrophobicity and in vitro intestinal membrane binding mediated by type 1 fimbriae. The isolation of a few strains of uropathogenic strains (40) that do not express high cell surface hydrophobicity indicates that this surface property may not always be critical in the initiation of urinary tract infection.

EIEC and *Shigella* spp.

Like shigellae, strains of EIEC cause a dysenterylike disease in humans. The two groups of organisms exhibit similar pathogenesis, and there is considerable homology among the virulence plasmids (94). Virulence is dependent on the ability of these pathogens to invade and multiply in intestinal cells (32). Their invasiveness is mediated by a high-molecular-weight plasmid (1). Stugard et al. (105) recently reported that this plasmid encodes for a 101-kilodalton protein responsible for Congo red or hemin uptake. They suggested that binding of heme compounds by the pathogen in vivo may disguise it as a desirable molecule for the intestinal epithelial cell. The consequence would be for the host cells to bind the heme-coated bacteria through heme receptors and then actively endocytose the bacterial cells. The possibility exists that the adhesion phase is aided by an increase in bacterial surface hydrophobicity. *Shigella flexneri* has been shown to express plasmid-associated hydrophobicity (87, 99), a property that was correlated with the ability of pathogen to bind to HeLa cells (84). The capability of EIEC strains to express high cell surface hydrophobicity has not been determined.

EPEC

Strains belonging to the EPEC group cause diarrhea but do not produce heat-labile or heat-stable enterotoxin or exhibit *Shigella*-like invasiveness (18). Cravioto et al. (12) reported the ability of EPEC strains to attach to HEp-2 cells. This attachment capability subsequently was demonstrated to be encoded on a 60-megadalton plasmid (2, 61, 78). Scaletsky et al. (97) observed two distinct patterns of attachment of EPEC strains to HeLa cells. The group exhibiting the localized adhesion (LA) pattern formed microcolonies on the HeLa cell surface, whereas the group exhibiting the diffused adhesion (DA) covered the HeLa cell uniformly. Nataro et al. (78) provided evidence that these adhesion patterns were due to at least two genetically distinct adhesins. Moreover, they observed that the EPEC strains with the DA pattern were more hydrophobic than those that exhibited the LA pattern. The EPEC strains that Wadström et al. (113) examined recently apparently consisted of both groups, since 63 of the 157 strains were hydrophobic. Whereas the LA adhesin remains largely uncharacterized, the DA adhesin was shown to be fimbrial in nature (61).

RDEC-1 is an EPEC strain that causes diarrhea specifically in rabbits, with lesions morphologically indistinguishable from those caused by human EPEC strains (7). This rabbit pathogen expresses a mannose-resistant fimbrial adhesin (designated AF/RI) that mediates species-specific adherence to the glycoproteins of the rabbit mucosa (9). Mutants lacking the ability to express the fimbriae produced less severe disease (115). Type 1 mannose-sensitive fimbriae which are also expressed by RDEC-1 may promote bacterial attachment to the mucosa to some extent (8, 101). Sherman et al. (101) reported that both fimbrial adhesins were associated with increase in the hydrophobicity of RDEC-1. All 32 strains of *E. coli* isolated from rabbits with diarrhea were recently shown to exhibit high cell surface hydrophobicity (4).

EHEC

EHEC is a recently recognized group of pathogenic *E. coli* strains that cause hemorrhagic colitis in humans characterized by abdominal cramping and bloody diarrhea but little or no fever (90). Although the disease has been attributed to a particular serotype (O157:H7), strains from other serogroups also fulfill the same pathological criteria (68).

The mechanism of virulence of EHEC strains is unknown. The

bacteriophage-mediated cytotoxins that are neutralized by Shiga toxin antiserum are putative virulence factors (104). These strains have been shown to express a plasmid-encoded fimbrial adhesin that mediates bacterial attachment to cultured Henle 407 intestinal cells (56). Unfortunately, subsequent studies did not lend support to the notion that these factors play significant roles in the virulence of EHEC strains. When infected with plasmid-bearing and cytotoxin-positive EHEC strains and their plasmidless or cytotoxin-negative derivatives, all gnotobiotic piglets were shown to be affected with equal severity of diarrhea and characteristic mucosal lesions (111). EHEC strains did not appear to express high cell surface hydrophobicity (52, 100).

SALMONELLAE

Salmonellae are enteroinvasive pathogens that require attachment to the luminal surface to counter the peristaltic cleansing motion of the intestine. However, little is known about the role hydrophobic interaction plays in the initial binding of the pathogens to the host cell. Recently, Baloda et al. (3) observed no association between the formation of mannose-sensitive fimbriae and cell surface hydrophobicity among strains of *Salmonella typhimurium* and *S. enteritidis*.

Jones (47) suggested that a plasmid-mediated mannose-resistant hemagglutinin (48, 49) may provide the means for *S. typhimurium* cells to attach to the intestinal wall. The chromosomally encoded mannose-sensitive type 1 fimbrial adhesin was much less effective than the plasmid-mediated mannose-resistant hemagglutinin in promoting adhesion or internalization of HeLa cells (49, 50). There was a reduced ability of the pathogen to colonize the mouse intestine, with the accompanying loss of virulence upon curing of the adhesin plasmid (48). Although a plasmid was indeed required for the virulence of the pathogen, Hackett et al. (31) observed no change in the ability of *S. typhimurium* to bind to HeLa cells with the loss of the plasmid. Moreover, both plasmid-bearing and plasmid-free strains invaded the Peyer's patches of the small intestine to the same extent. In a study of six classes of Tn*phoA* mutants of *S. choleraesuis* that were unable to bind or enter epithelial cells, none of the insertions were in the genes encoding type 1 fimbriae or mannose-resistant hemagglutinin (23).

Recently, Finlay et al. (24) provided evidence that the binding and invasive capabilities of *S. typhimurium* and *S. choleraesuis* toward epithelial cells required de novo synthesis of several new bacterial proteins. Induction of these proteins required specific glycoproteinlike surface receptors. Transposon mutants that were unable to synthesize these proteins became nonadherent and noninvasive to eucaryotic cells and were avirulent to mice. Whether de novo synthesis of the proteins has any effect on the surface hydrophobicity of the pathogens was not investigated.

YERSINIA ENTEROCOLITICA

The invasive enteric pathogen *Yersinia enterocolitica* requires the presence of a 42- to 48-megadalton plasmid for the pathogenesis of disease (11, 28, 118). This plasmid, whose precise role is unclear, encodes for 16 to 20 proteins (86, 102). A number of these proteins are located on the outer membrane of the bacteria, one of which is a 225-kilodalton protein (YOP1).

The ability of *Y. enterocolitica* grown at 37°C but not at 25°C to express plasmid-mediated cell surface hydrophobicity (65, 98) was an early indication of attachment to the host cell surface as a critical step in *Yersinia* infection. Heesemann et al. (33) provided evidence that the virulence plasmid directs the capability of *Y. enterocolitica* to bind to HEp-2 cells. Mantle et al. (71) recently demonstrated that the binding of *Y. enterocolitica* to brush border membranes isolated from a rabbit intestine is associated with the virulence plasmid. Moreover, they showed that attachment was greater in those regions of the gut most affected during yersiniosis, namely, the distal small intestine and the proximal colon (82, 83).

Heesemann and Grüter (34) provided evidence that the binding capability of *Y. enterocolitica* is mediated by YOP1. They succeeded in mobilizing a fragment of the virulence plasmid that encodes YOP1 into a nonadherent *Y. enterocolitica* strain, transforming it into a transconjugant that is adherent to HEp-2 cells. Kapperud et al. (55) demonstrated in orally infected mice that excretion of YOP1-positive cells was prolonged compared with excretion of YOP1-negative cells. YOP1 appears to be a structural component of a matrix of fibrillae that cover the bacterial surface (54, 66). The fibrillae impart some rather striking properties to the pathogen which significantly influence their binding potential. This surface structure appears to behave as an adhesin, enabling the bacteria to attach to each other (autoagglutination) and to guinea pig erythrocytes (hemagglutination) (54, 55). Fibrillae formation is associated with increased cell surface charge and hydrophobicity (66), forces that may play a decisive role in the initial phase of *Yersinia* infection.

Certain strains of *Y. enterocolitica* have also been shown to elaborate chromosomally mediated fimbriae at low temperatures but not at 37°C (21, 81). Their formation is associated with increased cell surface hydrophobicity and agglutination of various animal erythrocytes (21, 58). There was no correlation between the presence of the fimbriae and ability to bind to human epithelial cells (81). This surface structure does not appear to have any significance in the intestinal colonization by *Y. enterocolitica*.

VIBRIO CHOLERAE

Cholera is an infectious diarrheal disease caused by *Vibrio cholerae*, a noninvasive bacterium that colonizes the small intestinal epithelium and subsequently secretes a protein toxin. Whereas the action of cholera toxin is well understood (26, 35), the colonization phase of cholera is far from clear.

By analogy to the adhesin-receptor interactions observed with ETEC strains, the adherence capability of *V. cholerae* has been suggested to be associated with hemagglutinating activity (46, 67). Several cell-bound and soluble hemagglutinins have been reported (22, 46, 67).

Recently, Taylor et al. (107) identified a 20.5-kilodalton protein that is a major subunit of a *V. cholerae* fimbria. Its expression was associated with increased mouse intestine colonization, an enhanced cell-cell interaction (autoagglutination), hemagglutination, and a dramatic increase in the surface hydrophobicity of *V. cholerae*. Expression of the fimbrial protein was shown to be coordinatedly regulated with the formation of the cholera toxin. On the other hand, another hemagglutinating fimbrial structure consisting of a 16-kilodalton protein had no effect in the binding capacity of *V. cholerae* (39).

Kabir and Ali (53) suggested that nonspecific hydrophobic interaction mediated by outer membrane proteins may play a major role in the binding of *V. cholerae* to the intestinal mucosa. They noted no close correlation in the effect of growth conditions on the emergence of hydrophobic cells of *V. cholerae* and their hemagglutinating activity. This observation is consistent with a recent report by Teppema et al. (108) that strains of *V. cholerae* with or without hemagglutinating activity in vitro were equally pathogenic as determined in an adult rabbit ligated-gut model. Further studies are needed to determine what surface component(s) promotes increased hydrophobicity of *V. cholerae* and subsequent attachment to the intestinal wall.

MESOPHILIC AEROMONADS

Besides being an important cause of disease in fish, mesophilic aeromonads have been implicated in recent years as one of the major agents of gastroenteritis (1, 42). They also cause severe bacteremia and wound infection. In mouse studies (41), *Aeromonas sobria* strains tended to be the most virulent, whereas strains of *A. caviae* tended to be the least virulent; the category of *A. hydrophila* tended to fall between these two species. A similar virulence pattern was observed with rainbow trout (85). Although the virulence mechanism of the pathogens has not been established, a number of potential virulence factors such as hemagglutinins, enterotoxins, and invasiveness have been described (1, 42, 43).

The binding of mesophilic aeromonads to the intestinal wall would be considered a critical initial step in the development of gastroenteritis. It is therefore significant that Burke et al. (6) observed a strong association between certain hemagglutinating patterns and diarrhea-producing capability of these organisms. Recently, Clark et al. (10) reported the binding capability of these aeromonads to mouse adrenal cells in association with their piliated and nonpiliated attachment mechanisms. The more virulent species (*A. sobria* and *A. hydrophila*) tended to be more adherent than *A. caviae*. However, there was no correlation between adherence capability and hydrophobicity. In fact, avirulent *A. caviae* strains were shown recently to be more hydrophobic than the strains of *A. sobria* and *A. hydrophila*, which displayed a high degree of virulence in fish (85).

BORDETELLA PERTUSSIS

Bordetella pertussis is a respiratory tract pathogen that is the causative agent of pertussis, or whooping cough, in children. Components of *B. pertussis* that warrant consideration as possible virulence determinants include adenylate cyclase toxin, agglutinogens, dermonecrotic toxin, filamentous hemagglutinin, pertussis toxin, and tracheal cytotoxin (77, 114). The genes mediating these factors are coordinately and reversibly regulated by growth conditions (phenotypic modulation). Thus, the pathogen is in the so-called avirulent phase when none of these factors are produced at growth conditions such as incubation temperatures much less than 37°C or high concentrations of $MgSO_4$ (64) or nicotinic acid (74) in growth media. Under permissive conditions, *B. pertussis* reverts to the virulent phase when expression of the virulence factors resumes.

To counter the mucociliary clearance mechanism of the host, the pathogen needs to bind to ciliated epithelial cells. The agglutinogens (serotype 2 and 6 fimbriae), pertussis toxin, and filamentous hemagglutinin (FHA) have been implicated as mediators of attachment in assays with tissue culture cells (30, 88). However, only the last two hemagglutinating proteins look promising as attachment factors (109, 112). They have been shown to act in concert as adhesins to human ciliated epithelial cells (109). Evidence from in vitro studies as well as in vivo in animal models showed that FHA was essential for initial colonization of the upper respiratory tract (89, 109, 110). However, this surface structure was not found to be essential for colonization of the lower respiratory tract of mice (59, 110). Systemic immunization of mice with FHA provided significant protection against aerosol challenge of *B. pertussis* cell suspension (59).

Robinson et al. (92) reported high cell surface hydrophobicity among the virulent-phase cells of *B. pertussis* and a marked reduction when the culture was chemically induced to revert to the avirulent phase. Fish et al. (25) suggested FHA as the likely component responsible for the cell surface hydrophobicity of the pathogen. Their conclusion was based on the examination of a series of spontaneous mutants characterized with reference to the expression of hemolysin, pertussis toxin, adenylate cyclase, and FHA. Indeed, FHA has been characterized as a highly hydrophobic, self-aggregating protein (109).

PSEUDOMONAS AERUGINOSA

Pseudomonas aeruginosa is considered as a major respiratory pathogen of immunocompromised and immunosuppressed patients (91). Successful colonization of the respiratory system of a patient by this pathogen appears to be initiated upon attachment of the organism to the respiratory epithelium (45, 116). Several studies have indicated that the pilus adhesin provides the initial adhesion of *P. aeruginosa* to the respiratory epithelial surface (16, 17, 117). It has also been shown to be a virulence factor in burn wound infection (96). This adhesin is composed of a single monomer protein, called pilin (95), whose epithelial cell-binding domain was recently located at the highly conserved C-terminal region of the protein (15, 38).

Studies indicated that hydrophobic interaction does not seem to play a major role in the binding capability of *P. aeruginosa* (19, 27, 72). However, Garber et al. (27) cautioned that the procedures used for measuring the surface hydrophobicity of this organism may not be sensitive enough to detect hydrophobic areas of the pilus adhesin.

CONCLUSION

On the basis of the possibility that hydrophobic interaction plays a major role in the initial phase of pathogenesis, a survey (summarized in

Table 1. Association of Hydrophobicity and Adhesiveness with the Virulence
of Pathogenic Gram-Negative Bacteria

Bacteria	Hydrophobicity	Adhesin	Reference(s)
Escherichia coli			
ETEC	+	Fimbriae	13, 37, 73, 113
Uropathogenic	+	Fimbriae	70, 101, 106
EPEC			
LA	−	?	78, 97
DA	+	Fimbriae	61, 78, 97, 101
EHEC	−	Fimbriae (?)	52, 56, 100, 111
EIEC	?	Surface protein	105
Shigella spp.	+	Surface protein	84, 87, 99, 105
Salmonellae	?	Surface protein(s)	24
Yersinia enterocolitica	+	YOP1 (fibrillae)	34, 55, 66, 71
Vibrio cholerae	+	Fimbriae (?)	53, 107
Mesophilic aeromonads	−	?	6, 10, 43, 85
Bordetella pertussis	+	FHA	25, 59, 109, 110
Pseudomonas aeruginosa	−	Fimbriae	17, 19, 72, 96

Table 1) was made of pathogenic gram-negative bacteria that affect the enteric, urinary, and respiratory tracts. Six of the eleven pathogenic species or groups were reported to express high cell surface hydrophobicity that is associated with bacterial binding to the host cell. High cell surface hydrophobicity was not expressed by a subgroup of EPEC strains.

It may be premature to conclude at this time that hydrophobic interaction is not a universal phenomenon in bacterium-host interaction. As pointed out by Rosenberg and Kjelleberg (93), the participation of hydrophobic interaction in adhesion phenomenon is often overlooked in various studies. This is certainly true for salmonellae, which have been shown recently to require de novo synthesis of several proteins to bind to cultured epithelial cells (24). Further studies along this line should provide some indication as to whether cell surface hydrophobicity of bacteria in situ differs greatly from results obtained from growth media. Moreover, the methods used in measuring cell surface hydrophobicity may be inadequate. Hydrophobic interaction chromatography (103) and salt aggregation (69) have been the methods of choice for most studies on gram-negative bacteria. Recent studies reported a general lack of correlation of results from methods for measuring bacterial hydrophobicity (14, 76). Consequently, reliance on a single method may not indicate the true extent of hydrophobicity of the bacterial surface (see also Rosenberg and Doyle, this volume).

LITERATURE CITED

1. **Altwegg, M., and H. K. Geiss.** 1989. Aeromonas as a human pathogen. *Crit. Rev. Microbiol.* **16:**253–286.

2. **Baldini, M. M., J. B. Kaper, M. M. Levine, D. C. A. Candy, and H. W. Moon.** 1983. Plasmid-mediated adhesion in enteropathogenic *Escherichia coli. Pediatr. Gastroenterol. Nutr.* **2:**534–538.

3. **Baloda, S. B., A. Faris, and K. Krovacek.** 1988. Cell-surface properties of enterotoxigenic and cytotoxic *Salmonella enteriditis* and *Salmonella typhimurium*: studies on hemagglutination, cell-surface hydrophobicity, attachment to human intestinal cells and fibronectin. *Microbiol. Immunol.* **32:**447–459.

4. **Baloda, S. B., G. Froman, J. E. Peeters, and T. Wadström.** 1986. Fibronectin binding and cell-surface hydrophobicity of attaching effacing enteropathogenic *Escherichia coli* strains isolated from newborn and weanling rabbits with diarrhoea. *FEMS Microbiol. Lett.* **34:**225–229.

5. **Beachey, E. H.** 1981. Bacterial adherence: adhesin-receptor interactions mediating the attachment of bacteria in mucosal surfaces. *J. Infect. Dis.* **143:**325–344.

6. **Burke, V., M. Cooper, J. Robinson, M. Gracey, M. Lesmana, P. Echeverria, and M. Janda.** 1984. Hemagglutinating patterns of *Aeromonas* spp. in relation to biotype and source. *J. Clin. Microbiol.* **19:**39–43.

7. **Cantey, J. R.** 1981. The rabbit model of *Escherichia coli* (strain RDEC-1) diarrhea, p. 39–47. *In* E. C. Boedeker (ed.), *Attachment of Organisms to the Gut Mucosa*, vol. 1. CRC Press, Inc., Boca Raton, Fla.

8. **Cantey, J. R., L. R. Inman, and R. K. Blake.** 1989. Production of diarrhea in the rabbit by a mutant of *Escherichia coli* (RDEC-1) that does not express adherence (AF/R1) pili. *J. Infect. Dis.* **160:**136–141.

9. **Cheney, C. P., S. B. Schad, S. B. Formal, and E. C. Boedeker.** 1980. Species specificity of in vitro *Escherichia coli* adherence to host intestinal cell membranes and its correlation with in vivo colonization and infectivity. *Infect. Immun.* **28:**1019–1927.

10. **Clark, R. B., F. C. Knoop, P. J. Padgitt, D. H. Hu, J. D. Wong, and M. J. Janda.** 1989. Attachment of mesophilic aeromonads to cultured mammalian cells. *Curr. Microbiol.* **19:**97–102.

11. **Cornelis, G., Y. Laroche, G. Balligand, M.-P. Sory, and G. Wauters.** 1987. *Yersinia enterocolitica*, a primary model for bacterial invasiveness. *Rev. Infect. Dis.* **9:**64–87.

12. **Cravioto, A., R. J. Gross, S. M. Scotland, and B. Rowe.** 1979. An adhesive factor found in strains of *Escherichia coli* belonging to the traditional enteropathogenic serotypes. *Curr. Microbiol.* **3:**95–99.

13. **De Graaf, F. K., and F. R. Mooi.** 1986. The fimbrial adhesins of *Escherichia coli. Adv. Microb. Physiol.* **28:**65–143.

14. **Dillon, J. K., J. A. Fuerst, A. C. Hayward, and G. H. G. Davis.** 1986. A comparison of five methods for assaying bacterial hydrophobicity. *J. Microbiol. Methods* **6:**13–19.

15. **Doig, P., A. Parimi, P. A. Sastry, R. S. Hodges, K. K. Lee, W. Paranchych, and R. T. Irvin.** 1990. Inhibition of pilus-mediated adhesion of *Pseudomonas aeruginosa* to buccal epithelial cells by monoclonal antibodies directed against pili. *Infect. Immun.* **58:**124–130.

16. **Doig, P., N. R. Smith, T. Todd, and R. T. Irvin.** 1987. Characterization of the binding of *Pseudomonas aeruginosa* alginate to human epithelial cells. *Infect. Immun.* **55:** 1517–1522.

17. **Doig, P., T. Todd, P. A. Sastry, K. K. Lee, R. S. Hodges, W. Paranchych, and R. T.**

Irvin. 1988. Role of pili in the adhesion of *Pseudomonas aeruginosa* to human respiratory epithelial cells. *Infect. Immun.* **56**:1641–1646.

18. Edelman, R., and M. M. Levine. 1983. Summary of a workshop on enteropathogenic *Escherichia coli*. *J. Infect. Dis.* **147**:1108–1118.

19. Elsheikh, E. L., S. Abaas, and B. Wretlind. 1985. Adherence of *Pseudomonas aeruginosa* to tracheal epithelial cells of mink. *Acta Pathol. Microbiol. Immunol. Scand. Sect. B* **93**:417–422.

20. Falkowski, W., M. Edwards, and A. J. Schaeffer. 1986. Inhibitory effect of substituted aromatic hydrocarbons on adherence of *Escherichia coli* to human epithelial cells. *Infect. Immun.* **52**:863–866.

21. Faris, A., M. Lindahl, A. Ljungh, D. C. Old, and T. Wadstrom. 1983. Auto-aggregating *Yersinia enterocolitica* express surface fimbriae with high surface hydrophobicity. *J. Appl. Bacteriol.* **55**:97–100.

22. Finkelstein, R. A., and L. F. Hanne. 1982. Purification and characterization of the soluble hemagglutinin (cholera lectin) produced by *Vibrio cholerae*. *Infect. Immun.* **36**:1199–1208.

23. Finlay, B. B., and S. Falkow. 1989. Common themes in microbial pathogenicity. *Microbiol. Rev.* **53**:210–230.

24. Finlay, B. B., F. Heffron, and S. Falkow. 1989. Epithelial cell surfaces induce proteins required for bacterial adherence and invasion. *Science* **243**:940–943.

25. Fish, F., Y. Navon, and S. Goldman. 1987. Hydrophobic adherence and phase variation in *Bordetella pertussis*. *Med. Microbiol. Immunol.* **176**:37–46.

26. Foster, J. W., and D. M. Kinney. 1985. ADP-ribosylating microbial toxins. *Crit. Rev. Microbiol.* **11**:273–298.

27. Garber, N., N. Sharon, D. Shohet, J. S. Lam, and R. J. Doyle. 1985. Contribution of hydrophobicity to hemagglutination reactions of *Pseudomonas aeruginosa*. *Infect. Immun.* **50**:336–337.

28. Gemski, P., J. R. Lazere, and T. Casey. 1980. Plasmid associated with pathogenicity and calcium dependency of *Yersinia enterocolitica*. *Infect. Immun.* **27**:682–685.

29. Giannella, R. A. 1981. Pathogenesis of acute bacterial diarrheal disorders. *Annu. Rev. Med.* **32**:341–357.

30. Gorringge, A. R., L. A. E. Ashworth, L. I. Iron, and A. Robinson. 1985. Effect of monoclonal antibodies on the adherence of *Bordetella pertussis* to Vero cells. *FEMS Microbiol. Lett.* **26**:5–9.

31. Hackett, J., I. Kotlarski, V. Mathan, K. Francki, and D. Rowley. 1986. The colonization of Peyer's patches by a strain of *Salmonella typhimurium* cured of the cryptic plasmid. *J. Infect. Dis.* **153**:1119–1125.

32. Hale, T. L., and S. B. Formal. 1987. Pathogenesis of *Shigella* infections. *Pathol. Immunopathol. Res.* **6**:117–127.

33. Heesemann, J., B. Algermissen, and R. Laufs. 1984. Genetically manipulated virulence of *Yersinia enterocolitica*. *Infect. Immun.* **46**:105–110.

34. Heesemann, J., and L. Grüter. 1987. Genetic evidence that the outer membrane protein YOP1 of *Yersinia enterocolitica* mediates adherence and phagocytosis resistance to human epithelial cells. *FEMS Microbiol. Lett.* **40**:37–41.

35. Holmgren, J., and A.-M. Svennerholm. 1983. Cholera and the immune response. *Prog. Allergy* **33**:106–119.

36. Honda, T., M. M. A. Khan, Y. Takeda, and T. Miwatani. 1983. Grouping of enterotoxigenic *Escherichia coli* by hydrophobicity and its relation to hemagglutination and enterotoxin productions. *FEMS Microbiol. Lett.* **17**:273–276.

37. Honda, T., N. Wetprasit, M. Arita, and T. Miwatani. 1989. Production and character-

ization of monoclonal antibodies to a pilus colonization factor (colonization factor antigen III) of human enterotoxigenic *Escherichia coli. Infect. Immun.* **57**:3452–3457.

38. **Irvin, R. T., P. Doig, K. K. Lee, P. A. Sastry, W. Paranchych, T. Todd, and R. S. Hodges.** 1989. Characterization of the *Pseudomonas aeruginosa* pilus adhesin: confirmation that the pilin structural protein subunit contains a human epithelial cell-binding domain. *Infect. Immun.* **57**:3720–3726.

39. **Iwanaga, M., N. Nakasone, and M. Ehara.** 1989. Pili of *Vibrio cholerae* O1 biotype El Tor: a comparative study on adhesive and non-adhesive strains. *Microbiol. Immunol.* **33**:1–9.

40. **Jacobson, S. H., K. Tullus, and A. Brauner.** 1989. Hydrophobic properties of *Escherichia coli* causing acute pyelonephritis. *J. Infect.* **19**:17–23.

41. **Janda, J. M., R. B. Clark, and R. M. Brenden.** 1985. Virulence of *Aeromonas* species as assessed through mouse virulence studies. *Curr. Microbiol.* **12**:163–168.

42. **Janda, J. M., and P. S. Duffey.** 1988. Mesophilic aeromonads in human disease: current taxonomy, laboratory identification, and infectious disease spectrum. *Rev. Infect. Dis.* **10**:980–997.

43. **Janda, J. M., L. S. Oshiro, S. L. Abbott, and P. S. Duffey.** 1987. Virulence markers of the mesophilic aeromonads: association of the autoagglutination phenomenon with mouse pathogenicity and the presence of the peripheral cell-associated layer. *Infect. Immun.* **55**:3070–3077.

44. **Jann, K., G. Schmidt, E. Blumenstock, and K. Vosbeck.** 1981. *Escherichia coli* adhesion to *Saccharomyces cerevisiae* and mammalian cells: role of piliation and surface hydrophobicity. *Infect. Immun.* **32**:484–489.

45. **Johanson, W. G., Jr., J. H. Higuchi, T. R. Chaudhuri, and D. E. Woods.** 1980. Bacterial adherence to epithelial cells in bacillary colonization of the respiratory tract. *Am. Rev. Respir. Dis.* **121**:55–63.

46. **Jones, G. W.** 1980. The adhesive properties of *Vibrio cholerae* and other vibrio species, p. 219–249. *In* E. H. Beachey (ed.), *Bacterial Adherence. Receptors and Recognition*, series B, vol. 6. Chapman & Hall, Ltd., London.

47. **Jones, G. W.** 1984. Mechanisms of the attachment of bacteria to animal cells, p. 136–143. *In* M. J. Klug and C. A. Reddy (ed.), *Current Perspectives in Microbial Ecology.* American Society for Microbiology, Washington, D.C.

48. **Jones, G. W., D. K. Rabert, D. W. Svinarich, and H. J. Whitfield.** 1982. Association of adhesive, invasive, and virulent phenotypes of *Salmonella typhimurium* with autonomous 69-megadalton plasmids. *Infect. Immun.* **38**:476–486.

49. **Jones, G. W., and L. A. Richardson.** 1981. The attachment to, and invasion of HeLa cells by *Salmonella typhimurium*: the contribution of mannose-sensitive and mannose-resistant haemagglutinating activities. *J. Gen. Microbiol.* **127**:361–370.

50. **Jones, G. W., L. A. Richardson, and D. Uhlman.** 1981. The invasion of HeLa cells by *Salmonella typhimurium*: reversible and irreversible bacterial attachment and the role of bacterial motility. *J. Gen. Microbiol.* **123**:351–360.

51. **Jones, G. W., and J. M. Rutter.** 1972. Role of the K88 antigen in the pathogenesis of neonatal diarrhea caused by *Escherichia coli* in piglets. *Infect. Immun.* **6**:918–927.

52. **Junkin, A. D., and M. P. Doyle.** 1989. Comparison of adherence properties of *Escherichia coli* of O157:H7 and a 60-megadalton plasmid-cured derivative. *Curr. Microbiol.* **19**:21–27.

53. **Kabir, S., and S. Ali.** 1983. Characterization of surface properties of *Vibrio cholerae. Infect. Immun.* **14**:232–239.

54. **Kapperud, G., E. Namork, and H. Skarpeid.** 1985. Temperature-inducible surface

fibrillae associated with the virulence plasmid of *Yersinia enterocolitica* and *Yersinia pseudotuberculosis*. *Infect. Immun.* **47**:561–566.

55. **Kapperud, G., E. Namork, M. Skurnik, and T. Nesbakken.** 1987. Plasmid-mediated surface fibrillae of *Yersinia pseudotuberculosis* and *Yersinia enterocolitica*: relationship to the outer membrane protein YOP1 and the possible importance for pathogenesis. *Infect. Immun.* **55**:2247–2254.

56. **Karch, H., J. Heesemann, R. Laufs, A. D. O'Brien, C. O. Tackett, and M. M. Levine.** 1987. A plasmid of enterohemorrhagic *Escherichia coli* of O157:H7 is required for expression of a new fimbrial antigen and for adhesion to epithelial cells. *Infect. Immun.* **55**:455–461.

57. **Kass, E. H.** 1982. How important is bacteriuria? *Rev. Infect. Dis.* **4**:434–437.

58. **Kihlström, E., and K.-E. Magnusson.** 1983. Haemagglutinating, adhesive and physicochemical surface properties of different *Yersinia enterocolitica* and *Yersinia enterocolitica*-like bacteria. *Acta Pathol. Microbiol. Immunol. Scand. Sect. B* **91**:113–119.

59. **Kimura, A., K. T. Mountzouros, D. A. Relman, S. Falkow, and J. L. Cowell.** 1990. *Bordetella pertussis* filamentous hemagglutinin: evaluation as a protective antigen and colonization factor in a mouse respiratory infection model. *Infect. Immun.* **58**:7–16.

60. **Klemm, P.** 1985. Fimbrial adhesins of *Escherichia coli*. *Rev. Infect. Dis.* **7**:321–340.

61. **Knutton, S., D. R. Lloyd, and A. S. McNeish.** 1987. Adhesion of enteropathogenic *Escherichia coli* to human intestinal enterocytes and cultured human intestinal mucosa. *Infect. Immun.* **55**:69–77.

62. **Knutton, S., M. M. McConnell, B. Rowe, and A. S. McNeish.** 1989. Adhesion and ultrastructural properties of human enterotoxigenic *Escherichia coli* producing colonization factor antigens III and IV. *Infect. Immun.* **57**:3364–3371.

63. **Korhonen, T. K., V. Vaisanen, P. Kallio, E.-L. Nurmiaho-Lassila, H. Ranta, A. Siitonen, J. Elo, S. B. Svenson, and C. Svanborg-Eden.** 1982. The role of pili in the adhesion of *Escherichia coli* to human urinary tract epithelial cells. *Scand. J. Infect. Dis. Suppl.* **33**:26–31.

64. **Lacey, B. W.** 1960. Antigenic modulation of *Bordetella pertussis*. *J. Hyg.* **58**:57–91.

65. **Lachica, R. V., and D. L. Zink.** 1984. Plasmid-associated cell surface charge and hydrophobicity of *Yersinia enterocolitica*. *Infect. Immun.* **44**:540–543.

66. **Lachica, R. V., D. L. Zink, and W. R. Ferris.** 1984. Association of fibril structure formation with cell surface properties of *Yersinia enterocolitica*. *Infect. Immun.* **46**:272–275.

67. **Levine, M. M., J. B. Kaper, R. E. Black, and M. L. Clement.** 1983. New knowledge on pathogenesis of bacterial enteric infections as applied to vaccine development. *Microbiol. Rev.* **47**:510–550.

68. **Levine, M. M., J. G. Xu, J. B. Kaper, H. Lior, V. Prado, B. Tall, J. Nataro, H. Karch, and K. Wachsmuth.** 1987. A DNA probe to identify enterohemorrhagic *Escherichia coli* of O157:H7 and other serotypes that cause hemorrhagic colitis and hemorrhagic uremic syndrome. *J. Infect. Dis.* **156**:175–182.

69. **Lindahl, M., A. Faris, T. Wadström, and S. Hjertén.** 1981. A new test based on 'salting out' to measure relative surface hydrophobicity of bacterial cells. *Biochim. Biophys. Acta* **677**:471–476.

70. **Ljungh, Å., and T. Wadström.** 1982. Salt aggregation test for measuring surface hydrophobicity of urinary *Escherichia coli*. *Eur. J. Clin. Microbiol.* **1**:388–393.

71. **Mantle, M., L. Basaraba, S. C. Peacock, and D. G. Gall.** 1989. Binding of *Yersinia enterocolitica* to rabbit intestinal brush border membranes, mucus, and mucin. *Infect. Immun.* **57**:3292–3299.

72. **Marcus, H., and N. R. Baker.** 1985. Quantitation of adherence of mucoid and

nonmucoid *Pseudomonas aeruginosa* to hamster trachael epithelium. *Infect. Immun.* **47**:723–729.

73. **McConnell, M. M., P. Mullany, and B. Rowe.** 1987. A comparison of the surface hydrophobicity of enterotoxigenic *Escherichia coli* of human origin producing different adhesion factors. *FEMS Microbiol. Lett.* **42**:59–62.

74. **McPheat, W. L., A. C. Wardlaw, and P. Novotny.** 1983. Modulation of *Bordetella pertussis* by nicotinic acid. *Infect. Immun.* **41**:516–522.

75. **Morris, J. A., C. Thorn, A. C. Scott, W. J. Sojka, and G. A. Wells.** 1982. Adhesion in vitro and in vivo associated with an adhesive antigen (F41) produced by a K99 mutant of the reference strain *Escherichia coli* B41. *Infect. Immun.* **36**:1146–1153.

76. **Mozes, N., and P. G. Rouxhet.** 1987. Methods for measuring hydrophobicity of microorganisms. *J. Microbiol. Methods* **6**:99–112.

77. **Munoz, J. J., and R. K. Bergman.** 1977. *Bordetella pertussis*, immunological and other biological activities, p. 1–235. *In* N. Rose (ed.), *Immunology* series, vol. 4. Marcel Dekker, Inc., New York.

78. **Nataro, J. P., I. C. A. Scaletsky, J. B. Kaper, M. M. Levine, and L. R. Trabulski.** 1985. Plasmid-mediated factors conferring diffuse and localized adherence of enteropathogenic *Escherichia coli*. *Infect. Immun.* **48**:378–383.

79. **Öhman, L., K.-E. Magnusson, and O. Stendahl.** 1982. The mannose-specific lectin activity of *Escherichia coli* type 1 fimbriae assayed by agglutination of glycolipid-containing liposomes, erythrocytes, and yeast cells and hydrophobic interaction chromatography. *FEMS Microbiol. Lett.* **14**:149–153.

80. **Öhman, L., K.-E. Magnusson, and O. Stendahl.** 1985. Effect of monosaccharides and ethyleneglycol on the interaction between *Escherichia coli* bacteria and octyl-sepharose. *Acta Pathol. Microbiol. Immunol. Scand. Sect. B* **93**:133–138.

81. **Old, D. C., and J. Robertson.** 1981. Adherence of fimbriate and nonfimbriate strains of *Yersinia enterocolitica* to human epithelial cells. *Microbiol. Immunol.* **25**:993–998.

82. **O'Loughlin, E. V., G. Humphreys, I. Dunn, J. Kelly, C. J. Lian, C. Pai, and D. G. Gall.** 1986. Clinical, morphological, and biochemical alterations in acute intestinal yersiniosis. *Pediatr. Res.* **20**:602–608.

83. **Pai, C., and L. Destephano.** 1982. Serum-resistance-associated virulence in *Yersinia enterocolitica*. *Infect. Immun.* **35**:605–611.

84. **Pal, T., and T. L. Hale.** 1989. Plasmid-associated adherence of *Shigella flexneri* in a HeLa cell model. *Infect. Immun.* **57**:2580–2582.

85. **Paniagua, C., O. Rivero, J. Anguita, and G. Naharro.** 1990. Pathogenicity factors and virulence for rainbow trout (*Salmo gairdneri*) of motile *Aeromonas* spp. isolated from a river. *J. Clin. Microbiol.* **28**:350–355.

86. **Portnoy, D. A., A. H. Wolf-Watz, I. Bölin, A. B. Beeder, and S. Falkow.** 1984. Characterization of common virulence plasmids in *Yersinia* species and their role in the expression of outer membrane proteins. *Infect. Immun.* **43**:108–114.

87. **Qadri, F., S. A. Hossain, I. Čižnár, K. Haider, Å. Ljungh, T. Wadström, and D. A. Sack.** 1988. Congo red binding and salt aggregation as indicators of virulence in *Shigella* species. *J. Clin. Microbiol.* **26**:1343–1348.

88. **Redhead, K.** 1985. An assay of *Bordetella pertussis* adhesion to tissue-culture cells. *J. Med. Microbiol.* **19**:99–108.

89. **Relman, D. A., M. Dominighini, E. Tuomanen, R. Rappouli, and S. Falkow.** 1989. Filamentous hemagglutinin of *Bordetella pertussis*: nucleotide sequence and crucial role in adherence. *Proc. Natl. Acad. Sci. USA* **86**:2637–2641.

90. **Riley, L. W.** 1987. The epidemiologic, clinical, and microbiologic features of hemorrhagic colitis. *Annu. Rev. Microbiol.* **41**:383–407.

91. **Rivera, M., and M. B. Nicotra.** 1982. *Pseudomonas aeruginosa* mucoid strain. Its significance in adult chest diseases. *Am. Rev. Respir. Dis.* **126:**833–836.

92. **Robinson, A., A. R. Gorringe, L. I. Iron, and C. W. Keevil.** 1983. Antigenic modulation of *Bordetella pertussis* in continuous culture. *FEMS Microbiol. Lett.* **19:**105–109.

93. **Rosenberg, M., and S. Kjelleberg.** 1986. Hydrophobic interactions: role in bacterial adhesion. *Adv. Microb. Ecol.* **9:**353–393.

94. **Sansonetti, P. J., H. D'Hautville, C. Esobichon, and C. Porcel.** 1983. Molecular comparison of virulence plasmids in *Shigella* and enteroinvasive *Escherichia coli. Ann. Microbiol.* **134A:**295–318.

95. **Sastry, P. A., B. B. Finlay, B. L. Paloske, W. Paranchych, J. R. Pearlstone, and L. B. Smillie.** 1985. Comparative studies on the amino acid and nucleotide sequence of pilin derived from *Pseudomonas aeruginosa* PAK and PAO. *J. Bacteriol.* **164:**571–577.

96. **Sato, H., K. Okinaga, and H. Saito.** 1988. Role of pili in the pathogenesis of *Pseudomonas aeruginosa* burn infection. *Microbiol. Immunol.* **32:**131–139.

97. **Scaletsky, I. C. A., M. L. M. Silva, and L. R. Trabulski.** 1984. Distinctive patterns of adherence of enteropathogenic *Escherichia coli* to HeLa cells. *Infect. Immun.* **45:**534–536.

98. **Schiemann, D. A., M. R. Crane, and P. J. Swanz.** 1987. Surface properties of *Yersinia* species and epithelial cells *in vitro* by a method measuring total associated, attached and intracellular bacteria. *J. Med. Microbiol.* **24:**205–218.

99. **Seltman, G., T. Pál, and H. Tschäpe.** 1986. Surface hydrophobicity of plasmid-carrying virulent *Shigella flexneri* and their avirulent variants. *J. Basic Microbiol.* **26:**283–287.

100. **Sherman, P., R. Soni, M. Petric, and M. Karmali.** 1987. Surface properties of the Vero cytotoxin-producing *Escherichia coli* O157:H7. *Infect. Immun.* **55:**1824–1829.

101. **Sherman, P. M., W. L. Houston, and E. C. Boedeker.** 1985. Functional heterogeneity of intestinal *Escherichia coli* strains expressing type 1 pili somatic pili (fimbriae): assessment of bacterial adherence to intestinal membranes and surface hydrophobicity. *Infect. Immun.* **49:**797–804.

102. **Skurnik, M.** 1985. Expression of antigens encoded by the virulence plasmid of *Yersinia enterocolitica* under different conditions. *Infect. Immun.* **47:**183–190.

103. **Smyth, C. J., P. Honsson, E. Olsson, O. Soderlind, J. Rosengren, S. Hjertén, and T. Wadström.** 1978. Differences in hydrophobic surface characteristics of porcine enterotoxigenic *Escherichia coli* (ETEC) with or without K88 antigen as revealed by hydrophobic interaction chromatography. *Infect. Immun.* **22:**462–472.

104. **Strockbine, N. A., L. M. R. Marques, J. W. Newland, H. W. Smith, R. K. Holmes, and A. D. O'Brien.** 1986. Two toxin-converting phages from *Escherichia coli* O157:H7 strain 933 encode antigenically distinct toxins with similar biologic activities. *Infect. Immun.* **53:**135–140.

105. **Stugard, C. E., P. A. Daskaleros, and S. M. Payne.** 1989. A 101-kilodalton heme-binding protein associated with Congo red binding and virulence of *Shigella flexneri* and enteroinvasive *Escherichia coli* strains. *Infect. Immun.* **57:**3534–3539.

106. **Svanborg-Eden, C., A. Fasth, L. Hagberg, L. A. Hanson, T. Korhonen, and H. Leffler.** 1982. Host interaction with *Escherichia coli* in the urinary tract, p. 113–131. *In* L. Weinstein and B. N. Fields (ed.), *Seminars in Infectious Disease*, vol. 4. *Bacterial Vaccines*. Thieme-Stratton, New York.

107. **Taylor, R. K., V. L. Miller, D. B. Furlong, and J. J. Mekalanos.** 1987. Use of *pho* A gene fusions to identify a pilus colonization factor coordinately regulated with cholera toxin. *Proc. Natl. Acad. Sci. USA* **84:**2833–2837.

108. **Teppema, J. S., P. E. Guinée, A. A. Ibrahim, M. Pâques, and E. J. Ruitenberg.** 1987.

In vivo adherence and colonization of *Vibrio cholerae* strains that differ in hemagglutinating activity and mobility. *Infect. Immun.* **55**:2093–2102.

109. **Tuomanen, E., and A. Weiss.** 1985. Characterization of two adhesins of *Bordetella pertussis* for human ciliated respiratory-epithelial cells. *J. Infect. Dis.* **152**:118–125.

110. **Tuomanen, E., A. Weiss, R. Rich, F. Zak, and O. Zak.** 1985. Filamentous hemagglutinin and pertussis toxin promote adherence of *Bordetella pertussis* to cilia. *Dev. Biol. Stand.* **61**:197–204.

111. **Tzipori, S., H. Karch, K. Y. Wachsmuth, R. M. Robins-Brown, A. D. O'Brien, H. Lior, M. L. Cohen, J. Smithers, and M. M. Levine.** 1987. Role of a 60-megadalton plasmid and Shiga-like toxins in the pathogenesis of infection caused by enterohemorrhagic *Escherichia coli* O157:H7 in gnotobiotic piglets. *Infect. Immun.* **55**:3117–3125.

112. **Urisu, A., J. L. Cowell, and C. R. Manclark.** 1986. Filamentous hemagglutinin has a major role in mediating adherence of *Bordetella pertussis* to human WiDr cells. *Infect. Immun.* **52**:695–701.

113. **Wadström, T., R. A. Adegbola, S. B. Baloda, Å. Ljungh, S. K. Sethi, and Y. R. Yuk.** 1986. Non-haemagglutinating fimbriae of enteropathogenic *Escherichia coli* (EPEC). *Zentralbl. Bakteriol. Parasitenkd. Infektionskr. Hyg. Abt. 1 Reihe A* **261**:417–424.

114. **Weiss, A. A., and E. L. Hewlett.** 1986. Virulence factors of *Bordetella pertussis*. *Annu. Rev. Microbiol.* **40**:661–686.

115. **Wolf, M. K., G. P. Andrews, D. L. Fritz, R. W. Sjogren, Jr., and E. C. Boedeker.** 1988. Characterization of the plasmid from *Escherichia coli* RDEC-1 that mediates expression of adhesin AF/RI and evidence that AF/RI pili promote but are not essential for enteropathogenic disease. *Infect. Immun.* **56**:1846–1857.

116. **Woods, D. E., J. A. Bass, W. G. Johanson, Jr., and D. C. Straus.** 1980. Role of adherence in the pathogenesis of *Pseudomonas aeruginosa* lung infection in cystic fibrosis patients. *Infect. Immun.* **30**:694–699.

117. **Woods, D. E., D. C. Straus, W. G. Johanson, Jr., V. K. Kerry, and J. A. Bass.** 1980. Role of pili in the adhesion of *Pseudomonas aeruginosa* of mammalian buccal epithelial cells. *Infect. Immun.* **29**:1146–1151.

118. **Zink, D. L., J. C. Feeley, J. G. Wells, C. Vanderzant, J. C. Vickery, W. C. Roof, and G. A. O'Donovan.** 1980. Plasmid-mediated tissue invasiveness in *Yersinia enterocolitica*. *Nature* (London) **283**:224–226.

Microbial Cell Surface Hydrophobicity
Edited by R. J. Doyle and M. Rosenberg
© 1990 American Society for Microbiology, Washington, DC 20005

Chapter 11

Hydrophobic Characteristics of Staphylococci: Role of Surface Structures and Role in Adhesion and Host Colonization

Torkel Wadström

Staphylococci are gram-positive cocci and are among the most common of the pyogenic (pus-producing) bacteria (17). They produce local abscesses in almost any organ of the body, from the skin to the bone marrow. The organisms have a thick cell envelope and can survive for a long time on dry inanimate objects, and they are hard to eliminate from the human environment. Coagulase-positive staphylococci (*Staphylococcus aureus*) and several of the more than 30 species of coagulase-negative staphylococci (CNS) are normal inhabitants of the indigenous skin microbiota (46) (Table 1). Both *S. aureus* and some CNS are major pathogens in postoperative infections. Studies in recent years also have demonstrated foreign body biomaterial-associated infections (13, 97; T. Wadström, I. Eliasson, I. Holder, and Å. Ljungh, ed., *Pathogenesis of Wound and Biomaterial-Associated Infections*, in press).

The genus *Staphylococcus* contains several species able to cause pyogenic infections in humans and animals (27). *Staphylococcus saprophyticus* is a unique species in that it has the ability to colonize the urinary tract and cause urinary tract infections in young adult females (40). Staphylococci produce a number of toxins and enzymes involved in tissue

Torkel Wadström • Department of Medical Microbiology, University of Lund, Lund, Sweden.

Table 1. Clinical Importance of Members of the Genus *Staphylococcus*

Species group	Common pathogen	Common questionable pathogen
S. aureus	S. aureus	
S. epidermidis	S. epidermidis	S. haemolyticus, S. hominis, S. warneri, S. saccharolyticus
S. auricularis		
S. saprophyticus	S. saprophyticus	S. cohnii
S. simulans		S. simulans
S. intermedius	S. intermedius	
S. hyicus	S. hyicus subsp. hyicus[b]	S. hyicus subsp. chromogenes[b]
S. sciuri		
S. caseinolyticus		
S. gallinarum		

[a] Species group designation is questionable.
[b] Pathogenic for pigs and other animals but not humans.

degradation and killing of cells, including professional phagocytes (89, 94).

Even though staphylococci can colonize on various inert materials as well as on tissue-implanted prostheses in different organs, such as artificial hip joints, vascular grafts, and artificial cardiac valves, it was not until the last decade that research on surface properties of *S. aureus* and pathogens of CNS species began. Despite the knowledge gained in the early 1960s on the major components in the cell wall (peptidoglycan and teichoic acids) and the discovery that one cell surface protein (protein A) reacts with the Fc domain of immunoglobulins, it was not until the late 1970s and the 1980s that other cell surface proteins were reported to be involved in tissue colonization. In brief, our knowledge of the cell surface chemistry of *S. aureus* and pathogenic CNS as well as of nonpathogenic species of CNS is still quite poor compared with our knowledge of various pathogenic gram-negative species such as *Escherichia coli* and *Pseudomonas aeruginosa*. Species pathogenic for animals and humans are listed in Table 1.

THE STAPHYLOCOCCAL CELL SURFACE

The cell surface of both *S. aureus* and various CNS has a high negative net surface charge related to polymers such as ribitol and glycerol teichoic acids (61, 66, 87). However, it is not known whether lipoteichoic acid, a key polymer in membrane staphylococcal lipid cell metabolism (47), can be transported to the cell surface and expose its fatty acids. Lipoteichoic acid has been proposed as a surface amphiphile in group A streptococci (14, 15, 28). Because staphylococci do not bind radioactively labeled human or bovine serum albumin, it seems unlikely that lipoteichoic acid or other surface amphiphiles are exposed on the surfaces of *S. aureus* or various CNS (T. Wadström, S. Hjertén and M. Tylewska, unpublished data). It is possible that staphylococci produce other surface amphiphiles, as has recently been reported for other gram-positive bacteria such as *Corynebacterium diphtheriae* (48). However, it seems more likely that hydrophobic domains of such surface amphiphiles are buried in the deeper layers of the cell envelope and anchored in the cell membrane (98), as identified for staphylococcal LTA by Koch et al. (47).

In summary, the surface hydrophobicity common to *S. aureus* as well as strains of certain CNS species seems to involve protease- and heat-sensitive surface structures, not surface amphiphiles as proposed for group A streptococci and other common pathogens of wound and various surgical infections (5, 79).

DETERMINATION OF STAPHYLOCOCCAL CELL SURFACE HYDROPHOBICITY

In a first comparative study of cell surface properties of a limited number of strains of group A streptococci (*Streptococcus pyogenes*), *S. aureus*, and the strains of *S. saprophyticus* isolated from human urinary tract infections, we demonstrated that all three species commonly expressed high surface hydrophobicities as determined by hydrophobic interaction chromatography (HIC) on octyl-Sepharose and phenyl-Sepharose (41–43, 93, 95). This study was extended to investigate the relative surface hydrophobicities of numerous laboratory strains, protein A-negative mutants of *S. aureus* SA113(83A), and fresh clinical isolates from bovine mastitis (42, 43). The experiments were carried out both by the column procedure as described by Smyth et al. (81) and by a batch procedure in centrifuge tubes to avoid nonspecific trapping of bacteria between packed gel beads in the Pasteur pipette columns with glass wool in the bottom. Moreover, controls for both column and batch procedures also involved experiments with unsubstituted Sepharose CL-4B. Unlike group A streptococci, which commonly show a galactose-sensitive interaction with Sepharose (49, 96), nonspecific binding to unrelated gel beads was not observed with cells of various *S. aureus* and CNS strains. Interestingly, well-defined laboratory strains that are high producers of protein A (such as *S. aureus* Cowan 1 and Newman) expressed high surface hydrophobicity, whereas strain Wood 46 poorly expressed sur-

Table 2. Frequency of Autoaggregating *S. aureus* (SAT < 0.1)[a]

Source	No. of autoaggregating strains/total	%
Blood	123/135	91
Wound	54/60	90
Urine	12/14	86
Throat	22/35	61
Nose	9/23	39

[a] From Ljungh et al. (54).

face protein A as measured by ^{125}I-immunoglobulin binding to cells grown in various culture media. The strains interacted with octyl- and phenyl-substituted Sepharose gels only at extreme salt concentrations such as 4 M sodium chloride or 2 M ammonium sulfate. Furthermore, two mutants of *S. aureus* SA113(83A) (called SA113 U320 and SA113 U305) did not bind these hydrophobic gels even at high salt concentrations, suggesting a lower expression of cell surface hydrophobicity. A high-protein A-producing mutant of the same strain (SA113 *spaA*3) expressed a higher cell surface hydrophobicity, measured as binding to octyl-Sepharose at a lower salt concentration.

Screening of 72 clinical strains of *S. aureus* from bovine mastitis showed that all strains grown under similar conditions expressed high cell surface hydrophobicity as determined by HIC. These findings encouraged us to systematically study the cell surface hydrophobicities of both animal and human strains of *S. aureus* and various CNS (including type culture collection reference strains of various species). The HIC results correlated quite well with hydrophobicity tests as determined by "salting out" in the salt aggregation test (SAT) (43, 52). SAT was used to test several hundred strains through the years, and some observations on the cell surface hydrophobicities of *S. aureus* strains can be summarized as follows.

(i) Cell surface hydrophobicity, as determined by both HIC and SAT, was sensitive to heating cells at 80 or 100°C and sensitive to treatment with proteases such as *Streptomyces griseus* protease or proteinase K but was less sensitive to trypsin treatment.

(ii) Most strains isolated from both bovine mastitis and human infections (wound infections and septicemia) expressed autoaggregating properties more pronounced for blood agar-grown cells than for cells grown on a variety of other culture media (53, 54) (Tables 2 and 3). Since nonionic detergents such as Tween 80 (1%, vol/vol) and ethylene glycol (50%, vol/vol) as well as 2 M sodium isothiocyanate prevented autoag-

Table 3. Studies on Cell Surface Hydrophobicity Properties of *S. aureus*

Reference	Method	Surface hydrophobicity
85	HIC[a]	High[b]
63	TPP[c]	Low[d]
42	HIC	High[b]
43	SAT[e]	High
54	SAT	High[f]
57	AP[g]	High[h]
74	MATH	High[i]

[a] For a description of the method, see reference 43; also see Rosenberg and Doyle, this volume.
[b] *S. aureus* strains show higher relative surface hydrophobicities than do strains of *S. saprophyticus* and other CNS. See also references 53, 54, and 90.
[c] TPP, Two-phase partition system.
[d] In two-phase partition systems with dextran and polyethylene glycol, cells of strain Cowan 1 partition into the aqueous phase, whereas group A, C, and G streptococci with hydrophobic surface properties migrate to the other phase (group A, C, and G streptococci also show high surface hydrophobicities as studied by the SAT method (96; see below).
[e] For details, see references 52, 72, and 73; see also Rosenberg and Doyle, this volume.
[f] Strains isolated from septicemia and other infections show higher surface hydrophobicities than do strains isolated from the nose and skin (Table 2).
[g] AP, Adhesion to plastics.
[h] *S. aureus* strains show high binding to hydrophobic plastic surfaces and lower binding to hydrophilic surfaces (56–58).
[i] Strains isolated from pneumonia show higher surface hydrophobicities than do strains from normal skin and burns (74). Strains grown in defined media are more hydrophobic than strains grown in Trypticase soy broth and other complex media (see also references 1, 59, and 61).

gregation, this phenomenon was interpreted as an expression of extremely high surface hydrophobicity.

(iii) Repeated subcultures of such autoaggregating strains revealed that certain strains decreased in cell surface hydrophobicity after a few passages on blood agar. A similar phenomenon has been observed for certain oral streptococci associated with dental caries (J. Olsson, personal communication).

(iv) Competitive studies of some autoaggregating strains and other strains with high to moderately high cell surface hydrophobicity (aggregating in 0.1 to 0.5 M ammonium sulfate) revealed that sugar-supplemented agar media commonly caused a decline in surface hydrophobicity. Further studies on agar- and broth-grown cells revealed that cells grown on solid media showed higher cell surface hydrophobicity and that cells grown to exponential phase expressed higher hydrophobicity than did stationary-phase cells of the same strain. Growth at 20 or 42°C versus 37°C did not affect surface hydrophobicity, nor did anaerobic versus aerobic culturing (53).

More recent studies by Beck et al. (6) showed that stationary-phase cells expressed lower cell surface hydrophobicity, determined by binding to hexadecane as described by Rosenberg et al. (70). Comparative studies of binding of a great variety of *S. aureus* isolates to various plastic polymers showed that exponential-phase cells of hydrophobic strains of *S. aureus* bound most efficiently to a number of plastics compared with two encapsulated strains (strain M and strain Smith diffuse) (56, 57; A. Ludwicka, F. Rozgonyi, and T. Wadström, unpublished data).

CELL SURFACE HYDROPHOBICITY OF CNS

The cell surface hydrophobicities of four *S. aureus* strains and strains of 10 species of CNS were compared (53), with results as described below.

(i) Strains of all 10 species of CNS showed lower cell surface hydrophobicity than did *S. aureus* reference strains (Cowan 1 and Newman).

(ii) Cell surface hydrophobicity was generally more pronounced for blood agar-grown cells of CNS strains. Interestingly, addition of glucose to Trypticase soy broth caused an increase in surface hydrophobicity of strains Newman and Cowan 1 but did the opposite to prototype strains of the 10 CNS species. More recent studies have confirmed that strains of *Staphylococcus epidermidis* isolated from various human infections fall into two groups: strains expressing lower cell surface hydrophobicity when grown in sugar-supplemented Trypticase soy broth (group I strains) and strains not much affected by growth in sugar-supplemented media (group II). Strains of other CNS species commonly isolated from human infections, such as *Staphylococcus haemolyticus* from vascular grafts, are generally group II strains (Å. Ljungh, C. Edmiston, F. Rozgonyi, and T. Wadström, unpublished data). It seems likely that carbohydrate surface polymers, such as capsular substances and slime materials, are expressed under certain growth conditions and cause a decline in cell surface hydrophobicity despite the fact that by classical capsule-staining methods such as with India ink (72, 73; F. Rozgonyi, Å. Ljungh, W. Mamo, S. Hjertén, and T. Wadström, *in* T. Wadström, I. Eliasson, I. A. Holder, and Å. Ljungh, ed., *Pathogenesis of Wound and Biomaterial-Associated Infections*, in press), these surface materials cannot be demonstrated (Table 4).

CAPSULAR POLYSACCHARIDES, CELL SURFACE SLIME, AND EXPRESSION OF CELL SURFACE HYDROPHOBICITY OF *S. AUREUS* AND *S. EPIDERMIDIS*

Very few strains of *S. aureus* isolated from human and animal infections grow as mucoid colonies on blood agar and express a true

Table 4. Possible Cell Surface Virulence Determinants of *S. aureus* and CNS

Determinant	*S. aureus*	CNS
Capsule, slime, glycocalyx	Yes	Yes
Clumping factor[a]	Yes	No
Protein A	Yes	No
FNBP[b]	Yes	Yes
Collagen-binding protein[c]	Yes	Yes
FNBP	Yes	No

[a] The clumping reaction may be influenced by a high receptor density of *S. aureus*-binding fibrinogen. Also, mixing cells with a high fibronectin receptor density with fibronectin can induce cell clumping, whereas strains of CNS species do not clump, possibly because of a lower receptor density or because the receptors may be occluded by capsular material.

[b] Fibronectin binding to *S. aureus* was originally discovered by Kuusela et al. (50). The gene for FNBP was cloned (22, 78), and the gene for a second protein (FNBP II) was recently identified (M. Lindberg, in press). Fibronectin causes the classical clumping reaction of staphylococci suspended in a buffer. Another protease-sensitive surface protein involved has been called a clumping factor (86). However, extracellular coagulase under certain conditions is also surface located and referred to as cell-bound coagulase (10, 17).

[c] Collagen binding to *S. aureus* and various species of CNS was reported at the international staphylococcal conference in Warsaw in 1981 (82). Collagen-binding protein was purified by Holderbaum et al. (38) and more recently by Speziale et al. (82). The gene for collagen-binding protein has recently been cloned in *E. coli* (M. Lindberg, personal communication). We have recently shown that laminin and vitronectin binding occur to other heat-sensitive and protease-sensitive surface structures of *S. aureus*, whereas binding structures of *S. haemolyticus* and other CNS species are commonly more heat and protease resistant (M. Paulsson and T. Wadström, in press).

capsule as defined by classical capsule-staining methods such as with India ink (11, 20, 21). Two such famous classical strains of *S. aureus* are strains M and Smith diffuse. Studies by King and Wilkinson (45) have shown that expression of these surface capsules causes steric hindrance of binding of bacteriophages to teichoic acids and possibly binding of other cell surface receptors, such as immunoglobulin binding to protein A. We have also shown that binding of serum fibronectin to cells of strain Smith diffuse is lower than to cells of an acapsular variant (Smith compact), indicating that expression of a true capsule (44) causes steric hindrance of binding (T. Wadström, unpublished data) as in *Klebsiella* spp. (100). Comparative studies of numerous human clinical isolates of *S. aureus* (53, 54) for binding of fibronectin and collagen type II (molecular weight of 10^6) show that various media do not affect binding, suggesting that high densities of capsule or slime cell surface polymers are rare among fresh clinical isolates of *S. aureus* (62). In contrast, studies of class I and II strains of *S. epidermidis* reveal that binding of fibronectin, type 1 collagen, and vitronectin (molecular weight of 98,000) is much lower for cells grown in sugar-supplemented broth cultures, enhancing production of cell surface carbohydrate polymers commonly referred to as slime materials (15, 16).

These findings seem at first to contradict reports in the last few years that type 5 and 8 capsular polysaccharides, as well as several other capsular polymers, are among strains isolated from septicemia and other severe human infections (23) and strains isolated from cow, goat, and ewe milk (67). Moreover, type 5 and 8 capsular polysaccharides have been demonstrated among human strains isolated from septicemia and wound infections (53, 54) by use of type 5 and 8 specific antisera (J. M. Fournier and Å. Ljungh, unpublished data). Thus, it seems likely that various in vitro growth conditions may affect expression of these capsular polymers, but the findings also suggest that a true surface capsule is probably quite uncommon in S. aureus freshly isolated from human infections. However, recent studies by Lee et al. (51) on mutants of S. aureus SA1 with variations in capsular size show differences in mouse virulence. In summary, sometimes S. aureus isolates from clinical infections produce a true capsule, yielding cells with a high negative surface charge, hydrophilic characteristics, and antiphagocytic properties.

Preliminary studies in our laboratory showed that S. epidermidis isolates grown in sugar-supplemented broth media, such as Trypticase soy broth as described by Christensen et al. (13, 15, 16), suppress surface hydrophobicity (41). However, it should be emphasized that even slime-producing strains of S. epidermidis express moderate cell surface hydrophobicity when grown in iron-limited media which simulate in vivo growth conditions (Wadström, unpublished data). Moreover, studies by Williams et al. (99) showed that iron-limited S. epidermidis cells express other cell surface proteins in sodium dodecyl sulfate-gel electrophoresis compared with cells grown in conventional laboratory media. Preliminary studies in our laboratory have confirmed these findings and shown that blood agar-grown cells of S. aureus S396/81 and S1584/81 express more high-molecular-weight surface proteins as determined by sodium dodecyl sulfate-gel electrophoresis than do the same strains grown in nutrient broth. This preliminary study also revealed that the same strains grown in nutrient broth under iron limitation expressed higher cell surface hydrophobicity (T. Wadström and A. S. Naidu, unpublished data).

CELL SURFACE HYDROPHOBICITY OF S. AUREUS AND CNS IN RELATION TO BINDING TO PLASTIC BIOMATERIAL POLYMERS

Cells of S. aureus Cowan 1 expressing high cell surface hydrophobicity bind to plastic polymers at higher cell numbers than do cells of strain Wood 46 expressing low cell surface hydrophobicity (Fig. 1). Moreover, precoating of polystyrene surfaces with albumin causes a drastic decline in the binding of Cowan 1 cells, indicating a primary role

Figure 1. Percent attachment of staphylococcal cells from three *S. aureus* strains [Newman, SA113(83A), and SpA320] to three different surfaces. Bacteria were harvested in the stationary growth phase [Newman, 9.7 × 10^8; SA113(83A), 5.0 × 10^9; SpA320, 2.8 × 10^9; A_{540} = 0.20 to 0.22]. The surfaces were coated

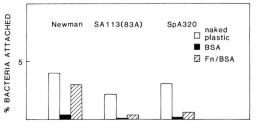

with bovine serum albumin (BSA) or fibronectin (Fn) and subsequently coated with bovine serum albumin. Levels are means of two independent determinations (accuracy between determinations was 10%).

of hydrophobic interactions in the binding process (62). Interestingly, even a precoating of these surfaces with fibronectin, collagen, laminin, or vitronectin to allow interactions between staphylococcal cell surface receptors for these proteins (Tables 4 and 5) did not cause higher bacterial binding per surface area than did uncoated hydrophobic plastic surfaces (39, 56–58).

A series of studies with various plastic polymers and polyethylene polymers treated to decrease surface hydrophobicity showed that such

Table 5. Possible Roles of Cell Surface Hydrophobicity and Charge in Development of Biomaterial-Associated Infections

Alternative I

Step 1. Staphylococci with high surface hydrophobicity (and charge?) bind to inert biomaterials such as various plastic polymers.[a]

Alternative II

Biomaterials introduced into the body are coated with serum and tissue proteins such as fibronectin, albumin, and immunoglobulins in the blood and with tissue proteins (fibronectin, laminin, collagen, etc.) in other organs. Strains of CNS with receptors for these proteins bind to the surface, using the proteins for binding.

Step 2. Surface-bound staphylococci multiply and develop microcolonies.

Step 3. Microbes lay down[b] a slime and capsular polymers, which interact with tissue proteins to develop a biofilm.[c] Cell surface proteins and nonidentified slime- or capsule-associated proteins are a major part of these host-parasite interactions.

[a] This binding is partly prevented by coating the surfaces with albumin or possibly also with other proteins (see Fig. 1 and reference 62).
[b] Phagocytes try to take up such surface-associated microbes in vain (stressed phagocytosis) (27).
[c] For further reading, see reference 30.

"hydrophilized" surfaces bind both *S. aureus* and *S. epidermidis* at lower cell numbers. Other studies by Hogt et al. (34, 35, 37) have shown that relative surface charge and hydrophobicity vary greatly among strains of *S. epidermidis* and *S. saprophyticus*. However, neither surface charge nor hydrophobicity was found to be directly correlated with binding to hydrophobic biomaterials (18). However, since culture conditions to control slime production (e.g., growth in different media to enhance and suppress slime production) were not defined, it is difficult to interpret these results. More recently, Hogt et al. (37) reported that nonencapsulated strains of CNS bound better to various plastic biomaterials such as fluorinated polyethylene-propylene and that the binding was blocked when staphylococci were pretreated with proteases.

FUTURE PROSPECTS

Both *S. aureus* and two CNS species (*S. epidermidis* and *S. haemolyticus*) are major pathogens in various biomaterial-associated infections (9, 18, 59, 90, 102). Many infections with CNS strains develop several months after implantation, such as in the case of eye lenses, joint prostheses, and vascular grafts. Both slime-producing and non-slime-producing strains seem to be common among isolates of *S. epidermidis* from such infections and infections associated with intravascular and intraperitoneal catheters in patients on chronic peritoneal dialysis (Å. Ljungh, N. Grefberg, and T. Wadström, unpublished data). For this reason, we propose a new model for infections associated with catheters and prostheses (77) modified from studies by Gristina et al. (30; A. Gristina, P. T. Naylor, and Q. N. Myrvik, *in* T. Wadström, I. Eliasson, I. A. Holder, and Å. Ljungh, ed., *Pathogenesis of Wound and Biomaterial-Associated Infections*, in press).

Gristina et al. have proposed "the race for the surface" as a general concept for the competition between the invading microbe and eucaryotic host cells to colonize and grow out on these biomaterial tissue implants. A hydrophobic plastic surface is more easily coated by tissue proteins such as fibronectin and collagens, which allow more rapid deposition of tissue fibroblasts and other host cells (102). Thus, it would appear difficult to develop antibacterial surfaces just by making biomaterial surfaces less hydrophobic and less charged (as an alternative to imparting a negative charge at physiological pH to decrease bacterium-material surface interactions). However, manipulation of biomaterial surfaces, including treatment with antibacterial substances and immunomodulators to strengthen the resistance to these infections, in experimental animal models (2) seems to be a fruitful area for future research.

Recently, Cheung and Fischetti (12) studied the role of cell surface proteins in staphylococcal adhesion to cotton and rayon fibers pretreated with serum proteins to simulate conditions in the vagina that may enhance staphylococcal colonization of menstrual tampons. High-molecular-mass (120- and 220-kilodalton) trypsin-sensitive proteins were identified in immunoblots and proposed as possible adhesins in binding to cotton fibers, which generally have an outer hydrophobic waxy layer. Precoating such fibers with fibronectin did not enhance binding, a result that may be due to albumin. The menstrual blood had first coated the fibers to block binding of the toxic shock toxin-producing strain used in this study. Further studies are now needed to better define how various surface proteins (Table 2) and not yet identified surface proteins of S. aureus and pathogenic CNS species are associated with colonization of cotton and other natural materials, such as various types of wound dressings. We have shown that removal of staphylococci from experimental skin wound infections in young pigs enhances the wound-healing processes (91). Recently, the first clinical trials with hydrophobic wound dressings have given promising results for treatment of staphylococcal and streptococcal skin and wound infections (24; G. Fröman, in T. Wadström, I. Eliasson, I. A. Holder, and Å. Ljungh, ed., Pathogenesis of Wound and Biomaterial-Associated Infections, in press). Because other skin and wound pathogens, such as Candida albicans, also bind fibronectin when grown under various conditions (32), it seems most likely that this pathogen, like group A, C, and G streptococci (60), uses similar strategies to cause wound and biomaterial-associated infections.

Effects of Antibiotics on Surface Hydrophobicity of Staphylococci: Possible Role of Hydrophobicity in Binding to Human Cell Connective Tissue Matrix

A number of antibiotics affecting cell wall synthesis (e.g., benzylpenicillin), protein synthesis (erythromycin), and nucleic acid synthesis (rifampin and nalidixic acid) cause a decrease in cell surface hydrophobicity of both S. aureus and group A streptococci (75, 84, 85, 93). Interestingly, studies by Proctor et al. (68) show that chloramphenicol and clindamycin decrease expression of fibronectin binding. Rifampin at subinhibitory concentrations causes a rapid decrease of surface hydrophobicity, of binding of group A streptococci to human pharyngeal cells (85), and of binding of S. aureus Cowan 1 to human nasal cells (93). The increased binding of S. aureus to both human keratinized nasal epithelial cells and human vulvar epithelial cells at low pH suggests that hydrophobic interactions are involved in the binding.

We have also recently shown that growth of *S. aureus* Cowan 1 in subinhibitory concentrations of rifampin and clindamycin causes a parallel decrease of cell surface hydrophobicity, immunoglobulin binding (to protein A), and fibronectin binding (to fibronectin-binding protein [FNBP]; Table 4) (19). It therefore seems most likely that antibiotics affect common regulatory mechanism(s) for several surface proteins in *S. aureus*. Interestingly, strain Wood 46, which is low in both protein A and FNBPs, exhibits low surface hydrophobicity (62). However, the low surface hydrophobicities of protein A and FNBP (22, 26, 78) (Table 4) indicate that other coregulated surface proteins may be the major hydrophobic proteins (or hydrophobins, in the nomenclature proposed by Rosenberg and Kjelleberg [71]). Influenza and other respiratory tract infections predispose to staphylococcal infections in the damaged epithelium (4). It seems most likely that exposure of collagen, fibronectin, vitronectin, and possibly other matrix components in virus-infected tissues exposes structures involved in binding.

Basic amino acids such as lysine and nonpolar amino acids such as leucine and isoleucine are likely involved in interactions between specific staphylococcal cell surface proteins and specific sequences of the connective tissue matrix proteins in exposed damaged tissues (29). However, the possibility remains that tissue lipids interact with surface lipids of *Mycobacterium leprae* (7), and these lipids also may be involved in interactions with staphylococcal cells. It seems unlikely that surface lipids contribute to staphylococcal cell surface hydrophobicity as they do in various mycobacteria and corynebacteria (7, 48). It cannot be excluded that lipid-modified surface proteins exist in staphylococci as in *Treponema pallidum* (76).

Cell Surface Hydrophobicity of *S. saprophyticus*

Pioneer studies by Gunnarsson et al. (31) showed that novobiocin-resistant CNS isolated from urinary tract infections (now called *S. saprophyticus* [40]) produce cell surface proteins which cause hemagglutination of sheep erythrocytes. Hemagglutination is now included in an early identification scheme for isolates from urine specimens in the diagnostic laboratory. Early comparative studies by HIC (92, 93) and later by SAT showed that *S. saprophyticus* strains generally express a hydrophobic cell surface upon growth in media that enhance hemagglutinin production, but hydrophobicity is suppressed in media that suppress synthesis of surface hemagglutinins (Wadström, unpublished data). Interestingly, unlike other CNS species such as *S. epidermidis*, fibronectin binding to *S. saprophyticus* seems to correlate with cell surface hydrophobicity and expression of surface hemagglutinin (95). However, since *S. saprophyticus* strains do not react with *S. aureus* FNBP gene probes, it is possible that binding of a nonspecific nature to the hydrophobic cell surface hemagglutinin or to other hydrophobic surface proteins or hydrophobins occurs. In the pathogenesis of *S. saprophyticus* infections, the hemagglutinin may allow colonization of the uroepithelium, whereas binding to fibronectin may allow the pathogen to colonize in subepithelial tissues after destruction of the surface epithelium by hemolytic (cytolytic) toxins (94, 97, 98).

Acknowledgments. I thank Martin Lindberg, Kristofer Rubin, Cecilia Rydén, and Ingmar Maxe for longtime fruitful collaborations on various aspects of bacterium-host tissue interactions.

Experimental portions of this chapter were supported by grants from the Swedish Medical Research Council (16 × 04723), Swedish Board for Technical Development, and Swedish Council for Forestry and Agricultural Research.

LITERATURE CITED

1. **Abbas, S.** 1983. Rapid labelling of bacteria with [14][C]-D-palmitic acid and its application in an aggregation assay. *FEMS Microbiol. Lett.* **18**:283–287.

2. **Adlam, C., J. C. Andersson, J. P. Arbuthnott, C. S. F. Easmon, and W. C. Noble.** 1983. Animal and human models of staphylococcal infections, p. 357–384. *In* C. S. F. Easmon and C. Adlam (ed.), *Staphylococci and Staphylococcal Infections*, vol. 1. Academic Press, Inc., New York.

3. **Aly, R., H. R. Shinefield, C. Litz, and H. I. Maibach.** 1980. Role of teichoic acid in the binding of *Staphylococcus aureus* to nasal epithelial cells. *J. Infect. Dis.* **141**:436–465.

4. **Babiuk, L. A., M. J. P. Lawman, and H. Bilefeldt Ohmann.** 1988. Viral bacterial synergistic interactions in respiratory diseases. *Adv. Virus Res.* **35**:219–249.

5. **Beachey, E. H.** 1981. Bacterial adherence: adhesin-receptor interactions mediating the attachment of bacteria to mucosal surfaces. *J. Infect. Dis.* **143**:325–344.

6. **Beck, G., E. Pucelle, C. Plotkowski, and R. Peslin.** 1988. Effect of growth on surface charge and hydrophobicity of *Staphylococcus aureus*. *Ann. Inst. Pasteur Microbiol.* **139**:655–664.

7. **Becky, M., R. Mukerjee, and N. H. Antia.** 1986. Adherence of *Mycobacterium leprae* to Schwann cells in vitro. *J. Med. Microbiol.* **22**:277–282.

8. **Bibel, D. J., L. Lahti, H. R. Shinefield, and H. I. Maibach.** 1987. Microbial adherence to vulvar epithelial cells. *J. Med. Microbiol.* **23**:75–82.

9. **Bisno, A. L., and F. A. Waldvogel (ed.).** 1989. *Infections Associated with Indwelling Medical Devices*. American Society for Microbiology, Washington, D.C.

10. **Bodén, M. K., and J. I. Flock.** 1989. Fibrinogen binding protein/clumping factor from *Staphylococcus aureus*. *Infect. Immun.* **57**:2358–2363.

11. **Butt, E. M., C. W. Bonynge, and R. L. Joyce.** 1936. The demonstration of capsules about hemolytic streptococci with India ink or A20 blue. *J. Infect. Dis.* **58**:5–9.

12. **Cheung, A. L., and V. A. Fischetti.** 1988. Role of surface proteins in staphylococcal adherence to fibers in vitro. *J. Clin. Invest.* **83**:2041–2049.

13. **Christensen, G. D., L. M. Baddour, D. L. Hasty, H. J. Lowrance, and A. L. Simpson.** 1989. Microbial and foreign body factors in the pathogenesis of medical device infections, p. 27–59. *In* A. L. Bisno and F. A. Waldvogel (ed.), *Infections Associated*

with Indwelling Medical Devices. American Society for Microbiology, Washington, D.C.

14. **Christensen, G. D., W. A. Simpson, and E. H. Beachey.** 1986. Binding of bacteria to animal tissues: complex mechanisms. p. 279–306. *In* D. C. Savage and M. Fletcher (ed.), *Bacterial Adherence.* Plenum Publishing Corp., New York.

15. **Christensen, G. D., W. A. Simpson, A. L. Bisno, and E. H. Beachey.** 1981. The production of slime by *Staphylococcus epidermidis* (SE): possible role in adherence to smooth surfaces, p. 427–434. *In* I. Phillips and F. P. Tally (ed.), *Resistance in Anaerobic Bacteria.* Academic Press, Inc. (London), Ltd., London.

16. **Christensen, G. D., W. A. Simpson, A. L. Bisno, and E. H. Beachey.** 1982. Adherence of slime-producing strains of *Staphylococcus epidermidis* to smooth surfaces. *Infect. Immun.* **37:**318–326.

17. **Cohen, J. O.** 1972. *The Staphylococci.* John Wiley & Sons, Inc., New York.

18. **Dankert, J., A. Hoght, and J. Feijen.** 1986. Biomedical polymers, bacterial adhesion, colonization and infection. *Crit. Rev. Biocompat.* **2:**219–300.

19. **Doran, J. E., and J. P. Rissing.** 1983. Influence of clindamycin on fibronectin-staphylococcal interactions. *J. Antimicrob. Chemother.* **12**(Suppl. C):75–83.

20. **Duguid, J. P.** 1951. The demonstration of bacterial capsules and slime. *J. Pathol. Bacteriol.* **63:**673–685.

21. **Faris, A., T. Wadström, and J. H. Freer.** 1981. Hydrophobic adsorptive hemagglutinating properties of Escherichia coli possessing colonization factor antigens (CFA/I or CFA/II), type 1 pili, or other pili. *Curr. Microbiol.* **5:**67–72.

22. **Flock, J. I., G. Fröman, B. Guss, K. Jönsson, M. Höök, T. Wadström, and M. Lindberg.** 1987. Cloning and expression of the gene for fibronectin binding protein from *Staphylococcus aureus. EMBO J.* **6:**2351–2357.

23. **Fournier, J. M., K. Hannon, M. Moreau, W. W. Karakawa, and W. F. Vann.** 1987. Isolation of type 5 capsular polysaccharide from *Staphylococcus aureus. Ann. Inst. Pasteur Microbiol.* **138:**561–567.

24. **Fröman, G.** 1987. A new hydrophobized wound dressing (Sorbact 10^5) in the treatment of infected wounds in patients. *Curr. Ther. Res.* **42:**88–93.

25. **Fröman, G., L. M. Switalski, A. Faris, T. Wadström, and M. Höök.** 1984. Binding of *Escherichia coli* to fibronectin. A mechanism of tissue adherence. *J. Biol. Chem.* **259:**4899–4905.

26. **Fröman, G., L. M. Switalski, B. Guss, M. Lindberg, M. Höök, and T. Wadström.** 1986. Characterization of a fibronectin binding protein of *Staphylococcus aureus*, p. 262–268. *In* S. Normark and D. Lark (ed.), *Protein-Carbohydrate Interactions in Biological Systems.* Academic Press, Inc. (London), Ltd., London.

27. **Gemmell, C. G.** 1986. Coagulase-negative staphylococci. *J. Med. Microbiol.* **22:**285–295.

28. **Ginsburg, I.** 1986. How are cell-wall components of pathogenic microorganisms degraded in infections and inflammatory sites? Facts and myths, p. 167–185. *In* P. H. Seidl and K. H. Schleifer (ed.), *Biological Properties of Peptidoglycan.* Walter de Gruyter, Berlin.

29. **Grinell, F., and M. Feld.** 1981. Adsorption properties of fibronectin in relationship to biological activity. *J. Biomed. Mater. Res.* **15:**363–381.

30. **Gristina, A.** 1987. Biomaterial-centered infections: microbial adhesion versus tissue integration. *Science* **228:**990–993.

31. **Gunnarsson, A., P. A. Märdh, A. Lundblad, and S. Svensson.** 1984. Oligosaccharide structures mediating agglutination of sheep erythrocytes by *Staphylococcus saprophyticus. Infect. Immun.* **45:**41–45.

32. **Hazen, B. W., and K. C. Hazen.** 1988. Dynamic expression of cell surface hydrophobicity during initial yeast cell growth and before germ tube formation of *Candida albicans*. *Infect. Immun.* **56:**2521–2525.

33. **Hjertén, S.** 1981. Hydrophobic interaction chromatography of proteins, nucleic acids, viruses, and cells of non-charged amphiphilic gels. *Methods Biochem. Anal.* **27:**89–108.

34. **Hogt, A. H., J. Dankert, and J. Feijen.** 1983. Encapsulation, slime production and surface hydrophobicity of coagulase-negative staphylococci. *FEMS Microbiol. Lett.* **18:**211–215.

35. **Hogt, A. H., J. Dankert, and J. Feijen.** 1985. Adhesion of *Staphylococcus epidermidis* and *Staphylococcus saprophyticus* to hydrophobic biomaterials. *J. Gen. Microbiol.* **131:**2485–2491.

36. **Hogt, A. H., J. Dankert, J. Feijen, and J. A. De Vries.** 1982. Cell surface hydrophobicity of Staphylococcus and adhesion onto biomaterials. *Antonie van Leeuwenhoek J. Microbiol. Serol.* **48:**496–498.

37. **Hogt, A. H., J. Dankert, C. E. Hulstaert, and J. Feijen.** 1986. Cell surface characteristics of coagulase-negative staphylococci and their adherence to fluorinated polyethylenepropylene. *Infect. Immun.* **51:**294–301.

38. **Holderbaum, D., G. S. Hall, and L. A. Erhart.** 1986. Collagen binding to *Staphylococcus aureus*. *Infect. Immun.* **54:**359–364.

39. **Höök, M., L. M. Switalski, T. Wadström, and M. Lindberg.** 1989. Interaction of pathogenic microorganisms with fibronectin, p. 295–308. *In* D. Mosher (ed.), *Fibronectin*. Academic Press, Inc., New York.

40. **Hovelius, B., and P. A. Mårdh.** 1984. Staphylococcus saprophyticus as a common cause of urinary tract infections. *Rev. Infect. Dis.* **6:**328–337.

41. **Jonsson, P., O. Kinsman, O. Holmberg, and T. Wadström.** 1981. Virulence studies on coagulase-negative staphylococci in experimental infections. A preliminary report. *Zentralbl. Bakteriol. Parasitenkd. Infectionskr. Hyg. Abt. 1 Orig. Reihe A* **10:**661–665.

42. **Jonsson, P., and T. Wadström.** 1983. High surface hydrophobicity of *Staphylococcus aureus* as revealed by hydrophobic interaction chromatography. *Curr. Microbiol.* **8:**347–353.

43. **Jonsson, P., and T. Wadström.** 1984. Cell surface hydrophobicity of *Staphylococcus aureus* measured by the salt aggregation test (SAT). *Curr. Microbiol.* **10:**203–210.

44. **Karakawa, W. W., and W. F. Vann.** 1982. Capsular polysaccharides of *Staphylococcus aureus*, p. 119–132. *In* J. B. Robbins, J. C. Hill, and J. C. Sadoff (ed.), *Bacterial Vaccines*. Georg Thieme Verlag, New York.

45. **King, B. F., and B. J. Wilkinson.** 1981. Binding of human immunoglobulin G to protein A in encapsulated *Staphylococcus aureus*. *Infect. Immun.* **33:**666–672.

46. **Kloos, W. E., and P. B. Smith.** 1980. Staphylococci, p. 83–87. *In* E. H. Lennette, A. Balows, W. H. Hausler, Jr., and J. P. Truant (ed.), *Manual of Clinical Microbiology*, 3rd ed. American Society for Microbiology, Washington, D.C.

47. **Koch, H. U., R. Hassa, and W. Fischer.** 1984. The role of lipoteichoic acid biosynthesis in membrane lipid metabolism of *Staphylococcus aureus*. *Eur. J. Biochem.* **138:**357–363.

48. **Kokeguchi, S., K. Kato, H. Ohta, K. Fukui, M. Tsujimoto, T. Ogawa, H. Takada, and S. Kotani.** 1987. Isolation and characterization of an amphiphilic antigen from *Corynebacterium diphtheriae*. *Microbios* **50:**183–199.

49. **Kuhnemund, O., J. Havlicek, K. H. Schmidt, T. Wadström, and W. Köhler.** 1982. Relationship of M protein to hydrophobic properties of streptococcal cells, p. 82–85. *In* S. Holm and P. Christensen (ed.), *Basic Concepts of Streptococci and Streptococcal Diseases*. Reedbooks Ltd., Chertsey, England.

50. **Kuusela, P., T. Vartio, M. Vuento, and E. Myhre.** 1985. Attachment of staphylococci and streptococci to fibronectin, fibronectin fragments, and fibrinogen bound to solid phase. *Infect. Immun.* **50:**77–81.

51. **Lee, J. C., M. J. Betlet, C. A. Hopkins, N. E. Perez, and G. B. Pier.** 1987. Virulence studies in mice of transposon-induced mutants of Staphylococcus aureus differing in capsule size. *J. Infect. Dis.* **156:**741–750.

52. **Lindahl, M., A. Faris, T. Wadström, and S. Hjertén.** 1981. A new test based on "salting out" to measure relative surface hydrophobicity of bacterial cells. *Biochim. Biophys. Acta* **677:**471–476.

53. **Ljungh, Å., A. Brown, and T. Wadström.** 1985. Surface hydrophobicity of coagulase-positive and coagulase-negative staphylococci determined by the salt aggregation test (SAT). *Zentralbl. Bakteriol. Parasitenkd. Infektionskr. Hyg. Abt. 1 Orig.* **S14:**157–161.

54. **Ljungh, Å., S. Hjertén, and T. Wadström.** 1985. High surface hydrophobicity of autoaggregating *Staphylococcus aureus* strains isolated from human infections studied with the salt aggregation test. *Infect. Immun.* **47:**522–526.

55. **Lopes, J. D., M. dos Reis, and R. R. Brentani.** 1985. Presence of laminin receptors in *Staphylococcus aureus*. *Science* **229:**275–277.

56. **Ludwicka, A., B. Jansen, T. Wadström, L. M. Switalski, G. Peters, and G. Pulverer.** 1984. Attachment of staphylococci to various synthetic polymers. *Zentralbl. Bakteriol. Parasitenkd. Infektionskr. Hyg. Abt. 1 Orig. Reihe A* **256:**479–489.

57. **Ludwicka, A., B. Jansen, T. Wadström, L. M. Switalski, G. Peters, and G. Pulverer.** 1984. Attachment of staphylococci to various synthetic polymers, p. 241–256. *In* S. W. Shalaby and A. S. Hoffman (ed.), *Polymers as Biomaterials*. Plenum Publishing Corp., New York.

58. **Ludwicka, A., L. M. Switalski, A. Lundin, G. Pulverer, and T. Wadström.** 1985. Bioluminescence assay for measurement of bacterial attachment to polyethylene. *J. Microbiol. Methods* **4:**169–177.

59. **Maki, D.** 1982. Infections associated with intravascular lines, p. 309–363. *In* J. S. Remington and M. N. Swartz (ed.), *Current Topics in Infectious Diseases*. McGraw-Hill Book Co., New York.

60. **Mamo, W. E., G. Fröman, A. Sundås, and T. Wadström.** 1987. Binding of fibronectin, fibrinogen and type II collagen to streptococci isolated from bovine mastitis. *Microb. Pathog.* **2:**417–422.

61. **Mamo, W., F. Rozgonyi, A. Brown, S. Hjertén, and T. Wadström.** 1987. Cell surface hydrophobicity and charge of *Staphylococcus aureus* and coagulase-negative staphylococci from bovine mastitis. *J. Appl. Bacteriol.* **62:**241–249.

62. **Maxe, I., C. Rydén, T. Wadström, and K. Rubin.** 1986. Specific attachment of *Staphylococcus aureus* to immobilized fibronectin. *Infect. Immun.* **54:**695–704.

63. **Miörner, H., E. Myhre, L. Björck, and G. Kronvall.** 1980. Specific binding of human albumin, fibrinogen, and immunoglobulin G on surface characteristics of bacterial strains as revealed by partition experiments in polymer phase systems. *Infect. Immun.* **29:**879–885.

64. **Mudd, S.** 1965. Capsulation, pseudocapsulation, and the somatic antigens of the surface of Staphylococcus aureus. *Ann. N.Y. Acad. Sci.* **128:**191–211.

65. **Myhre, E., and G. Kronvall.** 1980. Demonstration of specific binding sites for human serum albumin in group C and G streptococci. *Infect. Immun.* **27:**6–14.

66. **Oeding, P., and A. Grov.** 1972. Cellular antigens, p. 333–365. *In* J. O. Cohen (ed.), *The Staphylococci*. John Wiley & Sons, Inc., New York.

67. **Poutel, B., A. Boutonnier, L. Sutra, and J. M. Fournier.** 1988. Prevalence of capsular

polysaccharides type 5 and 8 among *Staphylococcus aureus* isolated from cow, goat, and ewe milk. *J. Clin. Microbiol.* **26**:38–40.

68. **Proctor, R. A., R. J. Hamill, D. F. Mosher, J. A. Textor, and P. J. Olbrantz.** 1983. Effects of subinhibitory concentrations of antibiotics on *Staphylococcus aureus* interactions with fibronectin. *J. Antimicrob. Chemother.* **C12**:85–95.

69. **Rosenberg, M.** 1981. Bacterial adherence to polystyrene: a replica method of screening for bacterial hydrophobicity. *Appl. Environ. Microbiol.* **42**:375–377.

70. **Rosenberg, M., D. Gutnik, and E. Rosenberg.** 1980. Adherence of bacteria to hydrocarbons: a simple method for measuring cell surface hydrophobicity. *FEMS Microbiol. Lett.* **9**:29–33.

71. **Rosenberg, M., and S. Kjelleberg.** 1986. Hydrophobic interactions: role in bacterial adhesion. *Adv. Microb. Ecol.* **2**:353–393.

72. **Rozgonyi, F., K. R. Szitha, S. Hjertén, and T. Wadström.** 1985. Standardization of the salt aggregation test for reproducible determination of cell surface hydrophobicity with special reference to Staphylococcus species. *J. Appl. Bacteriol.* **59**:451–457.

73. **Rozgonyi, F., K. R. Szitha, Å. Ljungh, S. B. Baloda, S. Hjertén, and T. Wadström.** 1985. Improvement of the salt aggregation test to study bacterial cell surface hydrophobicity. *FEMS Microbiol. Lett.* **30**:131–138.

74. **Sanford, B. A., and M. A. Ramsay.** 1987. Bacterial adherence to the upper respiratory tract of ferrets infected with influenza A virus (42525). *Proc. Soc. Exp. Biol. Med.* **185**:120–128.

75. **Schmidt, K. H., O. Kuhnemund, T. Wadström, and W. Köhler.** 1987. Binding of fibrinogen fragment D to group A streptococci causes strain dependent decrease in cell surface hydrophobicity as measured by the salt aggregation test (SAT) and cell clumping in polyethylene glycol. *Zentralbl. Bakteriol. Parasitenkd. Infectionskr. Hyg. Abt. 1 Orig. Reihe A* **264**:185–196.

76. **Schoulds, L. M., R. Mut, J. Dekker, and J. D. A. van Embden.** 1989. Characterization of lipid-modified immunogenic proteins of *Treponema pallidum* expressed in *Escherichia coli*. *Microb. Pathog.* **7**:175–188.

77. **Sheath, N. K., T. R. R. Fransson, and H. D. Rose.** 1983. Colonization of bacteria on polyvinyl chloride and Teflon intravascular catheters in hospitalized patients. *J. Clin. Microbiol.* **18**:1061–1063.

78. **Signäs, C., G. Raucci, K. Jönsson, P. E. Lindgren, G. M. Anantharamaiah, M. Höök, and M. Lindberg.** 1989. Nucleotide sequence of the gene for fibronectin-binding protein from *Staphylococcus aureus*: use of this peptide in the synthesis of biological active peptide. *Proc. Natl. Acad. Sci. USA* **86**:699–703.

79. **Simpson, W. A., I. Ofek, and E. H. Beachey.** 1980. Binding of streptococcal lipoteichoic acid to the fatty acid binding sites on serum albumin. *J. Biol. Chem.* **255**:6092–6097.

80. **Slomiany, B., and A. Slomiany.** 1984. Lipids in mucous secretion of the alimentary tract, p. 23–32. *In* E. Boedeker (ed.), *Attachment of Organisms to the Gut Mucosa*, vol. 1. CRC Press, Inc., Boca Raton, Fla.

81. **Smyth, C. J., P. Jonsson, E. Olsson, O. Söderlind, J. Rosengren, S. Hjertén, and T. Wadström.** 1978. Differences in hydrophobic surface characteristics of porcine enteropathogenic *Escherichia coli* with or without the K88 antigen as revealed by hydrophobic interaction chromatography. *Infect. Immun.* **22**:462–472.

82. **Speziale, R., G. Raucci, L. Visai, L. M. Switalski, R. Timpl, and M. Höök.** 1986. Binding of collagen to *Staphylococcus aureus* Cowan 1. *J. Bacteriol.* **167**:77–81.

83. **Świtalski, L. M., C. Rydén, K. Rubin, Å. Ljungh, M. Höök, and T. Wadström.** 1983. Binding of fibronectin to *Staphylococcus* strains. *Infect. Immun.* **42**:628–633.

84. **Tylewska, S. K., S. Hjertén, and T. Wadström.** 1979. Contribution of M protein to the

hydrophobic surface properties of *Streptococcus pyogenes*. *FEMS Microbiol. Lett.* **6**:249–253.

85. **Tylewska, S., S. Hjertén, and T. Wadström.** 1981. Effect of subinhibitory concentrations of antibiotics on the adhesion of *Streptococcus pyogenes* to pharyngeal epithelial cells. *Antimicrob. Agents Chemother.* **20**:563–566.

86. **Uaui, Y.** 1986. Biochemical properties of fibrinogen binding protein (clumping factor) of the staphylococcal cell surface. *Zentralbl. Bakteriol. Parasitenkd. Infektionskr. Hyg. Abt. 1 Orig. Reihe A* **262**:287–297.

87. **Van Oss, C., and C. F. Gillman.** 1972. Phagocytosis as a surface phenomenon. I. Contact angles and phagocytosis of nonopsonized bacteria. *RES J. Reticuloendothel. Soc.* **12**:283–292.

88. **Vercellotti, G. M., J. B. McCarthy, P. Lindholm, P. K. Peterson, H. S. Jacob, and L. T. Furcht.** 1985. Extracellular matrix proteins (fibronectin, laminin, and type IV collagen) bind and aggregate bacteria. *Am. J. Pathol.* **120**:13–21.

89. **Wadström, T.** 1983. Biological effects of cell damaging toxins, p. 671–704. *In* C. S. F. Easmon and C. Adlam (ed.), *Staphylococci and Staphylococcal Infections.* Academic Press, Inc. (London), Ltd., London.

90. **Wadström, T.** 1986. Bacterial toxic products and their effects on host cells and their tissues, p. 121–130. *In* M. Rye and J. Franek (ed.), *Bacteria and the Host.* Avicenum Czechoslovakia Medical Press, Prague.

91. **Wadström, T., S. Björnberg, and S. Hjertén.** 1985. Hydrophobized wound dressing in the treatment of experimental *Staphylococcus aureus* infections in the young pig. *Acta Pathol. Microbiol. Scand.* **93B**:359–364.

92. **Wadström, T., I. Haraldsson, P. Jonsson, C. Rydén, L. M. Switalski, L. M., M. Höök, and M. Lindberg.** 1981. Experimental infections with Staphylococcus aureus, p. 91–96. *In* G. T. Keusch and T. Wadström (ed.), *Animal Models for Experimental Bacterial and Parasitic Infections.* Elsevier, Amsterdam.

93. **Wadström, T., S. Hjertén, P. Jonsson, and S. Tylewska.** 1981. Hydrophobic surface properties of *Staphylococcus aureus, Staphylococcus saprophyticus* and *Streptococcus pyogenes*. A comparative study, p. 441–447. *In* J. Jeljascewicz (ed.), *Staphylococci and Staphylococcal Infections.* Gustav Fischer Verlag, Stuttgart.

94. **Wadström, T., and F. Rozgonyi.** 1986. Virulence determinants of coagulase-negative staphylococci, p. 123–130. *In* P. A. Mårdh and K. H. Schleifer (ed.), *Coagulase-Negative Staphylococci.* Almquist & Wicksell, Stockholm.

95. **Wadström, T., F. Rozgonyi, and Å. Ljungh.** 1989. Molecular pathogenesis of *Staphylococcus saprophyticus*, p. 214–218. *In* E. H. Kass and C. Svanborg-Edén (ed.), *Host-Parasite Interactions of Urinary Tract Infections.* Chicago University Press, Chicago.

96. **Wadström, T., K. H. Schmidt, O. Kuhnemund, J. Havlicek, and W. Köhler.** 1984. Comparative studies on surface hydrophobicity of streptococcal strains of group A, B, C, D, and G. *J. Gen. Microbiol.* **130**:657–664.

97. **Wadström, T., P. Speziale, F. Rozgonyi, Å. Ljungh, I. Maxe, and C. Rydén.** 1987. Interactions of coagulase-negative staphylococci with fibronectin and collagen as possible first step of tissue colonization in wounds and other tissue trauma. *Zentralbl. Bakteriol. Parasitenkd. Infectionskr. Hyg. Abt. 1 Orig.* **S16**:83–91.

98. **Wicken, A. J., and K. W. Knox.** 1980. Bacterial cell surface amphiphiles. *Biochim. Biophys. Acta* **604**:1–26.

99. **Williams, P., S. P. Denyer, and R. G. Finch.** 1988. Protein antigens of *Staphylococcus epidermidis* grown under iron-restricted conditions in human peritoneal dialysate. *FEMS Microbiol. Lett.* **50**:29–33.

100. **Williams, P., P. A. Lambert, and M. R. W. Brown.** 1988. Penetration of immunoglobulins through the Klebsiella capsule and their effect on cell surface hydrophobicity. *J. Med. Microbiol.* **26:**29–35.
101. **Yamada, K. M.** 1983. Cell surface interactions with extracellular materials. *Annu. Rev. Biochem.* **52:**761–799.
102. **Younger, J. J. U., L. M. Baddour, F. F. Barrett, D. M. Melton, E. H. Beachey, G. D. Christensen, and W. A. Simpson.** 1985. Adherence of coagulase-negative staphylococci to plastic tissue culture plates: a quantitative model for the adherence of staphylococci to medical devices. *J. Clin. Microbiol.* **22:**996–1006.

Microbial Cell Surface Hydrophobicity
Edited by R. J. Doyle and M. Rosenberg
© 1990 American Society for Microbiology, Washington, DC 20005

Chapter 12

Relative Importance of Surface Free Energy as a Measure of Hydrophobicity in Bacterial Adhesion to Solid Surfaces

H. J. Busscher, J. Sjollema, and H. C. van der Mei

The importance of hydrophobicity in bacterial adhesion to solid substrata is generally recognized and accepted (31). However, the extent to which hydrophobicity influences bacterial adhesion in comparison with other important factors, such as zeta potentials (24, 40, 44), absence or presence of surface appendages (53, 55), or production of biosurfactants by the adhering cells (7, 30), is still a subject of discussion.

Moreover, there is no universally accepted definition of "bacterial hydrophobicity," and various experimental methods, all of which claim to measure "hydrophobicity," do not yield consistent results (17, 25, 43). Most physicochemical approaches to determining bacterial adhesion are based on zeta potentials and surface free energies, estimated from contact angles of liquids (1, 11). From a purely physicochemical point of view, water contact angles on bacterial lawns seem to be an extremely suitable measure of hydrophobicity (22, 38; H. J. Busscher, M. N. Bellon-Fontaine, N. Mozes, H. C. van der Mei, J. Sjollema, A. J. Léonard, P. G. Rouxhet, and O. Cerf, submitted for publication).

Bacterial adhesion starts with the transport of microorganisms toward a surface, which can be mediated by gravity (sedimentation), diffusion, or convection. Once a microorganism is within the range of the interaction forces, the actual adhesion process commences. The first forces to become operative are Lifshitz-van der Waals forces, generally

H. J. Busscher, J. Sjollema, and H. C. van der Mei • Laboratory for Materia Technica, University of Groningen, Antonius Deusinglaan 1, 9713 AV Groningen, The Netherlands.

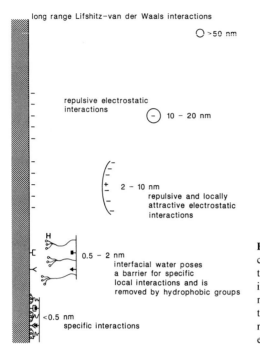

long range Lifshitz–van der Waals interactions

○ >50 nm

repulsive electrostatic
interactions
                    (−)   10 – 20 nm

(−
 −
 +   2 – 10 nm
 +   repulsive and locally
 −   attractive electrostatic
 −   interactions

H
0.5 – 2 nm
   interfacial water poses
   a barrier for specific
   local interactions and is
   removed by hydrophobic groups

<0.5 nm
   specific interactions

Figure 1. Various stages in the process of adhesion of a microorganism to a solid substratum. The distances indicated are approximate. "H" denote localized hydrophobic groups on the cell surface, acting as brooms to remove interfacial water (see reference 10 for details).

attractive and long range in character (Fig. 1). At closer approach, a microorganism will experience repulsive electrostatic interactions. Although most known microbial strains carry a net negative charge, yielding repulsive electrostatic interactions, localized positively charged domains on the cell surface may also yield attractive electrostatic interactions. However, these localized, positively charged domains are only recognizable by the interacting surfaces at even closer approach. Specific interactions between stereochemically complementary molecular groups require the removal of a thin interfacial water layer from in between the interacting surfaces. Previously, we have hypothesized that localized hydrophobic groups on appendages on the cell surface can be considered "water brooms" that remove this water film (Fig. 1) before specific, short-range interactions can start to become operative (see reference 10 for details).

Contact angles reflect the overall hydrophobic character of a microorganism (45) and are unlikely to be influenced greatly by the presence of a localized hydrophobic group. The same is true for zeta potentials. Therefore, we expect that these parameters will be useful mainly to describe the first two steps in the adhesion process (Fig. 1). Assessment of hydrophobicity by the microbial adhesion to hydrocarbons test (17, 25, 31, 43) or charge as measured by anion-exchange resin chromatography (17, 25, 43) probes the cell surface at a more microscopic level and will be more meaningful with regard to the later stages of the adhesion process.

Various methods are in vogue to determine bacterial adhesion, but it is seldom realized that it is extremely difficult to properly carry out bacterial adhesion experiments. The term "bacterial adhesion" is often used to denote either the number of adhering cells (39), their immobilization (24), or their strength of attachment (the shear force required for removal) (12, 13). However, these expressions refer to completely different parameters. Probably flow cell systems, preferably used in combination with real-time image analysis techniques (36), offer the only way to experimentally distinguish these various parameters (35). An additional advantage of flow cell systems is that the spatial arrangement of cells adhering to a solid substratum is preserved, opening new avenues for studying the role of cell-cell interactions and cooperative effects (14, 18, 19, 33, 37) in bacterial adhesion.

In this chapter, we will illustrate the relative importance of overall hydrophobicity in bacterial adhesion by describing the adhesion of *Streptococcus sanguis* 12, an oral streptococcal strain, to a hydrophobic and a hydrophilic solid substratum. The adhesion data presented have all been obtained in a flow cell system using real-time image analysis and will be interpreted in terms of numbers of adhering bacteria and their immobilization. The existence of possible (non)cooperative effects will be inferred from an analysis of the spatial arrangement of the adhering cells. Special attention will be given to the influences of (i) surface free energy as an overall measure of hydrophobicity, (ii) ionic strength of the suspension and zeta potentials, and (iii) shear rates acting during the experiments.

PHYSICOCHEMICAL BACKGROUND

In every adhesion process, new interfaces are created at the expense of old ones. Since every interface contains a certain amount of free energy, a free energy balance of all interfaces involved in adhesion can be used to tell whether adhesion is energetically favorable (1, 11).

In a thermodynamic approach, this free energy balance can be developed on the basis of the interfacial free energies γ_{sl}, γ_{bl}, and γ_{bs} (1, 11) (for nomenclature, see Fig. 2), yielding

$$\Delta F_{adh} = \gamma_{bs} - \gamma_{sl} - \gamma_{bl} \qquad (1)$$

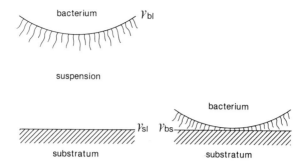

Figure 2. Schematic presentation of three interfaces involved in bacterial adhesion to a solid substratum from a liquid suspension. The three interfacial free energies γ_{sl}, γ_{bl}, and γ_{bs}, are indicated.

in which ΔF_{adh} is the free energy of adhesion. Thermodynamically, adhesion is favorable if ΔF_{adh} is negative.

The interfacial free energies γ_{ij} occurring in equation 1 can, among other methods (27, 47, 50), be evaluated from the concept of dispersion and polar surface free energy components by using the geometric mean equation (28)

$$\gamma_{ij} = \gamma_i + \gamma_j - 2(\gamma_i^d \, \gamma_j^d)^{1/2} - 2(\gamma_i^p \, \gamma_j^p)^{1/2} \qquad (2)$$

in which d and p denote the dispersion and polar components, respectively, of the surface free energy of the individual solid, bacterial, or liquid phase.

The surface free energy of the liquid phase is usually determined tensiometrically, whereas contact angle measurements can be carried out to estimate the surface free energies of the solid and bacteria. Being relatively easy for solids, contact angle measurements can present great difficulties when carried out on bacterial cell surfaces (48; Busscher et al., submitted). It is generally advocated that a bacterial lawn be produced on agar plates (48) or on membrane filters (11; Busscher et al., submitted), which should then be dried to evaporate free water. Although it may be argued that this procedure brings the surface to a state differing from the physiological one, it has been shown recently that surface free energies determined in this manner correlate well with electrophoretic data obtained under physiological conditions (41, 42).

No simple relationship exists between bacterial adhesion and ΔF_{adh}. Pratt-Terpstra et al. (29, 30) described strain-dependent linear relationships between the number of bacteria adhering in a stationary state n_s and ΔF_{adh} according to

$$n_s = a \cdot (\Delta F_{\text{adh}} - 23.1) + 1.2 \qquad (3)$$

in which a denotes a strain-dependent microbial factor, associated with the absence or presence of surface appendages (29), secretion of biosurfactants by the cells (30), or possibly charge properties. From a theoretical point of view, it may be surprising that a thermodynamic approach to bacterial adhesion works (4). Reversible adhesion, as required in a thermodynamic equilibrium, is encountered only as an exception (9), not as a general rule (34). Furthermore, the actual contact area between a cell and a substratum is very small compared with the area probed by a contact angle (45), and electrical charge interactions are not specifically included in a thermodynamic approach (24).

A physicochemical approach to bacterial adhesion that does include the influence of electrical charge interactions is based on the so-called DLVO theory (32). The DLVO theory, however, is used primarily to describe the kinetics of the adhesion process rather than the equilibrium situation. The theory accounts both for repulsive or attractive electrostatic interactions and for van der Waals forces. The interaction energy between a spherical particle and a flat solid substratum can accordingly be described as

$$V = V_{vdW} + V_{el} \qquad (4)$$

in which

$$V_{vdW} = - \frac{A_{slb}}{6} \left\{ \frac{2a\,(h + a)}{h(h + 2a)} - \ln \left(\frac{h + 2a}{h} \right) \right\} \qquad (5)$$

$$V_{el} = \pi \epsilon \epsilon_0 a (\zeta_s^2 + \zeta_b^2) \left\{ \frac{2\zeta_s\,\zeta_b}{\zeta_s^2 + \zeta_b^2} \ln \left[\frac{1 + \exp\,(-\kappa h)}{1 - \exp\,(-\kappa h)} \right] \right.$$

$$\left. + \ln\,[1 - \exp\,(-2\kappa h)] \right\} \qquad (6)$$

with A_{slb} the effective Hamaker constant, a the bacterial diameter, h the interaction distance, $\epsilon \epsilon_0$ the permittivity of the medium, ζ_s and ζ_b the zeta

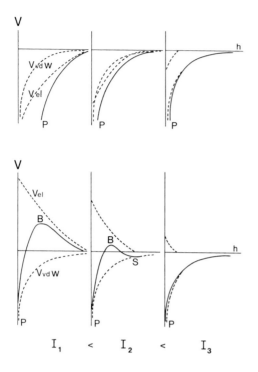

Figure 3. Interaction energies V, V_{vdW}, and V_{el}, calculated on the basis of the DLVO theory as a function of the distance h between an (inert) particle and a flat substratum surface in the cases of attractive (top) and repulsive (bottom) electrostatic interactions (adapted from reference 32). I denotes various ionic strengths of the suspension medium. P, Primary interaction minimum; B, interaction barrier; S, secondary interaction minimum. Note that because many strains possess surface appendages, the concept of distance loses its meaning at close approach between particle and substratum (10).

potentials of the substratum and bacteria, respectively, and κ the reciprocal Debye-Hückel length (21).

Figure 3 shows the interaction energies V, V_{vdW}, and V_{el} for various ionic strengths and for attractive and repulsive electrostatic interactions. Note that the importance of electrostatic interactions decreases as the ionic strength of the suspension increases. The depth of the primary minimum in the interaction energy versus distance curve cannot be calculated on the basis of the DLVO theory, and it is sometimes argued that the depth of the primary minimum can be inferred from the free energy of adhesion ΔF_{adh} (see equation 1). One of the problems associated with linking equations 1 and 4 is the fact that equation 1 is per unit area, whereas equation 4 is per particle. Hence, a linkage can be made only if the contact area of a cell with a given substratum is known. Although such estimates can be made, they always remain speculative (10).

Another linkage between the DLVO theory and the thermodynamic approach is based on the relationship between the Hamaker constant and the dispersion component of the free energy of adhesion ΔF_{adh} (46, 49, 50)

$$\Delta F_{adh}{}^d = -\frac{A_{slb}}{12 \cdot \pi \cdot h_{min}{}^2} \tag{7}$$

in which h_{min} is assumed to be 1.8 Å (0.18 nm) (49). ($\Delta F_{adh}{}^d$ can be easily evaluated from equations 1 and 2, accounting only for the dispersion components.)

All physicochemical approaches assume that bacterial cell surfaces are more or less smooth and homogeneous. This is never the case (53), and therefore simple physical chemistry as described above, neglecting microscopic details of the bacterial cell surface, will never be able to fully describe all aspects of bacterial adhesion. Nevertheless, it can be instructive to see how far one can get in explaining bacterial adhesion data on the basis of the principles above outlined, as will be illustrated in the next section.

AN EXPERIMENTAL DESIGN FOR STUDYING BACTERIAL ADHESION

Numerous experimental systems have been described for studying bacterial adhesion (see Sjollema et al. [35] for an overview). It is crucial to decide which parameters of the adhesion process one wants to determine before choosing and constructing a specific system. Among the parameters that can be determined are the following.

1. The number of cells adhering per unit area. These kinds of measurements should preferably be done as a function of time, yielding (34)

$$n_b\,(t) = n_s\,(1 - e^{-\beta \cdot t}) \tag{8}$$

in which $n_b(t)$ is the number of cells adhering per unit area at time t, n_s is the number of cells adhering per unit area in a stationary state, and β is the reciprocal adhesion time constant.

The initial rate of adhesion can be derived from n_s and β according to

$$j_0 = n_s \cdot \beta \tag{9}$$

in which β describes the way by which the adhesion process approaches a stationary state and includes blocking, desorption, and effects of biosurfactant production by the adhering cells.

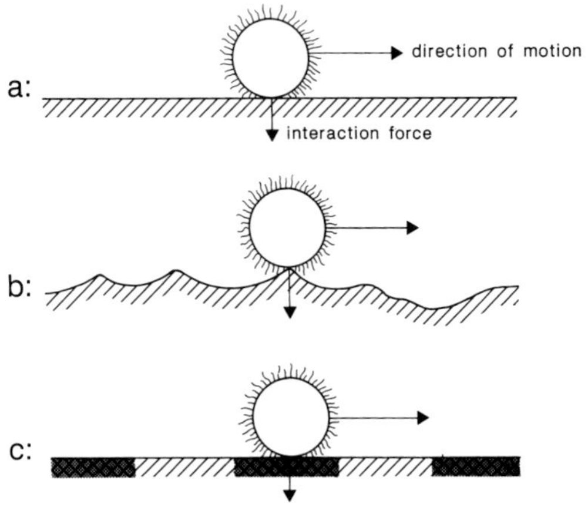

Figure 4. Difference between adhesion and immobilization of adhering cells. (a) Independent of the magnitude of the interaction forces acting perpendicularly to the substratum surface, motion of a cell over the surface remains possible (compare sliding a magnet over a metal plate). (b) Motion of an adhering cell over a rough substratum surface may require that interaction forces acting perpendicularly to the surface are balanced and overcome by a lift force. (c) Local areas of high attraction between an adhering cell and a substratum may cause an effective means of immobilization.

2. The degree of immobilization of adhering cells. The number of adhering cells, if not determined by real-time image analysis, is usually determined as a snapshot. This approach excludes the possibility of analyzing their motion over a surface. No matter how strong the adhesion forces are, motion of adhering cells over a surface remains always possible on perfectly smooth and chemically homogeneous substrata. Only surface roughness or chemical heterogeneity can induce real immobilization (Fig. 4). Therefore, it is of utmost importance to distinguish between adhesion and immobilization.

3. The actual force of attachment, acting perpendicularly to the substratum surface. This force, which need not be the same as the perpendicular force of detachment, is very difficult to measure. However, it should not be confused with the shear force operating during attachment or detachment. In addition, it has been suggested that incomplete immobilization of cells adhering to rough surfaces may stimulate detachment (23).

4. The number of cells desorbing per unit area per unit time and the residence time of cells before desorption. Particularly in thermodynamic approaches to measuring bacterial adhesion, it is necessary to establish whether adhesion is reversible. Real-time image analysis and fast microcomputers have made it possible to determine exactly the number of cells desorbing and their residence times before desorption by comparing successively stored images.

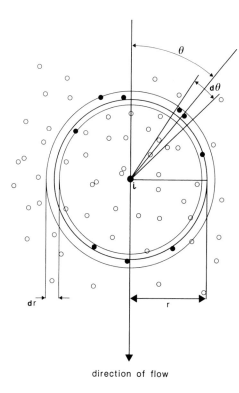

Figure 5. Schematic visualization of pair distribution functions between adhering particles. The radial distribution function $g(r)$ describes the relative density of adhering particles in a shell with thickness dr at a distance r from a given particle i, whereas the angular distribution function $g(\theta)$ describes the relative density in a segment of the shell dr with angular width $d\theta$ and at an angle θ with the direction of flow [see Sjollema and Busscher (33) for details concerning the exact calculations of $g(r)$ and $g(\theta)$].

A reciprocal desorption time constant α, with the same dimension as β in equation 8 can be calculated according to

$$\alpha = \frac{\Delta n_{21}}{n_b(t_1) \cdot (t_2 - t_1)} \tag{10}$$

in which t_2 and t_1 represent the times at which the images to be compared were stored, $n_b(t_1)$ is the number of bacteria in the image taken at t, and Δn_{21} is the number of cells that desorbed during the time interval $(t_2 - t_1)$.

5. The spatial arrangement of the adhering cells. This parameter can be analyzed to give the radial or angular distribution functions of adhering cells (Fig. 5). The angular distribution indicates the influence of the direction of flow on the adhesion of the cells compared with the influence of cell-cell interactions. The radial distribution indicates whether cells have a preference to adhere

close to each other (positive cooperativity) or far away from each other. Hitherto, the question of cooperativity in bacterial adhesion has been addressed only by analyzing the slopes appearing in Scatchard plots of adhesion data (14, 18, 19) (Fig. 6). Although it has not yet been proven that the information that can be obtained from an analysis of the spatial arrangement of adhering cells is the same as that from Scatchard plots, we expect that the two types of information are at least largely complementary. Finally, this type of analysis can be done only if it is well established that sample manipulation has not destroyed the spatial arrangement of the cells that existed during the experiment. A direct observation technique as described by Sjollema et al. (36) probably is the only method available to this end at present.

In the next section, we present some experimental results to illustrate the principles described above.

EXPERIMENTAL DATA FOR THE ADHESION OF S. SANGUIS 12 TO A HYDROPHOBIC AND A HYDROPHILIC SUBSTRATUM

Experiments were carried out in a rectangular parallel plate flow cell, using real-time image analysis. The dimensions of the flow cell were 0.06 by 3.8 by 5.5 cm, and a pulse-free flow was established by hydrostatic pressure (for details, see Sjollema et al. [34, 36]). All experiments described were done in duplicate with S. sanguis 12 cells suspended in potassium phosphate buffer (pH 7.0) to a density of 3.10^8 cells $\cdot$ ml^{-1}, using FEP (fluorethylenepropylene) and glass as extremely hydrophobic and hydrophilic substrata, respectively.

Initial Deposition Rates and Numbers of Cells Adhering in a Stationary State to a Hydrophobic and a Hydrophilic Substratum

Table 1 lists the experimental conditions under which the assays were done together with the initial deposition rates j_{0s} and the number of cells adhering in a stationary state n_{st}. The second subscript of each of these two parameters denotes whether the data refer to single cells (s) or also to higher-adhering multiple aggregates (t). Table 1 reveals a distinct preference for S. sanguis 12 to adhere to glass, a hydrophilic substratum, rather than to FEP, a hydrophobic substratum. Buffer concentrations and shear rates influence both the deposition rates and the number of cells adhering in a stationary state.

The initial deposition rates for single cells observed can be compared

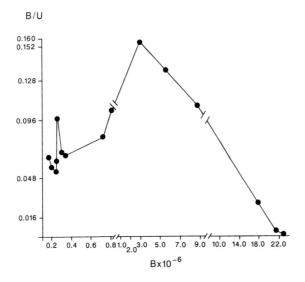

Figure 6. Scatchard analysis of data for adhesion of *S. sanguis* to an artificial pellicle plotted according to $B/U = K(N - B)$, in which B and U denote the numbers of bound and unbound cells, respectively, and K and N are an association constant and the number of available adhesion sites, respectively. In the absence of positive cooperative effects, a Scatchard plot has a negative slope. However, if positive cooperativity exists, the number of available adhesion sites N increases as more cells become bound, and the increase in the number of bound cells at the expense of the number of unbound cells is such that a positive slope arises. (Adapted from reference 14.)

with those calculated on the basis of the Smoluchowski-Levich approximation. In the Smoluchowski-Levich approximation, it is assumed that the increase in hydrodynamic drag which a particle experiences while approaching a surface (6) is balanced by attractive dispersion forces. All other particle-substratum interactions are neglected. Accordingly, the theoretical deposition rate j_0^* can be expressed as (2, 5)

$$j_0^* = \frac{D_\infty \cdot c}{0 \cdot 89a} \cdot \left(\frac{2\text{Pe}b}{9x}\right)^{1/3} \tag{11}$$

with

$$\text{Pe} = \frac{3V_m a^3}{2b^2 D_\infty} \tag{12}$$

Table 1. Adhesion of *S. sanguis* 12 to a Hydrophobic Substratum (FEP) and a Hydrophilic Substratum (Glass) under Various Experimental Conditions[a]

Substratum	Buffer concn (mM)	Shear rate (s^{-1})	j_{0s} $(cm^{-2} \cdot s^{-1})$	n_{st} $(10^6 \cdot cm^{-2})$	$Sh_0/Sh_0{}^*$
FEP	80	10	433	2.1	0.98
	80	90	302	4.4	0.31
	80	200	124	2.3	0.10
Glass	80	10	397	36.1	0.90
	80	90	666	19.8	0.69
	80	200	508	12.2	0.41
FEP	10	90	135	3.1	0.14
Glass	10	90	241	4.2	0.24

[a] Data were obtained in a flow cell system, using real-time image analysis. j_{0s}, Initial deposition rates for single cells; n_{st}, total number of cells adhering in a stationary state; $Sh_0/Sh_0{}^*$ deposition efficiency for single cells as compared with purely convective-diffusion controlled mass transport.

and

$$D_\infty = \frac{kT}{6\pi\mu a} \tag{13}$$

in which x is the longitudinal distance from the entrance of the cell (3.0 cm), b is the half-depth of the cell (0.03 cm), V_m is the mean liquid velocity in the cell, a is the bacterial cell radius (assumed to be $0.4 \ 10^{-4}$ cm), Pe is the dimensionless Péclet number, D_∞ is the Stokes-Einstein diffusion coefficient (5.5×10^{-9} cm$^2 \cdot$ s^{-1}), c is the bacterial cell density at the entrance (3×10^8 cm^{-3}), k is the Boltzmann constant (1.38×10^{-16} erg $\cdot$ K^{-1}), T is the absolute temperature (298 K), and μ is the viscosity of the medium (10^{-2} g $\cdot$ cm$^{-1} \cdot$ s^{-1}).

The ratio between the experimentally observed j_0 and the calculated initial deposition rate $j_0{}^*$ is often referred to as the deposition efficiency and is indicated in the literature as $Sh_0/Sh_0{}^*$.

The deposition efficiency of *S. sanguis* 12 thus calculated is shear rate dependent and generally low, especially in the case of deposition from a low-ionic-strength buffer in which electrostatic repulsions are high (Table 1).

Approach to a Stationary State and Reversibility of the Adhesion Process

In general, we were able to describe the adhesion kinetics of *S. sanguis* 12 during the first 4 to 6 h to both substrata and under all conditions by equation 8, yielding j_0, n_s, and β. β values for single cells are summarized in Table 2. In principle, the approach to a stationary state as

Table 2. Reciprocal Adhesion Time Constants β_s for Single
Cells and Reciprocal Desorption Time Constants α,
Evaluated Approximately 1,000 and 5,000 s
after the Start of an Experiment

Substratum and conditions	β_s $(10^{-4}\,s^{-1})$	α (1,000) $(10^{-4}\,s^{-1})$	α (5,000) $(10^{-4}\,s^{-1})$
FEP, 80 mM			
10 s^{-1}	5.9	4.1	1.0
90 s^{-1}	2.7	5.3	1.5
200 s^{-1}	0.7	5.8	1.1
Glass, 80 mM			
10 s^{-1}	0.3	5.7	2.9
90 s^{-1}	1.6	1.0	0.2
200 s^{-1}	0.4	1.4	0.2
FEP, 10 mM, 90 s^{-1}	0.7	2.2	2.4
Glass, 10 mM, 90 s^{-1}	0.5	2.0	0.4

expressed by β is governed by desorption, blocking, or secretion of biosurfactants by the cells.

If β is determined only by blocking and not by the other processes mentioned and if the stationary state obtained is a thermodynamic equilibrium, the reciprocal adhesion time constant β should equal the reciprocal desorption time constant α as expressed by equation 10.

Table 2 lists desorption constants α, evaluated shortly (approximately 1,000 s) and approximately 5,000 s after the onset of an experiment. In all cases, α values decrease after prolonged adhesion times.

A completely different measure for the reversibility of adhesion can be obtained by draining the flow cell while counting the numbers of adhering cells before and after draining. During draining, adhering cells have to withstand passage through a liquid-air interface once, which exerts a force parallel to the substratum surface of approximately 10^{-2} dynes. Such a force may cause spatial rearrangement or detachment of adhering cells by the mechanism shown in Fig. 4. The effect of one passage through a liquid-air interface upon the spatial arrangement and number of adhering cells is illustrated in Fig. 7. Figure 8 shows the percentage of adhering cells removed by draining the flow cell. It can be seen that only cells which have adhered from a suspension with a buffer concentration of 10 mM are able to successfully withstand shear forces thus exerted. The spatial arrangement of the adhering cells after draining was very different on the hydrophobic substratum (FEP) than on the hydrophilic substratum (glass): on FEP, cells had slid over the surface to form highly localized clumps, whereas on glass, a much more homogeneous detachment occurred (see also Fig. 7).

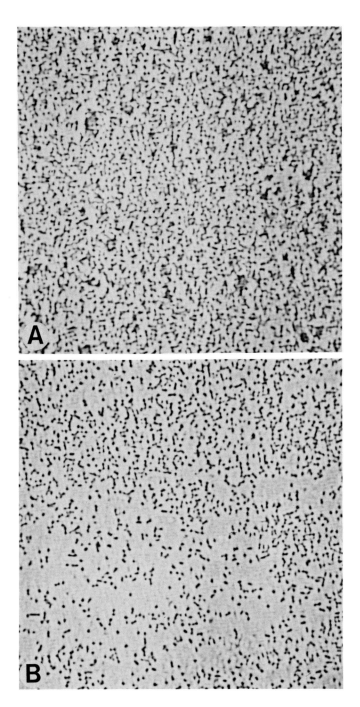

Figure 7. *S. sanguis* 12 adhering to glass (80 mM; 90 s^{-1}) in the flow cell before (A) and after (B) draining of the channel, i.e., one passage of the substratum through a liquid-air surface.

Analysis of the Spatial Arrangement of Adhering Cells

Figure 9 shows examples of radial distribution functions for *S. sanguis* 12 adhering to FEP and glass. The relative density $g(r)$ is smaller than unity (indicating the average density) within the so-called screening distance r_s. Particularly on FEP, bacteria have a great preference to adhere in each other's neighborhood (Fig. 9). In the example shown, this preferential distance r_p equals 2.5 diameters and $g(r_p)$ is 1.8 for FEP. Table 3 summarizes the features of the radial distribution functions obtained for *S. sanguis* 12 adhesion to FEP and glass under the various conditions.

Figure 9 also shows the angular distribution of the adhering cells. The

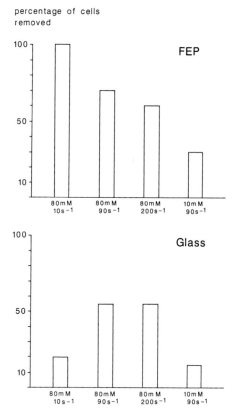

Figure 8. Percentage of adhering *S. sanguis* 12 removed from a hydrophobic substratum (FEP) and a hydrophilic substratum (glass) after adhesion under various conditions upon draining of the flow cell, i.e., one passage of a substratum with adhering cells through a liquid-air interface.

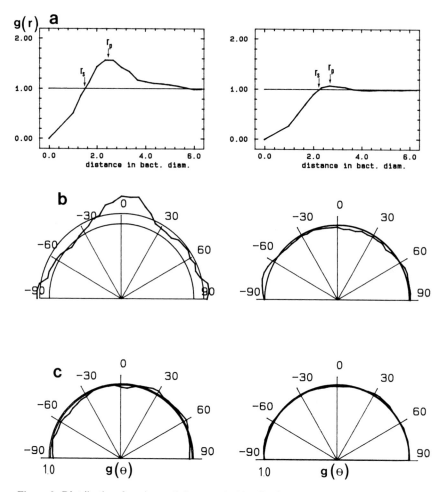

Figure 9. Distribution functions of *S. sanguis* 12 adhering to a hydrophobic substratum (FEP; left) and a hydrophilic substratum (glass; right) from a 80 mM potassium phosphate buffer at a shear rate of 90 s^{-1}. (a) Radial distribution functions $g(r)$ as a function of the bacterial cell diameter (bact. diam.). (b) Angular distribution functions $g(\theta)$ for near neighbors. (c) Angular distribution functions $g(\theta)$ for far neighbors.

relative density $g(\theta)$ is expressed in a polar plot for near and far neighbors and reveals a preferential adhesion for near neighbors in the direction of the flow on FEP. Because it cannot be determined on the basis of one image whether a cell is deposited in front of or behind another, the angular distributions are restricted to angular segments between $-90°$ and $+90°$ with respect to the direction of flow. Table 4 summarizes the features of the angular distribution functions of *S. sanguis* 12 to FEP and glass.

Table 3. Analysis by Radial Distribution of the Spatial Arrangement of *S. sanguis* 12 Adhering to a Hydrophobic Substratum (FEP) and a Hydrophilic Substratum (Glass) under Various Experimental Conditions[a]

Substratum and conditions	No. of particles involved	r_s	r_p	$g(r_p)$
FEP, 80 mM				
10 s^{-1}	4,940	2.0	2.7	1.2
90 s^{-1}	4,717	1.0	2.1	2.4
200 s^{-1}	3,893	1.3	2.3	2.3
Glass, 80 mM				
10 s^{-1}	13,538	1.9	2.3	1.2
90 s^{-1}	7,964	1.8	2.3	1.2
200 s^{-1}	3,836	2.0	2.0	1.1
FEP, 10 mM, 90 s^{-1}	4,745	1.2	2.4	2.7
Glass, 10 mM, 90 s^{-1}	5,407	0.8	2.3	1.5

[a] r_s, Screening distance; r_p, distance of preferential adhesion; $g(r_p)$, relative density at r_p. Both r_s and r_p are expressed in bacterial diameters.

ANALYSIS AND MECHANISM OF ADHESION

In this chapter, we have presented data on the adhesion of *S. sanguis* 12, one of the initial colonizers of human teeth (20). Substratum surface free energies used in this study encompass the value for ground and polished enamel, viz., γ_s = 88 erg · cm^{-2} (51); glass is considerably more hydrophilic ($\gamma_s \approx$ 120 erg · cm^{-2}), and FEP is much more hydrophobic (γ_s = 20 erg · cm^{-2}). Ionic strengths used are comparable to or lower than those under physiological conditions. Furthermore, it should be noted that the range of shear rates applied is thought to be representative for the wide spectrum of values encountered in the oral cavity, which range from 200 s^{-1} during swallowing and speech (3, 52) to 0.4 s^{-1} at rest (16). Although the system is also ideal for studying the effect of a pellicle on bacterial adhesion, time has allowed us to work only with uncoated, bare substrata.

S. sanguis 12, FEP, and glass surface characteristics have all been described in detail elsewhere (29, 43; J. Sjollema, H. C. van der Mei, H. M. Uyen, and H. J. Busscher, submitted for publication). Table 5 shows the parameters important for the interaction of the microorganism with the above two substrata, as calculated from their known surface characteristics. Rather than evaluating equation 6 for the exact calculation of the electrostatic interaction, we have taken the product $\zeta_s \cdot \zeta_b$ as a measure of the electrostatic interactions. It is obvious that a sizable electrostatic potential barrier is present at the low ionic strength for both substrata.

Table 4. Analysis by Angular Distribution of the Spatial Arrangement of *S. sanguis* 12 Adhering to a Hydrophobic Substratum (FEP) and a Hydrophilic Substratum (Glass) under Various Experimental Conditions

Substratum and conditions	Neighbor type	Distribution[a]						
		90°	60°	30°	0°	30°	60°	90°
FEP, 80 mM								
10 s⁻¹	Near	+	0	0	0	0	0	+
	Far	0	0	0	0	0	0	0
90 s⁻¹	Near	0	0	0	+ +	0	0	0
	Far	0	0	0	+	0	0	0
200 s⁻¹	Near	0	0	0	+	0	0	0
	Far	0	0	0	0	0	0	0
Glass, 80 mM								
10 s⁻¹	Near	0	0	0	0	0	0	0
	Far	0	0	0	0	0	0	0
90 s⁻¹	Near	0	0	0	0	0	0	0
	Far	0	0	0	0	0	0	0
200 s⁻¹	Near	0	0	0	0	0	0	0
	Far	0	0	0	0	0	0	0
FEP, 10 mM, 90 s⁻¹	Near	0	0	0	+ +	0	0	0
	Far	0	0	0	+	0	0	0
Glass, 10 mM, 90 s⁻¹	Near	0	−	0	+	0	−	0
	Far	0	0	0	0	0	0	0

[a] 0, Random distribution within the angular segment indicated; + and + +, preferential distributions of 10 and 20% or more, respectively. An analogous criterion holds for the − sign.

The interfacial free energies only yield conditions favorable for adhesion of *S. sanguis* 12 on glass. Two lengths of fibrils, concentrated toward the poles of the cells (54), may assist the microorganism in its adhesion to surfaces. Later in this discussion we will address the actual adhesion data as collected with the interaction parameters presented in Table 5.

Table 5. Free Energy of Adhesion of *S. sanguis* 12 from an Aqueous Suspension to a Hydrophobic Substratum (FEP) and a Hydrophilic Substratum (Glass) Calculated from Equation 1 with an Indicator for the Magnitude and Sign of the Electrostatic Interactions I_{el}, Defined as $\zeta_s \cdot \zeta_b$ (See Also Equation 6)

Substratum	Interaction parameter[a]		
	ΔF_{adh} (erg · cm⁻²)	I_{el}, 10 mM (mV²)	I_{el}, 80 mM (mV²)
FEP	+18	+440	→0
Glass	−7	+1452	→0

Deposition Kinetics and Stationary State

The initial deposition rates observed are rather low compared with theoretically calculated deposition rates on the basis of the Smoluchowski-Levich approximation, both for low and for high ionic strengths. This observation indicates that there is a sizable electrostatic repulsion which effectively reduces mass transport to the substratum surface. It has been shown that the Smoluchowski-Levich approximation adequately describes mass transport in the case of attractive electrostatic interactions (5). For a positively charged microorganism, *Streptococcus thermophilus* B, we have even observed deposition efficiencies above unity (7), although it is rare to encounter positively charged streptococci. Because theoretically the deposition efficiency should be independent of the prevailing hydrodynamic conditions, i.e., the shear rate applied, it can be concluded from the decrease in Sh_0/Sh_0^* with shear rate (and the high values occurring at 10 s^{-1}) that sedimentation contributes to the transport of cells toward the substrata. Nevertheless, the efficiency remains lower than unity. Calculation of the efficiency Sh_0/Sh_0^* as indicated in Table 1 is generally useful, since it yields an immediate indication of whether an observed deposition rate is to be considered high or low.

There is no relationship between the deposition rate j_0 and the number of cells adhering in the stationary state n_{st}. This proves that the origin of the stationary state is at least partly thermodynamic in nature and not kinetically controlled. The thermodynamic nature of the adhesion of *S. sanguis* 12 is confirmed by the observation of a relatively large number of cells desorbing per unit time and area (compare the α and β_s values in Table 2). The reciprocal desorption time constants α decreased markedly after prolonged adhesion times, demonstrating that adhesion changes gradually into a more irreversible state. This is in accordance with the observation that equation 8 could describe the adhesion kinetics only during the first 4 to 6 h. After approximately 6 h, an increase in the number of adhering cells sets in, considered to be the onset of a second, less reversible state in the adhesion process. Such an observation of a two-phase kinetics in the adhesion of *S. sanguis* has been described by Busscher et al. (8) and Cowan et al. (14).

The number of *S. sanguis* 12 cells is smaller on the hydrophobic substratum (FEP) than on the hydrophilic substratum (glass), obviously because of the unfavorable thermodynamic conditions (positive ΔF_{adh}; Table 5). The preference of these cells for glass exists at all ionic strengths and shear rates used, although it is most pronounced in the absence of electrostatic interactions (80 mM) and at the lowest shear rate. This observation confirms that the sensitivity of a given strain to interfacial free energy changes is dependent on its charge properties and the prevailing hydrodynamic conditions (29) as well as on the absence or presence of surface appendages and the secretion of biosurfactants.

Immobilization under Shear

In the experimental set-up described here, immobilization of adhering cells can be observed in real time from the screen, a time-consuming procedure. Nevertheless, slow motion of adhering cells was occasionally seen on FEP. Despite the fact that slow motion should theoretically also occur under more favorable adhesion conditions such as on glass, we seldom observed slow motion on glass.

Concurrent with the above observations, it was found that the destruction of the spatial arrangement of adhering cells upon draining of the channel was much more severe for FEP than for glass. Although the percentage of adhering cells removed upon draining is difficult to determine reproducibly, we have the impression that more cells were removed from FEP than from glass (see also Fig. 7), most likely as a consequence of the unfavorable thermodynamic conditions prevailing for this strain on a hydrophobic substratum such as FEP (Table 5).

The percentage of cells removed upon draining is lower in the case of adhesion from a 10 mM buffer than in the case of adhesion from a 80 mM buffer, even though the electrostatic repulsion between the negatively charged cell and substratum surfaces is greater in 10 mM buffer (21). Therefore, it can be concluded that $-/-$ interactions are not determinant for the strength of adhesion. More likely, $+/-$ interactions between oppositely charged domains on the interacting surfaces will determine the strength of adhesion in addition to attractive Lifshitz-van der Waals forces. For FEP, this conclusion is also confirmed by a comparison of the desorption rate constants α, which are considerably larger at 80 mM than at 10 mM, indicating a less reversible adhesion in the case of a 10 mM buffer.

Analysis of the Spatial Arrangement of Adhering Cells

It should first be emphasized that analysis of the spatial arrangement of adhering cells is meaningful only after it has been established that the spatial arrangement to be analyzed is representative for the actual adhesion process, i.e., fully preserved and not influenced by, for example, passage of a sample through a liquid-air interface (Fig. 7).

Positive cooperativity in the adhesion of S. sanguis seems to exist not only on artificial pellicles (14, 26) but also on bare solid substrata, as demonstrated here. Positive cooperativity is, however, much more pronounced on FEP than on glass, as indicated by the large relative densities $g(r_p)$ at nearest-neighbor distances r_p (Table 3). Also on FEP, there is a preference for cells to deposit in the direction of flow (Table 4). This tendency is completely contradictory to theoretical calculations on inert particles (15), predicting that an adhering particle casts a shadow zone in the direction of flow in which no particles can adhere. Thus, positive cooperativity between *S. sanguis* cells appears more influential than electrostatic and hydrodynamic effects predicting a shadow zone (15).

We believe that positive cooperativity can be an expression of the influence of lateral interactions between adhering cells and the force required to stimulate the motion of adhering cells. Thus, the observations of incomplete immobilization and larger removal of cells in the case of FEP are in agreement with the larger positive cooperativity on this hydrophobic substratum.

Positive cooperativity is slightly more pronounced at low than at high ionic strength, in accordance with previous findings for *Streptococcus salivarius* HB (37). For this strain, we attributed the lack of cooperativity at high ionic strength to the collapse of surface fibrils, thought to be responsible for the lateral attraction between adhering cells. At present, however, we have insufficient evidence that this explanation will also hold for *S. sanguis* 12, although it is known that this strain is also fibrillated (54).

Positive cooperativity on FEP is also stronger at higher shear rates. On the one hand, this finding is surprising since the area blocked by an adhering cell will be larger at higher shear rates. On the other hand, however, higher shear rates may enhance lateral motion of adhering cells.

We have no full explanation for the mechanisms of positive cooperativity. With the experimental set-up described and by systematically analyzing the factors involved in the spatial arrangement of adhering cells, we believe that major progress will be made in understanding these phenomena.

CONCLUDING REMARKS

Many factors are involved in bacterial adhesion to solid surfaces, and it is doubtful that this phenomenon can ever be completely described by one theory, valid for various strains. This chapter has presented examples of the types of data that can be obtained by using flow cell systems and real-time image analysis techniques.

Substratum hydrophobicity appears, along with zeta potentials and prevailing hydrodynamic conditions, to be a determinant factor in the adhesion of *S. sanguis* 12, even though these parameters are measured as overall cell surface characteristics. However, care should be taken in extrapolating the results for this strain to other microorganisms.

Acknowledgment. We are greatly indebted to Marjon Schakenraad-Dolfing for preparation of the manuscript.

LITERATURE CITED

1. **Absolom, D. R., F. L. Lamberti, Z. Policova, W. Zingg, C. J. van Oss, and A. W. Neumann.** 1983. Surface thermodynamics of bacterial adhesion. *Appl. Environ. Microbiol.* **46:**90–97.
2. **Adamczyk, Z., and T. G. M. van de Ven.** 1981. Deposition of particles under external forces in laminar flow-through parallel-plate and cylindrical channels. *J. Colloid Interface Sci.* **80:**340–356.
3. **Balmer, R. T., and S. R. Hirsch.** 1978. The non-Newtonian behaviour of human saliva. *AICHE Symp. Ser. Biorheol.* no. 181. **74:**125–129.
4. **Bellon-Fontaine, M. N., N. Mozes, H. C. van der Mei, J. Sjollema, O. Cerf, P. G. Rouxhet, and H. J. Busscher.** 1990. A comparison of thermodynamic approaches to predict the adhesion of dairy microorganisms to solid substrata. *Cell Biophys.*, in press.
5. **Bowen, B. D., and M. Epstein.** 1979. Fine particle deposition in smooth parallel-plate channels. *J. Colloid Interface Sci.* **72:**81–97.
6. **Brenner, H.** 1961. The slow motion of a sphere through a viscous fluid towards a plane surface. *Chem. Eng. Sci.* **16:**242–251.
7. **Busscher, H. J., M. N. Bellon-Fontaine, N. Mozes, H. C. van der Mei, J. Sjollema, O. Cerf, and P. G. Rouxhet.** 1990. Deposition of *Leuconostoc mesenteroides* and *Streptococcus thermophilis* to solid substrata in a parallel plate flow cell. *Biofouling* **2:**55–63.
8. **Busscher, H. J., H. M. Uyen, A. W. J. van Pelt, A. H. Weerkamp, and J. Arends.** 1986. Kinetics of adhesion of the oral bacterium *Streptococcus sanguis* CH3 to polymers with different surface free energies. *Appl. Environ. Microbiol.* **51:**910–914.
9. **Busscher, H. J., H. M. Uyen, A. H. Weerkamp, W. J. Postma, and J. Arends.** 1986. Reversibility of adhesion of oral streptococci to solids. *FEMS Microbiol. Lett.* **35:**303–306.
10. **Busscher, H. J., and A. H. Weerkamp.** 1987. Specific and nonspecific interactions in bacterial adhesion to solid substrata. *FEMS Microbiol. Rev.* **46:**165–173.
11. **Busscher, H. J., A. H. Weerkamp, H. C. van der Mei, A. W. J. van Pelt, H. P. de Jong, and J. Arends.** 1984. Measurement of the surface free energy of bacterial cell surfaces and its relevance for adhesion. *Appl. Environ. Microbiol.* **48:**980–983.
12. **Christersson, C. E., R. G. Dunford, P. O. Glantz, and R. E. Baier.** 1989. Effect of critical surface tension on retention of oral microorganisms. *Scand. J. Dent. Res.* **97:**247–256.
13. **Christersson, C. E., P. O. Glantz, and R. E. Baier.** 1988. Role of temperature and shear forces on microbial detachment. *Scand. J. Dent. Res.* **96:**91–98.
14. **Cowan, M. M., K. G. Taylor, and R. J. Doyle.** 1986. Kinetic analysis of *Streptococcus sanguis* adhesion to artificial pellicle. *J. Dent. Res.* **65:**1278–1283.
15. **Dabros, T.** 1989. Interparticle hydrodynamic interactions in deposition processes. *Colloids Surfaces* **39:**127–141.
16. **Dawes, C., S. Watanabe, P. Biglow-Lecomte, and G. H. Dibdin.** 1989. Velocity of the salivary film at different locations in the mouth. *J. Dent. Res.* **68:**920.
17. **Dillon, J. K., J. A. Fuerst, A. C. Hayward, and G. H. G. Davis.** 1986. A comparison of five methods for assaying bacterial hydrophobicity. *J. Microbiol. Methods* **6:**13–19.

18. **Doyle, R. J., W. E. Nesbitt, and K. G. Taylor.** 1982. On the mechanism of adherence of *Streptococcus sanguis* to hydroxylapatite. *FEMS Microbiol. Lett.* **15**:1–5.

19. **Doyle, R. J., J. D. Oakley, K. R. Murphy, D. McAlister, and K. G. Taylor.** 1985. Graphical analyses of adherence data, p. 109–113. *In* S. E. Mergenhagen and B. Rosan (ed.), *Molecular Basis of Oral Microbial Adhesion.* American Society for Microbiology, Washington, D.C.

20. **Gibbons, R. J., and J. van Houte.** 1975. Bacterial adherence in oral microbial ecology. *Annu. Rev. Microbiol.* **29**:19–44.

21. **Hiemenz, P. C.** 1977. Electrophoresis and other electrokinetic phenomena, p. 452–487. *In* J. J. Lagowski (ed.), *Principles of Colloid and Surface Chemistry.* Marcel Dekker, Inc., New York.

22. **Hoffman, A. S.** 1986. Letter to the editor: A general classification scheme for "hydrophilic" and "hydrophobic" biomaterial surfaces. *J. Biomed. Mater. Res.* **20**:IX–XI.

23. **Hubbe, M. A.** 1985. Detachment of colloidal hydrous oxide spheres from flat solids exposed to flow. *Colloids Surfaces* **16**:249–270.

24. **Mozes, N., F. Marchal, J. C. van Haecht, L. Reuliaux, A. J. Léonard, and P. G. Rouxhet.** 1987. Immobilization of microorganisms by adhesion: interplay of electrostatic and non-electrostatic interactions. *Biotechnol. Bioeng.* **30**:439–450.

25. **Mozes, N., and P. G. Rouxhet.** 1987. Methods for measuring the hydrophobicity of microorganisms. *J. Microbiol. Methods* **6**:99–112.

26. **Nesbitt, W. E., R. J. Doyle, K. G. Taylor, R. H. Staat, and R. R. Arnold.** 1982. Positive cooperativity in the binding of *Streptococcus sanguis* to hydroxylapatite. *Infect. Immun.* **35**:157–165.

27. **Neumann, A. W., R. J. Good, C. J. Hope, and M. Sejpal.** 1974. An equation of state approach to determine surface tensions of low-energy solids from contact angles. *J. Colloid Interface Sci.* **49**:291–304.

28. **Owens, D. K., and R. C. Wendt.** 1969. Estimation of the surface free energy of solids. *J. Appl. Polymer Sci.* **13**:1741.

29. **Pratt-Terpstra, I. H., A. H. Weerkamp, and H. J. Busscher.** 1988. On a relation between interfacial free energy dependent and non-interfacial free energy dependent adherence of oral streptococci to solid substrata. *Curr. Microbiol.* **16**:311–313.

30. **Pratt-Terpstra, I. H., A. H. Weerkamp, and H. J. Busscher.** 1989. Microbial factors in a thermodynamic approach of oral streptococcal adhesion. *J. Colloid Interface Sci.* **129**:568–574.

31. **Rosenberg, M., and S. Kjelleberg.** 1986. Hydrophobic interactions: role in bacterial adhesion. *Adv. Microb. Ecol.* **9**:353–393.

32. **Rutter, P. R., and B. Vincent.** 1980. The adhesion of microorganisms to surfaces: physico-chemical aspects, p. 79–93. *In* R. C. W. Berkeley, J. M. Lynch, J. Melling, P. R. Rutter, and B. Vincent (ed.), *Microbial Adhesion to Surfaces.* Ellis Horwood Ltd., Chichester, England.

33. **Sjollema, J., and H. J. Busscher.** 1990. Deposition of polystyrene particles in a parallel plate flow cell. 2. Pair distribution functions between deposited particles. *Colloids Surfaces* **47**:337–354.

34. **Sjollema, J., H. J. Busscher, and A. H. Weerkamp.** 1988. Deposition of oral streptococci and polystyrene particles onto glass in a parallel plate flow cell. *Biofouling* **1**:101–112.

35. **Sjollema, J., H. J. Busscher, and A. H. Weerkamp.** 1989. Experimental approaches for studying adhesion of microorganisms to solid substrata: applications and mass transport. *J. Microbiol. Methods* **9**:79–90.

36. **Sjollema, J., H. J. Busscher, and A. H. Weerkamp.** 1989. Real-time image analyses of

adhering microorganisms in a parallel plate flow cell using automated image analysis. *J. Microbiol. Methods* **9**:73–78.

37. Sjollema, J., H. C. van der Mei, H. M. Uyen, and H. J. Busscher. 1990. Direct observation of cooperative effects in oral streptococcal adhesion to glass by analysis of the spatial arrangement of adhering bacteria. *FEMS Microbiol. Lett.* **69**:263–267.

38. Tanford, C. 1979. Interfacial free energy and the hydrophobic effect. *Proc. Natl. Acad. Sci. USA* **76**:4175–4176.

39. Uyen, H. M., H. J. Busscher, A. H. Weerkamp, and J. Arends. 1985. Surface free energies of oral streptococci and their adhesion to solids. *FEMS Microbiol. Lett.* **30**:103–106.

40. Uyen, H. M., H. C. van der Mei, A. H. Weerkamp, and H. J. Busscher. 1988. Comparison between the adhesion to solid substrata of *Streptococcus mitis* and that of polystyrene particles. *Appl. Environ. Microbiol.* **54**:837–838.

41. van der Mei, H. C., A. J. Léonard, A. H. Weerkamp, P. G. Rouxhet, and H. J. Busscher. 1988. Surface properties of *Streptococcus salivarius* HB and nonfibrillar mutants: measurement of zeta potential and elemental composition with X-ray photoelectron spectroscopy. *J. Bacteriol.* **170**:2462–2466.

42. van der Mei, H. C., A. J. Léonard, A. H. Weerkamp, P. G. Rouxhet, and H. J. Busscher. 1988. Properties of oral streptococci relevant for adherence: zeta potential, surface free energy and elemental composition. *Colloids Surfaces* **32**:297–305.

43. van der Mei, H. C., A. H. Weerkamp, and H. J. Busscher. 1987. A comparison of various methods to determine hydrophobic properties of streptococcal cell surfaces. *J. Microbiol. Methods* **6**:277–287.

44. van Loosdrecht, M. C. M., J. Lyklema, W. Norde, G. Schraa, and A. J. B. Zehnder. 1987. Electrophoretic mobility and hydrophobicity as a measure to predict the initial steps of bacterial adhesion. *Appl. Environ. Microbiol.* **53**:1898–1901.

45. van Loosdrecht, M. C. M., J. Lyklema, W. Norde, and A. J. B. Zehnder. 1989. Bacterial adhesion: a physico-chemical approach. *Microb. Ecol.* **17**:1–15.

46. van Oss, C. J., D. R. Absolom, and A. W. Neumann. 1980. Applications of net repulsive van der Waals forces between different particles, macromolecular or biological cells in liquids. *Colloids Surfaces* **1**:45–56.

47. van Oss, C. J., M. K. Chaudhury, and R. J. Good. 1988. Interfacial Lifshitz-van der Waals and polar interactions in macroscopic systems. *Chem. Rev.* **88**:927–941.

48. van Oss, C. J., and C. F. Gillman. 1972. Phagocytosis as a surface phenomenon. I. Contact angles and phagocytosis of non-opsonized bacteria. *RES J. Reticuloendothel. Soc.* **12**:283–292.

49. van Oss, C. J., and R. J. Good. 1984. The "equilibrium distance" between two bodies immersed in a liquid. *Colloids Surfaces* **8**:373–381.

50. van Oss, C. J., R. J. Good, and M. K. Chaudhury. 1986. The role of van der Waals forces and hydrogen bonds in "hydrophobic interactions" between biopolymers and low energy surfaces. *J. Colloid Interface Sci.* **111**:378–390.

51. van Pelt, A. W. J., H. P. de Jong, H. J. Busscher, and J. Arends. 1983. Dispersion and polar surface free energies of human enamel (an *in vitro* study). *J. Biomed. Mater. Res.* **17**:637–643.

52. Vissink, A., H. A. Waterman, E. J. 's-Gravenmade, A. K. Panders, and A. Vermey. 1984. Rheological properties of saliva substitutes containing mucin, carboxymethylcellulose or polyethyleneoxide. *J. Oral Pathol.* **13**:22–28.

53. Weerkamp, A. H., P. S. Handley, A. Baars, and J. W. Slot. 1986. Negative staining and immunoelectron microscopy of adhesion-deficient mutants of *Streptococcus salivarius*

reveal that the adhesive protein antigens are separate classes of cell surface fibrils. *J. Bacteriol.* **165:**746–755.

54. **Willcox, M. D. P., J. E. Wyatt, and P. S. Handley.** 1989. A comparison of the adhesive properties and surface ultrastructure of the fibrillar *Streptococcus sanguis* 12 and an adhesion deficient non-fibrillar mutant 123 na. *J. Appl. Bacteriol.* **66:**291–299.

55. **Wyatt, J. E., L. M. Hesketh, and P. S. Handley.** 1987. Lack of correlation between fibrils, hydrophobicity and adhesion for strains of *Streptococcus sanguis* biotypes I and II. *Microbios* **50:**7–15.

Microbial Cell Surface Hydrophobicity
Edited by R. J. Doyle and M. Rosenberg
© 1990 American Society for Microbiology, Washington, DC 20005

Chapter 13

Hydrophobicity of Group A Streptococci and Its Relationship to Adhesion of Streptococci to Host Cells

Harry S. Courtney, David L. Hasty, and Itzhak Ofek

In the last decade, our perceptions of how the surface structures of streptococci are organized and how this organization is related to function have been dramatically altered. In the 1970s, many textbooks presented the rather simple view of bacterial surfaces as consisting of a layer of hyaluronic acid, followed by a layer of M protein, cell wall, and cell membrane. Today's models, though not complete, provide richer details of organization. These new models have come from ultrastructural analyses and from studies on the physical attributes of bacterial surfaces. For group A streptococci, the finding that the bacterial surface is hydrophobic led to new concepts of how some of its surface structures may be complexed to each other and to animal cells.

The impetus for studies on the surface properties of bacteria comes from a concept that is intuitively obvious: that the interaction of bacteria with host cells or other substrata is related to the molecular composition of the bacterial surface. The bacterial surface can be considered to be a mosaic of molecules possessing different physical attributes such as being positively or negatively charged, hydrophilic, hydrophobic, or amphipathic. It is the balance of these attributes that confers upon the bacterium its net surface charge or renders the organism hydrophobic.

Harry S. Courtney • Veterans Administration Medical Center and Department of Medicine, University of Tennessee, Memphis, Tennessee 38163. **David L. Hasty** • Veterans Administration Medical Center and Department of Anatomy, University of Tennessee, Memphis, Tennessee 38163. **Itzhak Ofek** • Department of Human Microbiology, Sackler School of Medicine, Tel-Aviv, Israel.

The hydrophobicity of an organism may depend on the surface exposure of one or more lipophilic components. To reflect this concept, the term "hydrophobin" was introduced by Rosenberg and Kjelleberg (49) and defined as a molecule that is exposed on the surface of a bacterium and contributes directly to the hydrophobicity of the organism. In this chapter, we discuss surface components of *Streptococcus pyogenes* that are potential hydrophobins and provide evidence that one of these components, lipoteichoic acid (LTA), serves as the major hydrophobin and also serves as an adhesin.

SURFACE COMPONENTS OF GROUP A STREPTOCOCCI

C Carbohydrate

The C carbohydrate is the group-specific antigen of *S. pyogenes*; it is composed of branched polymers of L-rhamnose and *N*-acetylglucos-amine. The latter sugar contains the antigenic epitope. The C carbohydrate has no known relationship with hydrophobicity.

M Protein

M protein is an alpha-helical coiled-coil dimer that can be seen as fibrillae which radiate outwardly from the cell wall (19). It is the type-specific antigen and exists in various sizes. M proteins are also highly variable in their N-terminal peptide, and antibodies against this region can provide immunity to streptococcal infections (30). M protein is the major virulence factor, since it confers resistance to phagocytosis in nonimmune hosts (19). M protein has been suggested to contribute to the hydrophobicity of streptococci, as will be discussed in more detail below.

LTA

LTA of group A streptococci is a highly negatively charged am-phiphile that is composed of a polyglycerolphosphate (PGP) covalently linked to a diglucosyl lipid moiety. The PGP backbone may be substituted to various degrees with D-alanine and exists in the fully acylated form as well as the deacylated form. It is found on the cell surface and is released into the growth medium (71). LTA can form complexes with M protein through ionic interactions between the negative charges of LTA and positive charges of M protein (Fig. 1), and it has been suggested that LTA complexed in this manner would have its lipid moiety in an orientation that permits it to interact with the external milieu (40). The evidence indicating that LTA is the major component responsible for the hydro-phobicity of streptococci (33) is discussed below.

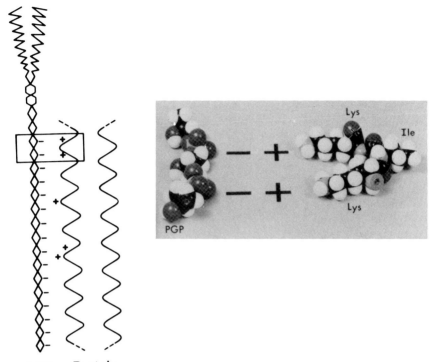

LTA   Protein

Figure 1. Proposed alignment of the PGP backbone of LTA with the alpha-helical coiled-coil region of type 24 M protein-containing clusters of positively charged amino acid residues. A space-filling model of the boxed region is shown on the right. It is proposed that the interactions between the positively charged lysine residues and the negatively charged glycerolphosphate residues can anchor LTA to the streptococcal surface. (From reference 40.)

Hyaluronic Acid Capsule

The hyaluronic acid capsule is composed of repeating units of the disaccharide glucuronic-β-1,3 N-acetylglucosamine. The presence of the capsule is related to virulence; unencapsulated organisms are more susceptible to phagocytosis. Data will be presented to support the concept that expression of the capsule is inversely related to the hydrophobicity of group A streptococci.

R Proteins

R proteins are surface proteins with molecular masses of 78 to 100 kilodaltons (kDa) found in group A, B, C, F, G, and L streptococci (24,

27, 72, 73). Antisera to R proteins are used to subclassify serotypes of group A streptococci. R proteins can be removed from the streptococcal surface by trypsin. The released protein is no longer sensitive to trypsin but is sensitive to pepsin. Certain antisera to R proteins were found to react with M protein. However, these antisera were not opsonic, and the presence of R protein does not confer resistance to phagocytosis. Since R and M proteins have physicochemical similarities, it was hypothesized that R proteins are defective forms of M protein in which the antiopsonic determinant was deleted (24). The biological role of R proteins and their relationship to hydrophobicity is unknown.

T Proteins

T proteins are located on the surface of many group A streptococci, and antisera to T proteins provide an important scheme for subclassification of the bacteria. The T protein is removed from whole cells by trypsin as a 160- to 200-kDa protein that is resistant to further cleavage by trypsin. It appears that T proteins can form complexes with LTA mediated by ionic interactions and by weak hydrophobic interactions (25). It has been suggested that T proteins bind to and mask the fatty acid moieties of LTA during transit, but because of the weakness of this interaction, the fatty acid may free itself at the surface. Because LTA also seems to undergo ionic interactions with T proteins, these antigens may also anchor LTA in a manner similar to that of M protein. The role of T proteins in adhesion of streptococci to host cells and to hydrocarbons has not been fully investigated.

TECHNIQUES FOR MEASURING SURFACE HYDROPHOBICITY OF STREPTOCOCCI

The methodologies for measuring hydrophobicity of cell surfaces are discussed in detail by Rosenberg and Doyle (this volume) and therefore will be mentioned here only as they pertain to group A streptococci. Five methods have been used to determine the surface hydrophobicity of group A streptococci: hydrophobic interaction chromatography (66), salt agglutination tests (28, 68), polymer two-phase partition tests (31, 34), enhanced electrophoretic mobility in sodium dodecyl sulfate solutions (22, 23), and interaction with hydrocarbons (12, 41, 48, 50; Fig. 2). Two of these techniques deserve further mention because in some cases the data may lead to false assumptions. Results derived from hydrophobic interaction chromatography must be interpreted with caution because in some cases group A streptococci may bind to the column matrix by nonhydro-

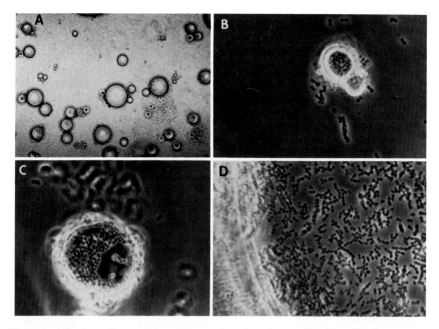

Figure 2. Adhesion of group A streptococci to hexadecane droplets. (A) Low-power magnification of streptococci attached to hexadecane droplets. (B to D) Higher magnifications of this attachment.

phobic interactions. For example, Wadström and Tylewska (69) found that M^+ streptococci bound to Sepharose beads and that the bound bacteria could be eluted with galactose. Data derived from the salt aggregation test may also be difficult to interpret because some strains of streptococci autoaggregate.

VARIATION IN THE HYDROPHOBICITY OF GROUP A STREPTOCOCCI

From perusal of the literature, it is clear that different strains of streptococci vary widely in hydrophobicity (31–34, 66–68). This section discusses some of the possible causes for phenotypic variations that may affect surface hydrophobicity.

Variations Due to Growth Conditions

The composition of the growth medium can have pronounced effects on the hydrophobicity of bacteria. Hill and co-workers (23) found that the addition of glycerol, acetate, or oleate to the growth media resulted in an increase in the lipid content of the streptococci, with a concomitant increase in hydrophobicity as measured by enhanced electrophoretic mobility of the bacteria in solutions of sodium dodecyl sulfate. Glycerol and acetate were presumed to increase lipid synthesis, since other parameters tested remained unchanged. Oleic acid (used at sublethal concentrations) was found to inhibit the lipase activity in the growth supernatant. It was suggested that a decrease in lipase activity could account for the increase in surface lipids and the resultant increase in hydrophobicity. The effect of these compounds on the expression of LTA or other surface components was not tested.

It is likely that addition of other substances to the growth media, such as serum containing substances known to bind to streptococci, will have dramatic effects on the hydrophobicity of bacteria. Although not tested as a growth supplement, human serum, when added to washed streptococci, altered their hydrophobic properties (32, 68). Other parameters, such as pH or temperature, may also affect hydrophobicity. For example, Stamp (60) found that the temperature, pH, and oxidation-reduction potential of media affected the expression and activity of the streptococcal protease. Because other proteases affect surface hydrophobicity, it is likely that the streptococcal protease may also affect these surface properties. The growth temperature also affected expression of the streptococcal protease.

Variations Due to Enzyme Expression

As described above, an inhibitor of the enzyme lipase caused an increase in surface lipids and hydrophobicity. Hill et al. (23) found that variations in the amount of lipase expressed can also affect hydrophobicity. Two groups of *S. pyogenes* that exhibited unusually high surface hydrophobicity were investigated. One group was tetracyline resistant, had a high rate of lipid synthesis, and had a normal amount of lipase activity. The other group, isolated from patients with impetigo, was tetracyline susceptible and produced no extracellular lipase. Thus, surface hydrophobicity can be modulated by regulation of expression of lipase and enzymes in the lipid pathway.

Another enzyme that can have an effect on streptococcal hydrophobicity is hyaluronidase. The hyaluronic acid capsule of streptococci accumulates on the bacterial surface during exponential growth and is subsequently lost as the organisms enter stationary phase. This loss of capsule is thought to be due to the accumulation of hyaluronidase in the growth medium as the organisms enter stationary phase so that the balance between degradation and synthesis is altered in favor of degradation (22). Ofek et al. (41) found that encapsulated, M^+ streptococci from the exponential phase of growth did not adhere to hexadecane, whereas those from the stationary phase did adhere to the hydrocarbon. Treatment of the encapsulated organisms with hyaluronidase converted these nonadhering bacteria to adhering bacteria. Hill et al. (23) also found that the electrophoretic mobility of streptococci was related to the amount of capsular material on the streptococcal surface. Thus, it appears that the degree of encapsulation is inversely proportional to the relative hydrophobicity of group A streptococci.

As mentioned above, proteases from group A streptococci may also affect hydrophobicity. The streptococcal protease is found in growth supernatants; it can cleave M protein and may affect other surface structures as well (17). Because other proteases that cleave M protein are known to alter streptococcal hydrophobicity, it is possible that this extracellularly secreted protease may also alter hydrophobicity.

These studies provide evidence that the enzymatic activity expressed by streptococci can vary from one strain to another and from one growth phase to another in the same strain and that these variations have a direct effect on the surface hydrophobicity of streptococci.

Variation in M Protein

Several investigators have suggested that M protein contributes to the hydrophobicity of group A streptococci (26, 66, 68). This idea is based on the observation that M^+ streptococci generally are more hydrophobic than M^- streptococci. However, there are M^+ strains of streptococci that are not hydrophobic, and there are M^- strains that are hydrophobic (68). Thus, the expression of M protein on the surface does not always coincide with the degree of hydrophobicity.

Attempts to correlate expression of M protein with hydrophobicity may be hampered by the criteria used to determine whether a strain is M^+ or M^-. At present, a strain is considered to be M^+ if it grows in whole blood from a nonimmune donor and reacts with absorbed typing sera. However, over half of the isolated strains of S. pyogenes are nontypable (46), and M protein can vary by as much as 40 kDa in size even among strains of the same serotype (19). Therefore, it is possible that some of these M^- strains of streptococci produce an altered version of M protein that has lost its function, i.e., resistance to phagocytosis. Scott et al. (53) screened 50 strains of group A streptococci with a probe from the conserved C terminus of the M protein gene and found that all 50 strains reacted with this probe, including those that were functionally M^-. It appears that these M⁻ strains have the M protein gene, but it was not determined whether these strains failed to express M protein or whether an altered version (nontypable?) of M protein was expressed. It is tempting to speculate that some of these of M⁻ strains are actually M⁺ strains in which the mutated M protein no longer confers resistance to phagocytosis but retains other functions of M protein, such as its ability to form complexes with LTA (see above).

As regards conversion of streptococcal strains from the M⁺ to M⁻ phenotype, Elliot (17) observed that of the 58 nontypable strains, 56 produced the streptococcal protease, whereas all 20 of the strains that were typable did not produce this extracellular protease. Because this protease cleaves M protein, it was suggested that the protease removes portions of the M protein and renders the organisms nontypable.

The amount of M protein on the surface can vary. Mouse passage of a strain of streptococci can increase the virulence of the organisms, and this effect is correlated with an increased production of M protein. Laboratory passage of strains in Todd-Hewitt broth usually leads to a decrease in the expression of M protein. The elaboration of the streptococcal protease may have a bearing on this observation, since growth of streptococci in Todd-Hewitt broth enhances expression of the protease, whereas growth in vivo leads to lack of expression of this enzyme (17).

Does the amount of M protein on the surface correlate with the degree of hydrophobicity of streptococci? Only one study has attempted to answer this question. Wadström et al. (68) measured the hydrophobicity of 20 strains of streptococci that were M⁺ by the salt aggregation test. The amount of M protein on these organisms was estimated by their ability to escape phagocytosis and multiply in whole, nonimmune human blood. In general, strains that grew well were also hydrophobic, but there were strains that grew well and were less hydrophobic. Thus, at present there is no clear correlation between the expression of M protein and the degree of hydrophobicity of streptococci.

Variation in LTA

LTA is expressed by all group A streptococci. No LTA-negative mutant has been found or created. It therefore seems that a mutation in LTA is lethal. The LTA expressed by streptococci is not homogeneous; it varies in the length of the PGP backbone, in the number of fatty acids bound to PGP, and in the number of alanine substitutions on the PGP backbone. These variations could have profound effects on the biologic activity of LTA. Nealon and Mattingly (37) found that virulent group B streptococci produced LTA with 30 to 35 glycerolphosphate units, whereas avirulent strains produced LTA with 10 to 12 glycerolphosphate units. It is possible that on streptococci which produce LTA with longer chains of glycerolphosphate, the lipid moiety of LTA is extended farther from the surface of the bacteria, thereby enhancing the ability of the organisms to interact with hydrophobic receptors.

The lipids of LTA also vary in the type of fatty acids that are substituted. Ofek et al. (39) found that LTA is composed mostly of fatty acids with the following composition: 16:0 (48%), 16:1 (10%), 18:0 (15%), 18:1 (16%), and 18:2 (8%). It is interesting to note that both group A streptococci and LTA bind preferentially to polyethylene glycol that is substituted with fatty acids containing 14 or more hydrocarbons (33). Since the fatty acids of LTA contain mostly fatty acids with 16 carbons, this finding provides supporting evidence that LTA is a hydrophobin.

The amount of LTA produced by different strains of streptococci also varies, and this variation correlates directly with the degree of hydrophobicity. Miörner et al. (33) quantitated the relative amounts of LTA by two methods. In the first method, LTA was extracted from the organisms with phenol, and the amount of LTA released was determined by rocket immunoelectrophoresis. In the second technique, streptococcal cells were reacted with anti-LTA serum, and the amount of bound antibodies was quantitated with radiolabeled protein A. In each case, there was a clear correlation between the amount of LTA measured and the degree of hydrophobicity.

The limitations of these assays are that (i) phenol will extract LTA that is both membrane associated and complexed with surface components and (ii) antibodies against LTA will detect both acylated LTA and deacylated LTA. It is presumed that LTA that has its lipid anchored in the membrane and that deacylated LTA cannot contribute to the surface hydrophobicity of streptococci. Nevertheless, LTA is the only surface molecule studied in which the degree of expression is clearly related to hydrophobicity.

The variations in surface hydrophobicity of group A streptococci seem to be related to expression of the hydrophobin itself, to its carrier molecule, or to its accessibility on the surface. In the first case, variation in the expression of LTA has been linked to hydrophobicity. In the second case, removal of the carrier molecule by proteases alters hydrophobicity. In the third case, the hyaluronic acid capsule can impede hydrophobic interactions.

PROBES FOR HYDROPHOBINS OF STREPTOCOCCI

The quantitation of hydrophobicity of streptococci involves interactions between the hydrophobin and lipophilic surfaces; therefore, agents and procedures that block these interactions can help to identify the chemical nature of the hydrophobin. The use of inhibitors as probes for determining the specificity of a reaction is a widely used technique. In many cases, the isolated ligand, receptor, or analogs thereof have been used. For example, sugar analogs of receptors have helped to determine the specificity of attachment of *Escherichia coli* and pneumococci to host cells (4, 18). Enzymatic, chemical, and physical treatments that are selective in their action can furnish corroborating evidence that a particular substance is a hydrophobin. These approaches have been applied to determine which surface component of group A streptococci contributes to the hydrophobicity of the organism.

Enzymatic Treatments

Enzymatic treatment of streptococci with pepsin at pH 5.8 selectively removes the amino-terminal half of M protein (8). When streptococci were treated with pepsin at this suboptimal pH, the organisms bound less fibrinogen, but adhesion to hexadecane and fibronectin-binding activity remained unchanged (12). Since fibrinogen binds to pepsin-extracted M protein, the data suggest that M protein, or at least its amino-terminal half, is not involved in adhesion of streptococci to hexadecane. On the other hand, trypsinization, which removes T proteins, R proteins, M protein, and LTA, reduced the hydrophobicity of streptococci.

As regards the hydrophobic nature of M protein, Kuhnemund et al. (26) found that M protein in bacteriophage-associated lysin extracts of group A streptococci was not very hydrophobic. M protein extracted under these conditions lacks the membrane anchor region (42). It would be anticipated that recombinant M protein that contains an intact membrane anchor would be much more hydrophobic in solution. In situ, however, this region would be buried in the membrane and not accessible for hydrophobic interactions at the streptococcal surface.

Streptococcal Binding Proteins

Among the various proteins tested, the most effective inhibitors were albumin and fibronectin. Since both albumin and fibronectin bind to LTA and this binding is mediated by the lipid moiety of LTA (13, 58), the data suggest that LTA is the major hydrophobin. Fibronectin was found to be a more potent inhibitor of adhesion to hexadecane than was albumin, in agreement with the finding that fibronectin has a higher affinity for LTA than does albumin (13).

Wheat germ agglutinin binds to the *N*-acetylglucosamine residues of the C carbohydrate of group A streptococci (44). The failure of wheat germ agglutinin to inhibit adhesion of streptococci to hexadecane suggests that the C carbohydrate is not involved in the hydrophobic interactions of streptococci.

Fibrinogen also inhibits hydrophobic interactions of group A streptococci (68). Fibrinogen is known to have an affinity for purified M proteins, but fibrinogen binds to both M^+ and M^- streptococci (9, 51) and can also bind to the lipid moiety of LTA (H. S. Courtney and E. Whitnack, unpublished results) and to T proteins (51, 52). It is also possible that fibrinogen, by binding to the streptococci, converts its surface from hydrophobic to hydrophilic. It is possible that fibrinogen blocks hydrophobic interactions of streptococci by both specific and nonspecific mechanisms. The specific mechanism would involve the binding of fibrinogen to the hydrophobin, thereby blocking the interaction of the hydrophobin with hydrophobic surfaces. Nonspecific blocking would involve the molecule's binding to a surface component other than the hydrophobin and sterically hindering the interaction between the hydrophobin and other surfaces.

Antibiotic Treatments

Antibiotics have also been used to investigate the surface properties of group A streptococci. Alkan and Beachey (3) were the first to show that penicillin can induce the release of LTA from streptococci. Tylewska et al. (64, 67) subsequently reported that penicillin and rifampin reduced the adhesion of M^+ and M^- streptococci to phenyl-Sepharose. Berberine sulfate was also found to induce the release of LTA and to reduce adhesion to hexadecane (62).

Other evidence in support of LTA as the major hydrophobin comes from studies in which fibronectin bound to streptococci was released when the organisms were treated with penicillin (36). Furthermore, the released fibronectin was precipitated with antibodies to LTA but not with antibodies to M protein or with antibody preparations to whole streptococcal cells that had been absorbed with LTA-sensitized erythrocytes. Because fibronectin inhibits adhesion of streptococci to hexadecane, these data suggest that fibronectin binds to the lipid moiety of LTA and inhibits the attachment of streptococci to hexadecane.

Chemical and Physical Treatments

Of all the treatments that have been used to garner information about the hydrophobins of streptococci, the chemical and physical treatments are the least selective in their actions. Nevertheless, these treatments do

Table 1. Effects of Various Agents on the Hydrophobicity of Group A
Streptococci and on Their Adhesion to Host Cells

Agent	Hydro-phobicity[a]	Adhesion to cells[a]	Reference(s)
Enzymes			
Trypsin	↓	↓	7, 12, 15, 33
Hyaluronidase[b]	↑	↑	41
Lipase	↓	↓	7, 16
Pepsin			
pH 4.5	↓	↓	7, 12, 34
pH 5.8	↔	↔	7, 12, 34
Streptococcal binding substances			
Albumin	↓	↓	34, 41, 68
Fibronectin	↓	↓	1, 12, 54
Fibrinogen	↓	↓	34, 68
Emulsan[c]	↓	↓	47
Immunoglobulin G	↓	NT[d]	34, 68
Wheat germ agglutinin	↔	NT	12
Physical agents			
Hot acid	↓	↓	7, 12, 41
Heat, 100°C	↔, ↓	↔, ↓	12, 16
Growth supplements			
Acetate, glycerol, oleate	↑	NT	23
Antibiotics, sublethal concn			
Penicillin	↓	↓	3, 64, 67
Rifampin	↓	↓	64, 67
Tetracycline	↓	↓	64, 67
Berberine	↓	↓	62

[a] Arrows indicate effects on the tests. ↑, Increased binding; ↓, decreased binding; ↔, no change.
[b] The effect of hyaluronidase is seen only on streptococci from the exponential phase of growth.
[c] Emulsan has not been shown to bind directly to streptococci.
[d] NT, Not tested.

provide information that can corroborate data generated from other approaches. Hot acid treatment of streptococci abolished their ability to bind to hydrophobic surfaces (7, 12, 41). This treatment removed many substances, including LTA and M protein, indicating that the hydrophobins are sensitive to this treatment.

The results from the use of heat treatment do not lead to a clear interpretation. Heat treatment in some cases resulted in no change in streptococcal hydrophobicity, whereas in other cases heat treatment caused a reduction in hydrophobicity (12, 33). Data gathered from heat treatment of streptococci should be interpreted with caution because heat treatment may expose sites that are cryptic in the native bacteria or may destroy native receptors. Heat treatment of streptococci did not alter the quantity of fibronectin bound, but fibronectin binding was no longer sensitive to inhibition with LTA, suggesting that heat-treated bacteria may express a different receptor for fibronectin (57). Heat treatment also abolished the ability of streptococci to be agglutinated by fibrinogen (63).

A common thread in all of the experiments with enzymes, streptococcal binding proteins, and antibiotic treatments is that in each case in which there was a reduction in the hydrophobicity, there was a parallel reduction in the exposure of LTA.

HYDROPHOBINS AS ADHESINS

Relationship between Hydrophobicity of Streptococci and Attachment to Host Cells

Several studies have found a relationship between the hydrophobicity of bacteria and their attachment to host cells (2, 49, 62, 64, 67). The adhesion of group A streptococci to host cells also appears to be related to hydrophobicity (Table 1). Using hydrophobic interaction chromatography, Tylewska et al. (64) reported that the antibiotics penicillin and rifampin reduced the attachment of streptococci to phenyl-Sepharose and to human pharyngeal cells. Penicillin induced the release of LTA from streptococci, and this loss of LTA was correlated with reduced attachment to host cells (3). Berberine sulfate also caused an eightfold increase in the release of LTA from streptococci and reduced the attachment of streptococci to hexadecane, fibronectin, and mucosal cells (62).

Enzymatic treatments that alter the hydrophobicity of streptococci also alter attachment of the bacteria to host cells. Trypsinization of streptococci reduced adhesion to both hexadecane and mucosal cells, whereas pepsin at pH 5.8 had little or no effect (7, 12, 15, 16, 33). Lipase decreases the hydrophobicity of streptococci (23) and diminishes the capacity of the bacteria to adhere to epithelial cells (16). Streptococci in the exponential phase of growth adhere poorly to hexadecane and to host cells. Treatment of these cells with hyaluronidase increased their surface hydrophobicity and their adhesion to host cells (41).

Substances that bind to streptococci and block adhesion to hexadecane also block adhesion to host cells. Albumin and fibronectin, both of which contain fatty acid-binding sites, inhibited attachment of streptococci to hexadecane and to host cells (12, 41). Emulsan, a polyanionic heteropolysaccharide with covalently linked fatty acids, prevented the adhesion of streptococci to octane and to epithelial cells (47).

Phospholipids and detergents inhibited the adhesion of group A streptococci to host cells (16), suggesting that hydrophobic interactions

Table 2. Summary of Evidence that LTA Is a Streptococcal Hydrophobin

LTA is hydrophobic; i.e., it contains fatty acids.

LTA is in the right location; it is found on fibrillae protruding from the streptococcal surface.

Both LTA and streptococci bind optimally to hydrocarbons with 14 or more carbons.

Sublethal concentrations of antibiotics that induce release of LTA cause a reduction in streptococcal hydrophobicity.

Proteins that bind to fatty acids of LTA also block streptococcal hydrophobic interactions.

Enzymatic treatments that release LTA cause a reduction in the surface hydrophobicity of streptococci.

The degree of hydrophobicity of streptococci is directly related to the amount of LTA expressed on the surface.

are involved in adhesion to host cells. As one would suspect, detergents also inhibited hydrophobic interactions of streptococci (68).

In each of the aforementioned cases, hydrophobicity was found to be related to adhesion to host cells; treatments that increased streptococcal hydrophobicity caused increased adhesion to host cells, and treatments that decreased streptococcal hydrophobicity caused a decrease in adhesion.

The evidence that LTA is the major hydrophobin is compelling (Tables 1 and 2). In each case in which hydrophobicity was altered, there was concomitant alteration in the exposure of LTA. The data are equally

Table 3. Relationship btween LTA Receptors and Streptococcal Adhesion

Buccal epithelial cells[a]	No. of LTA receptors/cell[b] (10^9)	% of cells with LTA receptors[c]	% Adhesion[d]
<1 day	2.1	17	22
1 day	2.5	29	36
2 days	3.3	53	46
3 days	5.2		
Adults	5.0		60

[a] Obtained from newborn infants and their mothers.
[b] Extrapolated from data of Simpson et al. (59).
[c] Data from Ofek et al. (38). Values for cells with LTA receptors were the percentages of cells demonstrating binding of LTA by immunofluorescence.
[d] Combined data from Ofek et al. (38) and Simpson et al. (59).

compelling for the correlation between hydrophobicity of streptococci and adhesion to host cells (Table 1). Evidence suggesting that LTA is also the major adhesin is presented below.

Role of LTA in Adhesion

The possible roles of various surface components of streptococci in attachment to host cells were investigated by purifying these components and assaying for their ability to block attachment. Of the substances tested (LTA, pepsin-extracted M protein, C carbohydrate, and peptidoglycan), only LTA inhibited attachment (6).

It was subsequently found that mucosal cells pretreated with LTA had fewer attached streptococci and that these mucosal cells had specific receptors for LTA (38, 59). The expression of receptors for LTA was age related; epithelial cells from infants less than 3 days old had fewer receptors for LTA, and streptococci attached in fewer numbers to these cells (Table 3). By day 3, the number of LTA receptors and the number of attached streptococci were equivalent to those of adult mucosal cells. The attachment of streptococci to neutrophils was also related to the number of available LTA receptors (10). The observation that not all of the oral epithelial cells from infants or their mothers exhibited a capacity to bind LTA deserves further mention. Only 17% of the cells from infants less than 1 day old were able to bind LTA. This finding suggests that LTA binds to a specific receptor(s) and does not simply intercalate its lipid moiety into the membrane. Furthermore, the percent adhesion of streptococci to these cells correlated with the percentage of cells with LTA receptors; as the percentage of oral cells with positive binding for LTA increased, the attachment of streptococci to these cells also increased.

Supporting evidence for LTA as the adhesin came from studies using penicillin. Alkan and Beachey (3) found that penicillin induced the release of LTA from streptococci and that these streptococci exhibited a diminished capacity to attach to host cells. Berberine sulfate also induced the release of LTA, resulting in reduced attachment of bacteria to host cells (62).

The nature of the binding of LTA to host cells was investigated by chemical deacylation of LTA. Treatment of LTA with ammonium hydroxide separates LTA into two parts: the lipid moiety, which is soluble in chloroform-methanol solutions, and the PGP moiety, which is soluble in aqueous solutions. Testing of these fractions in adhesion assays revealed that only the lipid fraction contained inhibitory activity. Esterification of PGP with palmitic acid restored inhibitory and binding activity, confirming that the lipid moiety was the biologically active part of LTA (39).

The data that we have presented strongly suggest that LTA is the major adhesin in the attachment process. However, at that time it was thought that the lipid moiety of LTA was always intercalated in the phospholipid membrane, with its PGP backbone protruding through the cell wall. If the lipid of LTA is buried in the membrane, how can it interact with receptors on host cells? It was known that LTA is released into the medium during growth and therefore is present, at least transiently, on the surface during its passage through the surface structures of streptococci and into the medium. This knowledge prompted investigations to determine whether any of the surface components could serve as an anchor for LTA during its transit. It was found that M protein and LTA could form complexes as a result of ionic interactions between the negatively charged backbone of LTA and positively charged residues of M protein (Fig. 1). Such an interaction could anchor LTA in an orientation that would free its lipid moiety and allow it to interact with the extracellular milieu. M proteins form fibrillae that project outwardly from the cell wall and are at an optimal location to immobilize LTA in order for its lipid moiety to interact with host cell receptors. It should be stressed that other surface structures may also serve as anchors for LTA. Thus, it appears that M protein may play a supportive role in the hydrophobic interactions of streptococci and in their adhesion to host cells.

Role of M Protein in Adhesion

M protein was first suspected to be an adhesin because M^+ strains of streptococci adhered in higher numbers to epithelial cells than did M^- strains (15, 16). Others have found no difference in the attachment of M^+ and M^- strains (3a). Because strains vary markedly in the amount of LTA expressed and probably also vary in the amount of M protein expressed, it is difficult to draw firm conclusions from these studies.

Data suggesting that M protein is not involved in adhesion came from studies by Beachey and Ofek (8). Pepsin treatment of streptococci at pH 5.8, which selectively removes the amino-terminal half of M protein, did not alter their ability to adhere to epithelial cells but did alter their ability to resist phagocytosis. Furthermore, the purified pepsin-extracted M protein did not inhibit adhesion, nor did it bind to epithelial cells (6, 7, 39).

More recently, pretreatment of epithelial cells with M protein was found to inhibit adhesion of streptococci to host mucosal cells (65). However, recombinant M protein from *E. coli* was used in these experiments, and it is possible that this recombinant molecule contains the membrane anchor region that is very hydrophobic and would be likely to inhibit hydrophobic interactions. In situ, this region would be buried in the membrane and would not be available for interactions with the host cells. Thus, if M protein is involved in adhesion, it would seem that the region involved would have to be from the pepsin cleavage site to the C terminus, since pepsin-extracted M protein neither binds to epithelial cells nor inhibits adhesion.

Role of Fibronectin in Adhesion

Fibronectins are a family of large glycoproteins (~440 kDa) that are found in body fluids, in extracellular matrix, in basement membranes, and on cell surfaces (35). The presence of fibronectin on mucosal surfaces and its ability to interact with bacteria led to investigations of its role in streptococcal adhesion (20).

As a first step, the binding of fibronectin to streptococci was investigated, and it was found to bind to various strains of streptococci (54). In adhesion assays, as little as 1 μg of fibronectin blocked attachment of streptococci to host cells (54). Streptococci were found to attach primarily to cells that contained fibronectin (1), and the level of attachment was directly related to the amount of fibronectin on host cells. For example, treatment of epithelial cells with mercaptoethanol or by heating at 90°C for 1 min reduced cell-associated fibronectin by 50% and caused a 50% reduction in attachment of streptococci. Conversely, pretreatment of epithelial cells with exogenous fibronectin resulted in a 50% increase in cell-associated fibronectin and a 50% increase in the adhesion of streptococci. Moreover, mouse and rabbit antibodies to fibronectin reduced adhesion to oral epithelial cells by 58 and 76%, respectively (11).

LTA-Fibronectin Interactions

The accumulated evidence suggests that LTA is a major adhesin and that fibronectin is a major receptor (Table 4). The missing link has been determination of whether LTA interacts directly with fibronectin. LTA was found to competitively inhibit the binding of fibronectin to streptococci, suggesting that LTA is the ligand on streptococcal surfaces capable of binding fibronectin (12). Direct binding of LTA to fibronectin was demonstrated by the enhanced electrophoretic mobility of fibronectin when bound to LTA (13). This binding was dependent on the lipid moiety of LTA, since deacylated LTA had no effect on the mobility of fibronectin. This finding was confirmed by the binding of acylated but not deacylated LTA to fibronectin-Sepharose. In addition, palmitic acid bound to fibronectin, a further indication that fibronectin has fatty acid-binding sites (13). Furthermore, the binding of a monoclonal antibody to fibronectin was inhibited by LTA and by palmitic acid, suggesting that LTA and palmitic acid bind to the same site on fibronectin (61).

Table 4. Summary of Evidence that LTA and Fibronectin Mediate Adhesion
of Streptococci to Epithelial Cells

LTA	Fibronectin
LTA binds to specific receptors on host cells.	Fibronectin is present on host cell surfaces.
LTA inhibits streptococcal adhesion.	Fibronectin inhibits adhesion.
The degree of streptococcal adhesion is related to the amount of LTA receptors.	The degree of streptococcal adhesion is related to the amount of fibronectin expressed on cells.
Anti-LTA blocks adhesion.	Antifibronectin blocks adhesion.
Antibiotics that induce release of LTA cause a reduction in adhesion.	Antibiotics that induce release of LTA cause a reduction in fibronectin binding to streptococci.
LTA competitively inhibits binding of fibronectin to streptococci.	Fibronectin binds to LTA.
Deacylated LTA does not bind to or inhibit adhesion to epithelial cells.	Deacylated LTA does not bind to fibronectin.
Proteins that have fatty acid-binding sites block adhesion.	Fibronectin has fatty acid-binding sites.

That fibronectin binds to LTA in situ is suggested by experiments in which penicillin induced the release of LTA-fibronectin complexes. Fibronectin bound to streptococci was released by penicillin, and the LTA-fibronectin complexes were precipitated by antibodies to LTA but not by antibodies to other surface components of group A streptococci (36).

The LTA-binding domain of fibronectin was localized to a 24-kDa peptide from the N terminus of fibronectin (14). Adsorption of a thermolysin digest of fibronectin with streptococci selectively removed a 24-kDa peptide that reacted with antibodies to a synthetic peptide from the N terminus of fibronectin. Furthermore, LTA completely blocked the binding of this peptide to streptococci. Electron microscopic studies using gold-labeled amino-terminal peptide of fibronectin confirms that this region binds to LTA on the surface of streptococci (70).

The amino terminus of fibronectin contains a "five-finger" construct conferred by intrachain disulfide bonds, and this region is rich in hydrophobic amino acids (43). It is proposed that the lipid moiety of LTA interacts with a hydrophobic pocket created by one or more of these

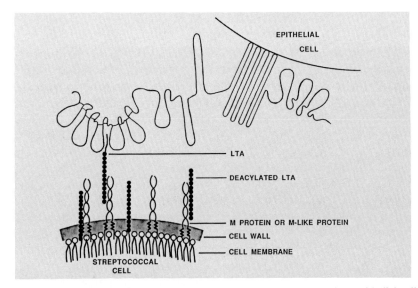

Figure 3. Hypothetical model of the adhesion of group A streptococci to epithelial cells. LTA is constantly released during growth of streptococci. During transit, some of these LTA molecules are bound via ionic interactions between the negatively charged PGP backbone of LTA and the positive charges of an LTA-binding protein. This arrangement allows an orientation in which the lipid moiety of LTA is exposed to the external milieu and therefore can bind to hydrophobic pockets located in the NH_2-terminal region of fibronectin on host cell surfaces. This type of interaction could serve as means to anchor streptococci to host epithelial cells. (From reference 14.)

fingers and that such a molecular interaction would serve as a mechanism for the adhesion of group A streptococci to epithelial cells (Fig. 3).

HYDROPHOBICITY OF *STREPTOCOCCUS PNEUMONIAE*

Most strains of pneumococci are encapsulated and bind poorly to hexadecane or to plastic (H. S. Courtney, unpublished data), suggesting that these organisms are not hydrophobic. These findings are similar to those for encapsulated group A streptococci, in which the capsule renders the organisms more hydrophilic. However, since noncapsulated pneumococci have not been tested for their hydrophobic properties, it is not known whether the capsule simply prevents hydrophobic interactions or whether the pneumococci are truly nonhydrophobic. Unlike group A streptococci, whose capsule interferes with adhesion to host cells, encapsulated pneumococci readily bind to epithelial cells (4), suggesting that for pneumococci, hydrophobicity may not be linked to adhesion.

CONCLUDING REMARKS

The data discussed above provide ample evidence that hydrophobic interactions play an integral part in the attachment of group A streptococci to host cells. Furthermore, the evidence supports the concept that LTA is the major surface component that contributes to these hydrophobic interactions. This does not mean, nor do we intend to imply, that other surface structures are not involved. M protein has been suggested to contribute directly to the hydrophobicity of streptococci (68). We have presented an alternative theory that M protein acts as an anchor protein for LTA. Other surface structures, such as T and R proteins, may be involved as well. The interactions between surface structures of streptococci are poorly understood, and additional studies are needed to determine how these components interact and what regulates their expression, with particular attention to the effect of environmental factors.

The types of environment that streptococci encounter during the course of an infection can change dramatically, from the oral cavity, where the pH varies from 5 to 7 and the quantity of nutrients varies (28a), to deeper tissues, where the pH is relatively constant (~7.4) and there is a richer source of nutrients. These two environments require the bacteria to respond in opposite ways. In the oral cavity the bacteria must attach to the host epithelium in order to avoid being swept away by the host's cleansing mechanisms, whereas in deeper tissues the bacteria must avoid attachment to phagocytic cells in order to survive. One of the ways in which bacteria accommodate these changing environments is by producing a heterogeneous population. In a population of bacteria, there will be a percentage of cells that have different surface properties with respect to electrical charge and hydrophobicity (45). Thus, when the population is exposed to a particular set of environmental conditions, there is a good probability that some of the bacteria will have a phenotype suitable for survival in that environment.

One such variation is the presence or absence of the hyaluronic acid capsule on group A streptococci. How this variation can affect survival of the bacteria in the host is hypothetically shown in Fig. 4. In the oral cavity, where the pH is comparatively low and there is competition for nutrients, the surviving bacteria would be noncapsulated, since the capsule impedes the attachment of streptococci to host cells and these unattached bacteria are removed by salivary flow. In the oral cavity the pH ranges from 5 to 7 (28a); this is the pH optimum for hyaluronidase activity (45), which may increase the percentage of bacteria that are

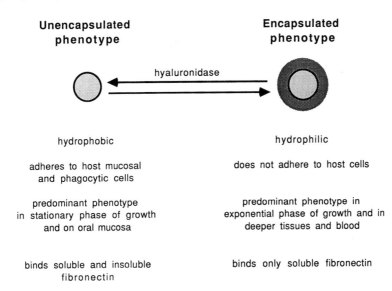

Figure 4. Hypothetical role of hyaluronic acid capsule in adhesion of streptococci to host cells in different environments. Encapsulated streptococci are not hydrophobic and do not adhere well to immobilized fibronectin or to host cells. Therefore, it would be expected that in deeper tissues encapsulated streptococci would be the predominant phenotype because they would be able to avoid attachment to phagocytes more effectively. Conversely, in the oral cavity the streptococci must attach to the host epithelium to avoid being swept away by the cleansing mechanisms of the host. Thus, in the oral cavity the prevailing bacteria would be of the noncapsulated phenotype because they would be able to attach to the host epithelium.

noncapsulated. In addition, there may be a limited amount of some nutrients that can affect expression of the capsule (28a).

Once the bacteria have invaded the deeper tissues, they encounter a different environment that requires a different phenotype for survival. The surviving bacteria would have the encapsulated phenotype because this phenotype allows the bacteria to avoid phagocytosis more effectively (29, 41). In the blood, the pH is not optimal for hyaluronidase activity, and the blood is a rich source of nutrients that enhance growth of the bacteria and expression of the capsule.

The components of these two environments may also affect the surface properties of the bacteria by coating their surfaces or by enzymatic modification of the bacterial surface (5, 20, 21). For example, saliva contains fibronectin, which can bind to streptococci and prevent both hydrophobic interactions and adhesion to host cells (12, 54–56). Saliva also contains glycosidases and proteases that may alter the surfaces of streptococci or the surfaces of the host cells (28a, 74).

In summary, we have discussed the surface hydrophobicity of group A streptococci, how this hydrophobicity is related to adhesion of streptococci to host cells, and how host factors can influence these interactions. In most bacteria, the function of the hydrophobin is to help the bacteria overcome repulsive forces between the bacteria and the host cells, which allows the bacteria to approach closer to the host cell and paves the way for the more specific interactions between the adhesin and the receptor. In group A streptococci, the hydrophobin performs both of these functions; LTA serves not only as a hydrophobin but also as an adhesin.

Dedication. We wish to dedicate this work to our late friend and colleague, Edwin H. Beachey. His lifelong contributions to the field of microbial pathogenesis have influenced many investigators, and his enthusiasm and insight were constant sources of inspiration. He will be deeply missed.

LITERATURE CITED

1. **Abraham, S. N., E. H. Beachey, and W. A. Simpson.** 1983. Adherence of *Streptococcus pyogenes*, *Escherichia coli*, and *Pseudomonas aeruginosa* to fibronectin-coated and uncoated epithelial cells. *Infect. Immun.* **41:**1261–1268.
2. **Absolom, D. R., C. J. van Oss, W. Zingg, and A. W. Neumann.** 1982. Phagocytosis as a surface phenomenon: opsonization by aspecific adsorption of IgG as a function of bacterial hydrophobicity. *RES J. Reticuloendothel. Soc.* **31:**59–70.
3. **Alkan, M. L., and E. H. Beachey.** 1978. Excretion of lipoteichoic acid by group A streptococci; influence of penicillin on excretion and loss of ability to adhere to human oral mucosal cells. *J. Clin. Invest.* **61:**671–677.
3a. **Alkan, M. L., I. Ofek, and E. H. Beachey.** 1977. Adherence of pharyngeal and skin strains of group A streptococci to human skin and oral epithelial cells. *Infect. Immun.* **18:**555–557.
4. **Anderson, B., J. Dahmen, T. Frejd, H. Leffler, G. Magnusson, G. Noori, and C. Svanborg-Eden.** 1983. Identification of an active dissaccharide unit of glycoconjugate receptor for pneumococci attaching to human pharyngeal cells. *J. Exp. Med.* **158:**559–570.
5. **Babu, J. P., E. H. Beachey, and W. A. Simpson.** 1986. Inhibition of the interaction of *Streptococcus sanguis* with hexadecane droplets by 55- and 60-kilodalton hydrophobic proteins of human saliva. *Infect. Immun.* **53:**278–284.
6. **Beachey, E. H.** 1975. Binding of group A streptococci to human oral mucosal cells by lipoteichoic acid. *Trans. Assoc. Am. Phys.* **137:**285–292.
7. **Beachey, E. H., G. L. Campbell, and I. Ofek.** 1974. Peptic digestion of streptococcal M protein. II. Extraction of M antigen from group A streptococci with pepsin. *Infect. Immun.* **9:**891–896.
8. **Beachey, E. H., and I. Ofek.** 1976. Epithelial cell binding of group A streptococci by lipoteichoic acid on fimbriae denuded of M protein. *J. Exp. Med.* **143:**759–771.
9. **Bjorck, L., S. K. Tylewska, T. Wadström, and G. Kronvall.** 1981. β₂ microglobulin is bound to streptococcal M protein. *Scand. J. Immunol.* **13:**391–394.
10. **Courtney, H. S., I. Ofek, W. A. Simpson, and E. H. Beachey.** 1981. Characterization of lipoteichoic acid binding to polymorphonuclear leukocytes of human blood. *Infect. Immun.* **32:**625–631.

11. **Courtney, H. S., I. Ofek, W. A. Simpson, D. L. Hasty, and E. H. Beachey.** 1986. Binding of *Streptococcus pyogenes* to soluble and insoluble fibronectin. *Infect. Immun.* **53:** 454–459.

12. **Courtney, H. S., I. Ofek, W. A. Simpson, E. Whitnack, and E. H. Beachey.** 1985. Human plasma fibronectin inhibits adherence of *Streptococcus pyogenes* to hexadecane. *Infect. Immun.* **47:**341–343.

13. **Courtney, H. S., W. A. Simpson, and E. H. Beachey.** 1983. Binding of streptococcal lipoteichoic acid to fatty acid-binding sites on human plasma fibronectin. *J. Bacteriol.* **153:**763–770.

14. **Courtney, H. S., L. Stanislawski, I. Ofek, W. A. Simpson, D. L. Hasty, and E. H. Beachey.** 1988. Localization of a lipoteichoic acid binding site to a 24-kilodalton NH_2-terminal fragment of fibronectin. *Rev. Infect. Dis.* **10:**s360–s362.

15. **Ellen, R. P., and R. J. Gibbons.** 1972. M protein-associated adherence of *Streptococcus pyogenes* to epithelial surfaces: prerequisite for virulence. *Infect. Immun.* **5:**826–830.

16. **Ellen, R. P., and R. J. Gibbons.** 1974. Parameters affecting the adherence and tissue tropisms of *Streptococcus pyogenes*. *Infect. Immun.* **9:**85–91.

17. **Elliot, S. D.** 1945. A proteolytic enzyme produced by group A streptococci with special reference to its effect on the type specific M antigen. *J. Exp. Med.* **81:**573–592.

18. **Firon, N., S. Ashkenazi, D. Mirelman, I. Ofek, and N. Sharon.** 1987. Aromatic α-glycosides of mannose are powerful inhibitors of the adherence of type 1 fimbriated *Escherichia coli* to yeast and intestinal epithelial cells. *Infect. Immun.* **55:**472–476.

19. **Fischetti, V. A.** 1989. Streptococcal M protein: molecular design and biological behavior. *Clin. Microbiol. Rev.* **2:**285–314.

20. **Hasty, D. L., E. H. Beachey, H. S. Courtney, and W. A. Simpson.** 1990. Interaction between fibronectin and bacteria, p. 89–112. *In* S. E. Carsons (ed.), *Fibronectin in Health and Disease*. CRC Press, Inc., Boca Raton, Fla.

21. **Hasty, D. L., and W. A. Simpson.** 1987. Effects of fibronectin and other salivary macromolecules on the adherence of *Escherichia coli* to buccal epithelial cells. *Infect. Immun.* **55:**2103–2109.

22. **Hill, M. J., A. M. James, and W. R. Maxted.** 1963. Some physical investigations of the behaviour of bacterial surfaces. VIII. Studies on the capsular material of *Streptococcus pyogenes*. *Biochim. Biophy. Acta* **66:**264–274.

23. **Hill, M. J., A. M. James, and W. R. Maxted.** 1963. Some physical investigations of the behaviour of bacterial surfaces. X. The occurrence of lipid in the streptococcal cell wall. *Biochim. Biophys. Acta* **75:**414–424.

24. **Johnson, R., and E. H. Beachey.** 1979. Purification and immunobiological properties of R antigen and its relation to M protein of type 3 group A streptococcus. *Infect. Immun.* **25:**1051–1059.

25. **Johnson, R., W. A. Simpson, J. B. Dale, I. Ofek, and E. H. Beachey.** 1980. Lipoteichoic acid-binding and biological properties of T protein of group A streptococcus. *Infect. Immun.* **29:**791–798.

26. **Kuhnemund, O., J. Havlicek, K. H. Schmidt, T. Wadström, and W. Köhler.** 1982. Relationship of M protein to hydrophobic properties of streptococcal cells, p. 182. *In* S. E. Holm and P. Christensen (ed.), *Basic Concepts of Streptococci and Streptococcal Diseases*. Reedbooks Ltd., Chertsey, England.

27. **Lancefield, R. C., and G. E. Perlmann.** 1952. Preparation and properties of A (R antigen) occurring in streptococci and other serological groups. *J. Exp. Med.* **96:**83–97.

28. **Lindahl, M., A. Faris, T. Wadström, and S. Hjertén.** 1981. A new test based on 'salting out' to measure surface hydrophobicity of bacterial cells. *Biochim. Biophys. Acta* **677:**471–476.

28a.**Mason, D. K., and D. M. Chisholm.** 1975. Saliva, p. 37–69. *In* D. K. Chisholm and D. M. Mason (ed.), *Salivary Glands in Health and Disease*. W. B. Saunders Co., London.
29. **Meynell, G. G.** 1961. Phenotypic variation and bacterial infection. *Symp. Soc. Gen. Microbiol.* **11**:174–195.
30. **Miller, L., L. Gray, E. Beachey, and M. Kehoe.** 1988. Antigenic variation among group A streptococcal M proteins. *J. Biol. Chem.* **263**:5668–5673.
31. **Miörner, H., P. A. Albertsson, and G. Kronvall.** 1982. Isoelectric points and surface hydrophobicity of gram-positive cocci as determined by cross-partition and hydrophobic affinity partition in aqueous two-phase systems. *Infect. Immun.* **36**:227–234.
32. **Miörner, H., J. Havlicek, and G. Kronvall.** 1984. Surface characteristics of group A streptococci with and without M protein. *Acta Pathol. Microbiol. Immun. Scand. Sect. B* **92**:23–30.
33. **Miörner, H., G. Johansson, and G. Kronvall.** 1983. Lipoteichoic acid is the major cell wall component responsible for surface hydrophobicity of group A streptococci. *Infect. Immun.* **39**:336–343.
34. **Miörner, H., E. Muhre, L. Björck, and G. Kronvall.** 1980. Effect of specific binding of human albumin, fibrinogen, and immunoglobulin G on surface characteristics of bacterial strains as revealed by partition experiments in polymer phase systems. *Infect. Immun.* **29**:879–885.
35. **Mosher, D. F. (ed.).** 1989. *Fibronectin*. Academic Press, Inc., San Diego, Calif.
36. **Nealon, T. J., E. H. Beachey, H. S. Courtney, and W. A. Simpson.** 1986. Release of fibronectin-lipoteichoic acid complexes from group A streptococci with penicillin. *Infect. Immun.* **51**:529–535.
37. **Nealon, T. J., and S. J. Mattingly.** 1985. Kinetic and chemical analysis of the biologic significance of lipoteichoic acid in mediating adherence of serotype III group B streptococci. *Infect. Immun.* **50**:107–115.
38. **Ofek, I., E. H. Beachey, F. Eyal, and J. C. Morrison.** 1977. Postnatal development of binding of streptococci and lipoteichoic acid by oral mucosal cells of humans. *J. Infect. Dis.* **135**:267–274.
39. **Ofek, I., E. H. Beachey, W. Jefferson, and G. L. Campbell.** 1975. Cell membrane-binding properties of group A streptococcal lipoteichoic acid. *J. Exp. Med.* **141**:990–1003.
40. **Ofek, I., W. A. Simpson, and E. H. Beachey.** 1982. Formation of molecular complexes between a structurally defined M protein and acylated or deacylated lipoteichoic acid of *Streptococcus pyogenes*. *J. Bacteriol.* **149**:426–433.
41. **Ofek, I., E. Whitnack, and E. H. Beachey.** 1983. Hydrophobic interactions of group A streptococci with hexadecane droplets. *J. Bacteriol.* **154**:139–145.
42. **Pancholi, V., and V. A. Fischetti.** 1989. Identification of an endogenous membrane anchor-cleaving enzyme for group A streptococcal M protein. *J. Exp. Med.* **170**:2119–2133.
43. **Petersen, T. E., H. C. Thogersen, K. Skorstengaard, K. Vibe-Pedersen, P. Sahl, L. Sottrup-Jensen, and S. Magnusson.** 1983. Partial primary structure of bovine plasma fibronectin: three types of internal homology. *Proc. Natl. Acad. Sci. USA* **80**:137–141.
44. **Pistole, P. G.** 1981. Interaction of bacteria and fungi with lectins and lectin-like substances. *Annu. Rev. Microbiol.* **38**:85–112.
45. **Plummer, D. T., A. M. James, and W. R. Maxted.** 1962. Some physical investigations of the behaviour of bacterial surfaces. V. The variation of the surface structure of streptococci during growth. *Biochim. Biophys. Acta* **60**:595–603.
46. **Quinn, R. W., R. Vander Zwaag, and P. N. Lowry.** 1985. Acquisition of group A streptococcal M protein antibodies. *Pediatr. Infect. Dis.* **4**:374–378.

47. **Rosenberg, E., A. Gottlieb, and M. Rosenberg.** 1983. Inhibition of bacterial adherence to hydrocarbons and epithelial cells by emulsan. *Infect. Immun.* **39:**1024–1028.

48. **Rosenberg, M., D. Gutnick, and E. Rosenberg.** 1980. Adherence of bacteria to hydrocarbons: a simple method for measuring cell-surface hydrophobicity. *FEMS Microbiol. Lett.* **9:**29–33.

49. **Rosenberg, M., and S. Kjelleberg.** 1989. Hydrophobic interactions: role in bacterial adhesion. *Adv. Microb. Ecol.* **1:**353–393.

50. **Rosenberg, M., A. Perry, E. A. Bayer, D. L. Gutnick, E. Rosenberg, and I. Ofek.** 1981. Adherence of *Acinetobacter calcoaceticus* RAG-1 to human epithelial cells and to hexadecane. *Infect. Immun.* **33:**29–33.

51. **Schmidt, K. H., D. Gerlach, O. Kuhnemund, and W. Köhler.** 1984. Quantitative differences in specific binding of fibrinogen fragment D by M-positive and M-negative group A streptococci. *Med. Microbiol. Immunol.* **173:**145–153.

52. **Schmidt, K. H., and W. Köhler.** 1981. Interaction of streptococcal cell wall components with fibrinogen. I. Adsorption of fibrinogen by immobilized T-proteins of *Streptococcus pyogenes*. *Immunobiology* **158:**330–337.

53. **Scott, J. R., W. M. Pulliam, S. K. Hollingshead, and V. A. Fishetti.** 1985. Relationship of M protein genes in group A streptococci. *Proc. Natl. Acad. Sci. USA* **82:**1822–1826.

54. **Simpson, W. A., and E. H. Beachey.** 1983. Adherence of group A streptococci to fibronectin on oral epithelial cells. *Infect. Immun.* **39:**275–279.

55. **Simpson, W. A., H. S. Courtney, and E. H. Beachey.** 1982. Fibronectin—a modulator of the oropharyngeal flora, p. 346–347. *In* D. Schlessinger (ed.), *Microbiology—1982*. American Society for Microbiology, Washington, D.C.

56. **Simpson, W. A., D. L. Hasty, and E. H. Beachey.** 1985. Binding of fibronectin to human buccal epithelial cells inhibits the binding of type 1 fimbriated *Escherichia coli*. *Infect. Immun.* **48:**318–323.

57. **Simpson, W. A., T. J. Nealon, D. L. Hasty, H. S. Courtney, and E. H. Beachey.** 1988. Binding of fibronectin to Streptococcus pyogenes: heat induction of fibronectin receptors, p. 349–354. *In* W. Strobrer, M. E. Lamm, J. R. McGhee, and S. P. James (ed.), *Mucosal Immunity and Infections at Mucosal Surfaces*. Oxford University Press, Oxford.

58. **Simpson, W. A., I. Ofek, and E. H. Beachey.** 1980. Binding of streptococcal lipoteichoic acid to the fatty acid binding sites on serum albumin. *J. Biol. Chem.* **255:**6092–6097.

59. **Simpson, W. A., I. Ofek, C. Sarasohn, J. C. Morrison, and E. H. Beachey.** 1980. Characteristics of the binding of streptococcal lipoteichoic acid to human oral epithelial cells. *J. Infect. Dis.* **141:**457–462.

60. **Stamp, L.** 1953. Studies on O/R potential, pH and proteinase production in cultures of *Streptococcus pyogenes*, in relation to immunizing activity. *Br. J. Exp. Pathol.* **34:**347–364.

61. **Stanislawski, L., H. S. Courtney, W. A. Simpson, D. L. Hasty, E. H. Beachey, L. Robert, and I. Ofek.** 1987. Hybridoma antibodies to the lipid-binding site(s) in the amino terminal region of fibronectin inhibits binding of streptococcal lipoteichoic acid. *J. Infect. Dis.* **156:**344–349.

62. **Sun, D., H. S. Courtney, and E. H. Beachey.** 1988. Berberine sulfate blocks adherence of *Streptococcus pyogenes* to epithelial cells, fibronectin, and hexadecane. *Antimicrob. Agents Chemother.* **32:**1370–1374.

63. **Tillet, W. S., and R. L. Garner.** 1934. The agglutination of hemolytic streptococci by plasma and fibrinogen. *Bull. Johns Hopkins Hosp.* **54:**145–149.

64. **Tylewska, S., S. Hjertén, and T. Wadström.** 1981. Effect of subinhibitory concentrations

of antibiotics on adhesion of *Streptococcus pyogenes* to pharyngeal cells. *Antimicrob. Agents Chemother.* **20**:563–566.

65. **Tylewska, S. K., V. A. Fischetti, and R. J. Gibbons.** 1988. Binding selectivity of *Streptococcus pyogenes* and M-protein to epithelial cells differs from that of lipoteichoic acid. *Curr. Microbiol.* **16**:209–216.

66. **Tylewska, S. K., S. Hjertén, and T. Wadström.** 1979. Contribution of M protein to the hydrophobic surface properties of *Streptococcus pyogenes*. *FEMS Microbiol. Lett.* **6**:249–253.

67. **Tylewska, S. K., S. Hjertén, and T. Wadström.** 1979. Hydrophobic properties of *Streptococcus pyogenes* related to M protein: decrease of surface hydrophobicity and epithelial cell binding by antibiotics, p. 788–790. *In* J. D. Nelson and C. Grassi (ed.), *Current Chemotherapy and Infectious Diseases.* American Society for Microbiology, Washington, D.C.

68. **Wadström, T., K. Schmidt, O. Kuhnemund, J. Havlicek, and W. Köhler.** 1984. Comparative studies on surface hydrophobicity of streptococcal strains of groups A, B, C, D, and G. *J. Gen. Microbiol.* **130**:657–664.

69. **Wadström, T., and S. Tylewska.** 1982. Glycoconjugates as possible receptors for *Streptococcus pyogenes*. *Curr. Microbiol.* **7**:343–346.

70. **Wagner, B., K. Schmidt, M. Wagner, and T. Wadström.** 1988. Localization and characterization of fibronectin-binding to group A streptococci, an electron microscopic study using protein-gold-complexes. *Zentralbl. Bakteriol. Parasitenkd. Infektionskr. Hyg. Abt. 1 Reihe A* **269**:479–491.

71. **Wagner, B., M. Wagner, M. Ryc, and R. Bicova.** 1984. Ultrastructural localization of lipoteichoic acid on group A streptococci, p. 192–194 *In* Y. Kimura, S. Kotami, and Y. Shiokawa (ed.), *Recent Advances in Streptococci and Streptococcal Diseases.* Reedbooks Ltd., Chertsey, England.

72. **Wiley, G. G., and P. N. Bruno.** 1970. Cross-reactions among group A streptococci. III. The M and R antigens of type 43 and serologically related streptococci. *J. Immunol.* **105**:1124–1130.

73. **Wilkinson, H. W.** 1972. Comparison of streptococcal R antigens. *Appl. Microbiol.* **24**:669–670.

74. **Woods, D. E., D. C. Straus, W. G. Johansson, Jr., and J. A. Bass.** 1981. Role of fibronectin in the prevention of adherence of *Pseudomonas aeruginosa* to buccal cells. *J. Infect. Dis.* **143**:784–790.

Microbial Cell Surface Hydrophobicity
Edited by R. J. Doyle and M. Rosenberg
© 1990 American Society for Microbiology, Washington, DC 20005

Chapter 14

Hydrophobicity of Oral Bacteria

R. J. Doyle, Mel Rosenberg, and David Drake

The oral environment serves as an ecological home for many different kinds of microorganisms. The microbes have evolved specialized structures or metabolic enzymes which enable them to occupy various oral sites. In many cases, oral microorganisms are found nowhere else in nature. The ability of microorganisms to adhere to various surfaces in the oral cavity has frequently been invoked as a major determinant for colonization. A variety of factors, such as genetics, age, and diet, combine to permit microbial colonization yet prevent symptoms of diseases from appearing. In some individuals, these same factors combine with an indigenous or new microbiota to permit a disease process. Some of the diseases encountered in the oral environment include dental caries, periodontal disease, thrush, and various mucosal and glandular infections. In some cases the determinants of the disease process are known, but with most oral diseases it has been difficult to precisely define the virulence traits of the microorganisms. In this chapter, we emphasize the role of surface hydrophobicity in oral microbial ecology in health and disease. Because much of the literature is concerned with the hydrophobicity of oral streptococci, most of the chapter will focus on the characteristics of these organisms and how they interact with the environment.

R. J. Doyle • Academic Health Center, University of Louisville, Louisville, Kentucky 40292. **Mel Rosenberg** • Maurice and Gabriela Goldschleger School of Dental Medicine and Department of Human Microbiology, Sackler School of Medicine, Tel Aviv University, Ramat Aviv, Israel 69978. **David Drake** • College of Dentistry, University of Iowa, Iowa City, Iowa 52242.

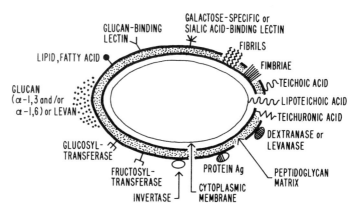

Figure 1. Surface structures of an oral streptococcus. Hydrophobins may be on lectins, fimbrial proteins, enzymes, lipoteichoic acid, lipids, or fatty acids. The oval shape more closely approximates streptococcal shapes than true coccal forms. Ag, Antigen.

SURFACE STRUCTURES OF ORAL STREPTOCOCCI

There are different degrees of complexity among bacterial cell surfaces. The oral streptococci have especially complex surfaces (Fig. 1), ranging from wall-bound enzymes to well-defined fibrillar or fimbriate structures (10, 21, 29, 39, 40–42, 46, 55, 81, 95, 115, 145–147, 149). The compositions and functions of some of the surface molecules and structures are known, whereas with others more information is needed. There is little doubt that the hydrophobicity of a streptococcus is due to cell wall-associated molecules. There is no evidence that the overall hydrophobicity of a streptococcus is due to the cytoplasmic membrane or membranous protrusions (7, 13, 97). Even when lipoteichoic acid has been considered to contribute to hydrophobicity of a bacterium, the polyanion is wall associated and not usually membrane bound (83). Some of the findings concerned with oral bacterial cell surfaces are summarized below.

Surface Enzymes

The idealized surface depicted in Fig. 1 reveals that several enzymes may be wall bound. These enzymes include invertase, which is capable of hydrolyzing sucrose into glucose and fructose; glucosyltransferase(s) (GFT[s]), capable of transferring the gluco moiety of a sucrose molecule to a glucan with the concomitant liberation of fructose, creating α-1,6 (soluble) or α-1,3 (insoluble) glucans or both; fructosyltransferases (FTFs,), capable of synthesizing a polyfructose (levan) from sucrose and liberating a glucose for each fructose added to the fructan; and glucanases and levanases, which may cleave previously synthesized glucan or levan. In some streptococci these enzymes are measurably bound to the cell wall, whereas in others the enzymes may be distributed in the culture medium. In general, streptococci synthesizing α-1,3 glucans tend to be more cariogenic, although some exceptions have been reported. The glucans may aid in the adhesion and colonization of some of the oral streptococci. The GTFs of *Streptococcus sobrinus* are highly hydrophobic, as evidenced by their strong association with octyl-Sepharose (S. Thaniyavarn and R. J. Doyle, unpublished data). It is likely that these enzymes possess hydrophobic domains capable of interacting with environmental hydrophobins. Other surface enzymes also probably possess hydrophobins.

Surface Lectins

Some oral streptococci have been reported to express surface lectins such as the glucan-binding lectin (GBL) (22). The GBL of *S. sobrinus* is capable of complexing with 6 to 10 internally-linked α-1,6 glucose residues (22). As shown later, the GBL may be related to the surface hydrophobicity of *S. sobrinus* or *S. cricetus*. Other lectins reported on some oral streptococci are said to be specific for galactose and sialic acids (33, 36, 69, 88). The probable function of the lectins is to aid in adhesion to pellicle.

Surface Polyanions

Most oral streptococci possess peptidoglycan-bound teichoic or teichuronic acids. These polymers probably do not contribute to cell surface hydrophobicity, but they do impart a net negative charge to the cells. The teichoic or teichuronic acids may function to bind oppositely charged molecules (1, 34, 38, 47, 53, 65, 88, 100, 126, 127) such as protons, metal ions, amino acids, or proteins (particularly autolysins). Their possible relationship to fibrils or fibrillae has not yet been resolved. On the basis of staining with ruthenium red, it has been suggested that fibrils may be at least partially composed of teichuronic acids (40).

Most oral streptococci seem to possess nonfimbrial protein antigens whose true functions are unknown. It seems likely that the antigens are adhesins in some of the streptococci. They do not possess known enzymatic activities. Almost certainly the wall-associated proteins exhibit hydrophobic regions capable of interaction with complementary hydrophobins. The reported glucan-binding domain of *Streptococcus mutans* GTF is clearly hydrophilic (5), however. It is unknown how various surface proteins remain cell associated. It is possible that proteins such as GTFs, FTFs, GBLs, and others possess a domain capable of binding to peptidoglycan or other surface structures. This binding is due to noncovalent interactions, because the proteins are extractable with denaturants. Some proteins, however, may be covalently bound to the cell wall, as suggested by a report of Nesbitt et al. (95), who found that exhaustive extraction of streptococcal cell walls did not solubilize all nonpeptidoglycan amino acids. Others have suggested that the cell walls of bacteria contain hydrophobic residues (3, 135).

As far as is known, the streptococcal cell surface does not turn over (shed into growth medium) during balanced growth (19, 89). Turnover refers to the loss of cell wall components into the growth medium, a process that requires the actions of autolysins (19). Normally, turnover is coupled to surface expansion of some bacteria (19). Turnover could cause the loss of surface lectins, enzymes, and hydrophobins into the environment, but all of these components would no doubt be replaced by new synthesis during division processes. There is a report that fluoride induces turnover in *S. mutans* (68). This finding suggests that streptococcal wall turnover may occur when the proton motive force has been dissipated (52) or when imbalanced growth occurs. The consequences of possible turnover events on surface hydrophobicity have yet to be considered by researchers, although there are numerous reports describing the effects of muramyl peptides on immune functions (reviewed by Doyle and Sonnenfeld [21]). If surface turnover occurs in oral streptococci, there will have to be a reassessment of the virulence traits of these organisms, since immune reactions can contribute to various kinds of oral disorders, such as periodontal disease and mucosal infections.

MECHANISMS OF ORAL MICROBIAL ADHESION

The initiation of oral microbial colonization usually begins with the adhesion of the organism to saliva-coated mucosal or tooth surfaces. Salivary components bind to the enamel surface, giving rise to pellicles. The pellicles in turn are substrata for bacteria, giving rise to adherent masses called dental plaques. Exposed (supragingival) plaque may be the source for localized proton production, resulting in demineralization of the tooth, or formation of dental caries. Members of the mutans streptococci (*S. mutans*, *S. sobrinus*, and *S. cricetus*) have been implicated in dental caries. *S. sanguis* has a high affinity for pellicle and may contribute to plaque formation (2, 14, 38, 72, 74, 75). *S. salivarius* and *S. mitis* seem to bind to saliva-coated mucosa, although it is likely that the bacteria would adhere to many kinds of saliva-coated surfaces. The apparent tropism of oral streptococci is probably a result of multiple factors rather than inherent surface characteristics of oral substrata. The same kinds of mechanisms involved in oral microbial adhesion also appear to be involved in environmental microbial adhesion (16).

Subgingival dental plaque, buried below the gingival margin, may lead to gingival inflammation. Certain microbial constituents of subgingival plaque (e.g., members of the genera *Bacteriodes* or *Porphyromonas*, *Actinobacillus*, *Actinomyces*, *Fusobacterium*, and *Treponema*) have been implicated in periodontal diseases. Whereas adhesive properties of purported periodontal pathogens have been studied, the role of adhesion in subgingival colonization and ensuing disease is less clear than in the case of exposed (supragingival) plaque and caries formation.

Some of the factors that have been reported to influence oral microbial adhesion include salivary flow rates, salivary composition (including mucins, immunoglobulins, and other proteins), bacterial proteases, lipoteichoic acids, phospholipids, endotoxins, bacterial cell wall turnover products, foodstuffs that may contain lectins or sucrose, metal ions, particularly Ca^{2+}, desquamated mucosal cells, competition between various microbial species, antibiotics, fluorides, ionophores, and metabolic poisons (31, 64, 73, 77, 78, 92, 104, 105, 113, 114, 125, 129, 141).

In this chapter, we are concerned primarily with the mechanism of adhesion of streptococci to saliva-coated surfaces and the contributions of hydrophobic interactions to the adhesion processes. It has become apparent in recent years that the hydrophobic effect (11) is a major factor in oral microbial adhesion. Some of the proposed mechanisms for streptococcal adhesion are summarized below. Only in the case of the multiple-site model is there compelling experimental evidence showing hydrophobic dependence on adhesion.

Glycoconjugate Binding, Lectin Mediated

In the glycoconjugate-binding, lectin-mediated mechanism, the bacteria possess surface lectins capable of binding to salivary glycoproteins or mucins. Lectins specific for sialic acids, amino sugars, and galactose have been proposed (33, 36). There is strong evidence that sialic acids contribute to the low desorption rate of *S. sanguis* from experimental pellicle (17). In other bacteria, hydrophobic sites are thought to be close to carbohydrate-binding sites (26, 27).

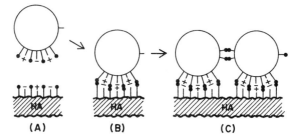

Figure 2. Multiple-site model for streptococcal adhesion. The model assumes that as a streptococcus (or possibly other bacteria as well) approaches a substratum surface, then either the streptococcal surface or the substratum changes so as to promote additional adhesion. This phenomenon is termed positive cooperativity and has been observed for various oral streptococci. Details of the model can be found in reference 20.

Glucan Binding, Lectin Mediated

The glucan-binding, lectin-mediated mechanism requires that α-1,6 glucan be present on pellicle. The glucan is a product of GTFs and sucrose. The glucan may bind directly to pellicle, or it may be formed by pellicle-bound streptococcal GTF (108, 120, 121). In addition, *S. sanguis* bound to pellicle can synthesize an α-1,6 glucan to which mutans streptococci can bind via the glucan-specific lectin. It is unknown whether the GBL has a hydrophobic site adjacent to the saccharide-binding site (70). Hydrophobic sites flanking saccharide-specific sites are common in lectins. Glucan-specific, lectin-mediated adhesion may be more important in the presence of sucrose. In the absence of sucrose, α-1,6-rich glucans are not found in plaques.

Polyionic, with Contributions from Multivalent Cations

The bacteria may possess polyelectrolytes such as teichoic acids which can interact with oppositely charged sites on the pellicle. Calcium ions may act to bridge negatively charged mucins and teichoic acids. This model does not take into account hydrophobic contributions to adhesion.

Multiple-Site Model

The multiple-site model model requires a contribution of hydrophobicity in order to stabilize cell-pellicle complexes (12, 20). The main features of the model are outlined in Fig. 2. The model states that a bacterium may weakly interact with complementary sites on the pellicle surface. These interactions may involve ion-dipole, dipole-dipole, cation bridging, direct hydrogen bonding, or a combination thereof. Lectins could be included in the model. All of the interactions are then stabilized by the hydrophobic effect. In this model, it is considered that some kind of change occurs on either the bacterial surface or the pellicle surface (or both) which results in increased hydrophobicity. This increased hydrophobicity could enhance additional adhesion, leading to positive cooperativity. Positive cooperativity, observed experimentally (18, 85, 94, 96), requires a change so as to promote additional adhesion. More will be said about cooperative effects later. The multiple-site model (20) predicts that (i) it would be easier to prevent adhesion than to cause desorption; (ii) the best desorbing agents would be those capable of disrupting hydrophobic interactions; (iii) adherent streptococci would be hydrophobic and hydrophilic mutants would tend to be nonadherent; (iv) the kinetics of adhesion would involve at least two distinct steps, one of which can be readily inhibited by hydrophobic effect-breaking agents; and (v) adhesion would be expected to occur in clusters of cells, although it would not rule out single-cell adhesion. In general, these predictions have been satisfied experimentally, although some interesting exceptions are known.

INVOLVEMENT OF HYDROPHOBIC INTERACTIONS IN ORAL MICROBIAL ADHESION

Table 1 summarizes some of the major findings concerning the role of hydrophobicity in oral microbial adhesion. Nesbitt et al. (93, 94) observed that *S. sanguis* could adhere to hexadecane. Nesbitt et al. (93) also found that tetramethyl urea was superior to urea in inhibiting the adhesion of *S. sanguis* to saliva-coated hydroxylapatite (HA). Cowan et al. (17, 18) showed that the initial adhesion of *S. sanguis* to saliva-coated HA was strongly entropy dependent but only weakly temperature dependent. Their results were derived from studies on the kinetics and equilibria of adhesion at different temperatures and at different cell densities. The results were interpreted to support a view that a combination of hydrophobic and electrostatic forces contributed to the initial adhesion process, a view consistent with the multiple-site model (Fig. 2). It should be kept in mind that Cowan et al. studied only the initial, readily reversible association between the bacteria and the artificial pellicle. Fujioka et al. (30) found that the association between a strain of *S. mutans* and various biomaterials was greater if the biomaterials possessed low surface free energies (see Rosenberg and Doyle [this volume] for a discussion of surface free energies and hydrophobicities). Similarly, Satou et al. (116) found that the adhesion of *S. sanguis* to biomaterials was related to the hydrophobicities of the bacteria. Van der Mei et al. (133) also recently

Table 1. Evidence Favoring Involvement of Hydrophobic Interactions
in Oral Microbial Adhesion

Results	Reference(s)
Adhesion of *Streptococcus sanguis* to saliva-coated HA was inhibited more readily by tetramethyl urea than by unsubstituted urea. Various oral streptococci bound to hexadecane.	93, 94
Initial adhesion of *S. sanguis* to saliva-coated HA was strongly entropy dependent but weakly temperature dependent, suggesting that a combination of hydrophobic and electrostatic forces contributes to the adhesion process.	18
Adhesion of *Streptococcus mutans* to biomaterials was inversely related to the surface free energies of the substrata, suggesting that hydrophobic forces contribute to the adhesion.	30
Adhesion of *S. sanguis* strains to various biomaterials was related to the hydrophobicities of the bacteria.	116
Adhesion of *Streptococcus salivarius* strains to polymethylmethacrylate and to saliva-coated HA was correlated with bacterial hydrophobicity.	133
There was a positive correlation between surface hydrophobicity and adhesion to saliva-coated HA.	29
When *S. sanguis* was trypsinized, the bacterium adhered poorly to octyl-Sepharose, saliva-coated HA, and hexadecane.	96
Adhesion of *S. salivarius* to saliva-coated HA and to HeLa cells was inhibited by protease, resulting also in lower bacterial hydrophobicity.	142
A mutant of *S. sanguis* lost its ability to adhere to salivary pellicles and to bind to hydrocarbons, suggesting a close relationship between adhesion and bacterial hydrophobicity.	35
Isolates from intraoral sites of *Macaca fascicularis* were usually hydrophobic.	9
Hydrophilic isolates tended to become hydrophobic upon exposure to saliva.	9
Most isolates from teeth were hydrophobic.	143
Hydrophilic strains of *S. mutans* were less cariogenic than hydrophobic strains.	99
Mutants of *S. mutans* were deficient in protein P1 (antigen I/II), were hydrophilic, and adhered less avidly to protein-coated HA than did the parent strain.	67
A 160,000-molecular-weight cell surface protein may contribute to hydrophobicity and adhesion to saliva-coated HA.	84

(*Continued on following page*)

Table 1—*Continued*

Results	Refer-ence(s)
Hydrophobic strains express cell surface proteins, whereas hydrophilic strains secrete the proteins into the culture medium.	80
A hydrophobic salivary protein prevented the adhesion of *S. sanguis* to hexadecane.	4
Human dental plaque microorganisms are generally able to bind to hydrocarbons.	112
Fresh isolates of oral streptococci were hydrophobic and adhered well to HA. Upon subculturing, the bacteria became hydrophilic and less adherent.	144
HA disks were coated with various proteins. The proteins with higher amounts of nonpolar residues created better substrata for adhesion of *S. mutans*, suggesting a hydrophobic involvement in the bacterium-surface union.	106
S. sanguis tended to adhere better to surfaces with low free energies.	136
Salivary deposits stick to hexadecane. Thermodynamic studies on plaque integrity suggested the hydrophobic effect as a stabilizing force.	66
Adhesion of *S. mutans* and *S. sanguis* was inhibited by fatty acids and by several proteins.	75
Mutants of *S. sanguis* lost surface hydrophobicity and adhered poorly to saliva-coated HA. Another mutant possessed increased hydrophobicity and demonstrated superior adhesion to saliva-coated HA.	51
Sublethal concentrations of tetracycline in the growth medium resulted in reduced hydrophobicity for *Actinomyces viscosus* and *Bacteroides gingivalis*. Antibiotic-grown cells also adhered very poorly to experimental pellicles.	102, 103

showed that in several strains of *S. salivarius*, there was a positive correlation between hydrophobicities and the tendency to bind to polymethylmethacrylate. It seems clear that the adhesion of various oral streptococci to artificial surfaces is dependent on the hydrophobic character of the bacteria.

The evidence favoring a role for hydrophobicity in streptococcal adhesion to saliva-coated HA is even more compelling. Fives-Taylor and Thompson (29) isolated several surface mutants of *S. sanguis* and found that the hydrophobic strain tended to adhere better than hydrophilic variants. Oakley et al. (96) observed that after trypsin treatment, *S.*

sanguis became more hydrophilic and lost its ability to adhere to hexadecane, octyl-Sepharose, and saliva-coated HA. Weerkamp et al. (142) also observed that proteolysis caused *S. salivarius* to lose its hydrophobicity and ability to adhere to HeLa cells or saliva-coated HA. Gibbons et al. (35) found that mutants of *S. sanguis* which had lost their ability to bind to hexadecane were also defective in adhesion to experimental pellicle.

Fresh isolates of oral streptococci from animals or humans are usually hydrophobic, as are dental plaque microorganisms scraped directly from the surfaces of teeth. Beighton (9) showed that streptococcal isolates from macaque monkeys were capable of binding to hydrocarbon. Similarly, Weiss et al. (143) have reported that human oral streptococci tend to be hydrophobic. Recently, Olsson and Emilson (99) concluded that hydrophobic and hydrophilic strains of *S. mutans* could be implanted with the same degree of ease in humans. The hydrophobic strains, however, were more cariogenic.

At present, the molecular basis for streptococcal hydrophobicity is poorly understood. Lee et al. (67) maintain that surface protein P1 (antigen I/II) of *S. mutans* may be a hydrophobin and adhesin. Mutants deficient in the protein were hydrophilic and adhered poorly to experimental pellicle. Their approach was to use recombinant techniques to construct the mutants and reinsert the gene coding for the protein back into appropriate recipients. There was convincing evidence to conclude that only protein P1 was altered. Morris et al. (84) found that a 160,000-molecular-weight protein on the surface of *S. sanguis* was probably responsible for, or at least contributed to, the ability of the bacterium to bind to hexadecane, aggregate with saliva, and adhere to saliva-coated HA. These results provide firm evidence for a role of hydrophobicity in streptococcal adhesion.

A study by McBride et al. (80) compared relatively hydrophobic strains of *S. mutans* with hydrophilic variants at the molecular level in terms of biochemical and immunological differences. High-molecular-weight proteins were isolated from purified cell walls of hydrophobic strains but were absent from identical extracts obtained from the hydrophilic *S. mutans*. However, analysis of culture supernatants of the hydrophilic strains revealed that high-molecular-weight proteins similar to those described from cell wall extracts of hydrophobic strains could be readily found in the culture milieu, possibly indicating that the hydrophilic strains were compromised in their ability to incorporate these proteins into the cell wall matrix. Treatment of whole cells of the hydrophobic strains with trypsin destroyed the proteins and concurrently substantially reduced the cell surface hydrophobicity. A prominent hydrophobin was found to have a molecular weight of 190,000 and was tentatively identified as antigen I/II of *S. mutans*. Because of the association of this high-molecular-weight protein with the relative hydrophobicity of the cell surface, the authors speculated that this hydrophobin (antigen I/II) may serve as an adhesin to hydrophobic regions of the acquired salivary pellicle.

Babu et al. (4) isolated from human saliva two salivary proteins that could inhibit the adhesion of *S. sanguis* to hexadecane. These proteins could serve as receptors for streptococcal hydrophobins. It may prove useful to employ the salivary proteins isolated by Babu et al. (4) as affinity probes for the isolation of adhesins of *S. sanguis*. These adhesins, in turn, may be good candidates for vaccine antigens. It is well known that *S. sanguis* can adhere to heart tissue and to platelets (45). Epitopes from the putative *S. sanguis* adhesins may also be of value in studies of the role of the bacterium in endocarditis or embolisms.

Table 1 summarizes additional experiments showing that adhesion of streptococci to various surfaces is coupled with hydrophobicity. Some of the results show that hydrophobic surfaces make superior substrata for streptococci. Others show that loss of streptococcal hydrophobicity results in reduced adhesion to surfaces, ranging from HA disks to other bacteria to plastics to hydrocarbons. The impressive array of approaches and diverse experimental procedures can lead only to the conclusion that streptococcal hydrophobicity is involved in adhesion to various substrata.

Although the evidence for a relationship between oral streptococcal adhesion and hydrophobicity is strong, the lack of molecular details impedes significant progress. More attention needs to be given to isolation and characterization of hydrophobin adhesins. The role of the hydrophobin in cariogenesis must be assessed, and the role of hydrophobins in cooperative phenomena must be resolved. Hydrophobin mimics (138) may prove to be useful in preventing streptococcal adhesion.

FACTORS INFLUENCING THE HYDROPHOBICITY
OF ORAL BACTERIA

Hydrophobicity of intact microorganisms requires the expression of surface hydrophobins. Thus, various factors can influence the synthesis of the hydrophobins as well as their behavior in aqueous media (Table 2). Westergren and Olsson (144) noted that when human isolates of *S. mutans* were subcultured many times, there was a loss of hydrophobic surface properties. Their findings could be very important, since the results suggest that the human oral environment may induce or select for

Table 2. Factors Influencing the Hydrophobicity and
Surface Structures of Oral Bacteria

Variable	Reference(s)
Growth medium	57, 59, 130
Growth rate	57, 107, 140
Proteases	96
Repeated subculture	101, 144
Buffer and/or ionic strength used in assay, pH	93, 94
Fatty acids and/or proteins bound to the cells	94
Antibiotics	102, 103
Saliva and/or salivary proteins	4, 9
Mutations in surface structures	29, 67, 146, 147

hydrophobicity. Very little is known about induction of surface proteins in streptococci. Unpublished work in our own laboratories has shown that repeated subculture of *S. sobrinus* 6715 does not alter its ability to adhere to hexadecane. Strain 6715, however, has been passed from laboratory to laboratory for many years, and it may have lost surface components in its early culturing history.

Knox et al. (57), using pH-controlled growth conditions, found that protein P1 (antigen I/II, previously discussed) was cell bound in some strains and secreted in the culture medium in others. The strains with cell-bound P1 were hydrophobic, whereas those deficient in the protein tended to be hydrophilic. The extracellular and surface protein profiles were different in comparisons of glucose- with fructose-grown bacteria.

Recently, D. Drake and R. J. Doyle have observed that the adhesion of *S. cricetus* AHT to hexadecane is dependent on the presence of elevated CO_2 in the culture gas phase (Fig. 3 and 4). In these studies, the adhesion of *S. cricetus* AHT to hydrocarbon was studied both as a function of time and as water/hexadecane ratio according to the method of Lichtenberg et al. (71). This approach is a major advance in that a quantitative assessment of bacterial hydrophobicity can be obtained from kinetic measurements. A series of rate constants is generated by varying the hexadecane/water ratio in the assay and measuring the association of the bacteria to the hexadecane layer over a defined period of time. Plotting these individual rates (k) versus the hydrocarbon (H)/water (W) volume (V) ratios [$(V_H/V_W) \times 10^3$] results in a linear relationship with a slope K, termed the removal coefficient. This value is best described as a measure of the change in rate of bacterial removal as a function of the hydrocarbon/water ratio in the assay (71, 123).

In the first series of experiments, the rate of association of *S. cricetus*

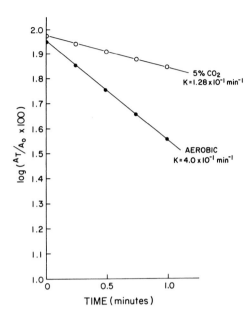

Figure 3. Adhesion of *S. cricetus* AHT to hexadecane. Cells were from a defined medium containing glucose (2 mg/ml) as the carbon source. Cells were suspended in a phosphate-saline buffer (pH 7.2) and mixed with hexadecane. The ratio of cell suspension to hexadecane was 1.0:0.2. A_T, Absorbance at time T; A_0, absorbance at time zero.

to hexadecane was determined at various hexadecane/buffer ratios (Fig. 3). Bacteria cultured in an atmosphere of 5% CO_2 were less hydrophobic than cells grown in an aerobic environment. Removal coefficients obtained from a plot of the individual rate constants versus the hexadecane/buffer ratio confirmed these initial findings and demonstrated that aerobically grown cells exhibit a higher affinity for hydrocarbons (Fig. 3 and 4). An interesting observation relative to these findings was that *S. cricetus* cultured aerobically aggregated in the presence of higher-molecular-weight glucan significantly less than did bacteria grown in an atmosphere containing 5% CO_2 (Fig. 5). It therefore appears that expression of the GBL and relative surface hydrophobicity are inversely related. One possible explanation for the results is that the lectin may sequester or cover hydrophobic sites on the surface of *S. sobrinus*, or the lectin may be in the growth medium of aerobically grown cells. The observations of Schilling et al. (120, 121) that glucan on saliva-coated HA surfaces makes a good substratum for streptococcal adhesion suggests that in the presence of sucrose (precursor of glucans), lectin sites are more important than hydrophobicity-dependent adhesion sites. In the absence of sucrose, the hydrophobic sites may be critical for adhesion. At present, the relationships between hydrophobicity, GBL, sucrose, and CO_2 are only empirically understood.

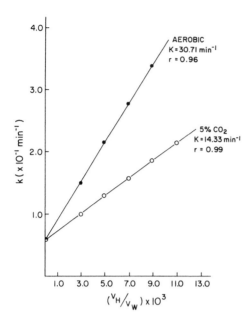

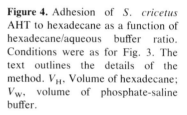

Figure 4. Adhesion of *S. cricetus* AHT to hexadecane as a function of hexadecane/aqueous buffer ratio. Conditions were as for Fig. 3. The text outlines the details of the method. V_H, Volume of hexadecane; V_W, volume of phosphate-saline buffer.

Proteases have been reported by several researchers to reduce oral streptococcal hydrophobicity. Presumably, the proteases release surface hydrophobins, resulting in decreased ability to bind complementary hydrophobin surfaces. Jenkinson (49, 50) has shown that hydrophobicity of *S. sanguis* may be related to its ability to coaggregate with *Actinomyces* spp. Protease-treated *S. sanguis* became more hydrophilic and concomitantly lost some of its coaggregation properties. These results are interesting because *Actinomyces* spp. possess a lectin capable of complexing with a polysaccharide on the cell wall of *S. sanguis* (61, 62). A possible explanation for the results is that a hydrophobin on the streptococcal surface interacts with a hydrophobin on the actinomycotic surface, stabilizing the lectin-polysaccharide interaction. Jenkinson isolated a 16,000-molecular-weight streptococcal surface polypeptide that was hydrophobic (49) and may contribute to the interaction between *S. sanguis* and *Actinomyces* spp.

Figure 6 shows how trypsin reduces the ability of radioactive *S. sanguis* to adhere to octyl-Sepharose. In control cells, water or saline eluted very little radioactivity, whereas sodium dodecyl sulfate eluted most of the cells in two major fractions. Trypsin-treated cells, in contrast, were partially eluted by saline. Elution of the remainder of the radioactivity by sodium dodecyl sulfate revealed a pattern different from that

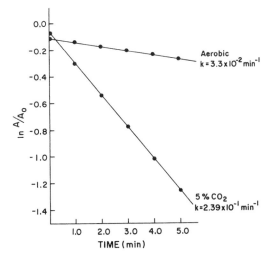

Figure 5. Aggregation of *S. cricetus* AHT by glucan. Conditions for the assay are found in reference 22. Results are for cells grown in the presence or absence of elevated CO_2. A, Absorbance at time of measurement; A_0, absorbance at time zero.

observed for the control cells. These results, along with others published by Oakley et al. (96), can be taken as evidence that cells in a given population are not homogeneous with respect to hydrophobicity. Oakley et al. (96) speculate that in a bacterial culture, hydrophobicity may be related to the cell cycle. To our knowledge, there have been no reports describing hydrophobicity in synchronized cultures. Another area that needs research is whether salivary proteases could modify oral microbial hydrophobicity.

When washed suspensions of *S. cricetus* were mixed with glucans or hexoses, there was no change in bacterial hydrophobicity (Table 3). Saliva and bovine serum albumin, when in the presence of the bacteria, caused a loss of binding to hexadecane. When the saliva or albumin was removed by centrifugation and washing in saline, hydrophobicity returned, as assayed by the microbial adhesion to hydrocarbons test. The results suggest that *S. cricetus* may possess complementary structures (hydrophobins?) for salivary proteins and bovine serum albumin. Nesbitt et al. (94) found several years ago that the adhesion of streptococci to albumin-coated HA beads followed binding equilibria much different from the equilibria involved in streptococcus-buffer-coated bead systems. The interaction between the *S. cricetus* and albumin or saliva must be relatively weak, since hydrophobicity of the bacteria returns upon washing the cells in saline. Eli et al. (23) recently showed that saliva tended to inhibit the adhesion of *Acinetobacter* spp. and *Serratia* spp. to polystyrene. Extraction of saliva with hexadecane resulted in loss of the ability

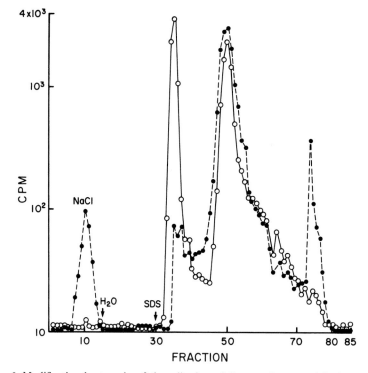

Figure 6. Modification by trypsin of the adhesion of *S. sanguis* to octyl-Sepharose. The column was sequentially eluted with 2.0 M NaCl, water, and 1% sodium dodecyl sulfate. Symbols: ○, control cells; ●, trypsin-treated cells. Washed cells (50 μl of 1.0 absorbance unit containing 8×10^5 cpm/ml) were added to the column after incubation with trypsin (50 μg/ml) at room temperature for 60 min. (From Oakley et al. [96] with permission.)

of the saliva to inhibit adhesion to the plastic. The results clearly demonstrate that salivary molecules contain hydrophobins. It is reasonable that the evolutionary history of oral streptococci would lead to surface structures capable of complexing with hydrophobins.

The work of Fives-Taylor and Thompson (29) convincingly demonstrates that surface fimbriae are responsible for the hydrophobicity of *S. sanguis*. Their hydrophilic mutants were devoid of fimbriae (surface structures that may appear to be ordered, or amorphous, depending on strain and fixation procedures). Fachon-Kalweit et al. (25) prepared antibodies against the fimbriae and found that the immunoglobulins would prevent adhesion to saliva-coated surfaces. Willcox et al. (146) have recently studied the ultrastructure of wild-type *S. sanguis* and an adhesion-defective strain. In the wild-type strain, fibrils of various lengths

Table 3. Effect of Solute Absorption on the Hydrophobicity of *S. cricetus* AHT[a]

Agent[b]	Coated, washed cells		Coated, unwashed cells	
	Saline	Hexadecane[c]	Saline	Hexadecane
Saline	0.63	0.03 (5)	0.38	0.01 (3)
Dextran T-10	0.66	0.03 (5)	0.66	0.05 (8)
Glucose	0.54	0.02 (4)	0.76	0.06 (7)
Fructose	0.62	0.02 (2)	0.70	0.05 (7)
Dextran sulfate	0.66	0.02 (3)	0.64	0.05 (7)
Bovine serum albumin (fatty acid free)	0.68	0.02 (3)	0.66	0.50 (71)
Saliva	0.74	0.03 (3)	0.90	0.35 (39)

[a] Unpublished results of J. David Oakley and R. J. Doyle.
[b] Saline = 0.15 M NaCl. Saliva was Parafilm stimulated and clarified by centrifugation at 25,000 × *g* for 15 min. All other agents were at concentrations of 100 μg/ml.
[c] Values represent reduction in turbidity of suspensions after standard mixing procedures with hexadecane (3 ml of cell suspension and 600 μl of hexadecane). Values in parentheses represent percentage of the original optical density. Suspensions were washed twice with and finally suspended in saline or the other agents shown.

were observed to be concentrated near cell ends. Some of the cells also carried fimbriae, distinct from the fibrils. The mutant strain did not possess any surface appendages. The two strains possessed similar hydrophobicities. These results can be taken as evidence that surface hydrophobicity is not directly related to adhesion. It should be kept in mind, however, that in any given bacterium there may be several hydrophobins. The hydrophobin near hydrophilic adhesins on the bacterium may be the most important in overall adhesion (20). Handley et al. (39) had earlier found that peritrichous fibrils on *S. sanguis* may be receptors (probably a peptidoglycan-associated polysaccharide-defective mutant rather than a surface protein mutant). Wyatt et al. (149) maintain that in *S. sanguis* there is no correlation between hydrophobicity, surface fibrils, and adhesion to saliva-coated surfaces. Their results are consistent with the premise that fibrils are polysaccharide and not protein. In another paper by the Handley group (41), it was shown that both fibrillar and fimbriate strains of *S. salivarius* were capable of adhering well to hexadecane. The fibrillar strains showed a greater tendency than did the fimbriate strains to adhere to saliva-coated HA and buccal epithelial cells and to coaggregate with *Veillonella* spp. Finally, Hesketh et al. (46) observed that when *S. sanguis* biotype strains I and II were treated with proteases, significant reductions in hydrophobicity and adhesion to saliva-coated HA occurred. In their work, the fibrils of biotype II were lost simultaneously with hydrophobicity. The biotype I strains lost their surface structures only slowly upon protease treatment.

Most studies describing hydrophobicity of oral streptococci have relied on batch culture methodology with a variety of different growth media. This type of approach may contribute significantly to the inherent variability observed in hydrophobicity experiments and the differences in streptococcal hydrophobicity data found in the literature, particularly since it is known that surface properties of oral streptococci are sensitive to growth conditions. Rogers et al. (107) showed that the hydrophobic nature of *S. mutans* was markedly affected by the composition of the growth medium and the growth rate in a continuous culture system. Slowly growing cells were substantially more hydrophobic than cells grown at rates similar to those encountered with batch cultures. More recently, Hogg and Manning (48) compared the hydrophobicity of *S. sanguis* grown in batch culture with that of chemostat-cultured cells. The percent hydrophobicity of *S. sanguis* AK1 grown in batch culture was 44%; however, this value varied considerably from experiment to experiment. In contrast, *S. sanguis* cells cultured in a chemostat were considerably more hydrophobic (63% hydrophobicity) and, importantly, were consistently hydrophobic over many generations, with little variability. These findings, along with the observations of Rosan et al. (109) that slowly growing cells adhere markedly differently to saliva-coated HA than do rapidly growing batch culture cells, provide further evidence that hydrophobicity may play an important role in mechanisms of bacterial adhesion. Future studies investigating changes in cell surface composition should be pursued in rigorously controlled continuous culture systems.

We believe that much more needs to be done to define the role of fibrils and fimbriae in bacteria. If fibrils are indeed polysaccharide, they would be expected to contribute little to hydrophobicity. Furthermore, mutations leading to deletions of polysaccharide fibril structures may or may not lead to changes in hydrophobicity. Mutations leading to the loss of proteinaceous fimbriae would probably cause a loss in the ability of the cells to bind complementary hydrophobins. There is a critical need to standardize growth conditions before ultrastructural results from various laboratories can be compared. This is especially true because changes in growth conditions can lead to changes in surface phenotypes in a single strain (Fig. 4 and 5).

HYDROPHOBIC ORAL STREPTOCOCCI IN VIVO

If the hydrophobic effect is important in adhesion to and accumulation of microorganisms on the oral surfaces, then it should be possible to make predictions concerning the nature of the microorganisms taken from plaques. First of all, plaque bacteria should bind to hydrocarbons or should demonstrate hydrophobicity as determined from other tests (see Rosenberg and Doyle, this volume). Second, oral bacteria with hydrophobic surfaces should be more easily implanted into suitable hosts. Both of these predictions have been confirmed experimentally but with different degrees of success.

Rosenberg et al. (112) obtained microorganisms directly from tooth surfaces and measured their ability to adhere to hexadecane and other hydrocarbons. After dispersion of plaque materials in buffer and admixture with the hydrocarbons, it was found that there was an almost quantitative recovery of the microorganisms at the interface between the two phases. These results show that most plaque microorganisms are hydrophobic. It would have been interesting to study the possible differences between plaques of caries-free and caries-active individuals. Weiss et al. (143) isolated 103 colonies from tooth surfaces and found that 82 adhered to hydrocarbon. Among the isolates were cocci, bacilli, coryneforms, and filamentous bacteria. Streptococci were the most frequently isolated. The results show that not only streptococci but also other oral bacteria adhere to hydrocarbons. Hogg and Manning (48) isolated various kinds of viridans streptococci from human volunteers and observed that most of the isolates were hydrophobic. Strains of *S. sanguis* and *S. salivarius* were more hydrophobic than the *S. mutans* isolates. Beighton (9) isolated 64 bacterial strains from *Macaca fascicularis* and found that most were hydrophobic, as assayed by adhesion to hexadecane. Interestingly, when bacteria isolated from mucosal sites were preincubated with saliva, the bacteria became more hydrophobic. We believe that these results suggest that the bacteria bound salivary glycoproteins or mucins via charge-charge interactions, but upon binding with the cells, hydrophobic groups were exposed. This phenomenon could reflect a type of cooperativity that leads to greater adhesion.

Svanberg et al. (128) studied the implantation into humans of strains of *S. mutans* with different degrees of hydrophobicity. They chose strains with antibiotic resistance markers so that recovery could be easily monitored. The hydrophobic strains were more easily implanted in various oral sites than were the hydrophilic variants. More recent work by Olsson and Emilson (99) revealed that in hamsters, hydrophobic and hydrophilic strains of *S. mutans* were nearly equally implanted. The hydrophobic strains, however, were superior in inducing caries when the animals were fed a sucrose-containing diet.

SUMMARY OF RESULTS ON THE HYDROPHOBICITY
OF ORAL MICROORGANISMS

An emphasis in this chapter has been the role of hydrophobicity in streptococcal adhesion to oral surfaces. Since 1982, there have been numerous papers describing inhibition of hydrophobic effect-dependent adhesion, attempts to define macromolecular hydrophobins, construction of hydrophobic mutants and studies of their interactions with various surfaces, conditions altering microbial hydrophobicity, and attempts to determine hydrophobic contributions to adhesion and colonization in vivo. Earlier sections in this chapter reviewed some of these papers. Discussion of members of the genera *Actinobacillus*, *Actinomyces*, *Bacteroides*, *Candida*, and others was limited because the literature is weighted toward oral streptococci. The hydrophobicities of other organisms, such as *Pseudomonas aeruginosa* (56, 82), *Streptococcus agalactiae* (79, 90, 91), *Streptococcus pyogenes* (7, 8, 97, 131), and *Serratia marcescens* (6), are discussed elsewhere in this volume. Table 4 summarizes results on oral microbial hydrophobicity not extensively discussed in this chapter. It appears that not only are the streptococci hydrophobic, but other oral microorganisms express surface hydrophobins as well. The role of hydrophobicity in pathogenesis is obscure. *Actinobacillus* spp., *Actinomyces* spp., and *Bacteroides* spp. have been reported to be involved in periodontal disease or root surface caries. These diseases, similar to dental caries on coronal surfaces, require plaque formation before the appearance of symptoms.

Clark et al. (15) examined 42 strains of *Actinomyces* spp. for hydrophobicity, using the salt aggregation test and hydrophobic interaction chromatography assays. Strains that adhered poorly also were relatively hydrophilic. The nonionic detergent Tween 80 was a good inhibitor of adhesion to phenyl-Sepharose but was a poor inhibitor of adhesion to saliva-coated HA. Because it is thought that *Actinomyces* lectins may mediate adhesion to salivary glycoproteins (reviewed by Kolenbrander [61]), it must be considered that the hydrophobic effect also participates in the interactions. It is unknown whether *Actinomyces* surface lectins possess hydrophobic sites near carbohydrate-specific sites (24). Such hydrophobic sites are known in *Escherichia coli* but not in *Salmonella* spp. (27).

Okuda et al. (98) found that *Bacteroides gingivalis*, a gram-negative anaerobe found in periodontal lesions, secreted a hemagglutinin into the culture medium. The hemagglutinin was not inhibited by carbohydrates, proteases, sialidase, or phospholipases A and C. The activity was, however, inhibited by phospholipase D and basic amino acids. Antibodies

Table 4. Summary of Results Demonstrating Hydrophobicity of Oral Microorganisms[a]

Organism(s)	Method for detecting hydrophobicity	Observations	Reference(s)
Actinobacillus actinomycetemcomitans	MATH	Several strains of the organism were weakly hydrophobic.	32
	MATH	Adhesion depended on culture medium and age.	63
Actinomyces viscosus, A. naeslundii	MATH	Both bacteria were highly hydrophobic.	32
	SAT, HIC	Hydrophobic strains tended to adhere well to saliva-coated HA. Tween 80 inhibited adhesion of the bacteria to the hydrocarbon substrata but not to the HA.	15
Bacteroides gingivalis	HIC	A surface hemagglutinin (adhesin?) was purified by chromatography.	98
Bacteroides spp.	MATH	Fibrils and fimbriae are not correlated with hydrophobicity.	42
	MATH	Wide variations (12–76%) in adhesion to xylene were observed for various strains.	137
Bacteroides gingivalis, B. melaninogenicus, B. intermedius	MATH	*B. gingivalis* was more hydrophobic than *B. melaninogenicus.*	32
Bordetella pertussis	HMA	Hydrophobicity was a trait independent of phase variation.	28
Branhamella catarrhalis	MATH	The bacterium exhibited a surface hydrophobicity greater than that of *E. coli* or *P. aeruginosa.*	37
Candida albicans	HMA	Hydrophobic interactions may be dominant in adhesion to epithelial cells.	43, 44

(Continued on following page)

Table 4—*Continued*

Organism(s)	Method for detecting hydropho-bicity	Observations	Refer-ence(s)
Fusobacterium necro-phorum	MATH	A virulent biovar was more hydro-phobic than an avirulent biovar.	124
Porphyromonas gingi-valis (*Bacteroides gingivalis*)	MATH	Nonpigmented strains exhibited reduced levels of hydrophobicity and protease.	122
Streptococcus sali-varius	MATH	Lipoteichoic acid does not confer hydrophobicity to the bacteria. Adhesion to hexadecane was de-pendent on surface fibrils.	142
	MATH	Strains with surface appendages were highly hydrophobic.	41
S. mutans	MATH	Colonization rates in hamsters were identical for hydrophobic and hy-drophilic strains, although in long-term experiments the hydrophobic strain was superior in inducing caries.	99
S. sanguis	MATH	Protein tufts on cell surface may be related to hydrophobicity.	46
S. sanguis, S. salivar-ius, S. mitior, Veil-lonella alcalescens	MATH, TPP	Hydrophobic strains tended to bind better to sulfated polystyrene than did hydrophilic strains.	134
S. sanguis, S. mutans	CAM	*S. sanguis* adhered better to hydro-phobic glass than did *S. mutans*. *S. mutans* tended to adhere bet-ter to positively charged glass surfaces than did *S. sanguis*.	117
S. sanguis, S. mutans, S. salivarius, S. mi-tis, S. milleri	MATH	Batch cultures revealed variability in hydrophobicity. *S. milleri* did not exhibit a tendency to bind to hexadecane.	48

(*Continued on following page*)

Table 4—*Continued*

Organism(s)	Method for detecting hydropho- bicity	Observations	Refer- ence(s)
S. mutans	MATH	Cell surface hydrophobic proteins were shed into the culture me- dium for some strains and when carbon source was changed.	57
S. sanguis, *S. mutans*, *S. salivarius*, *S. mi- tis*, *S. milleri*	Various methods	Observed no distinct correlation between adhesion to saliva- coated HA and hydrophobicity.	132
S. sanguis	MATH	Adhesion-proficient, fibrillar parent was similar in hydrophobicity to adhesion-deficient, fibrillar-defi- cient mutant.	146
	MATH	Hydrophobicity could not be corre- lated with surface fibrils, biotype, or adhesion to saliva-coated HA.	148
	SAT	Mutants were generally lower in adhesion to saliva-coated HA, in fimbriation, and in hydrophobic- ity.	29
S. oralis	MATH	Hydrophobicity appeared to be un- related to ability of strains to ad- here to saliva-coated glass.	147
S. mutans	MATH	Implantation of hydrophobic and hydrophilic mutants into human volunteers and subsequent recov- ery of the organisms revealed that their hydrophobicities had not changed.	128
S. sanguis	MATH	Salts enhanced adhesion. Acidic pH favored adhesion. Chaotropic agents inhibited adhesion of the bacteria to saliva-coated HA beads.	93, 94

(*Continued on following page*)

Table 4—*Continued*

Organism(s)	Method for detecting hydropho- bicity	Observations	Refer- ence(s)
	MATH, HIC	Trypsin reduced the ability of the cells to bind the hydrocarbons. Elution by sodium dodecyl sul- fate of cells bound to octyl-Seph- arose yielded complex profiles, suggesting that a cell population exhibits heterogeneity with re- spect to hydrophobicity.	96
	MATH	Hydrophobic proteins on cell sur- face may be required for coaggre- gation with *Actinomyces* spp.	50
Various oral strep- tococci	MATH, HIC	Most species were hydrophobic. A correlation between adhesion to hexadecane and chlorhexidine- mediated aggregation was ob- served.	148
S. mutans	MATH	Protein antigen may confer hydro- phobicity and promote adhesion to saliva-coated HA by a sero- type c strain.	60
Various noncoccal oral microorgan- isms	Adhesion to buccal cells	Hydrophobic effect disrupting agents inhibited adhesion.	110

[a] Abbreviations (see also Rosenberg and Doyle, this volume): MATH, microbial adhesion to hydrocar- bons; SAT, salt aggregation test; HIC, hydrophobic interaction chromatography; HMA, hydrophobic microsphere attachment (adhesion to polystyrene microspheres); TPP, two-phase polymer system; CAM, contact angle measurements. See also references 76, 87, 111, 118, 119, and 139 for discussions of methods for determining hydrophobicities of microorganisms. Rosenberg and Doyle (this volume) provide the most recent review on the characterization of surface hydrophobicities of microorganisms.

directed against the hemagglutinin bound to the *B. gingivalis* cell surface. The results suggest that the bacterium may possess a surface-associated hemagglutinin (probably a hydrophobin) and that the hemagglutinin may result from cell surface turnover (98). This view is tempered by the fact that surface turnover events usually are related to peptidoglycan (19). Shah et al. (122) found that pleiotropic mutants of *B. gingivalis* (now called *Porphyromonas gingivalis*) were unable to hemagglutinate and also had a lower hydrophobicity. Other studies had shown that the strains possessing reduced hydrophobicity tended to be avirulent. Hydrophobicity, hemagglutination, and virulence seem to be linked in this periodontopathogen.

Handley and Tipler (42) have studied the ultrastructure of *Bacteroides* spp. On the basis of adhesion to hexadecane, they concluded that most strains were hydrophilic. They found surface appendages (fibrils or fimbriae or both) on most of the strains. Van Steenbergen et al. (137) showed that a hydrophilic capsule surrounding some *Bacteriodes* spp. may prevent adhesion of the bacteria to hydrocarbon. It is unknown whether the "hydrophilic capsule" of van Steenbergen et al. was present on the surfaces of the strains examined by Handley and Tipler. Very little has been done to isolate and characterize hydrophobins from *Bacteroides* spp., although it does appear that hydrophobicity and virulence are coupled somehow.

Fusobacterium spp., also considered to be periodontopathogens, possess distinct hydrophobic characteristics. Shinjo et al. (124) found that biovar A strains of *Fusobacterium necrophorum* (common to cattle) were highly hydrophobic, whereas biovar B strains tended to be less hydrophobic, as determined by adhesion to octane. *F. nucleatum*, capable of intergenus coaggregation with *S. sanguis*, also expresses surface hydrophobins. Kaufman and DiRienzo (54) have isolated a 39,000-molecular-weight polypeptide thought to be the ligand for the *S. sanguis* surface polysaccharide. The polypeptide was hydrophobic, as determined by amino acid analysis. Antibody against the polypeptide inhibited coaggregation with *S. sanguis*. This paper makes a good beginning toward molecular characterization of hydrophobins that may be involved in coaggregation phenomena.

CONCLUDING REMARK

The foregoing results, taken together, provide compelling evidence that the hydrophobic effect is involved in oral microbial adhesion. When the adhesion has been defined on a more molecular basis, new inhibitors of the adhesins will no doubt emerge. It has been less than a decade since hydrophobic interactions were suggested to be involved in stabilizing the streptococcus-pellicle union. Another decade may yield a new generation of inhibitors based on the hydrophobic effect.

Acknowledgment. Research on streptococcal cell surfaces in the laboratory of R.J.D. has been supported by the NIH-NIDR.

LITERATURE CITED

1. **Abbott, A., and M. L. Hayes.** 1984. The conditioning role of saliva in streptococcal attachment to hydroxyapatite surfaces. *J. Gen. Microbiol.* **130:**809–816.

2. **Appelbaum, B., E. Golub, S. C. Holt, and B. Rosan.** 1979. In vitro studies of dental plaque formation: adsorption of oral streptococci to hydroxyapatite. *Infect. Immun.* **25:**717–728.

3. **Appelbaum, B., and B. Rosan.** 1984. Cell surface proteins of oral streptococci. *Infect. Immun.* **46:**245–250.

4. **Babu, J. P., E. H. Beachey, and W. A. Simpson.** 1986. Inhibition of the interaction of *Streptococcus sanguis* with hexadecane droplets by 55- and 66-kilodalton hydrophobic proteins of human saliva. *Infect. Immun.* **53:**278–284.

5. **Banas, J. A., R. R. B. Russell, and J. J. Ferretti.** 1990. Sequence analysis of the gene for the glucan-binding protein of *Streptococcus mutans* Ingbritt. *Infect. Immun.* **58:**667–673.

6. **Bar-Ness, A., N. Avrahamy, T. Matsuyama, and M. Rosenberg.** 1988. Increased cell surface hydrophobicity of a *Serratia marcescens* NS 38 mutant lacking wetting activity. *J. Bacteriol.* **170:**4361–4364.

7. **Beachey, E. H., T. M. Chiang, I. Ofek, and A. H. Kang.** 1977. Interaction of lipoteichoic acid of group A streptococci with human platelets. *Infect. Immun.* **16:**649–654.

8. **Beachey, E. H., J. B. Dale, W. A. Simpson, J. D. Evans, K. W. Knox, I. Ofek, and A. J. Wicken.** 1979. Erythrocyte binding properties of streptococcal lipoteichoic acids. *Infect. Immun.* **23:**618–625.

9. **Beighton, D.** 1984. The influence of saliva on the hydrophobic surface properties of bacteria isolated from oral sites of macaque monkeys. *FEMS Microbiol. Lett.* **21:**239–242.

10. **Bleiweis, A. S., R. A. Craig, S. E. Coleman, and I. V. De Rijn.** 1971. The streptococcal cell wall: structure, antigenic composition, and reactivity with lysozyme. *J. Dent. Res.* **50:**1118–1130.

11. **Burley, S. K., and G. A. Petsko.** 1988. Weakly polar interactions in proteins. *Adv. Protein Chem.* **39:**125–189.

12. **Busscher, H. J., and A. H. Weerkamp.** 1987. Specific and non-specific interactions in bacterial adhesion to solid substrata. *FEMS Microbiol. Rev.* **46:**165–173.

13. **Chiang, T. M., M. L. Alkan, and E. H. Beachey.** 1979. Binding of lipoteichoic acid of group A streptococci to isolated human erythrocyte membranes. *Infect. Immun.* **26:**316–321.

14. **Clark, W. B., L. L. Bammann, and R. J. Gibbons.** 1978. Comparative estimates of bacterial affinities and adsorption sites on hydroxyapatite surfaces. *Infect. Immun.* **19:**846–853.

15. **Clark, W. B., M. D. Lane, J. E. Beem, S. L. Bragg, and T. T. Wheeler.** 1985. Relative hydrophobicities of *Actinomyces viscosus* and *Actinomyces naeslundii* strains and their adsorption to saliva-treated hydroxyapatite. *Infect. Immun.* **47:**730–736.

16. **Costerton, J. W., R. T. Irvin, and K. J. Cheng.** 1981. The role of bacterial surface structures in pathogenesis. *Crit. Rev. Microbiol.* **8:**303–338.

17. **Cowan, M. M., K. G. Taylor, and R. J. Doyle.** 1987. Role of sialic acid in the kinetics of *Streptococcus sanguis* adhesion to artificial pellicle. *Infect. Immun.* **55:**1552–1557.

18. **Cowan, M. M., K. G. Taylor, and R. J. Doyle.** 1987. Energetics of the initial phase of adhesion of *Streptococcus sanguis* to hydroxylapatite. *J. Bacteriol.* **169:**2995–3000.

19. **Doyle, R. J., J. Chaloupka, and V. Vinter.** 1988. Turnover of cell wall in microorganisms. *Microbiol. Rev.* **52:**554–567.
20. **Doyle, R. J., W. E. Nesbitt, and K. G. Taylor.** 1982. On the mechanism of adherence of *Streptococcus sanguis* to hydroxylapatite. *FEMS Microbiol. Lett.* **15:**1–5.
21. **Doyle, R. J., and E. M. Sonnenfeld.** 1989. Properties of the cell surfaces of pathogenic bacteria. *Int. Rev. Cytol.* **118:**33–92.
22. **Drake, D., K. G. Taylor, A. D. Bleiweis, and R. J. Doyle.** 1988. Specificity of the glucan-binding lectin of *Streptococcus sanguis*. *Infect. Immun.* **56:**1864–1872.
23. **Eli, I., H. Judes, and M. Rosenberg.** 1989. Saliva-mediated inhibition and promotion of bacterial adhesion to polystyrene. *Biofouling* **1:**203–211.
24. **Ellen, R. P., E. D. Fillery, K. H. Chan, and D. A. Grove.** 1980. Sialidase-enhanced lectin-like mechanism for *Actinomyces viscosus* and *Actinomyces naeslundii* hemagglutination. *Infect. Immun.* **27:**335–343.
25. **Fachon-Kalweit, S., B. L. Elder, and P. Fives-Taylor.** 1985. Antibodies that bind to fimbriae block adhesion of *Streptococcus sanguis* to saliva-coated hydroxyapatite. *Infect. Immun.* **48:**617–624.
26. **Falkowski, W., M. Edwards, and A. J. Schaeffer.** 1986. Inhibitory effect of substituted aromatic hydrocarbons on adherence of *Escherichia coli* to human epithelial cells. *Infect. Immun.* **52:**863–866.
27. **Firon, N., S. Ashkenazi, D. Mirelman, I. Ofek, and N. Sharon.** 1987. Aromatic alpha-glycosides of mannose are powerful inhibitors of the adherence of type 1 fimbriated *Escherichia coli* to yeast and intestinal epithelial cells. *Infect. Immun.* **55:**472–476.
28. **Fish, F., Y. Navon, and S. Goldman.** 1987. Hydrophobic adherence and phase variation in *Bordetella pertussis*. *Med. Microbiol. Immunol.* **176:**37–46.
29. **Fives-Taylor, P. M., and D. W. Thompson.** 1985. Surface properties of *Streptococcus sanguis* FW213 mutants nonadherent to saliva-coated hydroxyapatite. *Infect. Immun.* **47:**752–759.
30. **Fujioka, Y., Y. Akagawa, S. Minagi, H. Tsuru, Y. Miyake, and H. Suginaka.** 1987. Adherence of *Streptococcus mutans* to implant materials. *J. Biomed. Mater. Res.* **21:**913–920.
31. **Gibbons, R. J., and I. Dankers.** 1983. Association of food lectins with human oral epithelial cells *in vivo*. *Arch. Oral Biol.* **28:**561–566.
32. **Gibbons, R. J., and I. Etherden.** 1983. Comparative hydrophobicities of oral bacteria and their adherence to salivary pellicles. *Infect. Immun.* **41:**1190–1196.
33. **Gibbons, R. J., I. Etherden, and E. C. Moreno.** 1983. Association of neuraminidase-sensitive receptors and putative hydrophobic interactions with high-affinity binding sites for *Streptococcus sanguis* C5 in salivary pellicles. *Infect. Immun.* **42:**1006–1012.
34. **Gibbons, R. J., I. Etherden, and E. C. Moreno.** 1985. Contribution of stereochemical interactions in the adhesion of *Streptococcus sanguis* C5 to experimental pellicles. *J. Dent. Res.* **64:**96–101.
35. **Gibbons, R. J., I. Etherden, and Z. Skobe.** 1983. Association of fimbriae with the hydrophobicity of *Streptococcus sanguis* FC-1 and adherence to salivary pellicles. *Infect. Immun.* **41:**414–417.
36. **Gibbons, R. J., and J. V. Qureshi.** 1979. Inhibition of adsorption of *Streptococcus mutans* strains to saliva-treated hydroxyapatite by galactose and certain amines. *Infect. Immun.* **26:**1214–1217.
37. **Gotoh, N., S. Tanaka, and T. Nishino.** 1989. Supersusceptibility to hydrophobic antimicrobial agents and cell surface hydrophobicity in *Branhamella catarrhalis*. *FEMS Microbiol. Lett.* **59:**211–214.

38. **Hamada, S., and H. D. Slade.** 1980. Biology, immunology, and cariogenicity of *Streptococcus mutans. Microbiol. Rev.* **44:**331–384.

39. **Handley, P. S., P. L. Carter, J. E. Wyatt, and L. M. Hesketh.** 1985. Surface structures (peritrichous fibrils and tufts of fibrils) found on *Streptococcus sanguis* strains may be related to their ability to coaggregate with other oral genera. *Infect. Immun.* **47:**217–227.

40. **Handley, P. S., J. Hargreaves, and D. W. S. Harty.** 1988. Ruthenium red staining reveals fibrils and a layer external to the cell wall in *Streptococcus salivarius* HB and adhesion deficient mutants. *J. Gen. Microbiol.* **134:**3165–3172.

41. **Handley, P. S., D. W. Harty, J. E. Wyatt, C. R. Brown, J. P. Doran, and A. C. Gibbs.** 1987. A comparison of the adhesion, coaggregation and cell-surface hydrophobicity properties of fibrillar and fimbriate strains of *Streptococcus salivarius. J. Gen. Microbiol.* **133:**3207–3217.

42. **Handley, P. S., and L. S. Tipler.** 1986. An electron microscope survey of the surface structures and hydrophobicity of oral and non-oral species of the bacterial genus *Bacteroides. Arch. Oral Biol.* **31:**325–335.

43. **Hazen, B. W., and K. C. Hazen.** 1988. Modification and application of a simple, surface hydrophobicity detection method to immune cells. *J. Immunol. Methods* **107:**157–163.

44. **Hazen, K. C.** 1989. Participation of yeast cell surface hydrophobicity in adherence of *Candida albicans* to human epithelial cells. *Infect. Immun.* **57:**1894–1900.

45. **Herzberg, M. C., K. K. Brintzenhofe, and C. C. Clawson.** 1983. Aggregation of human platelets and adhesion of *Streptococcus sanguis. Infect. Immun.* **39:**1457–1469.

46. **Hesketh, L. M., J. E. Wyatt, and P. S. Handley.** 1987. Effect of protease on cell surface structure, hydrophobicity and adhesion of tufted strains of *Streptococcus sanguis* biotypes I and II. *Microbios* **50:**131–145.

47. **Hogg, S. D., and G. Embery.** 1982. Blood-group-reactive glycoprotein from human saliva interacts with lipoteichoic acid on the surface of *Streptococcus sanguis* cells. *Arch. Oral Biol.* **27:**261–268.

48. **Hogg, S. D., and J. E. Manning.** 1987. The hydrophobicity of 'viridans' streptococci isolated from the human mouth. *J. Appl. Bacteriol.* **63:**311–318.

49. **Jenkinson, H. F.** 1986. Cell-surface proteins of *Streptococcus sanguis* associated with cell hydrophobicity and coaggregation properties. *J. Gen. Microbiol.* **132:**1575–1589.

50. **Jenkinson, H. F.** 1987. Novobiocin-resistant mutants of *Streptococcus sanguis* with reduced cell hydrophobicity and defective in coaggregation. *J. Gen. Microbiol.* **133:**1909–1918.

51. **Jenkinson, H. F., and D. A. Carter.** 1988. Cell surface mutants of *Streptococcus sanguis* with altered adherence properties. *Oral Microbiol. Immunol.* **3:**53–57.

52. **Jolliffe, L. K., R. J. Doyle, and U. N. Streips.** 1981. Energized membrane and cellular autolysis in *Bacillus subtilis. Cell* **25:**753–763.

53. **Jones, C., and C. Russell.** 1980. The effect of calcium and EDTA on initial adhesion of bacteria to teeth. *J. Dent. Res.* **60:**1105.

54. **Kaufman, J., and J. M. DiRienzo.** 1989. Isolation of a corncob (coaggregation) receptor polypeptide from *Fusobacterium nucleatum. Infect. Immun.* **57:**331–337.

55. **Kelstrup, J., J. Theilade, and O. Fejerskov.** 1979. Surface ultrastructure of some oral bacteria. *Scand. J. Dent. Res.* **87:**415–423.

56. **Klotz, S. A., S. I. Butrus, R. P. Misra, and M. S. Osato.** 1989. The contribution of bacterial surface hydrophobicity to the process of adherence of *Pseudomonas aeruginosa* to hydrophilic contact lenses. *Curr. Eye Res.* **8:**195–202.

57. **Knox, K. W., L. N. Hardy, L. J. Markeircs, J. D. Evans, and A. J. Wicken.** 1985. Comparative studies on the effect of growth conditions on adhesion, hydrophobicity,

and extracellular protein profile of *Streptococcus sanguis* G9B. *Infect. Immun.* **50**:545–554.

58. **Knox, K. W., L. N. Hardy, and A. J. Wicken.** 1986. Comparative studies on the protein profiles and hydrophobicity of strains of *Streptococcus mutans* serotype C. *J. Gen. Microbiol.* **132**:2541–2548.

59. **Knox, K. W., and A. J. Wicken.** 1985. Environmentally induced changes in the surface of oral streptococci and lactobacilli, p. 212–219. *In* S. E. Mergenhagen and B. Rosan (ed.), *Molecular Basis of Oral Microbial Adhesion.* American Society for Microbiology, Washington, D.C.

60. **Koga, T., N. Okahashi, I. Takahashi, T. Kanamoto, H. Asakawa, and M. Iwaki.** 1990. Surface hydrophobicity, adherence, and aggregation of cell surface protein antigen mutants of *Streptococcus mutans* serotype c. *Infect. Immun.* **58**:289–296.

61. **Kolenbrander, P. E.** 1989. Surface recognition among oral bacteria: multigeneric coaggregations and their mediators. *Crit. Rev. Microbiol.* **17**:137–159.

62. **Kolenbrander, P. E., and B. L. Williams.** 1981. Lactose-reversible coaggregation between oral actinomycetes and *Streptococcus sanguis*. *Infect. Immun.* **33**:95–102.

63. **Kozlovsky, A., Z. Metzger, and I. Eli.** 1987. Cell surface hydrophobicity of *Actinobacillus actinomycetemcomitans* Y4. *J. Clin. Periodontol.* **14**:370–372.

64. **Kuramitsu, H. K., and L. Ingersoll.** 1976. Differential inhibition of *Streptococcus mutans* in vitro adherence by anti-glucosyltransferase antibodies. *Infect. Immun.* **13**:1775–1777.

65. **Leach, S. A.** 1980. A biophysical approach to interactions associated with the formation of the matrix of dental plaque, p. 159–183. *In* S. A. Leach (ed.), *Dental Plaque and Surface Interactions in the Oral Cavity.* IRL Press, Oxford.

66. **Leach, S. A., and E. A. Agalamanyi.** 1984. Hydrophobic interactions that may be involved in the formation of dental plaque, p. 43–50. *In* J. M. ten Cate, S. A. Leach, and J. Arends (ed.), *Bacterial Adhesion and Preventive Dentistry.* IRL Press, Oxford.

67. **Lee, S. F., A. Progulske-Fox, G. W. Erdos, D. A. Piacentini, G. Y. Ayakawa, P. J. Crowley, and A. S. Bleiweis.** 1989. Construction and characterization of isogenic mutants of *Streptococcus mutans* deficient in major surface protein antigen P1 (I/II). *Infect. Immun.* **57**:3306–3316.

68. **Lesher, R. J., G. R. Bender, and R. E. Marquis.** 1977. Bacteriolytic action of fluoride ions. *Antimicrob. Agents Chemother.* **12**:339–345.

69. **Levine, M. J., M. C. Herzberg, M. S. Levine, S. A. Ellison, M. W. Stinson, H. C. Li, and T. Van Dyke.** 1978. Specificity of salivary-bacterial interactions: role of terminal sialic acid residues in the interaction of salivary glycoproteins with *Streptococcus sanguis* and *Streptococcus mutans*. *Infect. Immun.* **19**:107–115.

70. **Liang, L., D. Drake, and R. J. Doyle.** 1989. Stability of the glucan-binding lectin of oral streptococci. *J. Dent. Res.* **68**:1677.

71. **Lichtenberg, D., M. Rosenberg, N. Sharfman, and I. Ofek.** 1985. A kinetic approach to bacterial adherence to hydrocarbon. *J. Microbiol. Methods* **4**:141–146.

72. **Liljemark, W. F., and C. G. Bloomquist.** 1981. Isolation of a protein-containing cell surface component from *Streptoccus sanguis* which affects its adherence to saliva-coated hydroxyapatite. *Infect. Immun.* **34**:428–434.

73. **Liljemark, W. F., C. G. Bloomquist, and G. R. Germaine.** 1981. Effect of bacterial aggregation on the adherence of oral streptococci to hydroxyapatite. *Infect. Immun.* **31**:935–941.

74. **Liljemark, W. F., and S. V. Schauer.** 1975. Studies on the bacterial components which bind *Streptococcus sanguis* and *Streptococcus mutans* to hydroxyapatite. *Arch. Oral Biol.* **20**:609–615.

75. **Liljemark, W. F., S. V. Schauer, and C. G. Bloomquist.** 1978. Compounds which affect the adherence of *Streptococcus sanguis* and *Streptococcus mutans* to hydroxyapatite *J. Dent. Res.* **57**:373–379.

76. **Lindahl, M., A. Faris, T. Wadström, and S. Hjertén.** 1981. A new test based on 'salting out' to measure relative surface hydrophobicity of bacterial cells. *Biochim. Biophys. Acta* **677**:471–476.

77. **Lowrance, J. H., D. L. Hasty, and W. A. Simpson.** 1988. Adherence of *Streptococcus sanguis* to conformationally specific determinants in fibronectin. *Infect. Immun.* **56**:2279–2285.

78. **Magnusson, I., T. Ericson, and K. Pruitt.** 1976. Effect of salivary agglutinins on bacterial colonization of tooth surfaces. *Caries Res.* **10**:113–122.

79. **Mattingly, S. J., and B. P. Johnston.** 1987. Comparative analysis of the localization of lipoteichoic acid in *Streptococcus agalactiae* and *Streptococcus pyogenes. Infect. Immun.* **55**:2383–2386.

80. **McBride, B. C., M. Song, B. Krasse, and J. Olsson.** 1984. Biochemical and immunological differences between hydrophobic and hydrophilic strains of *Streptococcus mutans. Infect. Immun.* **44**:68–75.

81. **McCabe, M. M., E. E. Smith, and R. A. Cowman.** 1973. Invertase activity in *Streptococcus mutans* and *Streptococcus sanguis. Arch. Oral Biol.* **18**:525–531.

82. **Miller, M. J., and D. G. Ahearn.** 1987. Adherence of *Pseudomonas aeruginosa* to hydrophilic contact lenses and other substrata. *J. Clin. Microbiol.* **25**:1392–1397.

83. **Miörner, H., G. Johansson, and G. Kronvall.** 1983. Lipoteichoic acid is the major cell wall component responsible for surface hydrophobicity of group A streptococci. *Infect. Immun.* **39**:336–343.

84. **Morris, E. J., N. Ganeshkumar, and B. C. McBride.** 1985. Cell surface components of *Streptococcus sanguis*: relationship to aggregation, adherence, and hydrophobicity. *J. Bacteriol.* **164**:255–262.

85. **Morris, E. J., and B. C. McBride.** 1984. Adherence of *Streptococcus sanguis* to saliva-coated hydroxyapatite: evidence for two binding sites. *Infect. Immun.* **43**:656–663.

86. **Mozes, N., A. J. Leonard, and P. G. Rouxhet.** 1988. On the relations between the elemental surface composition of yeasts and bacteria and their charge and hydrophobicity. *Biochim. Biophys. Acta* **945**:324–334.

87. **Mozes, N., and P. G. Rouxhet.** 1987. Methods for measuring hydrophobicity of microorganisms. *J. Microbiol. Methods* **6**:99–112.

88. **Murray, P. A., M. J. Levine, L. A. Tabak, and M. S. Reddy.** 1982. Specificity of salivary-bacterial interactions. II. Evidence for a lectin on *Streptococcus sanguis* with specificity for a NeuAcα2, 3Galβ 1-3NAc sequence. *Biochem. Biophys. Res. Commun.* **106**:390–396.

89. **Mychajlonka, M., T. D. McDowell, and G. D. Shockman.** 1980. Conservation of cell wall peptidoglycan by strains of *Streptococcus mutans* and *Streptococcus sanguis. Infect. Immun.* **28**:65–73.

90. **Nealon, T. J., and S. J. Mattingly.** 1984. Role of cellular lipoteichoic acids in mediating adherence of serotype III strains of groups B streptococci to human embryonic, fetal, and adult epithelial cells. *Infect. Immun.* **43**:523–530.

91. **Nealon, T. J., and S. J. Mattingly.** 1985. Kinetic and chemical analyses of the biologic significance of lipoteichoic acids in mediating adherence of serotype III group B streptococci. *Infect. Immun.* **50**:107–115.

92. **Neeser, J.-R., A. Chambaz, S. E. Vedova, M.-J. Prigent, and B. Guggenheim.** 1988. Specific and nonspecific inhibition of adhesion of oral actinomyces and streptococci to

erythrocytes and polystyrene by caseinoglycopeptide derivatives. *Immun. Infect.* **56:**3201–3208.

93. **Nesbitt, W. E., R. J. Doyle, and K. G. Taylor.** 1982. Hydrophobic interactions and the adherence of *Streptococcus sanguis* to hydroxylapatite. *Infect. Immun.* **38:**637–644.

94. **Nesbitt, W. E., R. J. Doyle, K. G. Taylor, R. H. Staat, and R. R. Arnold.** 1982. Positive cooperativity in the binding of *Streptococcus sanguis* to hydroxylapatite. *Infect. Immun.* **35:**157–165.

95. **Nesbitt, W. E., R. H. Staat, B. Rosan, K. G. Taylor, and R. J. Doyle.** 1980. Association of protein with the cell wall of *Streptococcus mutans. Infect. Immun.* **28:**118–126.

96. **Oakley, J. D., K. G. Taylor, and R. J. Doyle.** 1985. Trypsin-susceptible cell surface characteristics of *Streptococcus sanguis. Can. J. Microbiol.* **31:**1103–1107.

97. **Ofek, I., E. Whitnack, and E. H. Beachey.** 1983. Hydrophobic interactions of group A streptococci with hexadecane droplets. *J. Bacteriol.* **154:**139–145.

98. **Okuda, K., A. Yamamoto, Y. Naito, I. Takazoe, J. Slots, and R. J. Genco.** 1986. Purification and properties of hemagglutinin from culture supernatant of *Bacteroides gingivalis. Infect. Immun.* **54:**659–665.

99. **Olsson, J., and C. G. Emilson.** 1988. Implantation and cariogenicity in hamsters of *Streptococcus mutans* with different hydrophobicity. *Scand. J. Dent. Res.* **96:**85–90.

100. **Olsson, J., P.-O. Glantz, and B. Krasse.** 1976. Surface potential and adherence of oral streptococci to solid surface. *Scand. J. Dent. Res.* **84:**240–242.

101. **Olsson J., and G. Westergren.** 1982. Hydrophobic surface properties of oral strepto-cocci. *FEMS Microbiol. Lett.* **15:**319–323.

102. **Peros, W. J., I. Etherden, R. J. Gibbons, and Z. Skobe.** 1985. Alteration of fimbriation and cell hydrophobicity by sublethal concentrations of tetracycline. *J. Periodont. Res.* **20:**24–30.

103. **Peros, W. J., and R. J. Gibbons.** 1982. Influence of sublethal antibiotic concentrations on bacterial adherence to saliva-treated hydroxyapatite. *Infect. Immun.* **35:**326–334.

104. **Pourdjabbar, F., and C. Russell.** 1979. Factors affecting adhesion of bacteria to a tooth *in vitro. Microbios* **26:**73–84.

105. **Pratt-Terpstra, I. H., A. H. Weerkamp, and H. J. Busscher.** 1987. Adhesion of oral streptococci from a flowing suspension to uncoated and albumin-coated surfaces. *J. Gen. Microbiol.* **133:**3199–3206.

106. **Reynolds, E. C., and A. Wong.** 1983. Effect of adsorbed protein on hydroxyapatite zeta potential and *Streptococcus mutans* adherence. *Infect. Immun.* **39:**1285–1290.

107. **Rogers, A. H., K. Pilowsky, and P. S. Zilm.** 1984. The effect of growth rate on the adhesion of the oral bacteria *Streptococcus mutans* and *Streptococcus milleri. Arch. Oral Biol.* **29:**147–150.

108. **Rolla, G., A. A. Scheie, and J. E. Ciardi.** 1985. Role of sucrose in plaque formation. *Scand. J. Dent. Res.* **93:**105–111.

109. **Rosan, B., B. Appelbaum, L. K. Campbell, K. W. Knox, and A. J. Wicken.** 1982. Chemostat studies of the effect of environmental control on *Streptococcus sanguis* adherence to hydroxyapatite. *Infect. Immun.* **35:**64–70.

110. **Rosenberg, E., A. Gottlieb, and M. Rosenberg.** 1983. Inhibition of bacterial adherence of hydrocarbons and epithelial cells by emulsan. *Infect. Immun.* **39:**1024–1028.

111. **Rosenberg, M.** 1981. Bacterial adherence to polystyrene: a replica method of screening for bacterial hydrophobicity. *Appl. Environ. Microbiol.* **42:**375–377.

112. **Rosenberg, M., H. Judes, and E. Weiss.** 1983. Cell surface hydrophobicity of dental plaque microorganisms in situ. *Infect. Immun.* **42:**831–834.

113. **Rosenberg, M., E. Rosenberg, H. Judes, and E. Weiss.** 1983. Bacterial adherence to hydrocarbons and to surfaces in the oral cavity. *FEMS Microbiol. Lett.* **20:**1–5.

114. Rosenberg, M., M. Tal, E. Weiss, and S. Guendelman. 1989. Adhesion of non-coccal dental plaque microorganisms to buccal epithelial cells: inhibition by saliva and amphipathic agents. *Microb. Ecol. Health Dis.* **2**:197–202.

115. Ryc, M., B. Wagner, M. Wagner, and R. Bicova. 1988. Electron microscopic localization of lipoteichoic acid on group A streptococci. *Zentralbl. Bakteriol. Mikrobiol. Hyg.* **269**:168–178.

116. Satou, J., A. Fukunaga, N. Satou, H. Shintani, and K. Okuda. 1988. Streptococcal adherence on various restorative materials. *J. Dent. Res.* **67**:588–591.

117. Satou, N., J. Satou, H. Shintani, and K. Okuda. 1988. Adherence of streptococci to surface-modified glass. *J. Gen. Microbiol.* **134**:1299–1305.

118. Savoia, D., T. Venesio, and M. G. Martinotti. 1986. Surface characteristics of streptococci: evaluation technics. *G. Batteriol. Virol. Immunol.* **79**:61–76.

119. Savoia, D., F. Vulcano, and M. G. Martinotti. 1987. Changes in surface characteristics of group A streptococci. *G. Batteriol. Virol. Immunol.* **80**:3–13.

120. Schilling, K. M., M. H. Blitzer, and W. H. Bowen. 1989. Adherence of *Streptococcus mutans* to glucans formed *in situ* in salivary pellicle. *J. Dent. Res.* **68**:1678–1680.

121. Schilling, K. M., and W. H. Bowen. 1988. The activity of glucosyltransferase adsorbed onto saliva-coated hydroxyapatite. *J. Dent. Res.* **67**:2–8.

122. Shah, H. N., S. V. Seddon, and S. E. Gharbia. 1989. Studies on the virulence properties and metabolism of pleiotropic mutants of *Porphyromonas gingivalis* (*Bacteroides gingivalis*) W50. *Oral Microbiol. Immunol.* **4**:19–23.

123. Sharon, D., R. Bar-Ness, and M. Rosenberg. 1986. Measurement of the kinetics of bacterial adherence to hexadecane in polystyrene cuvettes. *FEMS Microbiol. Lett.* **36**:115–118.

124. Shinjo, T., H. Hazu, and H. Kiyoyama. 1987. Hydrophobicity of *Fusobacterium necrophorum* biovars A and B. *FEMS Microbiol. Lett.* **48**:243–247.

125. Simpson, W. A., and E. H. Beachey. 1983. Adherence of group A streptococci to fibronectin on oral epithelial cells. *Infect. Immun.* **39**:275–279.

126. Stinson, M. W., D. C. Jinks, and J. M. Merrick. 1981. Adherence of *Streptococcus mutans* and *Streptococcus sanguis* to salivary components bound to glass. *Infect. Immun.* **32**:583–591.

127. Stinson, M. W., M. J. Levine, J. M. Cavese, A. Prakobhol, P. A. Murray, L. A. Tabak, and S. Reddy. 1982. Adherence of *Streptococcus sanguis* to salivary mucin bound to glass. *J. Dent. Res.* **61**:1390–1393.

128. Svanberg, M., G. Westergren, and J. Olsson. 1984. Oral implantation in humans of *Streptococcus mutans* strains with different degrees of hydrophobicity. *Infect. Immun.* **43**:817–821.

129. Torchilin, V. P., V. G. Omel'Yanenko, A. L. Kilibanov, A. I. Mikhailov, V. I. Gol'Danskii, and V. N. Smirnov. 1980. Incorporation of hydrophilic protein modified with hydrophobic agent into liposome membrane. *Biochim. Biophy. Acta* **602**:511–521.

130. Tylewska, S. 1983. Effect of culture conditions on the surface hydrophobic properties and adherence of streptococci. *Med. Dosw. Mikrobiol.* **35**:175–181.

131. Tylewska, S. K., S. Hjertén, and T. Wadström. 1979. Contribution of M protein to the hydrophobic surface properties of *Streptococcus pyogenes*. *FEMS Microbiol. Lett.* **6**:249–253.

132. van der Mei, H. C., A. H. Weerkamp, and H. J. Busscher. 1987. A comparison of various methods to determine hydrophobic properties of streptococcal cell surfaces. *J. Microbiol. Methods* **6**:277–287.

133. van der Mei, H. C., A. H. Weerkamp, and H. J. Busscher. 1987. Physico-chemical

surface characteristics and adhesive properties of *Streptococcus salivarius* strains with defined cell surface structure. *FEMS Microbiol. Lett.* **40**:15–19.

134. van Loosdrecht, M. C., J. Lyklema, W. Norde, G. Schraa, and A. J. Zehnder. 1987. Electrophoretic mobility and hydrophobicity as a measure to predict the initial steps of bacterial adhesion. *Appl. Environ. Microbiol.* **53**:1898–1901.

135. van Loosdrecht, M. C., J. Lyklema, W. Norde, G. Schraa, and A. J. Zehnder. 1987. The role of bacterial cell wall hydrophobicity in adhesion. *Appl. Environ. Microbiol.* **53**:1893–1897.

136. van Pelt, A. W. J., A. H. Weerkamp, M. H. W. J. C. Uyen, H. J. Busscher, H. P. deJong, and J. Arends. 1985. Adhesion of *Streptococcus sanguis* CH3 to polymers with different surface free energies. *Appl. Environ. Microbiol.* **49**:1270–1275.

137. van Steenbergen, T. J. M., F. Namavar, and J. de Graaff. 1985. Chemiluminescence of human leukocytes by black-pigmented *Bacteroides* strains from dental plaque and other sites. *J. Periodont. Res.* **20**:58–71.

138. Vecchio, G., P. C. Richetti, M. Zanoni, G. Artoni, and E. Gianazza. 1984. Fractionation techniques in a hydro-organic environment. 1. Sulfolane as a solvent for hydrophobic proteins. *Anal. Biochem.* **137**:410–419.

139. Verran, J., D. B. Drucker, and C. J. Taylor. 1980. Measurement of adherence to glass of *Streptococcus mutans* by image analysis. *J. Dent. Res.* **60**:1105.

140. Weerkamp, A. H., and P. S. Handley. 1986. The growth rate regulates the composition and density of the fibrillar coat on the surface of *Streptococcus salivarius* K$^+$ cells. *FEMS Microbiol. Lett.* **33**:179–183.

141. Weerkamp, A. H., H. M. Uyen, and H. J. Busscher. 1988. Effect of zeta potential and surface energy on bacterial adhesion to uncoated and saliva-coated human enamel and dentin. *J. Dent. Res.* **67**:1483–1487.

142. Weerkamp, A. H., H. C. van der Mei, and J. W. Slot. 1987. Relationship of cell surface morphology and composition of *Streptococcus salivarius* K$^+$ to adherence and hydrophobicity. *Infect. Immun.* **55**:438–445.

143. Weiss, E., M. Rosenberg, H. Judes, and E. Rosenberg. 1982. Cell-surface hydrophobicity of adherent oral bacteria. *Curr. Microbiol.* **7**:125–128.

144. Westergren, G., and J. Olsson. 1983. Hydrophobicity and adherence of oral streptococci after repeated subculture in vitro. *Infect. Immun.* **40**:432–435.

145. Wicken, A. J., and K. W. Knox. 1975. Lipoteichoic acids: a new class of bacterial antigen. *Science* **187**:1161–1167.

146. Willcox, M. D., J. E. Wyatt, and P. S. Handley. 1989. A comparison of the adhesive properties and surface ultrastructure of the fibrillar *Streptococcus sanguis* 12 and an adhesion deficient non-fibrillar mutant 12 na. *J. Appl. Bacteriol.* **66**:291–299.

147. Willcox, M. D. P., and D. B. Drucker. 1989. Surface structures, co-aggregation and adherence phenomena of *Streptococcus oralis* and related species. *Microbios* **59**:19–29.

148. Wilson, P. A. D., W. M. Edgar, and S. A. Leach. 1984. Some physical properties of oral streptococci that might have a role in their adhesion to oral surfaces, p. 99–112. *In* J. M. ten Cate, S. A. Leach, and J. Arends (ed.), *Bacterial Adhesion and Preventive Dentistry*. IRL Press, Oxford.

149. Wyatt, J. E., L. M. Hesketh, and P. S. Handley. 1987. Lack of correlation between fibrils, hydrophobicity and adhesion for strains of *Streptococcus sanguis* biotypes I and II. *Microbios* **50**:7–15.

INDEX